T0298362

Advanced Bioseparation of Industrial Wastes

Advanced Bioseparation of Industrial Wastes: Sustainable Recovery of High-Value Metal Ions examines resource recovery from a variety of industrial waste streams, including sludge and wastewater, with an emphasis on both the fundamentals and the more advanced concepts involved. Chemical leaching, waste treatment, and other processes for metal extraction are broken down into their component parts in great detail. Several important metals, such as lithium, copper, gold, platinum, nickel, zinc, chromium, uranium, cobalt, rhodium, and indium, could be salvaged from recyclables. This book presents the best practices for dealing with waste from industries such as those involved in the production of electronic goods, automobiles, batteries, as well as mining and electroplating. It provides readers with a comprehensive understanding of the many forms of industrial waste, including their composition, recycling processes, and the potential for recovery of essential metals, from the ground up.

Features:

- Provides updated occurrence and characteristics of a variety of high-value metal ions that can be recovered from different industrial wastes.
- Presents advanced chemical leaching technologies for those metal ions.
- Describes detailed accounts of physico-chemical-based reuse and recycle methodologies.
- Covers innovative approaches for the reutilization and management of industrial wastes.

Advanced Bioseparation of Industrial Wastes

Sustainable Recovery of High-Value Metal Ions

Edited by

Jayato Nayak, Sankha Chakrabortty, Suraj K. Tripathy, and Maulin P. Shah

CRC Press is an imprint of the
Taylor & Francis Group, an informa business

Designed cover image: Shutterstock

First edition published 2025
by CRC Press
385 NW Executive Center Drive, Suite 30, Boca Raton FL 3343

and by CRC Press
4 Park Square, Milton Park, Abingdon, Oxon, OX4 4RN

CRC Press is an imprint of Taylor & Francis Group, LLC

Library of Congress Cataloging-in-Publication Data
Names: Nayak, Jayato, editor. | Chakrabortty, Sankha, editor. |
Tripathy, Suraj, editor. | Shah, Maulin P., editor.
Title: Advanced bioseparation of industrial wastes : sustainable recovery
of high-value metal ions / edited by Jayato Nayak, Sankha Chakrabortty,
Suraj K. Tripathy, and Maulin P. Shah.
Description: First edition. | Boca Raton, FL : CRC Press, 2025. | Includes bibliographical references and index. | Identifiers: LCCN 2024034247 | ISBN 9781032541792 (hardback) |
ISBN 9781032541815 (paperback) | ISBN 9781003415541 (ebook)
Subjects: LCSH: Biomolecules—Separation. | Biochemical engineering. |
Factory and trade waste. | Recycling (Waste, etc.)
Classification: LCC TP248.25.S47 A38 2025 | DDC 660.6/3—dc23/eng/20241023
LC record available at https://lccn.loc.gov/2024034247

ISBN: 978-1-032-54179-2 (hbk)
ISBN: 978-1-032-54181-5 (pbk)
ISBN: 978-1-003-41554-1 (ebk)

DOI: 10.1201/9781003415541

Typeset in Times
by codeMantra

Contents

About the Editors

Dr. Jayato Nayak is a prominent researcher with more than nine years of experience in the field of novel design and applications embracing the theme of process intensification. He completed his MTech and PhD from NIT Durgapur on "Process intensification in acetic acid manufacture from waste material" and collaborated his research with CSIR-CMERI. His core expertise includes process intensification, bioprocess engineering, water and wastewater treatment, graphene-integrated photocatalyst development, and biosynthesis of value-added products. Currently, along with the national collaboration with IITs, NITs, and CSIR Labs, he has extended research collaborations with research professionals in Poland, Brazil, Malaysia, Vietnam, etc. By August 2022, he had published 25 international SCI/SCOPUS-indexed peer-reviewed journal papers and more than 20 book chapters and filed three patents. As of now, with a huge number of citations, he is having high h-index and i10-index. He provides expertise for several webinars and faculty development programs (FDPs); and he is a Senior Editor, Review Editor, Guest Editor, Editorial Board Member, and recognized reviewer of more than 15 international journals. He has received two best research awards and five distinguished awards from various national and international organizations.

Dr. Sankha Chakrabortty is an assistant professor at the School of Biotechnology/Chemical Technology, KIIT University, India. He is a renowned researcher with over 10 years of expertise in the domains of Environmental Engineering and Green Technology. He has had his research published in over 32 SCI/SCOPUS-indexed high-impact factor journals and has filed seven patents. He has garnered 14 awards/achievements from different organizations over the course of his 10-year research career. He received Springer's "Best Paper Award-2015" in 2015 and the "Best Researcher Award-2020" from an international organization in 2021, as well as the "NESA Young Scientist of the Year 2020" from the National Environmental Science Academy. In addition, he received an award from KIIT University in the field of "STE-Young Scientist Award (Faculty Category)" for his remarkable research effort.

Suraj Kumar Tripathy currently works as an associate professor and associate dean at the School of Chemical Technology and School of Biotechnology, KIIT University, where he heads the Chemical & Bioprocess Engineering Lab. His area of research is the processing of nanomaterials for water treatment, heterogeneous catalysis, and biomedical applications. He works closely with chemical and mineral industries to investigate the management of industrial wastes. He has published his research work in more than 60 SCI/Scopus journals and also filed eight patents based on his research development out of which two patents have been granted by the authority.

He has already finished four high-value projects as the principal investigator, and currently, three projects with a total value of 3 crores of Indian rupees are under his supervision. He is involved in a number of research partnerships with a variety of universities and research labs located in India as well as universities located in other countries. He is working on a couple of initiatives in other countries.

Maulin P. Shah has been an active researcher and scientific writer in his field for over 20 years. He received a BSc degree (1999) in Microbiology from Gujarat University, Godhra (Gujarat), India. He also earned his PhD degree (2005) in Environmental Microbiology from Sardar Patel University, Vallabh Vidyanagar (Gujarat), India. His research interests include biological wastewater treatment, environmental microbiology, biodegradation, bioremediation, and phytoremediation of environmental pollutants from industrial wastewaters. He has published more than 350 research papers in national and international journals of repute on various aspects of microbial biodegradation and bioremediation of environmental pollutants. He is the editor of 175 books of international repute (RSC, Wiley, De Gruyter, Elsevier, Springer, and CRC Press). He has edited 25 special issues specifically in industrial wastewater research, microbial remediation, and biorefinery of wastewater treatment area. He is associated as an Editorial Board Member in 25 highly reputed journals of Elsevier, Springer, Taylor & Francis, and Wiley.

List of Contributors

Anoar Ali Khan
Department of Chemical Engineering
Haldia Institute of Technology
Haldia, India

Shirsendu Banerjee
School of Chemical Technology
KIIT Deemed to be University
Bhubaneswar, India

Aradhana Basu
School of Sustainability
XIM University
Puri, India

Sujoy Bose
Indian Institute of Chemical Engineers
Kolkata, India

Sujoy Chattaraj
Department of Chemical Engineering
Indian Institute of Technology Madras
Chennai, India
and
Department of Chemical Engineering
Vignan's Foundation for Science, Technology & Research (Deemed to be University)
Guntur, Andhra Pradesh, India

Raj Kumar Das
Chemical Engineering Department
Jadavpur University
Kolkata, India

Tripti De
V. S. B. Engineering College
Karur, India

Sankha Chakrabortty
School of Chemical Technology
Kalinga Institute of Industrial Technology (KIIT)
Bhubaneswar, India

Trisha G
Department of Biotechnology
Rajalakshmi Engineering College
Chennai, India

Mercy Jacquline B
Centre for Industrial Safety
Anna University
Chennai, India

Neetha John
Department of Chemical Engineering
Saintgits College of Engineering
Ernakulam, India

Lakshmi Kanthan Bharathi A
Department of Mechanical Engineering, CEG Campus
Anna University
Chennai, India
and
Centre for Industrial Safety
Anna University
Chennai, India

Harish Kumar U
Department of Mechanical Engineering, CEG Campus
Anna University
Chennai, India
and
Centre for Industrial Safety
Anna University
Chennai, India

Bhuvanesh M
Department of Mechanical Engineering, CEG Campus
Anna University
Chennai, India
and
Centre for Industrial Safety
Anna University
Chennai, India

G. Damaru Mani
Mahindra University
Hyderabad, India

Angel Mathew
Greenamor Ventures Pvt Ltd
Ernakulam, India

Ananya Namala
Mahindra University
Hyderabad, India

Jayato Nayak
Centre for Life Science
Mahindra University
Hyderabad, India

Anna Nova
CIPET: IPT
Kochi, India

Reddem Poojitha Reddy
Mahindra University
Hyderabad, India

Chandrima Roy
Centre for Environment
University College of Engineering, Science and Technology, JNTUH
Hyderabad, India

Kalaiselvam S
Department of Mechanical Engineering, CEG Campus
Anna University
Chennai, India
and
Centre for Industrial Safety
Anna University
Chennai, India

Pandya, Shivamkumar N.
Department of Chemical Engineering
SKJP
Bharuch, India

Javvadi K J N S Thanishka
Mahindra University
Hyderabad, India

Anitha Thulasisingh
Department of Biotechnology
Rajalakshmi Engineering College
Chennai, India

Suraj K. Tripathy
School of Chemical Technology
KIIT Deemed to be University
Bhubaneswar, India

Somanchi Venkata Ramalakshmi
School of Sustainability
XIM University
Puri, India

Himabindu Vurimindi
Centre for Environment
University College of Engineering, Science and Technology, JNTUH
Hyderabad, India

Preface

This book describes how biological approaches can be utilized to recover metal ions from a range of industrial waste streams, such as sludge and wastewater, by focusing on both the fundamentals and the more sophisticated concepts that are involved in the process. Everything the target audience need to know about bioleaching is broken down in great detail, and every concept associated with waste treatment for metal extraction is explored in great detail as well. Lithium, copper, gold, platinum, nickel, zinc chromium, uranium, cobalt, rhodium, and indium are just a few examples of the essential metals that are among the byproducts that could be collected during the recycling process. Due to the in-depth research that was conducted, the editors have an in-depth understanding of the procedures that are the most efficient at dealing with waste from businesses such as those that are involved in the production of electronic goods, mining, electroplating, automobile, battery manufacturing, and so on. The readers of this book will get an in-depth grasp of the many different types of industrial waste as well as their compositions, recycling procedures, and recovery of key metals, beginning with the ground up.

The uniqueness of this book vis-à-vis others in its field is the detail with which it tackles the topic of metal ion recovery from industrial wastes. As a result, it helps to boost up confidence in the reader that they have a good grasp on the ideas that are presented. In addition, the editors demonstrate a great deal of promise in a variety of processes that take place further down the production line. These processes include membrane filtering, hybrid techniques, chemical leaching, electrochemical techniques, and a multitude of other contemporary recovery methods, all of which would be presented in an effective manner. In order to guarantee quick and reliable results, a focus will be placed on the nitty-gritty elements of the analysis, such as which instruments were used and how measurements were made in accordance with the set protocols.

It is indeed a fact that the process of bioleaching is easier and, hence, cheaper to run and maintain than conventional methods. In addition, this process is safer for the environment than conventional extraction techniques, which means it can be lucrative for the corporations involved. Because of the conductivity and chemical structures, metals including copper, nickel, gold, platinum, zinc, and others used to be trapped inside the matrices of industrial wastes and effluents. Metal recovery from industrial wastes is necessary due to the economic benefits of recycling and reusing waste as a secondary source of rare and expensive metals as well as the toxicity of some metals. Each year, humanity produces 800 million tons of solid trash. Rare and precious metals are becoming increasingly scarce in the natural world, but they can be recovered from electronic waste through bioleaching. Because of this, mining and other industries now have the attention of academic and industrial sectors. While there are current technologies for metal removal from industrial wastes, such as pyrometallurgical and hydrometallurgical processes, they are not environmentally sustainable due to their high cost, high risk, and management issues. These methods aren't preferred because of their severe heat treatment conditions, limited social acceptability, harmful effects on the environment, and expensive capital and operational costs. Due to the energy crisis and the need to reduce greenhouse gas emissions, businesses are looking for new methods to operate.

In the process of bioleaching, microorganisms are used to extract metals from the trash. Recovering metals with microorganisms is targeted, economical, and ecologically sound. Greater efficiency and safety, lower operating costs and energy consumption, easier management, working at atmospheric pressure and room temperature, less environmental effect, and few industrial requirements are only a few of the bio-hydrometallurgical benefits of bioleaching. The initial investment for traditional methods is more than that for bioleaching. Enzymatic oxidation-reduction, proton-promoted processes, ligand and complex formation, etc. are all driven by the microorganisms responsible for bioleaching. Recent research in the field of bioleaching of industrial wastes is summarized, and the basic concepts of this process are outlined in this book. Comprehensive reviews of the literature on the topic of bioleaching of industrial wastes classify these studies according to waste type, microbe, and target metal and discuss topics including bioleaching mechanism, microorganism, bioleaching methods, and process.

The focus of this book is on the recycling of industrial byproducts such as solid and liquid wastes in order to extract trace metals such as copper, gold, platinum, nickel, zinc, chromium, uranium, cobalt, rhodium, and indium. Bioleaching rare earth metal ions is just one example of a method that goes beyond the conventional technique and is described experimentally in our book. The authors of this book have taken a holistic approach toward bioleaching for the recovery of precious metal ions. This book covers all the important facets of waste reutilization for metal recovery through biological methodologies and promises to be of great use to the appropriate communities. Every section concludes with a discussion of where that metal ion recovery research could go in the future. This is useful information for the researcher in determining the best course of action.

The extraction of trace metal ions from industrial waste sludge and effluents is one of the most vital processes that must be carried out in order to guarantee the successful recycling and recovery of materials that have had value added to them. These recovered metals might have additional use in catalysis, the synthesis of nanoparticles, the fabrication of solar cells, the degradation of dyes, electroplating, and other processes. By recycling metals that have previously been mined, rather than engaging in the time-consuming and resource-intensive practice of continuously mining for metals from the interior of the Earth, we may be able to achieve tremendous economies of scale. As a consequence of the industrial revolution, a surge in population, and the spread of urbanization over the course of the past several centuries, the amount of waste that has accumulated on Earth has skyrocketed. In addition, although shifts are taking place in production strategies, these shifts will only be gradually implementable in the current economy because conventional production methods are still required for the majority of production. This book provides the reader with the information necessary to recover valuable metals from a wide variety of industrial wastes by utilizing a wide range of physicochemical technologies. This book is an alternative strategy for the utilization of resources and equips the reader with this information. The purpose of this book is to show readers how to extract useful metal ions from industrial waste by utilizing the most cutting-edge technology and most accurate analytical procedures that are now available.

This book:

1. Provides updated occurrence and characteristics of a variety of high-valued metal ions in different industrial wastes.
2. Presents advanced biological leaching technologies for those metal ions.
3. Describes a detailed account of bio-based reuse and recycle methodologies.
4. Covers innovative approaches for the reutilization and management of industrial wastes.

1 Introduction
Environmentally Friendly Recovery of High-Value Metal Ions

Tripti De

1.1 INTRODUCTION

Naturally occurring toxic metals polluting the nature have high atomic weight and their density is also almost five times higher than water. In spite of the advantages of some heavy metal ions like copper, zinc, and iron in various biological processes, trace amounts of other elements such as lead, mercury, and cadmium are toxic to the environment and have many hazardous health effects to humans in small concentrations. The trace of heavy metals in contaminated air is a significant environmental and public health concern, largely stemming from various industrial processes.

Industries contribute to atmospheric heavy metal pollution through a variety of operations, including mining, metal processing, power generation, battery manufacturing, and chemical processing (Forruque Ahmed, 2022; Nicomel et al., 2015). Each of these activities is responsible for emissions of heavy metals into the air, often as part of particulate matter, which can then be transported over long distances by wind currents before being deposited on the ground or water bodies.

Wastewater, a by-product of our daily lives and industrial activities, carries more than just organic matter. Lurking within this effluent are often invisible threats – heavy metals. These metallic elements, with densities exceeding water, create a remarkable challenge by their persistence, toxicity, and bioaccumulation potential. This chapter delves into the existence of various heavy metals in wastewater, exploring their origins, potential benefits (surprisingly few), detrimental effects, and the demand for fruitful mitigation policies (Hussain et al., 2021; Zhao et al., 2022).

The escalating industrialization and urbanization in recent decades have dramatically influenced the quality of water bodies around the globe. Among the myriad of pollutants, toxic metals in waste effluent have garnered remarkable attention because of their persistent, bioaccumulative, and toxic nature. The metals, such as lead, arsenic, nickel, mercury, chromium, and cadmium, originate from a variety of industrial, agricultural, and domestic sources, posing a severe warning to aquatic lifestyle, wildlife, and also to human lifestyle (Mitra et al., 2022). This introduction delves into the presence of various heavy metals in wastewater, highlighting their origin, healthiness, ecological impact, and the pressing need for effective management strategies.

DOI: 10.1201/9781003415541-1

The occurrence of heavy metal ions in waste effluent is not a recent phenomenon, but the scale and scope of pollution have intensified with the advent of manufacturing production. Industries such as mining, metal plating, battery construction, tanneries, and cloth manufacturing are primary contributors to heavy metal discharge into water bodies (Vardhan et al., 2019). Moreover, agricultural practices, including the use of fertilizers and pesticides containing metals like arsenic and cadmium, further exacerbate the contamination levels. Urban runoff, leaching from landfills, and improper disposal of electronic waste also participate effectively in initiating heavy metals into the wastewater stream.

The environmental ramifications of heavy metal pollution in wastewater are profound and multifaceted. Toxic metals persist in the environment and can build up in aquatic organisms, resulting in bioaccumulation and biomagnification throughout the food web. This accumulation can devastate aquatic ecosystems, resulting in decreased biodiversity and the disruption of aquatic life cycles (Sonone et al., 2020). Furthermore, heavy metals can sediment in water bodies, altering the physicochemical properties of the water and the sediment quality, which affects the overall health of the aquatic environment.

The health implications of heavy metal pollution are equally alarming. Humans can be exposed to heavy metals through various pathways, including direct consumption of contaminated water, ingestion of contaminated seafood, and dermal contact. The toxicity of heavy metals is well-documented, with metals like lead and mercury affecting the nervous system, cadmium causing renal failure, chromium leading to skin rashes and lung cancer, and arsenic causing skin lesions and cardiovascular diseases. The World Health Organization (WHO) and other health agencies have set strict rules for heavy metal occurrence limit in drinking water, emphasizing the need to control their levels in wastewater (Makuza et al., 2021; Schippers et al., 2014).

The concentration of toxic metals in waste sludge is a pressing ecological and public health issue that demands immediate and sustained attention. The sources of heavy metal impurity are diverse, ranging from factory discharges to agricultural runoff, and their impacts are wide-ranging, influencing water ecosystem and human health. Effective management and treatment of heavy metals in wastewater are crucial to mitigate their adverse effects and protect water quality. This requires a comprehensive approach, involving the adoption of advanced treatment technologies, stringent regulatory frameworks, and public awareness and education (Vidu et al. 2020). As the world continues to contend with the difficulties of pollution and environmental degradation, the issue of heavy metals in wastewater stands as a testament to the need for concerted efforts to safeguard our water resources for future generations. In general, metal ions are toxic to the mammalian environment as they chemically react with the cellular framework proteins, catalysts, and membrane network. This leads to the exposure to chemical compounds of the metal based on the metal's oxidation state, volatility, lipid solubility, etc.

The withdrawal of toxic elements that are hazardous to the environment from the mining low-grade ores is achieved by the techniques of pyrometallurgy or hydrometallurgy. Both these techniques are based on the requirement of high energy consumption and the production of major by-products for the usage of chemicals in hydrometallurgy which further needs the treatment processes for disposal and also generates many toxic gases. These toxic processes lead to human health exposure to cancer, neurological disorders, organ disorders, immunodeficiency, etc. This exploitation demands the introduction of promising environmentally friendly, low power usage, cost-effective, and high removal effectiveness process called bioleaching which is also known as biohydrometallurgy (Liu et al., 2023). This bioleaching technique bridged the leaching process of heavy metals from mining ores which are dangerous for the environment with the environmentally friendly bioleaching approach with the usage of autotrophic and heterotrophic bacteria, fungi, etc. Some investigations revealed that bioleaching of vanadium by both the *Acidithiobacillus* sp. and the *Pseudomonas* sp. has a maximum removal efficiency of 90%, whereas *Aspergillus* sp. had a removal efficiency of 92% (Mirazimi et al., 2015). Comparing the bioleaching removal of zinc and lead by 96.36% and 95.34%, respectively, to the remaining debris concentration of zinc, copper, and chromium with 0.34%, 0.64%, and 0%, respectively, this analysis reveals low toxicity and environmental risks (Liao et al., 2022; Ye et al., 2021). Different harmful effects of each individual metal are presented briefly in Table 1.1.

TABLE 1.1
Clinical Aspects of Chronic Toxicities

Metal	Target organs	Primary sources	Clinical effects
Arsenic	Pulmonary nervous system, skin	Industrial dusts, medicinal uses of polluted water	Perforation of nasal septum, respiratory cancer, peripheral neuropathy: dermatomes, skin, cancer
Cadmium	Renal, skeletal pulmonary	Industrial dust and fumes and polluted water and food	Proteinuria, glucosuria, osteomalacia, aminoaciduria, emphysema
Chromium	Pulmonary	Industrial dust and fumes and polluted food	Ulcer, perforation of nasal septum, respiratory cancer
Manganese	Nervous system	Industrial dust and fumes	Central and peripheral neuropathies
Lead	Nervous system, hematopoietic system, renal	Industrial dust and fumes and polluted food	Encephalopathy, peripheral neuropathy, central nervous disorders, anemia
Nickel	Pulmonary, skin	Industrial dust, aerosols	Cancer, dramatis
Tin	Nervous, pulmonary system	Medicinal uses, industrial dusts	Central nervous system disorders, visual defects and EEG changes, pneumoconiosis
Mercury	Nervous system, renal	Industrial dust and fumes and polluted water and food	Proteinuria

Source: Reproduced with copyright from Ref. Mahurpawar (2015).

1.2 SOURCES AND CAUSES OF VARIOUS HEAVY METAL POLLUTION

1.2.1 Contamination in Water

Heavy metal pollution continues to be one of the most urgent environmental problems, presenting severe threats to ecosystems and human health. Heavy metals, including lead (Pb), mercury (Hg), cadmium (Cd), chromium (Cr), and arsenic (As), are naturally occurring elements characterized by high atomic weights and densities much greater than water (Afzal et al., 2017). Despite their natural origins, anthropogenic (human-made) activities have exponentially increased their concentrations in the environment to levels that can cause toxicity. This chapter explores the origin and causes of heavy toxic element contamination, providing insights into the origins and pathways through which these pollutants enter our ecosystems. Akpor (2014) reviewed the negative impact of the heavy metals such as causing death to aquatic animals in water, reducing plant's growth in soil, causing cancer, nervous system damage, and death in animals. So, remediation of heavy metal ions in waste effluent through chemical method or biological (by microbial remediation or by phytoremediation) method reduces the negative impact on the ecosystem.

1.2.1.1 Industry Source

Industries are the primary contributors to heavy metal contamination. Metal mining and smelting operations release a significant amount of metals directly into the environment. These activities not only disturb the geological formations but also generate large volumes of waste known as tailings, which often leach heavy metals into the surrounding water supplies and soil. The manufacturing sector, including electronics, batteries, paints, and textiles, utilizes various heavy metals in production processes. For instance, lead and cadmium are used in battery manufacturing, chromium in metal plating, and mercury in the chlor-alkali industry. The effluents discharged from these industries contain a large concentration of toxic elements, which pollute rivers, lakes, and groundwater. The development of industrial sector and urban centers has increased the potentiality of water contamination by the emission of heavy metals which causes health risk to all living beings due to the toxicity and nondegradability features of heavy metal ions (Odumbe et al., 2023).

1.2.1.2 Agricultural Practices

Agriculture contributes to heavy metal contamination through applying manure, pesticides, and garbage silt. Phosphatic fertilizers are known to contain trace amounts of cadmium, while pesticides may contain metals such as mercury and arsenic. When applied to crops, these chemicals can leach into the soil and subsequently into water bodies. Moreover, the application of sewage sludge as a fertilizer, a common practice intended to improve soil fertility, can introduce various heavy metals into agricultural soils, given that sludge often contains contaminants from industrial and domestic wastewater.

1.2.1.3 Urban Runoff and Waste Disposal

Urban runoff, resulting from rainwater flowing over streets and other urban surfaces, can pick up heavy metals from vehicle emissions (lead and cadmium), tire wear (zinc), and building materials (copper and lead). This contaminated runoff then drains into sewer systems or directly into water bodies. Landfills and improper waste disposal practices are also culpable. Electronic waste, for example, contains a plethora of heavy metals, including lead, mercury, and cadmium. When disposed of in landfills, these metals can leach into the ground and contaminate soil and groundwater (Müller et al., 2020).

1.2.1.4 Atmospheric Deposition

Heavy metals emitted from various sources like industrial waste gas, vehicle emissions, and gas emitted from burning of fossil fuel can travel long distances and settle on the surface of the Earth. This process, known as atmospheric deposition, can contaminate soils and water bodies far removed from the original source of pollution. Mercury is particularly noteworthy for its ability to undergo atmospheric deposition, leading to widespread environmental contamination.

1.2.2 Contamination in Air

1.2.2.1 Mining and Refinery Activities

Mining and refinery operations are the primary origin of airborne heavy metals. These activities involve the extraction and processing of ores to obtain metals, which release a significant amount of dust and fumes containing metals like lead, arsenic, cadmium, and chromium into the environment (Harrison et al., 1981). Smelting, which involves heating and melting metal ores to extract the pure metal, emits vast amounts of metal-laden dust and fumes. For example, lead smelters have been identified as one of the largest sources of atmospheric lead emissions. The adverse health impacts of these emissions are well-documented, ranging from respiratory problems to severe neurological damage in humans (Verner et al., 1996).

The exposure of heavy metals like cadmium through air mainly occurs through inhalation of cigarette smoke, consumption of contaminated food, or exposure to cadmium-polluted workplaces. This toxic cadmium is carcinogenic and causes many respiratory problems. These types of production and use of these heavy metals have a direct impact on human exposure (Tchounwou et al., 2012).

1.2.2.2 Manufacturing Industries

The manufacturing of batteries, electronics, and other metal-containing products can emit heavy metals such as cadmium, nickel, and lead. During the manufacturing process, these metals can become airborne from activities such as welding, soldering, and other thermal and mechanical processes. For instance, cadmium, used in nickel-cadmium batteries, is a potent carcinogen and can also cause kidney and bone damage upon chronic exposure (Ishchenko, 2018; Fishben, 1981).

1.2.2.3 Power Plants and Combustion of Fossil Fuels

Coal-fired power plants are significant sources of airborne mercury emissions. Coal contains trace amounts of mercury, which, when burned, is released into the atmosphere. Mercury in the atmosphere can undergo transformation into methylmercury, a highly toxic form that accumulates in aquatic food chains, posing significant health risks to wildlife and humans, particularly affecting the nervous system and brain development in infants (Ghosh et al., 2022).

1.2.2.4 Chemical Industry

The chemical industry, including the production of plastics, pesticides, and other chemicals, can also emit toxic elements such as mercury and cadmium. These emissions occur during various chemical reactions and processing stages, where metals are used as catalysts or occur as impurities in chemical feedstocks. Chimneys are the main sources of environmental air pollutions.

1.2.2.5 Agricultural Practices

While not an industrial process per se, agricultural practices contribute to the existence of toxic metals in the air, particularly through the use of pesticides and fertilizers that contain metals like arsenic and cadmium. These substances can become airborne during application or through volatilization and dust generation (Wuana & Okieimen, 2011).

1.2.2.6 Vehicle Emissions

Vehicles also contribute to airborne heavy metal pollution, particularly through the wear and tear of parts such as brake pads and tires, which can release copper, zinc, and other metals. Moreover, internal combustion engines emit metals like nickel and chromium, which are present in fuel additives and lubricants. Besides vehicle emissions, radioactive aerosols and refrigerator pollutions are also indeed the sources of heavy metal environmental emissions (Masindi and Muedi, 2018).

1.2.2.7 Natural Sources

Although human activities are the main contributors to heavy metal pollution, natural processes also play a role in the presence of these metals in the environment. Volcanic eruptions, weathering of rocks, and forest fires can release heavy metals into the atmosphere, water, and soil. These natural sources, however, generally contribute to a lesser extent compared to human activities (Okorondu et al., 2022).

1.3 DEMERITS OF OCCURRENCE OF VARIOUS HEAVY METALS IN WASTEWATER

The transboundary nature of heavy metal pollution adds another layer of complexity to this issue. Heavy metals can be carried across miles by water currents and atmospheric winds, leading to the contamination of regions far from the original source. This global movement poses significant challenges to managing and mitigating heavy metal pollution, requiring international cooperation and policy-making.

The effects of heavy metal contamination on health and the environment are profound. Heavy metals can accumulate in the food chain, leading to biomagnification and posing serious health risks to wildlife and humans. For example, mercury can accumulate in fish, leading to neurological disorders in humans who consume them. Lead exposure is associated with developmental issues in children, and arsenic can cause skin lesions and cancers. In ecosystems, heavy metals can disrupt biological processes, reduce biodiversity, and degrade habitat quality.

The topic of heavy metals encompasses a wide range of elements, including lead (Pb), cadmium (Cd), chromium (Cr), mercury (Hg), copper (Cu), nickel (Ni), arsenic (As), platinum, and palladium. These elements have high atomic weight and high density than water. While often discussed in the context of their environmental and health risks (Azimi et al. 2017), it's important to recognize that many of these metals play significant roles in various industries, technology, and even medicine. This chapter explores both the advantages and disadvantages of these heavy metals, offering a balanced perspective on their impact on society and the environment. Copper (Cu) is renowned for its excellent electrical conductivity, making it indispensable in the manufacturing of electrical wires, motors, and other components. It's also used in plumbing and roofing due to its resistance to corrosion. Chromium (Cr) is used extensively in the production of stainless steel, which is prized for its resistance to rust and tarnishing. Chromium plating also provides a decorative and protective layer for automotive and furniture products. Nickel (Ni), like chromium, is crucial in the production of stainless steel. Its properties allow it to withstand extreme temperatures, making it vital in the manufacturing of turbines and aircraft engines. Platinum and palladium are used in catalytic converters, which reduce harmful emissions from vehicles. These metals also play crucial roles in various chemical reactions and are integral in the field of renewable energy, particularly in hydrogen fuel cells. Lead (Pb), despite its toxicity, has applications in batteries, radiation shielding, and in some types of glass and ceramics. Certain heavy metals have found unique applications in medicine. For instance, platinum compounds are used in chemotherapy drugs to treat various types of cancer. These drugs, including cisplatin, are effective in killing cancer cells.

The mining, processing, and disposal of heavy metals can contribute to significant environmental degradation. Heavy metals can accumulate in water bodies, soil, and living organisms, causing harm to ecosystems and biodiversity. Mercury (Hg) is particularly notorious for its environmental impact. Its application in artisanal and small-scale gold mining emits significant amounts of mercury into the environment, which can bioaccumulate in fish and enter the human food chain, leading to neurological and developmental problems (Gibb and O'Leary, 2014). Lead (Pb) and cadmium (Cd) are extremely poisonous and can pollute water sources and soil, posing health hazards to humans and wildlife. Exposure to lead is especially detrimental to children, impairing cognitive development and causing behavioral issues. Heavy metals are associated with a range of health issues. Prolonged contact to arsenic (As) through drinking water has been linked to skin lesions, cardiovascular diseases, and an increased risk of cancer. Occupational exposure to chromium (Cr), especially its hexavalent form, can cause lung cancer, while nickel (Ni) exposure is associated with respiratory illnesses and allergic reactions. Cadmium exposure, primarily through smoking and dietary sources, can lead to kidney damage and bone fragility (Satarug and Moore, 2004).

The cleanup and remediation of sites contaminated by heavy metals require significant financial investment. The long-term healthcare costs associated with exposure to heavy metals also place a burden on economies, particularly in regions with inadequate regulation and oversight (Tchounwou et al., 2012).

Recently, worldwide, the development of urbanization and increased industrial belts progress toward the environmental emissions of toxic heavy metal ions into the terrestrial environment and aquatic environment. Mining fields increase the production of high concentration of heavy metal wastes into the environment which cause the contamination of the ecosystem.

The metals are dispersed into the environmental water or wind in an uncontrolled manner by leaching process. The levels of heavy metal ions in the environment cause consequential human health effects because of their nondegradable characteristics that make them persevering for affecting human health and ecosystem for a longer period of time. Communities living near mines or industrial facilities that process heavy metals often bear the brunt of pollution, facing health risks and environmental degradation. This raises ethical questions about environmental justice. The method for the extraction of low-grade ores like cyanidation process generates hydrogen cyanide that is emitted into the environment and increases global warming and the production of huge amounts of tailings which is directly a source of heavy metals (HMs) (Fashola et al., 2016).

1.4 VARIOUS TREATMENT PROCESSES FOR HEAVY METAL REMOVAL

Various heavy metal ions such as lead (Pb), mercury (Hg), cadmium (Cd), nickel (Ni), chromium (Cr), copper (Cu), arsenic (As), platinum (Pt), and palladium (Pd) are well-known environmental pollutants due to their harmful effects on living organisms and ecosystems. The removal of these pollutants from wastewater has been a major focus of extensive research, focusing on the development of different treatment methods. This review delves into the prominent treatment strategies, discussing their procedures, advantages, and disadvantages, alongside necessary improvements highlighted in the literature.

1.4.1 Chemical Precipitation

Chemical precipitation (also called coagulation precipitation) involves adding precipitating agents to wastewater to convert dissolved metals into insoluble forms. Common agents include calcium hydroxide ($Ca(OH)_2$), sodium hydroxide (NaOH), and sodium sulfide (Na_2S). The precipitates are subsequently removed from the water via sedimentation and filtration. It is a simple and well-established method that can treat large volumes of wastewater efficiently. It requires relatively low operational skills and equipment. The generation of large volumes of sludge, which requires further treatment and disposal, poses environmental and economic challenges.

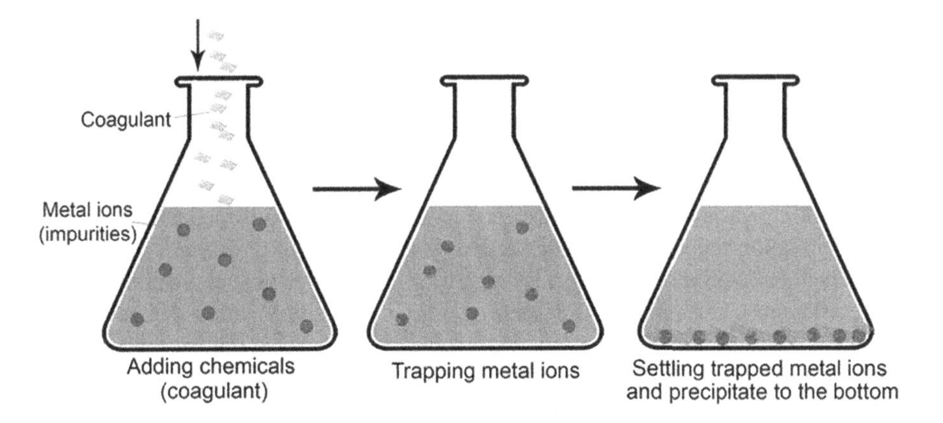

FIGURE 1.1 Schematic process of chemical precipitation. Reproduced with copyright from Ref. Qasem et al. (2021).

The process is also less effective for metals at low concentrations. The chemical precipitation process is represented in Figure 1.1.

Studies have explored the use of novel precipitating agents and process optimization to reduce sludge volume and improve metal removal efficiency. For example, Brboot et al. (2011) studied single-component and multi-component magnesium hydroxide, lime, and caustic soda precipitation in various dose ranges from 1 to 5.0 g/L for removal of different heavy metals like iron, chromium, copper, lead, and cadmium and obtained results of about 99% recovery in the pH range of 9.5–10 (with MgO precipitant) and 11.5–12 (with CaO precipitant).

Hydroxide precipitation is added to wastewater with stirring to generate metal hydroxide precipitates that are insoluble. This characteristic of hydroxide is used as it is relatively less expensive and simple and has an adjustable pH. High pH values of precipitates are disadvantageous as they require a high dosage of precipitates (Park et al., 2014). Calcium oxide or calcium hydroxide (CaO or $Ca(OH)_2$) is mostly used for high limits of heavy metal ions of 1,000 mg/L. Benalia et al. (2021) used lime, caustic soda, and soda ash for the withdrawal of simultaneous heavy metals like Cu and Zn in a laboratory scale with a reagent dose of 10–400 mg/L and removal efficiency of 90% and obtained $Zn(OH)_2$ and $Cu(OH)_2$ as precipitates.

For higher removal of toxic metal ions, sulfide precipitation is recommended for execution at neutral pH (Anotai et al., 2017).

$$\text{Metal}^{n+} + \text{S}^{2-} \leftrightarrow \text{Metal}_n\text{S} \downarrow \tag{1.1}$$

Besides sulfide precipitation, carbonate precipitation shows good effectiveness in heavy metal ion removal at lower pH values.

$$\text{Metal}^{n+} + n\text{NaCO}_3 \leftrightarrow n\text{Metal(CO}_3) \downarrow + n\text{Na}^+ \tag{1.2}$$

Li et al. (2020) studied the co-precipitation of Fe^{2+}, Cu^{2+}, Zn^{2+}, Cd^{2+}, and Ni^{2+} by a mechanochemical reaction with $CaCO_3$ with efficiencies of 99%, 98.4%, and 93.8%

with lower filter residue humidity (less than 50%). Pohl (2020) showed that sodium dimethyldithiocarbamate (SDTC), 1,3-benzenediamidoethanethiol (BDETH2), 2,6-pyridinediamidoethanethiol (PyDET), a pyridine-based thiol ligand (DTPY), or ligands with extended sulfur chains can extract harmful heavy metals while producing toxic by-products, and that the efficiency of metal precipitation can be enhanced by using a higher dose of the precipitating agent. According to the studies by Álvarez et al. (2006), in today's world, research is trending toward the advancement of chemical precipitation together with other recovery techniques like reverse osmosis and photochemical oxidation for effective withdrawal of heavy toxic elements from the waste effluent. Charerntanyarak (1999) investigated the elimination of heavy metals like zinc, cadmium, manganese, and magnesium ions from the wastewater with a pH value of 1.9 by the usage of lime water with a concentration of greater than 9.5, followed by secondary treatment using sodium sulfide with a concentration of 250 mg/L. In some studies, it was shown that the removal efficiency of more than 90% of toxic heavy metals from Pikeville mine by the introduction of aqueous solution of 1,3 benzenediamidoethanethiol dianion from 194 to 0.009 ppm (Brboot., 2002). So, from the above investigations, it can be concluded that the chemical precipitation process by the use of an aqueous solution of chemicals aids in the elimination of toxic elements from mining ore or waste sludge to a high efficiency, but the reagent cost attaches to the expensiveness of the precipitation process and also to the toxicity of the process condition due to large amount of reagent dose.

1.4.2 Adsorption

Adsorption demands the accumulation of metal elements on the exterior of adsorbents such as activated carbon, biochar, or nanomaterials. The choice of adsorbent and conditions like pH and temperature significantly influence the process efficiency. It is highly effective for removing low concentrations of metals and can sometimes be reversible, allowing for the recovery of both the adsorbent and the metals. The process is versatile and can be tailored to target specific metals. A schematic diagram of the process of adsorption is shown in Figure 1.2. The cost of adsorbent materials and the potential for secondary pollution due to adsorbent disposal are significant drawbacks (Upadhyay et al., 2021). The process efficiency is also heavily dependent on the characteristics of the wastewater and the metals present.

Recent studies have focused on developing cost-effective and high-efficiency adsorbents from waste materials. For instance, S.V. Renge et al. (2012) examined various low-cost, safe, and economical adsorbents and their potential applications for agricultural waste by-products like seaweed, algae, chitosan, eggshells, and sawdust for the removal of heavy metal ions from contaminated water.

Gedda et al. (2024) successfully investigated the removal of heavy metal ions such as cadmium, lead, and chromium from industrial wastewater, achieving removal efficiencies of 95.6%, 99.5%, and 99.5%, respectively, using a novel diphenylamine-coordinated cobalt complex (Co-DPA). Additionally, they optimized adsorption parameters, including dosage, pH, initial metal concentration, and adsorption time. Moreover, Ali et al. (2023) investigated and found that zinc and iron metal ions can be removed by using activated carbon as an adsorbent at different flow

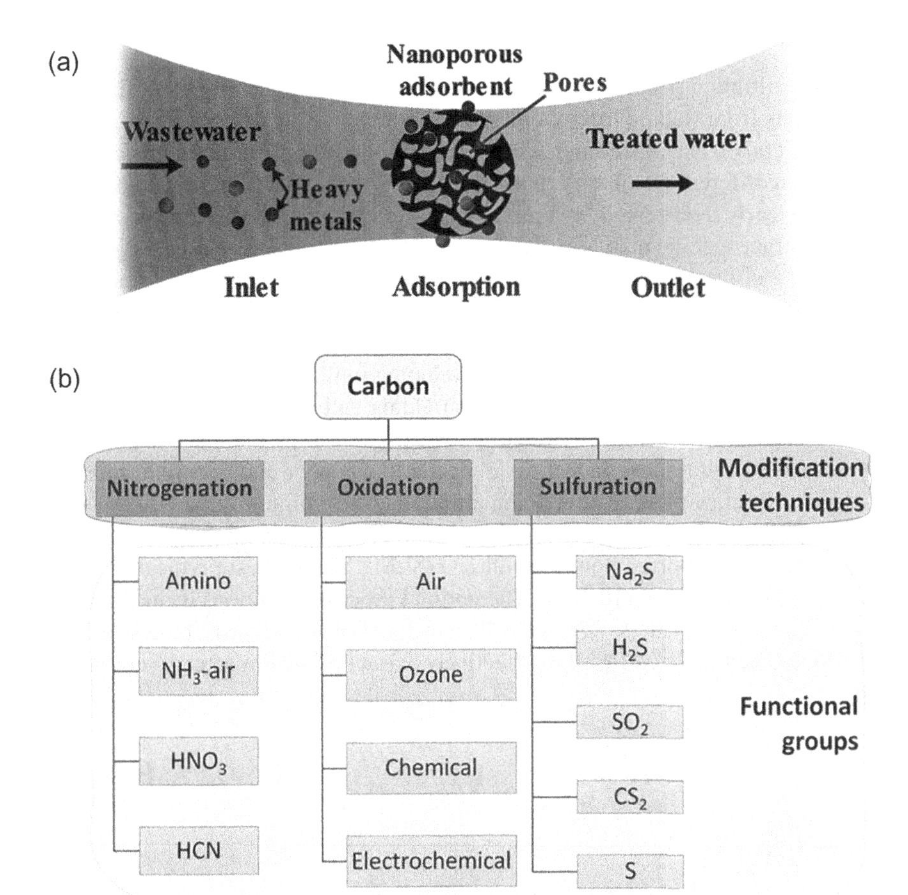

FIGURE 1.2 Adsorption process for water treatment. Reproduced with copyright from Ref. Qasem et al. (2021).

discharges and at different hydraulic parameters. The results concluded that zinc metal ion removal has an efficiency of 95% at pH 7, 20 L/h, and 5h and iron metal ions reached to 96.67% removal efficiency at pH 6, 20 L/h, and 5h.

In some investigations, it was shown that low concentrations of aqueous solution of amino functional mesoporous silica SBA-15 materials can remove heavy metals like copper, nickel, lead, and zinc from wastewater at a constant temperature of 25°C for removal efficiency of more than 90% (Aguado et al., 2009). In some observations, it is studied that the removal efficiency of copper, nickel, and chromium ions reached 90%, 68%, and 91%, respectively, with the help of bioadsorbents from *Moringa aptera* Gaertn (MAG) (Matouq, 2015). Thus, based on the findings from the above investigations, it can be concluded that the high cost of preparing adsorbents and maintaining process conditions highlights the limitations of the adsorption process, leading to the emergence of a more eco-friendly and cost-effective alternative for removal.

1.4.3 Membrane Filtration

Membrane filtration, through the usage of semi-permeable membranes, eliminates heavy metals from water. Different membrane separation techniques for elimination of heavy metals from wastewater like reverse osmosis, ultrafiltration, nanofiltration are based on different membrane characteristic like pore size, operation pressures etc which are clearly represented in Figure 1.3. It offers high removal efficiencies and the ability to target specific metals based on membrane selection. The process generates high-quality effluent that is suitable for reuse. High operational costs, membrane coagulation, and the need for pre-treatment to remove particulates are significant challenges. Energy consumption is also a concern for processes like reverse osmosis.

Research efforts are directed toward developing fouling-resistant membranes and energy-efficient processes. A study by Abu-Qdais and Moussa (2004) highlighted that treatment of industrial wastewater by reverse osmosis membrane and nanofiltration membrane can remove heavy toxic metals like copper and cadmium ions with an overall efficiency of 99.4% (reducing initial ion concentration from 500 to 3 ppm). P. Zaheri et al. (2015) developed a method for extracting europium (Eu) from a nitrate medium using a supported liquid membrane (SLM). The highest permeability coefficient recorded was 3.16×10^{-5} m/s. The optimal process conditions included 0.60 M Cyanex272 (carrier) with 2 mg/mL CNTs, a feed pH of 6, and a stripping solution of 1 M HNO_3. The SLM continued to effectively extract europium after 10 operational cycles.

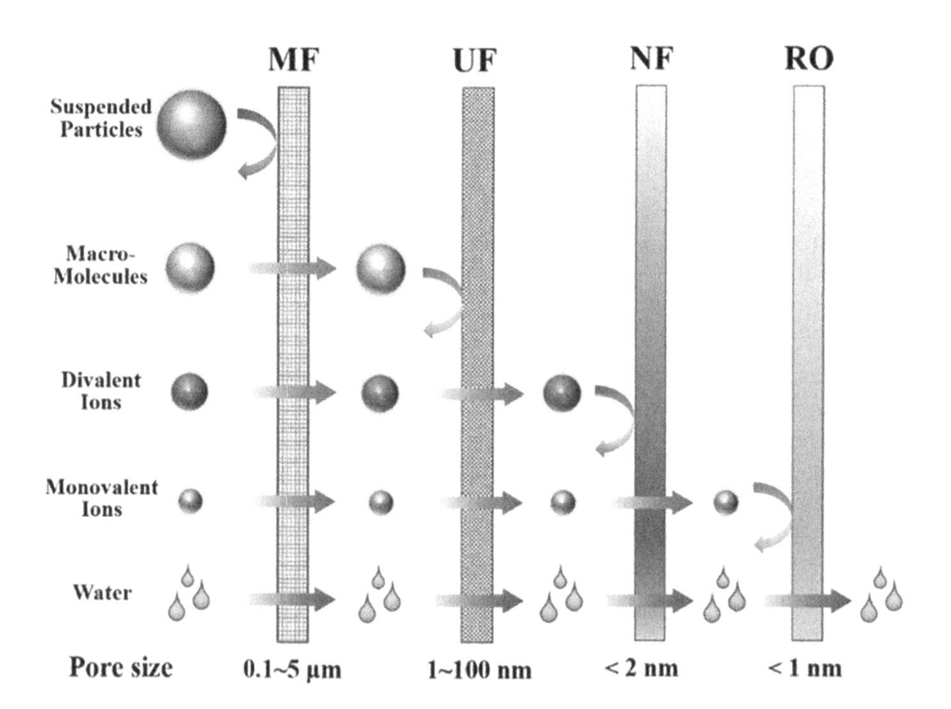

FIGURE 1.3 Elimination of heavy toxic metals by the membrane filtration method. Reproduced with copyright from Ref. Xiang et al. (2022).

Functional groups of arsenic metals like arsenite (As(III)) and arsenate (As(V)) are removed by the direct contact membrane distillation (DCMD) method. This experiment was investigated by Luan et al. (2009) with polyvinylidene fluoride (PVDF) membranes. The experimental results explain that permeate containing traces of As(III) and As(V) in the concentration range of 10 μg/L was obtained with a feed solution concentration of 40 and 2,000 mg/L at 450 h and the membrane performance limit was set at 0.5 mg/L. The traces of arsenic concentration in the permeate solution prove the arsenic removal efficiency. In some investigations, it is observed that the removal efficiency of copper, nickel, and zinc from initial concentrations of 474, 3.3, and 167 mg/L, respectively, to a final concentration of 0.05 mg/L, was achieved with a stable membrane flux concentration of 80 L/m^2hr and pressure of less than 100 mbar in a submerged microfiltration module. Therefore, from the investigations, it can be concluded that the membrane separations process can be disadvantageous due to several factors such as high operational costs, membrane fouling, selective rejections, concentration polarization, limited pH range, and complexity in manual handling which leads to the creation of the bioleaching process as a clean technology for heavy metal removal.

1.4.4 Ion Exchange

Ion exchange involves exchanging ions between a solution and an ion exchange resin to remove metals. The resins can be specific to certain metals, making the process highly selective. It is effective for both high and low concentrations of metals and allows for the recovery of metals and the reuse of resins after regeneration. The initial cost of resins and the need for periodic regeneration, involving the use of chemicals, limit the process's sustainability and economic feasibility. Figure 1.4 clearly illustrates the ion exchange method for elimination of different metal ions.

Innovations in resin technology focus on enhancing capacity, selectivity, and durability. For example, research by Dabrowski et al. (2004) presented elimination of toxic substances like Pb, Hg, Cd, Ni, V, Cr, Cu, and Zn from waterbodies and industrial wastewater by different types of ion exchangers.

Al-Enezi et al. (2004) studied the elimination of toxic elements using ion exchange resin on a laboratory scale from municipal wastewater sludges collected from the Ardiya plant in Kuwait, achieving a high efficiency of 99% across the concentration range found in wastewater effluents and sludges. M.S. Chauhan et al. (2023) explored the combined process of removing toxic metals from waste effluent of the Swarnamukhi River in Tirupati using ion exchange and membrane filtration. The process achieved removal efficiencies of 95% for total suspended solids (TSS) and 93.33% for total dissolved solids (TDS), as well as 76.25% for Biological Oxygen Demand (BOD) and 85% for Chemical Oxygen Demand (COD). J.A.S. Tenório (2001) used commercial resins for withdrawal of chromium metal ions from chromium plating companies. The rinse tank from where chromium ions are recovered is 150 mL capacity of cationic resin and another tank of same 150 mL capacity for the anionic resin. The feed solution is to be at a rate of 10 mL/min, with solutions containing 2% H_2SO_4 for the cationic resins and 4% NaOH for the anionic resin. The result shows chromium traces of less than 0.25 mg/L. S.-Y. Kang et al. (2004) experimented and

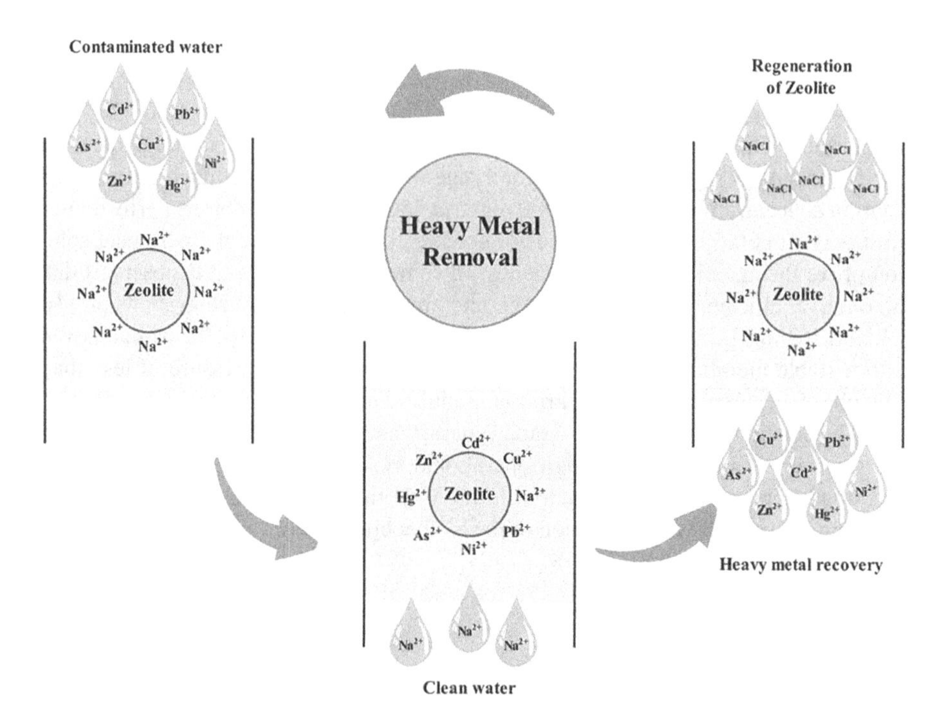

FIGURE 1.4 Ion exchange method for the elimination of metal ions. Reproduced with copyright from Ref. Aziz et al. (2023).

studied adsorption characteristics of many cation exchange resins like Co^{2+}, Ni^{2+}, and Cr^{3+} on an Amberlite IRN-77 cation in batch process system. The results were shown on adsorption isotherms like Langmuir isotherms. The experiment showed stability after 1 h. Co^{2+} and Ni^{2+} ions present in the solution showed similar results of adsorption efficiency on the resin after removal. With the above pros of ion exchange process, there are many cons of the process like high operational cost, resin fouling, selectivity issues, and water quality sensitivity nature which lead to the creation of the biohydrometallurgy process for heavy metal removal.

1.4.5 Electrochemical Treatment

The demand for hydrogen gas has been increasing due to its potential as a clean energy source. Electrolysis which is a key method for hydrogen production, where water is split into hydrogen and oxygen using an electric current indirectly facilitates the removal of these metals from waste water through various mechanisms. Further differential pulse voltammetry with scanning electron microscopy measurements investigates the proficiency of this electrochemical treatment method. This method is a profound green technology for the elimination of toxic metals from waste sludge by utilizing the selective electrode. The method requires high expenditure costs for using the high electricity source. The electrochemical method is shown in Figure 1.5.

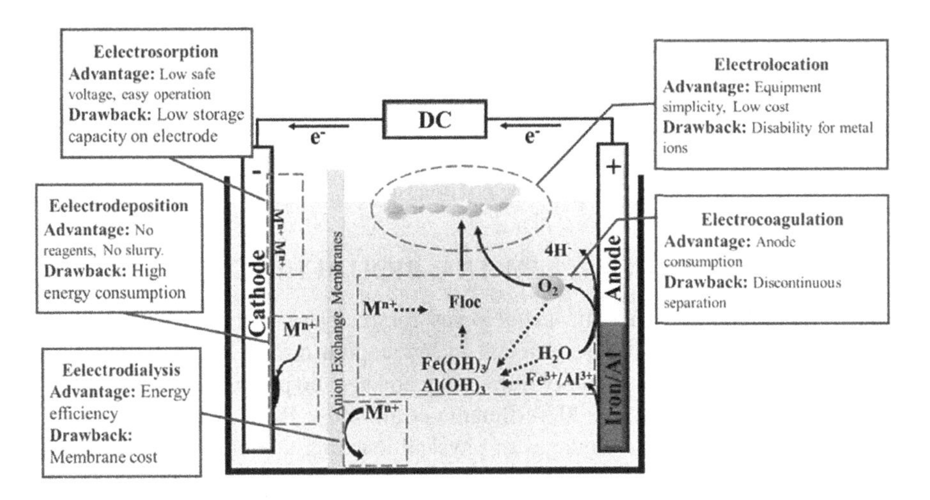

FIGURE 1.5 Electrochemical method process for heavy metal removal. Reproduced with copyright from Ref. Xu et al. (2023).

Further research improvements are increasing the EC process efficiency with AC/DC current using a selective electrode. For example, A.L.M.L. Mata et al. (2012) investigated the applicability and efficiency of the electrochemical process by using Fe electrodes in synthetic wastewater for elimination of toxic ions with a removal efficiency of 43%–100%. Yi Cui et al. (2009) researched on both low-value limits and high-value limits of heavy metal ion concentration like Cu, Cd, and Pb using both AC and DC current on CF-DO electrode with removal of more than 99.9%. Cristina Veronica Gherasim et al. (2014) investigated the withdrawal of lead from hydrated solutions using a batch electrodialysis system: increasing the feed value limit from 500 mg Pb/L led to a highest increase in the CE value, and vice versa. The experimental results demonstrated a fivefold increase in the concentrate section under optimal operating conditions, including an applied potential of 10V, a flow rate of 70 L/h, a temperature of 25°C, and a feed solution value of 1–2 mg Pb/L. In some investigations, the heavy metal removal leads to COD removal from the wastewater by more than 67% by the usage of electrocoagulation or electrochemical Fenton process for the wastewater treatment process at a total time of 4 min at a flux density of current of 24.2 mA/cm^2, whereas the efficiency rises to more than 99% removal at a fluid flow rate of 26, 35, 46, 53, and 61 mL/min and current density of 1, 2, and 3 A/dm^2 (Li et al., 2018; Basha et al., 2011). However, electrochemical methods that consist of electrocoagulation, electrodeposition, and electroflotation process have some disadvantageous factors like high electricity cost, electrode issues, electrochemical reaction conditions, and sludge generation which make this process a negative aspect for metal removal.

1.5 BIOLOGICAL TREATMENT

Biological treatment leverages microorganisms or plants to bioaccumulate or biotransform heavy metals. Techniques include bioremediation, phytoremediation, and the use of constructed wetlands. It also includes bioleaching for the elimination of toxic elemental ions from the surroundings. This approach is environmentally friendly, cost-effective, and capable of treating low concentrations of metals. It also offers the potential for metal recovery and the restoration of ecosystems. The process is slower than physical and chemical methods, with effectiveness dependent on the biological agents and environmental conditions. Controlling and optimizing the conditions for maximum efficiency can be challenging.

1.6 ACHIEVEMENT OF A NEW ERA: BIOLEACHING

One of the green eco-friendly technologies for recovery of heavy metals is the bioleaching process. It is more efficient in low capital cost investment (setup and operational objectives), low energy utilization, and also no poisonous waste material production (Dąbrowski et al., 2004; Benalia et al., 2021; Pohl, 2020; Anotai et al., 2007). It resolves the disadvantages of physical leaching and chemical leaching in terms of energy consumption, pollution, and reagent/adsorbent requirement for the process (Nguyen et al., 2021; Pathak et al., 2009; Liapun and Motola, 2023; Pathak et al., 2021). It has gained attraction in different chemical industries like mineral waste sludge and solid industrial by-products (e.g., galvanic sludge, electronic waste, sewage sludge, fly ash, residual slag, spent petrochemical catalysts, medical waste, and spent batteries). In these types of manufacturing industries, the mass of toxic substances is low but the trace of specific substances is very high which cause the environmental pollution risk. This tends to move the world toward the favorable biological treatment process which is represented by a schematic diagram in Figure 1.6.

The bioremediation or bioleaching process has the ability to eliminate toxic metals from solid wastes by converting them into an aqueous mixture using microbes, namely bacteria (autotrophic and heterotrophic) and also fungi. In this process, heavy metal ions can be removed by microorganisms in two methods: one is the indirect method, and another is the direct bioleaching method (Semerci et al., 2019). In the

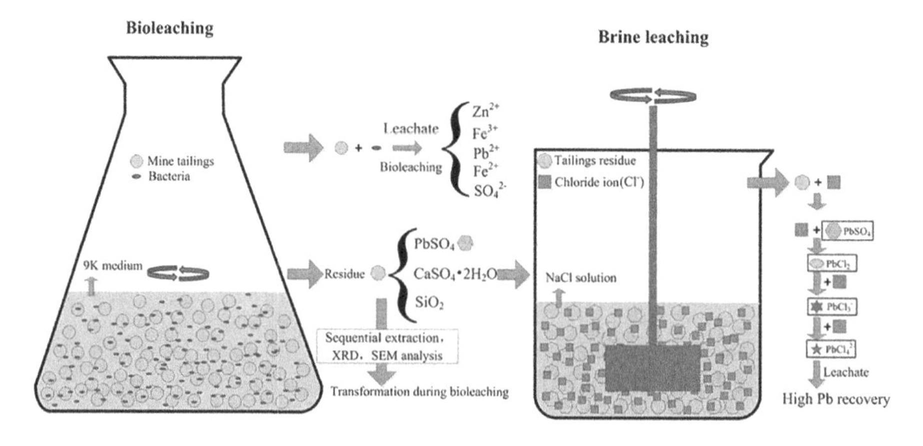

FIGURE 1.6 Bioleaching treatment process for heavy metal removal. Reproduced with copyright from Ref. Ye et al. (2017).

indirect method process, the bacteria have no direct contact with sewage but can convert indirectly the waste materials through their metabolic activities, whereas in the direct method, bacteria directly metabolize the solid wastes. As a result of which, the metals are soluble due to the electrochemical reaction process, Naseri et al. (2023) studied and invented that growth of autotrophic bacteria depends on CO_2 fixation from the atmosphere and O_2 becomes the terminal electron acceptor in the growth metabolism of bacteria. The mechanism of the reaction is illustrated below:

$$O_2 + 4H^+ + 4e \rightarrow 2H_2O$$

Autotrophic bacteria are directly related to the photosynthesis process due to the presence of pigment-containing vesicles called chromatophores, whereas heterotrophic bacteria are parasitic, saprophytic, or symbiotic. Some common examples of autotrophic bacteria for recovering heavy metal ions from electronic/industrial wastes are gram-negative bacteria which consist of *Acidithiobacillus* microorganism (aerobic bacteria) (Peter et al. 2022; Misra and Rhee, 2010). However, heterotrophic microorganisms need organic materials as the carbon source for their activity in microbial growth metabolism which includes organic acids like oxalic acid, citric acid, and nitric acid which are secreted during their metabolism in the substrate medium, that is, culture medium. In this type of microorganisms, protein is also catabolized which forms non-acidic complexes resulting in an alkaline medium in the leaching substance. Such heterotrophic microorganisms that are responsible for heavy metal recovery from different types of wastes are *Aspergillus* sp. and *Penicillium* sp. which constitute the fungi group, whereas strains of *Gluconobacter* sp., *Bacillus* sp., and *Chromobacterium* sp. constitute the bacteria group. In addition to fungi, cyanobacteria that secrete cyanides in the medium as the secondary metabolite constitute *Pseudomonas* strains, *Bacillus megaterium*, and *Chromobacterium violaceum* that follow the same mechanisms like the fungal bioleaching process (Vakilchap et al., 2020; Wang et al., 2021; Liang et al., 2014).

The fungal metabolism in the bioleaching process is linked with the generation of organic acids (Arwidsson et al., 2009; Brisson et al., 2020). The three mechanisms that are associated with fungal bioleaching processes are acidolysis, complexolysis, and redoxolysis. Acidolysis is akin to the acid leaching process, where protons from organic acids dissolve metals, and the protonation of oxygen atoms covers the surface of the solid waste. In the complexolysis mechanism, organic acid-metal complexes are formed through the carboxyl and hydroxyl groups of the organic acid. In the redoxolysis mechanism, metal mobilization from the solid waste occurs due to oxidation-reduction reactions. During the process mechanism, microbial growth metabolism generates high energy by electron transfer. Table 1.2 depicts the recovery of individual metal ions from different solid wastes by the help of microbes.

The biorecovery efficiency of heavy metal ions is influenced by different parameters, including individual concentration of the substrate constituents, concentrations of O_2 and CO_2, pH value, temperature value, volume density of inoculum volume density, particle size in the waste, solid-to-liquid flow rate ratio, bioleaching duration, and mixing speed. Challenges such as slow reaction kinetics, economic considerations, and safety concerns need to be addressed before large-scale commercialization or

TABLE 1.2
Microbes Used for Bioleaching Process of Toxic Metals from Different Types of Waste Sewage

Microorganism	Solid wastes	Target metals
Autotrophic		
Acidithiobacillus thiooxidans	TPCBs	Cu and Au
	LEDs	Cu, Ni, and Ga
	Tannery sludge	Cr
	Carbide slag	Zn, Ba, Ni, and Li
	Refinery spent catalyst	Ni, V, Mo, and Al
	SCCs	Li, Co, and Mn
	MPPCBs	Ni and Cd
Acidithiobacillus ferrooxidans	Mine tailings	Te
	PCBs	Cu, Ni, and Fe
	PCBs	Cu and Ni
	WLED	Cu, Ni, and Ga
	SCCs	Li, Co, and Mn
	LED	Cu, Ni, and Ga
	PCBs	Cu
	Low-grade ore	Cu
Heterotrophic (fungi)		
Penicillium citrinum	LIBs	Li and Mn
Aspergillus niger	Zinc plant purification residue	Zn, Co, and Mn
Aspergillus niger	LCD	In
A. niger, *Pseudomonas putida*, *Pseudomonas koreensis* and *P. bilaji*	Iron-rich laterite ore	Co and Ni
A. niger	Phosphorites	U, Sm, Th, and La
A. niger	Waste - Printed Circuit Boards (WPCBs) which are 3-6 wt % of the total electronic wastes	Ni, Cu, and Zn
A. niger, *Penicillium simplicissimum*	Vanadium-rich power plant residual ash	Ni and V
Heterotrophic (bacteria)		
Bacillus foraminis	AMOLED displays	Ag, Mo, and Cu
KB3B1 strain	Pyrolusite	Mn
Bacillus megaterium	Sulfide concentrate	Ni and Co

Source: Reproduced with copyright from Ref. Naseri et al. (2023).

industrial application or industrialized. Further research should be clear on optimizing the process variables which are the backbone of the bioleaching process (Dong et al., 2023). Although many studies have been published on the bioleaching process (use of microorganisms) to recover heavy metal ions, there is a relative deficiency of research studies on toxic elemental ion removal from metal tailing or mining resources, and there is a deviation in the development of the methods required for bioleaching processes. However, still bioleaching has many approaches in the usage of tailings bioremediation to increase the retrieval efficiency of metal substances in bioleaching technology (Nguyen et al., 2021). Bioleaching has achieved greater attention for its decontaminating nature which removes the toxic metals from the soil which is contaminated by the heavy metallic elements (Yang et al., 2018).

According to the studies of Srichandan et al. (2013), the biohydrometallurgy is a fascinating and increasingly technique, which leverages microbial processes to extract valuable metals from ores, concentrates, and waste materials, offering several benefits, especially in terms of environmental sustainability. The bioleaching process also extracts heavy metals from low-grade sulfide ores and concentrates that cannot be treated by conventional methods. Recently, bioleaching highlighted the process of decontaminating solid wastes like eliminating toxic heavy metals from any contaminated soils (Yang et al., 2018). It is the method of extracting heavy metal substances from mining ores, industrial sewage, or polluted soil. It occurs by the process of converting insoluble metal compounds into soluble forms by leaching methods with the help of microbes like bacteria and fungi. Bioleaching offers several advantages. It is environmentally friendly, as it requires less energy and produces fewer pollutants than traditional metal extraction methods. In conclusion, bioleaching so far is the efficient recovery treatment process for heavy metals than any other conventional processes and the rate of leaching can be improvised by genetic mutation of bioleaching microorganisms on the development of genetic engineering in the future (Bosecker, 1997).

REFERENCES

Abu-Qdais, H., H. Moussa (2004), Removal of heavy metals from wastewater by membrane processes: a comparative study, *Desalination*, 164(2), 105–110.

Afzal, A.M., M.H. Rasool, M. Waseem, B. Aslam (2017), Assessment of heavy metal tolerance and biosorptive potential of *Klebsiella variicola* isolated from industrial effluents, *AMB Express*, 7, 184. DOI: 10.1186/s13568-017-0482-2.

Aguado, J., J.M. Arsuaga, A. Arencibia, M. Lindo, V. Gascón (2009), Aqueous heavy metals removal by adsorption on amine-functionalized mesoporous silica. *Journal of Hazardous Materials*, 163(1), 213–221.

Akpor, O. (2014), Heavy metal pollutants in wastewater effluents: sources, effects and remediation, *Advances in Bioscience and Bioengineering*, 24(4), 11. DOI: 10.11648/j.abb.20140204.11

Al-Enezi, G., M.F. Hamoda, N. Fawzi (2004), Ion exchange extraction of heavy metals from wastewater sludges, *Journal of Environmental Science and Health*, 39(2), 455–464.

Ali, G.A., N.Q.M. Salih1, G.A. Faroun, R.F.C. Al-Hamadani (2022), Adsorption technique for the removal of heavy metals from wastewater using low-cost natural adsorbent, *IOP Conference Series: Earth and Environmental Science*, 1129, 012012. DOI: 10.1088/1755-1315/1129/1/012012.

Álvarez, M.T., C. Crespo, and B. Mattiasson (2007). Precipitation of Zn (II), Cu (II) and Pb (II) at bench-scale using biogenic hydrogen sulfide from the utilization of volatile fatty acids. *Chemosphere*, 66(9), 1677–1683

Anotai, J., P. Tontisirin, P. Churod (2007), Integrated treatment scheme for rubber thread wastewater: sulfide precipitation and biological processes, *Journal of Hazardous Materials*, 141, 1–7.

Arwidsson, Z., E. Johansson, T. von Kronhelm, B. Allard, P. van Hees (2010), Remediation of metal contaminated soil by organic metabolites from fungi I – production of organic acids. *Water, Air, and Soil Pollution*, 205, 215–226.

Azimi, A., A. Azari, M. Rezakazemi, M. Ansarpour (2017), Removal of heavy metals from industrial wastewaters: a review, *ChemBioEng Reviews*, 4(1), 37–59.

Aziz, K.H.H., F.S. Mustafa, K.M. Omer, S. Hama, R.F. Hamarawf, K.O. Rahman (2023), Heavy metal pollution in the aquatic environment: efficient and low-cost removal approaches to eliminate their toxicity: a review. *Royal Society of Chemistry*, 13, 17595.

Basha, C.A., M. Somasundaram, T. Kannadasan, C.W. Lee (2011), Heavy metals removal from copper smelting effluent using electrochemical filter press cells. *Chemical Engineering Journal*, 171(2), 563–571.

Benalia, M.C., Y. Leila, M.G. Bouaziz, A. Samia, M. Hayet (2021), Removal of heavy metals from industrial wastewater by chemical precipitation: mechanisms and sludge characterization, *Arabian Journal for Science and Engineering*, 47, 5587–5599. DOI: 10.1007/s13369-021-05525-7.

Bosecker, K. (1997), Bioleaching: metal solubilization by microorganisms, *FEMS Microbiology Reviews*, 20(3–4), 591–604.

Brboot, M.M., B.A. Abi, N.M. Al-Shuwaik (2011), Removal of heavy metals using chemical precipitation, *Engineering and Technology Journal*, 29(3), 595–612.

Brisson, V.L., W.Q. Zhuang, L. Alvarez-Cohen (2020), Metabolomic analysis reveals contributions of citric and citramalic acids to rare earth bioleaching by a Paecilomyces fungus. *Frontiers in Microbiology*, 10, 3008.

Charerntanyarak, L. (1999), Heavy metals removal by chemical coagulation and precipitation. *Water Science and Technology*, 39(10–11), 135–138.

Chauhan, M.S., A.K. Rahul, S. Shekhar, S. Kumar (2023) Removal of heavy metal from wastewater using ion exchange with membrane filtration from Swarnamukhi river in Tirupati, *Materials Today: Proceedings*, 78(1), 1–6.

Dąbrowski, A., Z. Hubicki, P. Podkościelny, E. Robens (2004), Selective removal of the heavy metal ions from waters and industrial wastewaters by ion-exchange method, *Chemosphere*, 56(2), 91–106.

Dong, Y., J. Zan, H. Lin (2023), Bioleaching of heavy metals from metal tailings utilizing bacteria and fungi: mechanisms, strengthen measures, and development prospect, *Journal of Environmental Management*, 344, 118511.

Fashola, M.O., V.M. Ngole-Jeme, O.O. Babalola (2016), Heavy metal pollution from gold mines: environmental effects and bacterial strategies for resistance, *International Research of Environmental Research and Public Health*, 13(11), 1047. DOI: 10.3390/ijerph13111047.

Fishbein, L. (1981), Sources, transport and alterations of metal compounds: an overview. I. Arsenic, beryllium, cadmium, chromium, and nickel. *Environmental Health Perspectives*, 40, 43–64.

Forruque Ahmed, S., P. Senthil Kumar, M. Rodela Rozbu, A.T. Chowdhury, S. Nuzhat, Nazifa Rafa, T.M.I. Mahlia, H. Chyuan Ong, M. Mofijur (2022), Heavy metal toxicity, sources, and remediation techniques for contaminated water and soil, *Environmental Technology and Innovations*, 25, 102114.

Gherasim, C.V., J. Křivčík, P. Mikulášek (2014), Investigation of batch electrodialysis process for removal of lead ions from aqueous solutions, *Chemical Engineering Journal*, 256, 324–334.

Ghosh, S., A. Othmani, A.Malloum, O. KeChrist, H. Onyeaka, S.S. AlKafaas, N.D. Nnaji, C. Bornman, Z.T. Al-Sharify, S. Ahmadi, M.H. Dehghani, N.. Mubarak, I. Tyagi, R.R. Karri, J.R. Koduru (2022), Removal of mercury from industrial effluents by adsorption and advanced oxidation processes: a comprehensive review, *Journal of Molecular Liquids*, 367, 120491.

Gibb, H., K.G. O'Leary, (2014), Mercury exposure and health impacts among individuals in the artisanal and small-scale gold mining community: a comprehensive review. *Environmental Health Perspectives*, 122(7), 667–672.

Harrison, R.M., C.R. Williams, I.K. O'Neill (1981), Characterization of airborne heavy metals within a primary zinc-lead smelting works. *Environmental Science & Technology*, 15(10), 1197–1204.

Hussain, A., S. Madan, R. Madan (2021), Removal of heavy metals from wastewater by adsorption. *Heavy Metals - Their Environmental Impacts and Mitigation*, IntechOpen, London. DOI: 10.5772/intechopen.95841

Ishchenko, V.A. (2018), Environment contamination with heavy metals contained in waste. *Environmental Problems*, 3(1), 21–24.

Kang, S.-Y., J.-U. Lee, S.-H. Moon, K.-W. Kim (2004), Competitive adsorption characteristics of Co2+, Ni2+, and Cr3+ by IRN-77 cation exchange resin in synthesized wastewater, *Chemosphere*, 56(2), 141–147.

Li, X., Q. Zhang, B. Yang (2020), Co-precipitation with CaCO3 to remove heavy metals and significantly reduce the moisture content of filter residue, *Chemosphere*, 239, 124660.

Liang, C.J., J.Y. Li, C.J. Ma (2014), Review on cyanogenic bacteria for gold recovery from E-waste, *Advanced Materials Research*, 878, 355–367.

Liao, X., M. Ye, J. Liang, S. Li, Z. Liu, Y. Deng, Z. Guan, Q. Gan, X. Fang, S. Sun (2022), Synergistic enhancement of metal extraction from spent Li-ion batteries by mixed culture bioleaching process mediated by ascorbic acid: Performance and mechanism, *Journal of Cleaner Production*, 380, 134991.

Liapun, V., M. Motola (2023), Current overview and future prospective in fungal biorecovery of metals from secondary sources, *Environmental Management*, 332, 117345. DOI: 10.1016/j.jenvman.2023.117345.

Liu, C., P. Hsu, J. Xie, J. Zhao, K. Liu, J. Xu, J. Tang, Z. Ye, D. Lin, Y. Cui (2019), Direct/alternating current electrochemical method for removing and recovering heavy metal from water using graphene oxide electrode, *ACS Nano*, 13(6), 6431–6437.

Mahurpawar, M. (2015), Effects of heavy metals on human health, *International Journal of Research*, 3(9SE), 1–7.

Makuza, B., Q. Tian, X. Guo, K. Chattopadhyay, D. Yu (2021), Pyrometallurgical options for recycling spent lithium-ion batteries: a comprehensive review. *Journal of Power Sources*, 491, 229622.

Masindi, V., K.L. Muedi (2018), Environmental contamination by heavy metals, *Heavy Metals*, 10(4), 115–133.

Matouq, M., N. Jildeh, M. Qtaishat, M. Hindiyeh, M.Q. Al Syouf (2015), The adsorption kinetics and modeling for heavy metals removal from wastewater by Moringa pods. *Journal of Environmental Chemical Engineering*, 3(2), 775–784.

Mavrov, V., T. Erwe, C. Blöcher, H. Chmiel (2003), Study of new integrated processes combining adsorption, membrane separation and flotation for heavy metal removal from wastewater. *Desalination*, 157(1–3), 97–104.

Mirazimi, S.M.J., Z. Abbasalipour, F. Rashchi (2015), Vanadium removal from LD converter slag using bacteria and fungi, *Journal of Environmental Management*, 153, 144–151.

Mishra, D., Y.H. Rhee (2010), Current research trends of microbiological leaching for metal recovery from industrial wastes. *Current Research, Technology and Education Topics in Applied Microbiology and Microbial Biotechnology*, 2, 1289–1292.

Mitra, S., A.J. Chakraborty, A.M. Tareq, T.B. Emran, F. Nainu, A. Khusro, A.M. Idris, M.U. Khandaker, H. Osman, F.A. Alhumaydhi, J. Simal-Gandara (2022), Impact of heavy metals on the environment and human health: novel therapeutic insights to counter the toxicity, *Journal of King Saud University*, 34(3), 101865

Müller, A., H. Österlund, J. Marsalek, M. Viklander (2020), The pollution conveyed by urban runoff: a review of sources, *Science of the Total Environment*, 709, 136125.

Naseri, T., V. Beiki, S.M. Mousavi, S. Farnaud (2023), A comprehensive review of bioleaching optimization by statistical approaches: recycling mechanisms, factors affecting, challenges, and sustainability, *RSC Advances*, 13(34), 23570–23589.

Nguyen, T.H., S. Won, M.-G. Ha, D.D. Nguyen, H.Y. Kang (2021), Bioleaching for environmental remediation of toxic metals and metalloids: a review on soils, sediments, and mine tailings, *Chemosphere*, 282, 131108.

Nicomel, N.R., K. Leus, K. Folens, P. VanDer Voort, G. Du Laing (2015), Technologies for arsenic removal from water: current status and future perspectives, *International Journal of Environmental Research and Public Health*, 13, 1–24. DOI: 10.3390/ijerph13010062.

Odumbe, E., S. Murunga, J. Ndiiri (2023), Heavy metals in wastewater effluent: causes, effects, and removal technologies. *Trace Metals in the Environment*, Springer Nature, Berlin, Germany. DOI: 10.5772/intechopen.1001452.

Okorondu, J., N.A. Umar, C.O. Ulor, C.G. Onwuagba, B.E. Diagi, S.I. Ajiere, C. Nwaogu (2022), Anthropogenic activities as primary drivers of environmental pollution and loss of biodiversity: a review, *Journal of Trend in Scientific Research and Development*, 6(4), 621–643.

Park, J.-H., G.-J. Choi, S.-H. Kim (2014), Effects of pH and slow mixing conditions on heavy metal hydroxide precipitation, *Journal of the Korea Organic Resource Recycling Association*, 22, 50–56.

Pathak, A., M.G. Dastidar, T.R. Sreekrishnan (2009), Bioleaching of heavy metals sewage sludge: a review, *Journal of Environmental Management*, 90(8), 2343–2353.

Pathak, A., R. Kothari, M. Vinoba, N. Habibi, V.V. Tyagi (2021), Fungal bioleaching of metals from refinery spent catalysts: a critical review of current research, challenges, and future directions, *Journal f Environmental Management*, 280(15), 111789.

Peter, D., L. Shruti Arputha Sakayaraj, T.V. Ranganathan (2022), Recovery of precious metals from electronic and other secondary solid waste by bioleaching approach. *Biotechnology for Zero Waste: Emerging Waste Management Techniques*, 10, 207–218.

Pohl, A. (2020), Removal of heavy metal ions from water and wastewater by sulphur coating precipitation agents, *Water, Air and Soil Pollution*, 231, 503.

Qasem, N.A.A., R.H. Mohammed, D.U. Lawal (2021), Removal of heavy metal ions from wastewater: a comprehensive and critical view, *Clean Water*, 4, 36. DOI: 10.1038/s41545-021-00127-0.

Qu, D., J. Wang, D. Hou, Z. Luan, B. Fan, C. Zhao (2009), Experimental study of arsenic removal by direct contact membrane distillation, *Journal of Hazardous Materials*, 163, 874–879.

Renge, V.C., S.V. Khedkar, S.V. Pande (2012), Removal of heavy metals from wastewater using low cost adsorbents: a review, *Scientific Reviews and Chemical Communications*, 2(4), 580–584.

Satarug, S., M.R. Moore, (2004), Adverse health effects of chronic exposure to low-level cadmium in foodstuffs and cigarette smoke. *Environmental Health Perspectives*, 112(10), 1099–1103.

Schippers, A., S. Hedrich, J. Vasters, M. Drobe, W. Sand, S. Willscher (2014), Biomining: metal recovery from ores with microorganisms. In: Schippers, A., Glombitza, F., Sand, W. (eds) *Geobiotechnology I. Advances in Biochemical Engineering/Biotechnology*, vol 141. Springer, Berlin, Heidelberg.

Semerci, N., B. Kunt, B. Calli (2019), Phosphorus recovery from sewage sludge ash with bioleaching and electrodialysis, *International Biodeterioration & Biodegradation*, 144, 104739.

Sonone, S.S., S.V Jadhav, M.S. Sankhla, R. Kumar (2020), Water contamination by heavy metals and their toxic effect on aquaculture and human health through food chain, *Letters in Applied NanoBioScience*, 10(2), 2148–2166.

Souza, K.R., D.R. Silva, W. Mata, C.A. Martinez-Huitle, A.L.M.L. Mata (2012), Electrochemical technology for removing heavy metals present in synthetic produced water, *Latin American Applied Research*, 42, 141–147.

Srichandan, H., D.J. Kim, C.S. Gahan, A.Akcil (2013), Microbial extraction metal values from spent catalyst: mini review, *Advances in Biotechnology*, 19, 225–239. https://www.researchgate.net/publication/236236619.

Tchounwou, P.B., C.G. Yedjou, A.K. Patlolla, D.J. Sutton (2012), Heavy metals toxicity and the environment, *Clinical and Environmental Toxicology*, 3, 133–164.

Tenório, J.A.S., D.C.R. Espinosa (2001), Treatment of chromium plating process effluents with ion exchange resins, *Waste Management*, 21(7), 637–642.

Upadhyay, U., I. Sreedhar, S.A. Singh, C.M. Patel, K.L. Anitha (2021), Recent advances in heavy metal removal by chitosan based adsorbents, *Carbohydrate Polymers*, 251, 117000.

Vakilchap, F., S.M. Mousavi, M. Baniasadi, S. Farnaud, (2020), Development and evolution of biocyanidation in metal recovery from solid waste: a review. *Reviews in Environmental Science and Bio/Technology*, 19, 509–530.

Vardhan, K.H., P.S. Kumar, R.C. Panda (2019), A review on heavy metal pollution, toxicity and remedial measures: current trends and future perspectives, *Journal of Molecular Liquids*, 290, 111197.

Verner, J.F., M.H Ramsey, E. Helios-Rybicka, B. Je˛drzejczyk, (1996), Heavy metal contamination of soils around a PbZn smelter in Bukowno, Poland. *Applied Geochemistry*, 11(1–2), 11–16.

Vidu, R., E. Matei, A.M. Predescu, B. Alhalaili, C.Pantilimon, C. Tarcea, C. Predescu (2020), Removal of heavy metals from wastewaters: a challenge from current treatment methods to nanotechnology applications, *Toxics*, 8(4), 101.

Wang, J., F. Faraji, J. Ramsay, A. Ghahreman (2021), A review of biocyanidation as a sustainable route for gold recovery from primary and secondary low-grade resources. *Journal of Cleaner Production*, 296, 126457.

Wuana, R.A., & F.E. Okieimen (2011), Heavy metals in contaminated soils: a review of sources, chemistry, risks and best available strategies for remediation Indian scholarly research network, *ISRN Ecology*, 2011, 2090–4614. DOI: 10.5402/2011/402647.

Xiang, H., X. Min, C.-J. Tang, M. Sillanpää, F. Zhao (2022), Recent advances in membrane filtration for heavy metal removal from wastewater: a mini review, *Journal of Water Process Engineering*, 49, 103023.

Xu, Y., Z. Zhong, X. Zeng, Y. Zhao, W. Deng, Y. Chen (2023), Novel materials for heavy metal removal in capacitive deionization, *Applied Sciences*, 13, 5635.

Ya, V., N. Martin, Y.H. Chou, Y.M. Chen, K.H. Choo, S.S. Chen, C.W. Li (2018) Electrochemical treatment for simultaneous removal of heavy metals and organics from surface finishing wastewater using sacrificial iron anode. *Journal of the Taiwan Institute of Chemical Engineers*, 83, 107–114.

Yang, Z., W. Shi, W. Yang, L. Liang, W. Yao, L. Chai, et al. (2018b), Combination of bioleaching by gross bacterial biosurfactants and flocculation: a potential remediation for the heavy metal contaminated soils, *Chemosphere*, 206, 83–91.

Ye, M., P.Yan, S. Sun, D. Han, X. Xiao, L. Zheng, S. Huang, Y. Chen, S. Zhuang (2017), Bioleaching combined brine leaching of heavy metals from lead-zinc mine tailings: transformations during the leaching process, *Chemosphere*, 168, 1115–1125.

Zaheri, P., T. Mohammadi, H. Abolghasemi, M.G. Maraghe (2015), Supported liquid membrane incorporated with carbon nanotubes for the extraction of Europium using Cyanex272 as carrier, *Chemical Engineering Research and Design*, 100, 81–88.

2 Global Strategies and Statistics of Biological Separation of Metal Ions From Industrial Waste

Sujoy Bose and Raj Kumar Das

2.1 INTRODUCTION

The metallic elements, well known as toxic and carcinogenic agents, and heavy metals have an intensity as daily-life usage of a few things like decorative arts, jewellery, coins, building structures, ships, vehicles, bridges, aircraft, batteries, etc. (Jadhav and Hochen, 2012; Krishnan et al., 2021; Razzak et al., 2022). Over the past two decades, environmental pollution has been raised by the larger population due to the regular usage of heavy metals eccentrically (Olaniran et al., 2013). These toxic heavy metals are regularly and consistently discharged into our environment by several industries including mining, metallurgical, electronic, batteries, pesticides and herbicides, electroplating, petrochemical, coal mining, refineries, tanning industries, paper and pulp industries, and metal finishing (Bazrafshan et al., 2015; Razzak et al., 2022).

According to the physicochemical features, the metals found in the waste can certainly be divided into four types oriented by the periodic table: (1) hazardous metals (Ag, As, Cr, Hg, Pb, Sr, Si, Ti); (2) risky radioactive metals (e.g., Am, Ra, Rn, Tc, Th, U); (3) metabolism-effective metals (Ca, Cu, Fe, K, Mo, Ni, Zn); and (4) metals detection of biologic effectiveness (B, Ge, Po, Sb, Te) (Das and Poater, 2021; Kumar and Nagendran, 2007). Based on the periodic table, 25 elements of heavy metals from natural and industrial sources are given as follows (Table 2.1):

2nd period: B (Group IIIA)
3rd period: Na (Group IA), Mg (Group IIA), Al (Group IIIA), Si (Group IVA)
4th period: K (Group IA), Ca (Group IIA), V (Group VB), Cr (Group VIB), Mn (Group VIIB), Fe (Group VIIIB), Co (Group VIIIB), Ni (Group VIIIB), Cu (Group IB), Zn (Group IIB), As (Group VA), Se (Group VIA)
5th Period: Pd (Group VIIIB), Ag (Group IB), Cd (Group IIB), Sb (Group VA)
6th Period: Hg (Group IIB), Tl (Group IIIA), Pb (Group IVA), Bi (Group VA)

The presence of these non-biodegradable and carcinogenic heavy metals in disproportionate amounts could cause critical human health issues. Table 2.2 summarises major

 DOI: 10.1201/9781003415541-2

TABLE 2.1
Presence of Metals in Various Types of Industrial Wastes (Jadhav and Hochen, 2012)

Waste Type	Metals in Waste
Electronic waste	Ag, Al, Au, Cu, Ni, Sn, Zn
Metal finishing industrial wastes	Ag, Au, Cd, Cu, Cr, Ni, Zn
MSW fly ash	Al, Cu, Cr, Ni, Pb, Zn
Petroleum spent catalyst	Co, Mo, Ni
Waste batteries	Ag, Cd, Ni
Waste X-ray films	Ag

TABLE 2.2
Sources of Heavy Metals with Health Impacts (Vidu et al., 2020; Barakat, 2011)

Heavy Metals	Sources of Heavy Metals	Health Effects in Organ
Arsenic (As)	Electronics industries and glass industries	Skin manifestations, brain injury, lungs exposition, kidney injury, metabolism, cardiovascular system, immune system, endocrine, visceral cancers, vascular disease
Cadmium (Cd)	Batteries, plastic and steel industries, metal refineries, paints	Bones, liver or kidney damage, lungs/brain problems
Copper (Cu)	Electronic and electrical industry, laboratory apparatus	Liver damage, kidney injury, brain, cornea, gastrointestinal distress system, lungs, immune system, and haematological system
Chromium (Cr)	Steel mills, tanneries, pulp mills	Skin infections, lungs and brain problems, kidney and liver damage, pancreas, allergic dermatitis, gastrointestinal system
Mercury (Hg)	laboratory apparatus, agriculture, electrical usages, refineries	Brain and lungs exposition, liver and kidney damage, nervous system, endocrine, and reproductive organism
Nickel (Ni)	Stainless steel production, nickel alloy manufacturing	Lung/kidney damage, gastrointestinal suffering, pulmonary fibrosis, skin exposition, nausea and coughing
Lead (Pb)	Lead-based batteries, alloys, ammunition, cable sheathing pigments, glazes, plastic stabilisers	Bones, brain, kidneys and liver problems, lungs, cardiovascular system, high blood pressure, reproductive system, spleen
Zinc (Zn)	Brass coating, rubber products, some cosmetics, and aerosol deodorants	Neurological signs, stomach cramps, skin irritations

sources, toxic impacts on health, and the permitted (by World Health Organisation (WHO)) quantity in drinking water of a few heavy toxic metals.

It is true that serious environmental issues as well as the health of living organisms are currently enhanced by pollution of low-concentration heavy metals due to the persistence of metal ions in the global environment due to their natural non-degradability, bioaccumulation tendency, and toxicity. Mani and his co-workers (2014) depicted that this kind of toxic metal can easily reach living organisms through the food chain and damage global ecosystems. Unlike organic contaminants, neither chemical nor biological processes can degrade these kinds of heavy metals; they can only transfer into less toxic species.

It has been observed that in the last 10 years, toxic heavy metals have been treated from industrial wastewater using suitable conventional treatment techniques that include physical techniques (coagulation and/or flocculation, adsorbents, etc.), chemical techniques (chemical precipitation, ion exchange, ion flotation, and photocatalytic processes and cementation), and biochemical techniques (bioleaching, bioremediation, bioaccumulation, and biosorption) (Razzak et al., 2022). In some cases, this method is inappropriate for the proper extraction of metals owing to some limitations, such as large toxic and hazardous sludge creation and high demands for energy. Later, overcoming those limitations of conventional techniques, a few noble and innovative techniques have been widely fabricated and developed since the modern era, followed by the membrane-based separation process, adsorbents, and electrochemical-related techniques. In these techniques, membranes and adsorbents have been studied to enhance removal efficiency in the last few decades, along with their limitations and benefits. Figure 2.1 delivers a synopsis of several techniques for heavy metal removal in overall experimental conditions. During the extraction of heavy metals, the drawbacks of conventional, chemical, and physical methods exposed were the following: (1) the large amount of toxic metal-sludge formation

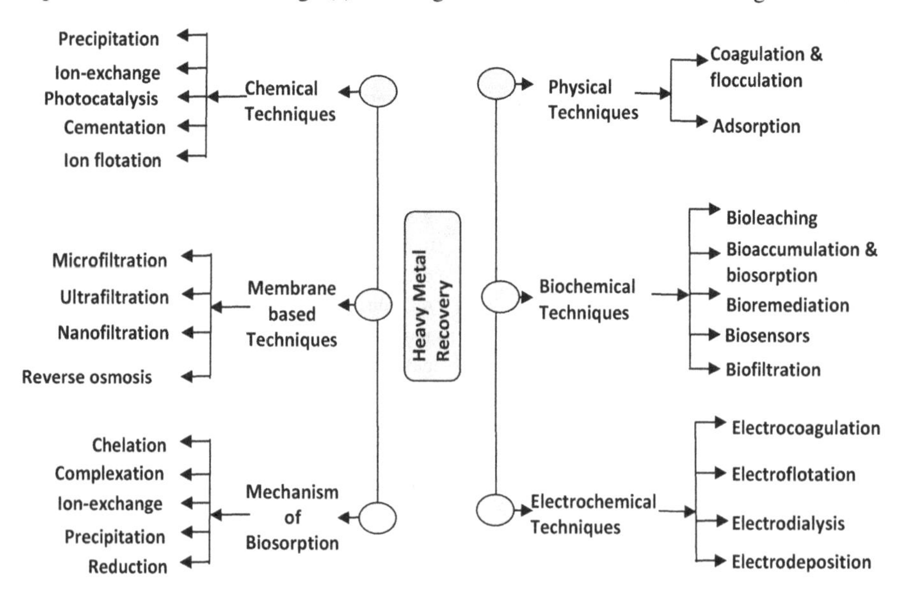

FIGURE 2.1 Overview of different technologies for heavy metal recovery.

and the solidity treatment of the large amount of low-concentration metals in wastewater by coagulation-precipitation; and (2) more operationally expensive for some methods like electrolysis and ion-exchange resins (Ahluwalia and Goyal, 2007). Over the severe limitations, the eco-friendly techniques exert the natural biological mechanisms to annihilate the toxic and hazardous microorganisms and plants (Ahmed et al., 2022). The major advantage of the biological method is that it is less expensive and more effective in recovering low metal concentrations compared to the conventional method (Olaniran et al., 2013). For example, during the metal recovery, the biochemical-like bioremediation process has been used to decrease the metal bioavailability through the speciation increment (Mani and Kumar, 2014).

In this chapter, the techniques in the different classes have been illuminated for the removal of toxic heavy metals from industrial wastewater. Universal strategies and statistics for the biological separation of metal ions from industrial waste have also been discussed.

2.2 CONVENTIONAL RECOVERY OF INDUSTRIAL WASTE

Based on the requirements of the desired metals, a few suitable conventional methods like coagulation and flocculation, adsorption, ion exchange, chemical precipitation, etc. are discussed for removing metals from industrialised wastewater.

2.2.1 Physical Techniques

2.2.1.1 Coagulation and Flocculation

Earlier, in the separation industry, the most competent method was coagulation and flocculation, in which metals or contaminants could be easily treated or removed from industrial wastewater or polluted water by the use of nanostructured materials, following the detaching process of physical and/or chemical techniques (sedimentation, filtration, or straining) for the collection of treated contaminants or metals from solution. Later, both processes will require sedimentation and filtration to collect the foam from the solution. The coagulation-flocculation process, known as the flocculation process, is preceded by coagulation. The substances, namely, coagulants and flocculants, are to be inserted into solutions for operating both processes. Coagulation is the destabilisation of suspended negative electrically charged colloidal particles offered by the addition of a chemical reagent (known as a coagulant) in solution as a negatively charged suspended solid particle moves around inside the solution in Brownian motion (constant motion). In contrast, flocculation refers to the chemical process of the agglomeration (like the use of polymers) of these destabilised suspended particles into microflocs that can be decanted as floc (like foam) and then detached by filtration, flotation, or sedimentation subsequently.

Several types of coagulants, like aluminium, ferric chloride, ferric chloride sulphate, ferrous sulphate, polyaluminium chloride (PAC), polyamides and polytannines, etc., and flocculants such as cationic, anionic, and non-ionic, like PAC, polyacrylamide (PAM), or polyferric sulphate (PFS), are employed for metal removal in this method (Vidu et al., 2020).

2.2.1.2 Adsorption

The easiest and most reliable reversible process, i.e., adsorption, has an inexpensive working function and great metal discharge strength. It is a facile physicochemical method in which, generally, heavy metals (known as adsorbate) are accumulated on the effective surface of the solid adsorbent under different operating conditions like solution pH, initial dosage of metals, temperature, adsorbent amount, and adsorption time.

Adsorbents with great adsorption strength and massive surface areas will be favourable for the metal recovery process. Based on the mass transport of the bulk solution (liquid phase containing contaminants or metals) and the solid phase (adsorbent), the three major stages are generated for heavy metal sorption onto a solid adsorbent: (1) the entrance of metal ions from the wasted solution onto the solid adsorbent surface; (2) the certain adsorption of metals onto the adsorbent surface; and (3) the entrance of metals in the adsorbent structure.

Various types of solid adsorbents for metal recovery from industrial water are discussed below.

2.2.1.2.1 Carbon-Based Adsorbents

Adsorbents should be heterogeneous in nature, and their surfaces should be rough to enhance metal ion adsorption. A few modified methods like oxidation, nitrogenation, and sulphuration are used to control the active parameters such as surface area and pore structure of adsorbents, adsorption capacity, thermal stability, and mechanical strength by adjusting their surface functional groups (e.g., carboxyl, phenyl, and lactone groups). These various functional groups significantly improve the surface carbon charge, which can increase heavy metal uptake.

Based on the adsorption capacity and specific surface area, a few carbon-based nanoporous adsorbents, e.g., activated carbons (ACs), carbon nanotubes (CNTs), graphene (GN), and sawdust, are employed largely in the application of metal recovery.

2.2.1.2.1.1 Activated Carbon (AC) Due to its highly microporous structure (an amorphous solid consisting of microcrystalline crystallites with a graphite lattice) and large interior surface area, AC is used as an adsorbent, either in the form of pellets or a powder. The external surface of AC holds several functional groups, such as carbonyl, carboxyl, lactone, quinone, and others, which can be attributed to its high adsorption efficiency. Even the active free oxidation states of AC and temperature can play a role in generating adsorption activity.

2.2.1.2.1.2 Carbon Nanotubes (CNTs) In this method, a common intricate reaction mechanism is the sorption of heavy metal ions onto CNTs in solution due to their remarkable properties. The heavy metal ions are adsorbed on various CNTs, such as single-walled carbon nanotubes (SWCNTs), multiwalled carbon nanotubes (MWCNTs), oxidised CNTs, and charcoal types like activated charcoal (granular activated charcoal, GAC), and powder activated charcoal (PAC).

2.2.1.2.1.3 Graphene (GN) Due to the chemical, physical, and distinctive mechanical characteristics, better-specified surface area, developed surface functional groups, notable active sites of GN surface, and tremendous chemical stability, GN is most useful in heavy metal removal. It is the hexagonal lattice structure of a single layer that is characterised by two-dimensional (2D) sp2 hybridisation. Some conventional GN, like graphene oxides (GO) and reduced graphene oxides (rGO), can also be used as GN-based materials in heavy metal removal.

2.2.1.2.1.4 Sawdust Because of its lignocellulosic composition (composed of 45%–50% cellulose and 23%–30% lignin), solid wood sawdust is used largely as an inexpensive and simple adsorbent of metals. It is obtained from mechanical wood processing. These adsorbents easily bind the heavy metal cations owing to the presence of hydroxyl (OH), carboxylic (COOH), and phenolic groups (C_6H_6O) in their respective structures. In this process, other agricultural materials can replace the sawdust, such as tree bark, wool, and nut waste.

2.2.1.2.2 Chitosan-Based Adsorbents

Chitosan is a natural, effective adsorptive polymer. It is the main derivative of chitin and has a polymer structure with suitable properties like biocompatibility, biodegradability, and non-toxicity, making it widely used for heavy metal ion recovery. The deacetylation of chitin is commonly involved in the formation of chitosan-based adsorbents under alkaline conditions. It can be applied to adsorb the toxic heavy metal ions from an aqueous solution. It has a strong affinity towards metal ions in wastewater due to having many chelation sites and other functional groups like amino ($-NH_2$) and hydroxyl (–OH) groups, where the metals are certainly disposed through coordination bond or ion exchange. In the modern era, chitosan derivatives are also used as metal ion sorbents. Chitosan is synthesised by either physical or chemical modifications, or both, to improve its properties for adsorption. The adsorption capacities of these adsorbents are affected by optimal operating situations like the primary amount of adsorbents, solution pH, time length of interaction, degree of adsorbent in solution, presence of competitive ions, etc.

2.2.1.2.2.1 Mineral Adsorbents The common mineral adsorbents like clay, diatomite, silica, zeolite, etc. offer low-priced operation for heavy metal recovery treatment through several well-known mechanisms like chemical and physical adsorption and ion exchange. The effective parameters, including temperature, solution pH, adsorption time, and adsorbent dosage, lead to higher adsorption removal efficiency.

Because of the fine silicate morphology, cation-charged toxic metals are adsorbed onto the negatively charged clay species and extracted by them. These adsorbents have remarkable ion-exchange capacities with selectivity, surface morphology, hydrophilicity, etc. Clay has strong availability in a few fundamental classes, such as kaolinite, micas (for example, illite), and smectites (for instance, montmorillonite).

A tetrahedral structure of zeolites is called a molecular sieve and is composed of potassium, calcium, and sodium, surrounded by oxygen atoms. These kinds of ion exchangeable groups act as vital adsorbents to adsorb the various shapes, sizes, and polarities of the toxic heavy metals (adsorbates). Zeolites can be either found naturally via silicate minerals or synthesised through artificial methods.

2.2.1.2.2.2 Magnetic Adsorbents Adsorption using magnetic adsorbents is another low-cost and easy-synthesis method. Having a specific material matrix is mostly employed in metal recovery as well as water treatment. A material matrix provides iron particles (generally magnetic nanoparticles (NPs), such as Fe_3O_4) and notable surface charge. Within the generated material structure, the fused force between solid adsorbents and magnetic components is the leading part of metal removal. This kind of adsorption during metal extraction is significantly influenced by some suitable effective parameters, like surface charge and the mobility nature of the redox reaction under the magnetic field. As base materials (carbon, Carbondisulfide (CS), polymers, starch, or biomass), magnetic adsorbents can be classified based on two major characteristics: the synthesis method with obtained material structures and the attaching force between magnetic materials and adsorbents. With the effective adsorption performance, numerous techniques (co-precipitation, thermal decomposition, solvo-thermal, sol-gel, hydrothermal, oxidation-polymerisation, direct precipitation, ultrasonication, etc.) have been employed to synthesise the various magnetic adsorbents (strong reducer zero-valent iron nanoparticles (ZVI NPs), iron oxide NPs (haematite (α-Fe_2O_3), maghemite (γ-Fe_2O_3), magnetite (Fe_3O_4)), spinel ferrites NPs, P zeolite-Fe_3O_4, amino-functionalised silica–Fe_3O_4, AC–Fe_3O_4/γ-Fe_2O_3, zeolite–chitosan–Fe_3O_4, Chitosan–Fe_3O_4, etc.)

The presence of parameters such as surface morphology, adsorbent magnetic behaviour, solution pH, temperature, the dosage of adsorbents, the initial concentration of pollutants, and irradiation time significantly lead to the reaction kinetics of this process with the proper mechanism. It is true that the presence of iron particles in adsorbents plays a crucial role in the heavy metal treatment of industrial effluents. Table 2.3 provides a summary of studies on the removal of heavy metals using physical techniques.

2.2.2 Chemical Techniques

2.2.2.1 Chemical Precipitation

The coagulation-precipitation, also known as chemical precipitation, is a process that changes from a simple ionic equilibrium to a dissolved hydroxide precipitate in which the reactive agents of the chemical precipitant can consume the metal ions at a higher pH. Consequently, the soluble metal ions are turned into the insoluble solid stage through the yields of the insoluble metal complex. These solid particles are detached from the solution through simple physical techniques such as sedimentation, flocculation, or filtration. The solution pH is used as a controlled variable for chemical precipitation to enhance the removal of metal. Table 2.4 summarises different metal removal methods via various chemical precipitation methods.

2.2.2.2 Ion Exchange

The common ions can be reversibly exchanged between two phases—the solid and liquid phases. Due to insoluble salt-acid or salt-base interactions, ion exchanges help

TABLE 2.3
Short Studies of Dismissal of Heavy Metals by Physical Techniques Method

Treatments	Heavy Metals	Operational Parameters	Separation Efficiency (%) or Removal Capacity (mg/g)	Sources
Coagulation-flocculation treatment	Sn^{2+}, Pb^{2+}, Fe^{2+}	pH: 9 Initial concentration: Coagulant (dosage, mg/La): Polyelectrolytes polydadmac Flocculant (dosage, mg/Lb): Trident 27,506	>99.0 each	López-Maldonado et al. (2014)
	Ag^{2+}	pH: 8.68, 8.00 Initial concentration: Coagulant (dosage, mg/La): Poly aluminium chloride (113.6) and poly-acrylated aluminium chloride (70) Flocculant (dosage, mg/Lb): Anionic polyelectrolyte (3.5) and anionic polyelectrolyte (3.68)	>98.0 each	Folens et al. (2017)
Flotation	Ca^{2+} (using precipitation flotation)	pH: 10.0 Initial concentration: 2,650 mg/L Added chemicals (dosage): Flue gas desulphurisation waste (FGD gypsum, > 20 mg/L)	95.0	Kang et al. (2018)
	Ni^{2+} (using ion flotation)	pH: 9.7 Initial concentration: 10 mg/L Added chemicals (dosage): Sodium dodecyl sulphate (135.1 mg/L) as collector and nickel nitrate (20.0 mg/L) as frother	100	Hoseinian et al. (2019)

(Continued)

TABLE 2.3 (*Continued*)
Short Studies of Dismissal of Heavy Metals by Physical Techniques Method

Treatments	Heavy Metals	Operational Parameters	Separation Efficiency (%) or Removal Capacity (mg/g)	Sources
		Adsorption		
Carbon-based adsorbents	Co^{2+}	Adsorbents: Nanoscaled zero-valent iron/graphene (0FG) Initial concentration: 10 mg/L Initial pH: 5.7 Temperature: 30 Adsorption characteristics: NA	Adsorption capacity (mg/g): 134.27	Xing et al. (2016)
	Hg^{2+}	Adsorbents: MWCNTs-Fe_3O_4 nanocomposite Initial concentration: 50 mg/L Initial pH: 2.0 Temperature: 30 Adsorption characteristics: As = 92 m^2/g	Adsorption capacity (mg/g): 238.78	Sadegh et al. (2018)
Magnetic adsorbents	Cr^{6+}	Adsorbents: Magnetite NP into Lagerstroemia speciose bark (MNPLB) Initial concentration: 100 mg/L each Initial pH: 2.05 Temperature: 35 Adsorption characteristics: dp = 5.269 nm Vp = 0.070 cc/g As = 2.791 m^2/g	Adsorption capacity (mg/g): 434.78	Srivastava et al. (2017)
	As^{3+} and As^{5+}	Adsorbents: Magnetic graphene oxide (MGO-IL) Initial concentration: 250 mg/L each Initial pH: 2.0 Temperature: 45 Adsorption characteristics: NA	Adsorption capacity (mg/g): 160.65 and 104.13	Zhang et al. (2019)

(*Continued*)

TABLE 2.3 (*Continued*)
Short Studies of Dismissal of Heavy Metals by Physical Techniques Method

Treatments	Heavy Metals	Operational Parameters	Separation Efficiency (%) or Removal Capacity (mg/g)	Sources
Mineral adsorbents	Ni^{2+}	Adsorbents: Mixed layer clay Initial concentration: 100 mg/L Initial pH: 5.6 Temperature: 25 Adsorption characteristics: NA	Adsorption capacity (mg/g): 6.25	Es-sahbany et al. (2019)
	U^{6+}	Adsorbents: Thermally activated montmorillonite Initial concentration: 23.7 mg/L Initial pH: 5.01 Temperature: 30 Adsorption characteristics: As = 63.2 m^2/g.	Adsorption capacity (mg/g): 3.93	Qianru et al. (2017)
Chitosan-based adsorbents	Cu^{2+}	Adsorbents: Cross-linked chitosan/ melamine-conjugated poly(hydroxyethy L methacrylate) (PHEMA) Initial concentration: 140 mg/L each Initial pH: 5.0 Temperature: 55 Adsorption characteristics:	Adsorption capacity (mg/g): 299	Yuan et al. (2018)
	Cd^{2+} and Pb^{2+}	Adsorbents: Arginine cross-linked chitosan–carboxymethyl cellulose (CS-ag- CM) Initial concentration: 350 mg/L each Initial pH: 6.5 Temperature: 40 Adsorption characteristics:	Adsorption capacity (mg/g): Adsorption capacity (mg/g): 182.5 and 168.5	Manzoor et al. (2019)

TABLE 2.4
Chemical Methods for Removing Heavy Metals and Related Procedures

Treatments	Heavy metals	Operational Parameters	Separation Efficiency (%) or Removal Capacity (mg/g)	Sources
Hydroxide participation	Cu^{2+}	Initial concentration: 25 mg/L Initial pH: 12–13 Precipitant: $Ca(OH)_2$	>99.0	Jiang et al. (2008)
	Cr^{3+}	Initial concentration: 5,363 mg/L Initial pH: 8 Precipitant: CaO or MgO	>99.0	Guo et al. (2006)
Sulphide participation	Zn^{2+}	Initial concentration: 800–5,800 mg/L Initial pH: 6.5 Precipitant: Na_2S	>99.9	Veeken et al. (2003)
	Cu^{2+}, Ni^{2+}, and Zn^{2+}	Initial concentration: 100 mg/L (for each) Initial pH: 1.9–2 Precipitant: CaS	>99.0	Soya et al. (2010)
Carbonate participation	Mn^{2+}	Initial concentration: 100,000 mg/L Initial pH: 8.0–10.0 Precipitant: Na_2CO_3	>99	Liu et al. (2018)
	Cu^{2+}	Initial concentration: 300 mg/L Initial pH: 7.4 Precipitant: $CaCO_3$	>99.9	Zhang et al. (2018)
Ion exchange	Mn^{2+}	Initial concentration: 25–450 mg/L Initial pH: 6 Ion-exchange particle: Manganese oxide coated zeolite	13.44	Taffarel and Rubio (2010)
	Cs^{+}	Initial concentration: 1,000 mg/L Initial pH: 5.7 Ion-exchange particle: Cross-linked persimmon waste gel	122.3	Pangeni et al. (2014)

TABLE 2.5
Recovery of Metal via Ion-Exchange Method (Krishna et al., 2021)

Resin Category	Types of Metal	Surfactants
Anionic	In, Ni, Pt, Sn	AMBERT JET, Amberlite IRA-400 AR, Purolite NRW-100, Resin D201
Cationic	Ce, Cr, Fe, Ni, Pb, Pd, Zn	Amberlite IRC86, Purolite C100, Chelite S, Duolite GT 73

to exchange either positively charged ions or negatively charged ions. The physical adsorption of toxic heavy metal ions occurs onto the surface of the solid adsorbent during initiative ion-exchange reactions, and consequently, through ion exchange, the positively charged ions of toxic metal are mannerly adsorbed on the negatively charged surface of solid adsorbents that manipulate the adsorption capacity. Here, the surfactant is used as a synthetic organic ion-exchange insoluble resin due to having an effective group of either a hydrophobic part or a hydrophilic part. The hydrophobic part refers to the uncharged carbohydrate group, while the hydrophilic part commonly depends on the nature of the surfactant. For example, a resin is used as an ion exchange by removing ions from the solution and discharging the other equivalent and uniformly charged ions. This method is manipulated by a few experimental parameters, like solution pH, temperature, the initial dosage of adsorbents and adsorbates, contact time, and anions. The cationic, anionic, and non-ionic resins are the types of resins shown in Table 2.5.

2.2.2.3 Photocatalysis Process

Due to its few properties, like its facile motive, the inexpensive operational process, and its large stability, it is auspicious for heavy metal removal. The non-toxic semiconductor materials highly convey a vital role as a photocatalyst through a particular wavelength of visible light. The great achievement is that no chemicals are required for the removal of heavy metals. The materials mostly use TiO_2, ZnO, CeO_2, WO_2, CdS, ZnS, and WS_2.

There are five stages to conducting the photocatalytic process in solution for metal removal treatment from industrial wastewater. The first stage is to secure the transmission of heavy metal ions to the catalyst surface from the aqueous phase. The slow penetration of metal ions onto the solution's surface is followed by their adsorption onto the photocatalyst's surface. During the second stage, the particular wavelength of the visible light favours the simultaneous photocatalytic reaction that occurs on the absorbed surface of the semiconductor. The next stage includes the convection of heavy metals into the target products, as well as the transfer of those metals into a common boundary between the catalyst and the solution surface, known as the interfacial section. Under a specific visible light wavelength, the semiconductor's radiation initiates the photon excitation process. In this case, a photo-hole (h−) is produced in the conduction band (CB) when the electron is prompted from the valence band (VB) to the CB. Therefore, an electron-hole pair is generated by leaving a photo-electron (e−) in the CB and a photo-hole (h−) in the VB, respectively. This electron-hole pair handles the oxidation-reduction reaction of metal ions by transforming them into target products.

The metal deposition at the boundary between the catalyst and solution surface is controlled by a few processes. During metal deposition, the mechanical and thermal processes can help detach the deposited heavy metals. Some drawbacks are found in the coalescence of electron-hole pairs and the formation of side products during photoelectrolysis under a particular wavelength of light. Most researchers use this method in advanced oxidation processes (AOP) for wastewater treatment.

2.2.2.4 Ion Flotation

It is a distinct and promising dissociation technique for the liquid phase for heavy metal removal due to its many advantages like simplicity and low cost, its small space and low energy requirements, fast function, fitness for different motive ions at several levels, a small quantity of sludge, a lower excess dosage, etc. In this technique, the oppositely charged substances to those of the motive ions, like surfactants or collectors, are added to generate a solid surfactant complex in the solution. The sufficient air (gas) is purged continuously through the solution to lead the micro-bubbles that extract the heavy metal ions, and the low-density agglomerates produced at once enhance the flocks via the solution. A small concentration of a hydrophobic coagulated slug (that attains the concentrated desired metal ions) presented at the upper surface can be easily detached. The process involves leading parameters such as solution pH, temperature, water flow rate, bubble size, feed concentration, the dosage of any surfactants, and ionic strength. Based on the reclamation of metal ions as well as water from industrialised water, ion flotation can be classified into several techniques, such as dissolved air flotation (DAF), biological flotation, dispersed-air flotation, electroflotation (EF), and vacuum air flotation.

2.2.2.5 Cementation

A special type of precipitation, cementation operates mostly in batch mode, which can recover the target valuable metals from the industrialised water (Guerra and Dreisinger, 1991) through a few methods like solution tailings, stabilisers, and stop baths. The cementation rate can be varied through various parameters, including the solution pH, the temperature, the cementation agent consumption, and the time and in different ways during the extraction of copper, lead, and silver. Table 2.5 shows different chemical methods for removing heavy metals and related procedures.

2.2.3 Biochemical Techniques

2.2.3.1 Bioleaching

Simple, effective non-conventional bioleaching is a naturally occurring solubilisation process and a microbial leaching method that is used for metal recovery from industrial effluents as well as low-grade ores and mineral concentrates because of organic acids, chelating, and complex compounds discharged into the global environment. The global bioleaching microorganisms (heterotrophic microorganisms, fungi, etc.) require competent conditions for growth, which easily convert the insoluble heavy metal sulphides into water-soluble metal sulphates through the oxidation reaction of bioleaching. Even these kinds of heterotrophic bacteria and fungi remarkably operate on non-sulphide ores and minerals (Bosecker, 1997). It needs the fitted situations established for the natural growth of ubiquitous bioleaching microorganisms such as heterotrophic microorganisms and fungi, *leptospirillum*, thermophilic bacteria, and *thiobacillus*, by which the heavy metal sulphides are transformed into water-soluble metal sulphates via biochemical oxidation reactions (Bosecker, 1997). There are two types of reaction steps for metal extraction: direct and indirect bacterial leaching. The growth efficiency of the microorganisms highly manipulates the leaching effectiveness based on leaching conditions such as pH, temperature, and heavy metal dosage, Nutrients, O_2, and CO_2 supply.

In the extraction industry, only two types of reaction steps are found for the extraction of toxic metals from any kind of sulphide mineral: direct and indirect bacterial leaching. The direct bacterial leaching provides the physical contact of microorganisms (bacterial cells) with the insoluble mineral sulphide surface. In the presence of oxygen, pyrite oxidation occurs to produce water-soluble iron(III) sulphate through either various enzymatically catalysed steps (Eqs. 2.1 and 2.2) (Silverman, 1967) or only one-step Eq. (2.3).

$$4FeS_2 + 14O_2 + 4H_2O \xrightarrow{\text{Bacteria}} 4FeSO_4 + 4H_2SO_4 \tag{2.1}$$

$$4FeSO_4 + O_2 + 2H_2SO_4 \xrightarrow{\text{Bacteria}} 2Fe_2(SO_2)_3 + 2H_2O \tag{2.2}$$

$$4FeS_2 + 15O_2 + 2H_2O \xrightarrow{\text{Bacteria}} 2Fe_2(SO_2)_3 + 2H_2SO_4 \tag{2.3}$$

However, the biological oxidation of a mineral sulphide is expressed through the following generalised reaction mechanism involved in direct leaching (Eq. 2.4),

$$MeS + 2O_2 \xrightarrow{\text{Bacteria}} MeSO_4 \tag{2.4}$$

where MeS(metal sulfide) is the metal sulphide.

But the few authors, (Torma, 1977), showed that in the direct interaction, *T. ferrooxidans oxidised* some non-iron metal sulphides such as cobaltite (CoS), chalcocite (Cu_2S), covellite (CuS), and galena (PbS).

Whereas, in indirect leaching, the bacteria are not required to be in contact with the mineral surface, and it mainly uses the ferric-ferrous cycle (indirect mechanism), in which a strong oxidising agent (ferric sulphate) largely assists in dissolving several metal sulphide minerals, and the ferrous sulphate is formed in the absence of both oxygen and viable bacteria.

$$MeS + Fe_2(SO_2)_3 \rightarrow MeSO_4 + 2FeSO_4 + S^0 \tag{2.5}$$

Tables 2.6 and 2.7 summarise the fundamental appliances of microorganisms through various metabolisms such as autotrophs, heterotrophs, mixotrophs, facultative autotrophs (FA), facultative heterotrophs (FH), and facultative mixotrophs (FM) in the bio leaching process, depending on the optimum pH and temperature (Sarkodie et al., 2022).

2.2.3.2 Bioremediation

Bioremediation refers to the toxic heavy metal removal treatment of industrialised water and/or polluted environments using three major elements: nutrients, food, and microorganisms. This method depends on animating the growth of certain microbes, which generally utilise toxic metals as a particular source of food and energy. These kinds of toxic metals are gradually consumed by microbes that convert them into water and non-toxic gases like carbon dioxide under the comfortable conditions of

TABLE 2.6
Several Applications of Microorganisms in Bioleaching (Sarkodie et al., 2022)

Metabolism		Optimum pH/Temp(C)	IO	SO
Microorganism—Bacteria				
Autotrophs	Acidiferro Bacteria Thiooxidans	2/38	+	+
	Acidithiobacillus albertensis	3.5–4.0/25–30	+	–
	Acidithiobacillus caldus	2.0–2.5/45	–	+
	Acidithiobacillus ferrooxidans	1.4–2.5/28–35	+	+
	Acidithiobacillus thiooxidans	2.0–3.0/Oct-37	–	+
	Leptospirillum ferriphilum	1.0–3.5/<45	+	–
	Leptospirillum ferrooxidans	1.5–3.0/28–30	+	–
Heterotroph	*Ferrimicrobium acidiphilum*	>1.4/<37	+	–
	Pseudomonas putida	7.0–8.0/30	–	–
Mixotroph	*Sulfobacillus benefaciens*	>0.8/<47	+	+
	Sulfobacillus thermosulfidooxidans	1.5–5.5/20–60	+	+
	Thiomonas cuprina	1.5–7.2/20–45	–	+
FA/FM	*Acidimicrobium ferrooxidans*	~2.0/45–50	+	–
	Alicyclobacillus disulfidooxidans	1.5–2.5/35	+	+
	Alicyclobacillus GSM	1.5–2.0/40–47	+	+
FH	*Acidiphilium cryptum*	3.0/35–40	+	–
Microorganism—Achaea				
Autotrophs	*Acidianus infernus*	~2.0/~90	+	+
	Acidianus sulfidivorans	0.35–3.0/45–83	+	+
Mixotroph	*Acidianus brierleyi*	1.5–2.0/45–75	+	+
	Acidiplasma cupricumulans	0.4–1.8/22–63	+	+
	Ferroplasma acidiphilum	1.3–2.2/15–45	+	–
	Sulfolobus solfataricus	3.0–4.5/85	–	–
	Sulfolobus thermosulfidooxidans	1.9–2.4/50–55	+	+
FA/FM	*Metallosphaera hakonensis*	3.0/70	–	+
	Metallosphaera prunae	2.0–3.0/~75	+	+
	Metallosphaera sedula	2.0–3.0/75	–	+
	Sulfolobus metallicus	2.0–3.0/65	+	+
FA	*Sulfolobus acidocaldarius*	2.0–3.0/70–75		–
Microorganism—Fungi				
Heterotroph	*Aspergillus fumigatus*	3.7–7.6/55–70	–	–
	Aspergillus niger	1.5–9.8/35–37	–	–
	Penicillium chrysogenum	3.5–5.5/35–37	–	–
	Penicillium funiculosum	3.5–4.0/45–55	–	–

FA—facultative autotroph; FH—facultative heterotroph; FM—facultative mixotroph; IO—iron oxidising; SO—sulphur oxidising.

TABLE 2.7
Studies of Metal Ion Extractions by the Bioleaching Process and Its Related Methods

Organisms	Heavy Metals	Operational Parameters	Adsorption Capacity (mg/g)	Sources
A. thiooxidans	Cd^{2+}, Cu^{2+}, Pb^{2+} and Zn^{2+}, Cr^{3+}	Adsorbents: NaOH-carob shells Inoculum: 10% (v/v) Solid ratio: 3% (w/v) Initial pH: 3.0–7.0 Temperature: 30 150 rpm shaking for 28 days	90–98	Kumar and Nagendran (2017)
Acidithiobacillus sp.	Cd^{2+}, Cu^{2+}, Zn^{2+},	Solid ratio: 3% (w/v) Initial pH: 4.0–6.3 Temperature: 28°C 150 rpm shaking for 25 days	80–98	Fang et al. (2013)
Geotrichum sp. G1 and *Bacillus* sp. B2	Cr^{3+}	Solid: 200 g soil Flow rate: 5 mL/min pH: 2–10	95	Qu et al. (2018)

the right temperature, nutrients, and food. Based on water media and soil, bioremediation processes are also divided into two kinds in order to implement the policy, i.e., in situ bioremediation and *ex situ* bioremediation (such as biopiles, bioreactors, land farming, and windrows). The two bio-techniques, bioaccumulation and biosorption, can remediate toxic heavy metals in wastewater (Ahmed et al., 2022). These bio-techniques have proven to be effective in removing toxic heavy metals from wastewater by utilising the natural abilities of microorganisms to accumulate or absorb these contaminants. This approach offers a sustainable and environmentally friendly solution for treating industrial effluents and reducing the harmful impact of heavy metal pollution on ecosystems.

2.2.3.3 Bioaccumulation and Biosorption

The microbial type makes a major difference between the biosorption and bioaccumulation processes. The activity of non-living microbial organisms is essential to operating biosorption, which is a metabolism-driven inert method, whereas bioaccumulation is carried out through the subsequent stages of biosorption using living organisms. The accumulation process of metal ions occurs in living organisms during bioaccumulation, which is the metabolically effective functional process that deals with the absorption of metal ions by adsorbents into living cells intercellularly. For example, aquatic fluid plants (*Nasutrium* officinale, *Typha latifolia,* and *Thelypteris palustria*) are known for the efficiency of bioaccumulation of biomass for the desired metal.

Biosorption is a physicochemical pathway and another class of adsorption that employs certain biomass of biological origin (called a biological matrix) as a type of biosorbent and passively accumulates heavy metal ions by binding them from solutions to its cellular structure (that represents functional groups onto the surface of certain biomass) on the external surface of the bio-absorbent in a fast, reversible process. It occurs through a few major reaction processes, such as chemical or physical interaction, reduction, chelation, coordination, or complexation. For the reversible biosorption process, several microorganisms, biopolymers, materials obtained from plants, and different wastes are used.

Two major parameters, such as the locus of metal recovery and the microbial cell metabolism, can affect the biosorption methods. In addition, depending on the mobility of certain biomass, biosorption can be classified as metabolism-dependent or non-metabolism-dependent. According to the source of metal recovery, biosorption of metal ions can certainly be achieved through effective methods such as (1) the acquisition of intracellular phase; (2) the biosorption of cell surface and/or precipitation; and (3) the acquisition of extracellular phase and/or precipitation. Overall, biosorption is mostly studied because of the unlimited presence of biosorbents that can be reutilised and the low concentration of sludge during the process. In some cases, due to the unavailability of cellular metabolism, only a few reaction methods like ion exchange, physical adsorption, and complexation can be achieved.

2.2.3.3.1 Mechanism of Biosorption

There are various metabolically dependent reaction processes for the biosorption of metal removal (see Table 2.8); typically, the intracellular phase is acquired first, and then the metal is transported across the cell membrane through interaction with effective functional groups.

2.2.3.3.2 Complexation

Complexation is a type of biosorption in which a toxic metal chemically reacts with a ligand to create a metal complex. In complexation, a minimum of one atom can be linked to an unshared pair of electrons in this reaction. The metal charges can be positive, negative, or neutral, depending on the concentration of the binding ligand

TABLE 2.8
Brief Studies of Metal Ion Removal from Wastewater Using Biosorption Techniques and the Processes Involved

Treatments	Heavy Metals	Operational Parameters	Adsorption Capacity (mg/g)	Sources
Biosorbents	Cd^{2+} and Co^{2+}	Adsorbents: NaOH-carob shells Initial concentration: 100 mg/L each Initial pH: 7.0 Temperature: 25 Adsorption characteristics: NA	49.62 and 30.04	Farnane et al. (2017)
	Pb^{2+}	Adsorbents: Cow dung Initial concentration: 60 mg/L Initial pH: 2.0 Temperature: 25 Adsorption characteristics:	37.35	Ojedokun and Bello (2016)

present in this reaction mechanism. The two-type metal complex is generated. A mononuclear complex is made by either only a metal ion or a ligand with the metal at the centre. A polynuclear complex is also created when more than one metal atom is used at the centre.

2.2.3.3.3 *Chelation*

In chemistry, chelation is a preceding structure of complexation in which the molecules are reversibly enclosed to metal ions as binder. In this process, the chelant (known as the binding of the ligand) to a heavy metal provides the chelate, which is a metal complex formation. The higher number of binding sites improves the chelate stability. At many loci, a stable encircling structure is created owing to its many bindings via the combination of chelant and heavy metal ions.

2.2.3.3.4 *Coordination*

In one of the chemical reactions, a coordination compound acts as a chemical structure for metal removal. A non-metal encompasses the central atom during a chemical reaction, and it is classified as a ligand or complexing agent.

2.2.3.3.5 *Ion Exchange*

The filtration-dependent reaction method involves the exchange of ions when one substance reacts with others (each composed of positively and negatively charged metal ions), resulting in the interchange of one or more ionic materials. In biosorption, the metal ions are to be exchanged with ions present on the biosorbent during the chemical reaction process.

2.2.3.3.6 *Precipitation*

In a different type of chemical reaction known as precipitation, which uses a biosorbent, the metal ions form a bond with the functional groups on the surface of a microbial cell to produce a complex compound.

2.2.3.3.7 *Reduction*

The reduced state of a few components, like chromium, gold, etc., may play a crucial role in metal removal. These metals may be reduced while they are exposed to a solid biosorbent in particular spaces. The growth of crystals is determined by the interaction between functional groups and heavy metal reduction.

2.2.3.4 Biosensors

Recently, biosensors have gained more attention in the detection of heavy metal presence in water with quantity due to their remarkable properties like inexpensiveness, quick and shorter time response, and superior sensitivity when limited resources are available (Hossain and Mansour, 2019). Based on the several pathways, only two common types of biosensors exist: (1) a group of transduction systems (that includes optical, electrochemical, thermal, and piezoelectric biosensors) and (2) another group of biorecognition (e.g., bioreceptors and biocomponents). In order to biorecognise, biosensors can be classified into antibody-based, enzyme-based, protein-based, and whole cell-based biosensors. A paper-based biosensor is also used as a biosensor to control low detection limits for heavy metals in a water body.

2.2.3.5 Biofiltration

In the biofiltration method, a particular biofilter comprises a porous medium whose surface is surrounded by water and various microorganisms. This method is used to remove heavy metals present in wastewater by using microorganisms and converting the heavy metals (contaminants) into biomass, carbon dioxide, water, and metallic by-products. The biodegradation of heavy metals is carried out in the biofilter through three major biological processes: (1) bonding of microorganisms; (2) growth of microorganisms; and (3) decay and detachment of microorganisms. The bio-reaction between microorganisms and heavy metals generally takes place on the surface of filter media through various reaction mechanisms, like (1) the movement of microorganisms to the porous surface of filter media, (2) primary adhesion, (3) firm closeness, and (4) colonisation. The movement of microorganisms is manipulated by a few processes, like diffusion (Brownian motion), natural convection, and sedimentation, owing to the effective mobility of microorganisms and gravity. It is assured that the active microorganisms can either oxidise or reduce the non-biodegradable and water-soluble heavy metals through the formation of less-soluble species.

2.3 MEMBRANE-BASED SEPARATION TECHNOLOGY

Recently, the membrane technique has received great attention in heavy metal removal as well as wastewater treatment processes because of its inexpensive and effortless work and fast and lower vacancy requirements. In the membrane filtration industry, based on the features of the pore structure of the membrane (e.g., the size and size distribution of the pore and its porosity), along with the ability of the permeable membrane and applied operating pressures, the membrane filtration technique can be divided into four categories despite the same working principle of membrane filtration: microfiltration (MF), ultrafiltration (UF), nanofiltration (NF), and reverse osmosis (RO).

Naturally, MF and UF are the two low-pressure-driven porous membranes that have the same working principle. The key dissimilarities between these processes are that the solutes present in solutions that are eliminated by MF are greater than those refused by UF. The membrane structure can divide MF and UF into symmetric or asymmetric; usually, only two membranes, such as MF and UF, extensively employ crystalline polymers as raw materials. When they utilise one kind of hydrophilic membrane, the mechanical resistance is highly dominated by water molecules, along with thermal resistance. Besides, the opposite work is also observed during the exploitation of hydrophobic membranes in reducing the major adsorption sites of solute on the active surface of the membrane that assist in reducing the permeate flux. The MF and UF processes are both conducted by two filtration modes (dead-end mode and cross-flow mode) to identify the segregation mechanism. The metal removal efficiency is manipulated by parameters like membrane porosity, structure, and material, which assist in selecting the membrane. Furthermore, the pH of the solution and the applied pressure can also affect this efficiency.

The dissociation range of NF has specific properties between UF and RO membrane processes. The polymer-made NF is composed of a multiple-layer thin film of chemical groups that are negatively charged. It is prompted via a few major techniques, e.g., interfacial polymerisation (IP), which is a polyamide thin film composite membrane, NPs, and ultraviolet (UV) treatments. For heavy metal removal, three major stages of the NF technique are generated: first, a pre-treatment step (such as pre-filtration, coagulation and filtration, flocculation and filtration, and ion exchange). The second stage is the utilisation of the NF membrane, in which the separation mechanism consists of three effects (steric hindrance (i.e., sieving), electrical (i.e., Donnan), and dielectric effects) (Krishana et al., 2021). That event facilitates the rejection of charged solutes, massive neutral solutes, and salt that is smaller than the membrane pore size. Finally, post-treatment belongs to the third and last stage.

The low-pressure-driven non-porous RO process operates a semipermeable membrane with pore sizes in the 0.5–1.5 nm range to remove monovalent and multivalent particles. In this auspicious method of metal removal treatment, only water can easily penetrate through the membrane while extracting heavy metal ions from water effluents, despite the major drawback of membrane fouling and degradation of the RO method (Wang et al., 2011). Under the hydraulic operating pressure (20–70 bar), which is higher than the osmotic pressure of the feed solution, the RO membrane functions oppositely to the normal osmosis process; the transition occurs by diffusion as the particles do not easily flow through any distinct pores. The membrane selectivity, size of the solute, etc. are the key parameters that affect metal removal efficiency (see Table 2.9).

TABLE 2.9
Short Studies of Filtration Process for Metal Ions Dismissal from Wastewater

Treatments	Heavy Metals	Operational Parameters	Separation Efficiency (%)	Sources
Ultrafiltration (UF)	Pb^{2+}	Membrane: Cellulose pH: 7.75 Initial concentration: 4.4–7.6 mg/L Surfactant/complexing agent (concentration): Dodecylbenzenesulpho nic acid (DSA, 10^{-5} M) and dodecylamine (10^{-6} M)	>99.0	Ferella et al. (2007)
	Zn^{2+}, Cu^{2+}, Cd^{2+}, and Pb^{2+}	Membrane: Polyethersulfone pH: 1–12 Initial concentration: 50, 150, and 300 mg/L Surfactant/complexing agent (concentration): SDS fixed at 8 mM (1 CMC)	>99.0 each	Huang et al. (2017)
Nanofiltration (NF)	Co^{2+}	Membrane Material: DL, NF4, NF6, and NF8 pH: Pressure: 4.0–8.0 bar Initial feed concentration: 0.006–0.03 mol/L (Ammoniacal solutions)	>99.0	Wu et al. (2016)
	Zn^{2+}	Membrane Material: Polyamide film pH: 7.0 Pressure: 5.0–30.0 bar Initial feed concentration: 50.0–200.0 mol/L (Zinc sulphate and Zinc nitrate)	>98.0	Cuhorka et al. (2020)

(Continued)

TABLE 2.9 (*Continued*)
Short Studies of Filtration Process for Metal Ions Dismissal from Wastewater

Treatments	Heavy Metals	Operational Parameters	Separation Efficiency (%)	Sources
Microfiltration (MF) (cross-flow)	Co^{2+}, Cu^{2+}, and Cd^{2+}	Membrane: Millipore membrane filter pH: 7.2 Initial concentration: 0.2 mmol/L each	>99	Brady and Duncan (1993)
	Cr^{3+}, Cu^{2+}, Pb^{2+}, and Cd^{2+}	Membrane: Polypropyle ne pH: 7.2 Initial concentration: 0.2 mmol/L each	60–71	Brady et al. (1994)
Reverse osmosis (RO)	Cu^{2+} and Ni^{2+}	Membrane Material: Thin film polyamide pH: 7.8 Pressure: 5.03 bar Initial feed concentration: 500.0 mg/L (Single-salt and mixed-salt systems)	99.5	Mohsen-Nia et al. (2007)
	B^{3+}	Membrane Material: Thin film composite polyamide pH: 7–11 Pressure: 5.16–20.64 bar Initial feed concentration: 70.0 mg/L (H_3BO_3-boric acid solution)	>99.0	Dydo et al. (2012)

2.4 ELECTROCHEMICAL TECHNIQUES FOR SEPARATION

Depending upon the nature of pollutants (heavy metal ions), wastewater treatment as well as heavy metal removal are driven by a few common techniques, such as primary, secondary, and tertiary treatment and conventional biological methods. But to overcome the limitations of that method, electrochemistry is widely used to either recover heavy metal ions from treated water from food, oil, electroplating, and textiles or dismiss the heavy metals from industrialised water, owing to their great separation efficiency, operation cost, selectivity, and reliability (Jin and Zhang, 2020). In both heavy metal recovery and wastewater treatment, the recently advanced techniques are electrooxidation (EO) and electro-reduction (ER), electrocoagulation (EC), EF, and electrodeposition (ED).

2.4.1 Electrochemical Oxidation (EO)

Electrochemical oxidation (EO) plays a vital role in decomposing and mineralising strong organic compounds present in industrialised water effluent for metal extraction from polluted water (Oliveira et al., 2018). In EO reactions, two types of reactions may occur: (1) heavy metal anions generate the direct electrons on the anode surface into the electrolytic solution, and/or (2) the active side radicals (OH, etc.)

and oxidants generated in the electrolyte induce the electrons indirectly. The direct oxidation method, in which metal ions lose electrons and can be oxidised from low-valence to high-valence states, may be classified into three types of oxidation methods: direct EO, synergistic electrochemical oxidation, and indirect EO (Ntagia and Co-workers, 2019).

In the case of direct EO, the metals generate direct electrons to the anode surface, but the potentially effective radicals and strong oxidants propagated in situ erode the electrons when indirect E-oxidation is applied. The specified oxidation outcomes are also linked to the electrode surface via nano-adsorbents in order to modify the electrode (Ntagia et al., 2019).

The direct EO of high-valence metal ions occurs on the anode surface of the electrode material (the treating electrode) to get heavy metals such as As, Sb, and Pb. The anode materials were employed as Boron-doped diamond (BDD), carbon fibres, stainless steel, metal oxide electrodes, porous carbon felt, glassy carbon, reticulated vitreous carbon (RVC), etc. (Carlos and Sergio, 2006). The direct switch of electrons (generally oxidising substances, two electrons) is exposed to the reducers from the anode. The metal ions are oxidised on the treated anode surface, and then the electrons are transferred directly to the anode. The direct electron transfer is achieved to the anode surface during the oxidation of metal ions on the treated anode surface (Yang et al., 2021). The activity of direct EO is significantly manipulated by several parameters, including the type of electrolytic cell, the electrode material (anode and cathode), electrolytic composition, flow rate, and current density. The future goal is to generate higher efficiency at a lower current density.

In contrast, when the EO process is applied, the oxidative forms created by in situ electrogeneration can achieve the oxidation of low-valence metal ions. Here, the oxidative forms are of two types: one is a strong oxidising agent like H_2O_2, Cl_2, and others are highly effective free radicals (OH, O_2^-, and HO2–) (Carlos and Sergio, 2006). The continuous electrogeneration of H_2O_2 can be performed either via the reduction of oxygen on a cathode surface (Eq. 2.1) or the oxidation of H_2O on an anode surface (Eq. 2.2). Moreover, in the presence of an electrolyte, the electrogeneration of chlorine (Cl_2) can be carried out at the anode surface via the oxidation of free Cl_2 (Eq. 2.3). Yang and his team (2021) examined that the effective free radicals performed in a very short time in the EO method are stronger than the oxidising agents.

Oxidising Agents	**Free Radicals**
$O_{2+}2H^+ + 2e^- \rightarrow H_2O_2$ (2.6)	$MC\ (L) + H_2O_2 + H^+ \rightarrow M\ (H) + H_2O + OH$ (2.9)
$H_2O - 2e^- \rightarrow H_2O_2 + 2H^+$ (2.7)	$H_2O - e^- \rightarrow OH\cdot + H^+$ (2.10)
$2Cl^- + 2e^- \rightarrow Cl_2$ (2.8)	$O_2 + MC\ (L) \rightarrow O_2^- + M\ (H)$ (2.11)
	$H^+ + O_2^- + e^- \rightarrow OH + HO_2^-$ (2.12)
	$H_2O + O_2 + 2\,e^- \rightarrow OH^- + HO_2^-$ (2.13)

MC (L): Metal catalyst (low-valence electron), M (H): Metal (high-valence electron)

In fact, like direct EO, the formation of the active substance (oxidation products) is strongly influenced by the electrode material and other functional parameters like electrolytic composition, flow rate, current density, solution pH, and convection (Vidu et al., 2020). The efficiency of EO is enhanced by the higher active substance.

It is well known that electrons are obviously transferred between the metal ions and the anode surface when direct EO is applied. However, lower oxidation efficiency may be achieved, and competitive active sites may limit the EO process. In contrast, the numerous strong oxidising agents generated by indirect EO in electrolytes perform the oxidation of the metal ions in little time. The maximum oxidation rate was ascending to direct E-oxidation, indirect E-oxidation, and synergistic E-oxidation, as shown by Yang and co-workers (2021). During the oxidation process, limited side reactions are achieved if (1) a higher potential is applied at the anode and (2) the voltage is increased. The higher potential at the anode encourages unwanted side products from H_2O_2 as it is converted to O_2 or directly converted to H_2O. On the other hand, the enhancement in voltage would generate hydrogen evolution at the cathode via the consumption or inhibition of the active substances OH and HO_2^-. The oxidation efficiency is reduced by some limitations of the direct EO process, manipulated by a few factors such as diffusion resistance and electrostatic repulsion between metal ions (Yang et al., 2021). They also revealed that because of the proximity of oxidising agents and the target metals, the oxidation efficiency was also enhanced, while numerous potent oxidising agents (formed in the electrolyte solution) can perform the oxidation of metal ions in a short response time.

2.4.2 Electrochemical Reduction

In the electrochemical reduction process, heavy metal ions are deposited as metallic elements by accepting electrons from the cathode surface (Yang et al., 2021). Under the influence of external energy sources, this reaction occurred at the cathode surface, confirming a decrease in the valence states of oxidisers at the cathode (Devda et al., 2021). The metal recovery can be processed through two reduction methods: metal electroplating and electrochemical denitrification (Martin et al., 2016).

2.4.3 Electrocoagulation (EC) and Electroflotation

The major policy of EC is a process of in situ generation of coagulants at the sacrificial solid metallic anode electrodes instead of metal salts (used in physical coagulants) under a power source to remove suspended pollutants (heavy metals) in solutions (Bazrafshan et al., 2015). This technique mainly comprises three reaction stages: (1) electrogeneration of coagulant agents (known as destabilising agents), (2) destabilisation of pollutants present in solution, and (3) agglomerated flocs creation (Aleboyeh et al., 2008).

When the electrical current initiates to flow through the immersed electrodes, the coagulated ions are created at the metallic anode (which is made by destabilising agents), and then the ions are dissolved into a solution. At the opposite cathode electrode, hydroxide (OH–) ions and hydrogen gas (H_2) are generated. The coagulants ($M(OH)_2/M(OH)_3/M(OH)_n$) are formed inside the solution by the reaction between the coagulated ions and OH– ions (Huang et al., 2020). These agents destabilise pollutants as well as heavy metals because they generate an opposite electrostatic charge. i.e., metals can also be accumulated as hydroxides (coagulants) via EC methods in which the electrogenerated coagulants react with heavy metals. A few operating parameters like electrode material, solution pH, solution conductivity, applied current, interelectrode distance, and electrolysis time can affect the EC process.

On the other hand, Balmer and Foulds (1986) clearly explained the concept of electro-flocculation, which is the aggregation of EF with electro-precipitation in the electro-flocculation reactor, where the dual process is operated at the same time. The electro-flocculation surely requires a flocculation process that generates insolublely precipitated metal ions as the flocculating agent from the anode material, and the products of metal ions are exempt from the electrolyte. That kind of metal ion is later adsorbed onto the outer surface of colloidal particles present in the solution. During this experiment, hydrogen gas bubbles (H_2) are created in the cathode material. The heavy metals are captured by the bubbles formed at the cathode; consequently, the metal ions have engaged with H_2 gas. These established gases mainly help them to float to the upper surface of the water as a stable floc (Zhang et al., 2013). The toxic metal initiates the coalesce process, creating stable flocs that can be detached from the bulk solution once charged.

Anode (+), Oxidation	**Cathode (−), Reduction**
$2H_2O \rightarrow O_2\ (g) + 4H^+ + 4e-$ (2.14)	
$M\ (s) \rightarrow Mn^+(aq) + ne-$ (2.15)	$2H_2O\ (l) + ne- \rightarrow H_2\ (g) + 2OH-\ (aq)$ (2.16)

In bulk solution,

$Mn^+(aq) + 2OH-\ (aq) \rightarrow M(OH)_2/M(OH)_3/M(OH)_n$ (Coagulants) (2.17)

$M(OH)_2/M(OH)_3/M(OH)_n$ + Pollutants (Heavy metals) → Coagulation→ (2.18)

Flocculation → Precipitation

Precipitation → H_2 (g) → Flotation

In the special case of a dilute solution, EF is the process of electrogenation of hydrogen bubbles, which can float heavy metals to the water surfaces (Manikandan and Saraswathi, 2023). Generally, it replaces bubbles generated from dissolved air. It is highly noticed that only H_2 gas bubbles are established at cathode surfaces in the EC process, but in the EF process, two bubble formations are created: one H_2 bubble is created at cathodes and other O_2 bubbles at anode surfaces (Mota et al., 2015). The O_2 bubbles are produced by H_2O decomposition with the help of higher current density. In addition, the cathodic OH ion is evaluated to control the electrolytic pH. Only two major parameters, such as the electrode surface area and the applied voltage that is used between the anode and cathode, smoothly dominate the current density (Chen, 2004). The formation of gas bubbles is improved through three major steps: (1) nucleation, (2) growth, and (3) detachment from the electrode surface (Khosla et al., 1991). In another study, it was investigated that the micro-roughness of electrode surfaces provided several energy-favourable places (e.g., pits and scratches) instead of ideal flat surfaces, generating a higher current density (Mickova et al., 2015).

The two mechanisms, such as supersaturation and consolidation of gas in the electrolytic solution, can begin growth, where two parameters—mass transport of soluble gas through the gas or liquid interference and high internal pressure—cause high propagation. Coalescence is created by either a process of single gas bubble formation when two or more gas bubbles get together or by one gas bubble surrounding the active surface of the cathode and consolidating with each gas bubble. A smaller diameter gas bubble is more effective with a flotation rate than a larger diameter gas bubble, which is affected by solution pH (Mansour et al., 2007).

On the other hand, two parameters, such as bubble size and the contact angle of gas bubbles, could potentially impact the separation ability of the gas bubbles from the electrode surface. The detachment efficiency can be highly enhanced if the addition of surface-active substances in the electrolyte reduces the surface tension between the electrolyte, the electrode surface, and the gas bubbles (Koren and Syversen, 1995).

Metal anode (+), Oxidation		Cathode (−), Reduction	
$2H_2O \rightarrow O_2\ (g) + 4H^+ + 4e^-$	(2.19)	$2H_2O\ (l) + ne- \rightarrow H_2\ \ (g) + 2OH-\ (aq)$	(2.21)
$M\ (s) \rightarrow Mn^+(aq) + ne-$	(2.20)		

In bulk solution,

$M\ (s) + H_2O \rightarrow H_2\ \ (g) + OH^-(aq) + Mn^+(aq)$ (2.22)

2.4.4 Electrodeposition (ED) (Electrowinning)

The ED process is a highly selective and low cost, one-step clean method in which no substances are required to recover heavy metals and no slurry will be formed during the process (Scott and Paton, 1993). The removal of heavy metals in non-aqueous solutions or solutions containing citrate, Ethylenediaminetetraacetic acid (EDTA), and Nitrilotriacetic acid (NTA) is preferable to that in aqueous solutions because of the availability of some technical issues such as the release of hydrogen gas molecules, low thermal solidity, and thin electrochemical gap. In some cases, chelating agents are employed in binding the cations of toxic metals to minimise the yield of insoluble salts and improve the recovery effectiveness because of the building of stronger complexes with metal ions (Chen and Lim, 2005; Oztekin and Yazicigil, 2006). A few factors, such as reduced current efficiency, heat stability of the cell, and degradation of cell components, hinder the process.

In the electrochemical cell (composed of one cathode and anode, electrolytic cell, and power source), this method leads to the conversion of dissolved heavy metal ions into solid elements on the cathode and anode electrodes (called ionic conductors) (cathode and anode), avoiding corrosion of metals (Chen and Lim, 2005; Jayakrishnan, 2012). During the redox reaction, heavy metals are commonly condensed and electroplated onto the cathode surface. In three-electrode chemical cells, the cathode electrode should be the metal cathode, and the anode should be an insoluble or inert material because anode materials strongly participate in the removal treatment of heavy metals (Jayakrishnan, 2012).

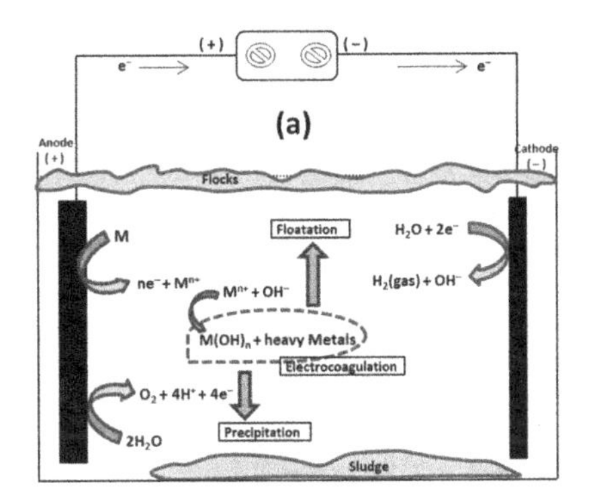

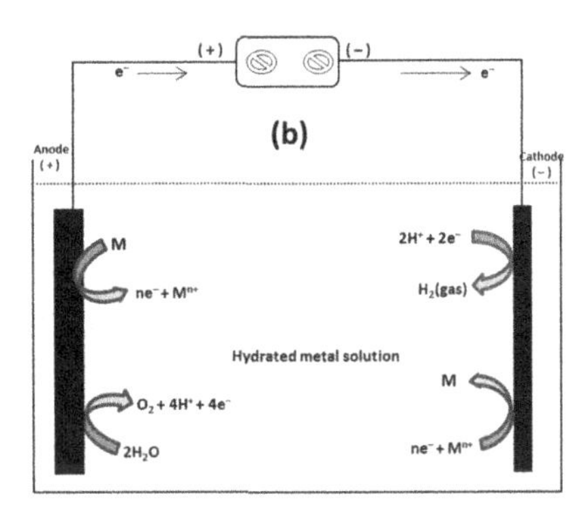

FIGURE 2.2 Reaction mechanism for (a) EC and (b) ED process.

In the cell, the cathode works as the conducting substrate on which the ED process takes place, and the anode may be either soluble or inert (insoluble). During electrolysis, the overall reaction occurs in solution, represented as Eq. (2.18) at soluble, Eq. (2.19) at insoluble, and Eq. (2.21) at cathode (Jayakrishnan, 2012). The figure depicts the transportation of hydrated ions towards the cathode, along with overall reactions. In simple salt solutions, metal ions are present in bulk solutions as hydrated ions (Figure 2.2).

The overall reaction is as follows:

Anode (+), Oxidation	Cathode (−), Reduction
$M \rightarrow Mn^{+}(aq) + ne^{-}$ (soluble anode) (2.23)	$Mn^{+}(aq) + ne^{-} \rightarrow M$ (2.25)
$2H_2O \rightarrow O_2\,(g) + 4H^{+} + 4e^{-}$ (Insoluble) (2.24)	$2H^{+} + 2e^{-} \rightarrow H_2\,(g) + 2$ (2.26)

2.4.5 Electrodialysis (ED)

A process of membrane separation technique, electrodialysis, can be employed in the removal of aqueous metal ions through electrically driven ion-exchange membranes, in which anions and cations are transported due to an electrical potential or concentration gradient across an ion-exchange membrane (Chen, 2004). When the polluted solution (that contains ion species) penetrates through the separation membranes, positively charged ions are attracted to the cathode side, and negatively charged ions are relocated to the opposite side of the anode (Ebbers et al., 2015).

In some cases, electrodialysis reversal (EDR) can provide better performance than ED during the reversion of the polarity of ED electrodes due to the ability to achieve certain higher recovery rates. EDR can also minimise scaling and fouling despite the high requirement for pipes, fixtures, and electrical controls (Arribas et al., 2015).

2.4.6 Microbial Electrochemical Technology

It occurs between microbes and electrodes through two types of processes (primary and secondary microbial electrochemical technologies) depending on the nature and level of interaction (Devda et al., 2021). Primary microbial electrochemical technologies are the part of microbial electrochemistry in which these connections expose the relocation of extracellular electrons at the system level.

Secondary microbial electrochemical technology, the non-part of microbial electrochemistry, deals with indirect interactions like observing or regulating the microbial response environment using electrochemical methods through the ionic bond between the electrochemical system and the microbial system. Table 2.10 summarises various electrochemical methods for removing ions from wastewater.

2.5 TYPES OF BIOSORBENTS

Biosorbents are economical because of their rich sources and reasonable preparation conditions. The metal-sequestering property of biosorbents is widely employed to reduce the amount (from ppm to ppb) of heavy metal ions in solution. The heavy metals are to be treated from industrial wastewater by the biological removal of naturally made biosorbents. Microorganisms such as algae, bacteria, fungi, and yeast are ideal sources of these biosorbents, as shown in Table 2.11.

2.5.1 Algae as Biosorbents

An aquatic plant, algae, is used as a favourable biosorbent material in heavy metal ion recovery for its large size, great sorption capacity, and lack of any formation

TABLE 2.10
Brief Investigations on the Electrochemical Method of Removing Ions from Wastewater

Treatments	Heavy Metal	Operating Conditions	Removal Efficiency (%)	Sources
Electrochemical oxidation (EO)	Cu^{2+}	Initial concentration: 65 mol/L Initial pH: 4.1 Anode/Cathode material: BDD/graphite $j = 50\,A/m^2$, $V = 4.5\,V$, $S = 3{,}000$ mS/cm, $t = 24$ min, and FV = 0.5 L/h	99.6	Raptopoulou et al. (2016)
	Ni^{2+}	Initial concentration: 2,156 mg/L Initial pH: 9.0 Electrodes material: RuO2-Ti/SS $j = 320\,A/m^2$, $t = 180$ min, and $T = 60°C$	99	Guan et al. (2017)
Electrochemical reduction	Cu^{2+}	Initial concentration: 200 mol/L Electrode material: Ti-$Ti_{0.7}Ru_{0.3}O_2$/SS, $I = 6$ A and $t = 1$ h	99.95*	Tonini & Ruotolo (2017)
	Cd^{2+} and Pb^{2+}	Initial concentration: 500 mg/L for both Electrodes material: SnS-Bi_2O_3/ SnS- Bi_2O_3 $t = 40$ min and $V = -1.5$ V pH: 5.0	99.8 and 100	Jin et al. (2020)

(Continued)

TABLE 2.10 (*Continued*)
Brief Investigations on the Electrochemical Method of Removing Ions from Wastewater

Treatments	Heavy Metal	Operating Conditions	Removal Efficiency (%)	Sources
Electrocoagulation	As^{3+} and As^{5+}	Initial concentration: 1–30 mol/L Initial pH: 2.86–4.14 Anode/Cathode material: Carbon steel (CS)/CS $j = 136–208\,A/m^2$, $V = 6.68–12.8\,V$, $t = 5\,min$, $T = 20°C$, $d = 0.5\,cm$, and $S = 0.95\,mS/cm$	99.9	Parga et al. (2014)
	Cr^{3+}	Initial concentration: 500 mg/L Initial pH: 5.83 Electrodes material: Al/Fe $V = 30\,V$, $t = 1\,h$, Stirring rate = 100 rpm, and $T = 25°C$	99.6	Shahriari et al. (2014)
Electroflotation	Ni^{2+} and Cu^{2+}	Initial concentration: 100 mol/L each Initial pH: 11–12 Anode/Cathode material: Ti-RuO_2/SS Na_2SO_4 (1 g/L), $I = 80\,mA$, $t = 300\,min$, and $d = 1\,cm$	>97	Brodskiy et al. (2015)
	Cr^{3+}	Initial concentration: 2,481 mg/L Initial pH: 4.4 Electrodes material: RuO_2/TiO_2- Ti/Ti Na_2SO_4 (0.01 N), $j = 1\,kA/m^2$, $t = 3\,h$, and $d = 0.5\,cm$	98	Selvaraj et al. (2018)

I—electric current; V—electric potential; j—current density; S—salinity; t—time consumed; and d—distance between electrodes.

TABLE 2.11
Classification of Local Biomass Used for Living Microorganisms Biosorbents Formation

Microorganism Category	Example
Phytoplankton	Microphytes—*Clorella sp.*, *Chlamydomonas sp.*, and others Macrophytes Brown algae—*Sargassum sp.*, *Ecklonia sp.*, etc. Green algae—*Enteromorpha sp.*, *Codium sp.*, and others Red algae—*Gelidium sp.*, *Porphyra sp.*, etc.
Bacteria	Gram-positive bacteria—*Bacillus* sp., *Corynebacterium* sp., etc. Gram-negative bacteria—*Escherichia* sp., *Pseudomonas* sp., etc. Cyanobacteria—*Anabaena* sp., *Synechocystis* sp., etc.
Fungal biomass	Moulds—*Aspergillus* sp., *Penicillium* sp., *Saccharomyces cerevisiae*, etc. Mushrooms—*Agaricus* sp., *Trichaptum* sp., etc.
Yeasts	*Saccharomyces* sp., *Candida* sp., etc.

of harmful poisonous materials (He and Chen, 2014). Moreover, algae type, ionic charge of metal, and chemical composition of metal ion solution are the significant parameters that assist in binding the heavy metal ions to the surface of algae (which contain amine, carboxyl groups, OH, imidazole, sulphydryl, phosphate, and sulphate) (Oyedepo, 2011). The internalisation process offers the metal uptake mechanism of algae, which helps detect the attachment of metal ions to algae. However, two algal-metal binding mechanisms are involved in the algal biosorption process: the ion-exchange method and the complexation between functional groups and metal ions. Carboxyl and sulphate groups form ionic binding, whereas covalent bonding between metal ions and functional groups is formed by amino and carboxyl groups.

Algae are divided into two main types: the multicellular plant microalgae and the photosynthetic unicellular plant macro-algae (known as seaweeds). Both grow in freshwater or saltwater. The pigmentation of microalgae causes three categories: brown, red, and green algae. Similarly, three notable characteristics like colouration, preparation of photosynthetic membranes, and other morphological structures are responsible for dividing the microalgae into four categories, such as diatoms, golden algae, green algae, and blue-green algae (cyanobacteria) (Abbas et al., 2014).

2.5.2 Bacteria as Biosorbents

Bacterial cell walls have great potential for metal ion removal during biosorption. The metal ions are attracted to the various functional groups (oxygen, phosphate, nitrogen, carboxyl, sulphur or phosphorus, amine, OH, and sulphate) of the bacterial cell wall. For example, various bacterial species like *Bacillus*, *Pseudomonas*, and *Escherichia* enhance biosorption properties due to their tiny size and capability to mature in diverse environmental circumstances (Ghosh et al., 2022). Generally, the cell wall is responsible for surface-binding sites. It also provides binding strength to metal ions that rely on different binding mechanisms. Based on bacterial cell wall composition and thickness, bacteria are categorised into two groups: one is gram-positive bacteria and the other is gram-negative bacteria (Zyoud et al., 2019). The cell wall of gram-positive bacteria contains thick peptidoglycan (90%) layers that are attached by amino acid bridges, also known to hold polyalcohols and teichoic acids. These teichoic and teichuronic acids are also connected by phosphodiester bonds linked to the peptidoglycan of the cell wall. The presence of this kind of teichoic and teichuronic acid causes a notable electronegative charge density in gram-positive bacteria, which is responsible for the easy elimination of heavy metal cations from wastewater. In contrast, the gram-negative bacterial cell wall holds an additional outer membrane composed of 10%–20% peptidoglycan, lipopolysaccharides, and phospholipids. Lipopolysaccharides, teichoic acids, and teichuronic acids cause a negative charge in gram-negative bacteria.

2.5.3 Fungi as Biosorbents

The cell structure of eukaryotic organisms, such as fungi, offers excellent metal binding properties as a biosorbent material. Examples of fungi are yeasts, mushrooms, moulds, etc. The fungal cell wall is composed of chitin, chitosans, OH groups, cellulose, glycoproteins, etc. (Dhankhar and Hooda, 2011). Contrarily, the functional groups used for the metal-fungus binding are carboxyl, phosphate, uranic acids, proteins, nitrogen-based ligands, chitin, or chitosan (Abbas et al., 2014).

Bioaccumulation (cell metabolism dependency) and biosorption are the two methods involved in the metal-fungi binding mechanism (Gadd and Rome, 1988). The biosorption ability can be controlled by a few parameters, like metabolic inhibitors, temperature, etc. The binding of metal ions with functional groups can offer ion exchange, complexation, or physical adsorption.

2.6 FEATURES AFFECTING BIOSORPTION

The following factors influence the biosorption capacity:

2.6.1 pH

Under the biosorption process, pH plays a vital role in biosorption capacity, binding sites of biomass, and the solubility of metal ions. It was found that the protonated form (hydronium ions) appeared at a lower pH, but the hydroxide form (protonated form) appeared at a higher pH. The pH of metal acceptance varies between 2.5 and 6 (Bilal et al., 2018).

2.6.2 At Lower pH

At low pH, the biosorption capacity of metal ions gets reduced by virtue of the competition between the metal cations and hydronium ions for the binding sites of the biosorbent. The functional groups of the biosorbent favour hydronium ions for binding sites. This happens as the functional groups exist in protonated form and as repulsive forces between the functional groups and metal cations.

2.6.3 At Optimum pH

The biosorption capacity often intensifies with a rise in pH. Maximum biosorption capacity can be achieved at the optimum pH. However, with the increase in pH, deprotonation occurs, and as a result, functional groups such as carboxyl, OH, and phosphate groups undergo negative charges. Due to the negative charge of the functional groups, the binding of metal cations increases and thus increases the biosorption capacity and rapidity of biosorption.

2.6.4 At Higher pH

At higher pH, the unrestricted heavy metal cations can initiate precipitation as metal hydroxides or hydroxide anionic complexes, and the efficiency of metal removal decreases due to the deprotonation of all acidic groups.

2.6.5 Temperature

Nature can manipulate the biosorption process for metal ion removal. The higher temperature can reduce the metal ion removal for the exothermic biosorption process because of protein damage (45°C), whereas the metal ion removal is increased at a lower temperature for the endothermic biosorption process (Redha, 2020). Few papers examined the temperature of the biosorption process being performed between 20°C and 35°C (Sao, 2014).

2.6.6 Initial Metal Concentration

The biosorption capacity against the mass transport of metal between the solid phase and aqueous solution is largely dependent on the initial metal concentration (Zouboulis et al., 1997). The higher the initial metal concentration, the higher the metal acceptance at a given biomass concentration, implying that the increased amount of metal is highly adsorbed by biomass. Therefore, the biting site of a biosorbent potentially offers metal removal. On the other side, lower metal concentrations provide slower metal removal due to the vacancy of metal binding sites.

2.6.7 Biomass Concentration

The biomass dosage offers a driving force that can impact the uptake of heavy metal ions, depending on binding sites. The increase in biomass concentration can limit the entrance of heavy metal ions to binding locations (functional groups) of biomass. Even larger amounts of metal ions also provide similar outputs (Nuhoglu and Malkoc, 2005).

2.6.8 Contact Time

Furthermore, a limiting factor, contact time, which is the time allotted for the biosorption process to occur, can exaggerate the biosorption capability. Increased contact time of the biosorbent material can result in extreme biosorption ability with fully saturated binding sites (Ati et al., 2021).

2.7 REGENERATION OF BIOSORBENTS

The regeneration of the adsorbent is vital towards reducing the operational cost of the recovery of valued metal ions after the biosorption process. Generally, regeneration is an important factor in developing the economy of the adsorption process, reducing global environmental issues, and accessing the potential of the adsorbent for commercial applications. Regeneration can generally be used to minimise the requirement for a new adsorbent and suppress the disposal issues of the used adsorbent, as the disposal of biosorbent is a major issue attached to the desorption process (Bayuo et al., 2020). In the case of heavy metal removal, the whole biosorption method commonly uses sorption followed by desorption, focusing on the solute. The biosorption method is biotechnologically utilised in heavy metal removal, depending on the regeneration efficiency of the biosorbent after the metal desorption method.

Several regeneration methods have been utilised with different degrees of success. For example, Kanamrlapudi and his team workers (2018) used eluents (solvents) like dilute mineral acids, organic acids, and complexing agents (EDTA, thiosulphate) to recover the biosorbent and metal, depending on the category of biosorbent and the biosorption mechanism, based on four requirements: (1) unaltered to the biomass, (2) less costly, (3) ecologically friendly, and (4) affectivity of eluents for desorption (Wang and Chen, 2009). Briefly, regeneration refers to the recycling of adsorbent and the retrieval of solute, and these are some major aspects of adsorption that can make this wastewater management more economical, valuable, and important.

2.8 APPLICATION OF BIOSORPTION FOR INDUSTRIAL WASTE

The biosorption process is favourable over conventional techniques due to the greater performance of metal exclusion processes at lower process expenses and oil costs. A group of favourable biosorbents may need to be improved for specific effluent varieties. The cost and feasibility in terms of large-scale applications may be evaluated. Eco-friendly solid biomass (or strains) or degradable agricultural, domestic, or industrial wastes can be used as biosorbents for wastewater treatment or metal removal. In this case, the biosorption process may be widely applied to the contaminated effluents before they go into rivers, lakes, seas, and so on.

2.9 CURRENT ADVANCEMENT AND COMMERCIALISATION OF BIOSEPARATION

The biosorbents formed naturally seek a lower removal capacity than conventional sorbents like AC, rain, or zeolites. Under the same experimental conditions, limited studies are carried out between less effective biosorbents and commercial sorbents because of the undefined capacity of these materials and the availability of several physicochemical non-contribution properties for enhancing biosorption. Those facts make the biosorbents lack attractive removal efficiency compared to the commercial ones and highly need modification to earn a high removal efficiency. The convenience of nanomaterials and microorganisms (effective living biomass rather than dead biomass) contributes to the tremendous outlook for metal removal. So, several studies should also be going on.

2.10 CONCLUSIONS

In summary, toxic heavy metals from industrial waste and other wastes significantly contribute to harm to both humans and the environment. A certain removal method has mostly been reviewed through proper mechanisms, outlooks, and optimisation studies of several parameters with removal efficiency. It was observed that each removal process was potentially affected by several experimental parameters to achieve higher removal efficiency, and outlooks were found separately for each method. To get better removal efficiency along with lower process costs, a few methods, such as EF and EC, RO, NF, and adsorption, have been studied under optimised conditions over conventional methods. The appropriate techniques are mostly selected based on several issues, such as solution pH, initial concentration of metal ions, and temperature. Most studies delivered 80%–90% ion removal efficiency through experimental conditions using conventional methods. More research is being explored on eco-friendly biosorbents used in membranes, biosorption and nanotechnology, bioleaching, and biofiltration over pollution of the environment. The major advantage of biological processes is their low cost and ease of operation compared to conventional methods based on different microorganisms.

In order to reduce the toxicity of the contaminated sites of wastewater, several conventional techniques have been frequently and largely used to extract the harmful metal ions from wastewater when ions are present in contaminated water at a low concentration.

Most studies achieved a separation efficiency of 80%–90% by utilising a wide range of relevant parameters. Over certain limitations of conventional methods like expensive operations, massive energy requirements, medium separation efficiency, great pH verity, overpotential (for electrochemical) with different cathodes and anodes, etc., the promising eco-friendly biological techniques (bioleaching, biosorption, bioaccumulation, biofiltration, etc.) provide the toxic as little as possible through common crucial factors such as temperature, pH, and activity of microorganisms.

REFERENCES

Abbas S.H., Ismail I.M., Mostafa T.M., Sulaymon A.H. Biosorption of heavy metals: a review, *Journal of Chemical Science and Technology* 3(4) (2014) 74–102.

Ahluwalia S.S., Goyal D. Microbial and plant derived biomass for removal of heavy metals from wastewater, *Bioresource Technology* 98 (2007) 2243–2257.

Ahmed R.A.E., ElKhawaga A.A., Hameed I.M.A. Toxic heavy metal ions removal techniques from wastewater in Saudi Arabia: a comprehensive review, *SINAI Journal of Applied Sciences* 11(5) (2022) 835–858. DOI: 10.21608/SINJAS.2022.154410.1138

Aleboyeh A., Daneshvar N., Kasiri M.B. Optimization of C.I. acid red 14 AZO dye removal by electrocoagulation batch process with response surface methodology, *Chemical Engineering and Processing: Process Intensification,* 47(5) (2008) 827–832.

Arribas P., Khayet M., Garcia-Payo M.C., Gil L. Novel and emerging membranes for water treatment by electric potential and concentration gradient membrane processes. In *Advances in Membrane Technologies for Water Treatment* (Eds: A. Basile and A. C. K Rastogi), Woodhead Publishing, Cambridge (2015) 287–325.

Ati E.M., Abbas R.F., Latif A.S., Ajmi R.N. Factors affecting biosorption of heavy metals by powder pomegranate and corn peels, *Natural Volatiles and Essential Oils* 8(4) (2021) 7020–7030. DOI: 10.1016/B978-1-78242-121-4.00009-5

Balmer L., Foulds A. Electroflocculation/electroflotation for the removal of oil from oil-in-water emulsions. *Filtration & Separation* 23(6) (1986) 366–372.

Barakat M.A. New trends in removing heavy metals from industrial wastewater, *Arabian Journal of Chemistry* 4(4) (2011) 361–377. DOI: 10.1016/j.arabjc.2010.07.019

Bazrafshan E., Mohammadi L., Ansari-Moghaddam A., Mahvi A.H. Heavy metals removal from aqueous environments by electrocoagulation process – a systematic review, *Journal of Environmental Health Science & Engineering* 13 (2015) 74. DOI: 10.1186/s40201-015-0233–8

Bilal, M., Rasheed, T., Sosa-Hernandez, J., Raza, A., Nabeel, F., Iqbal, H. Biosorption: an interplay between marine algae and potentially toxic elements – a review, *Marine Drugs* 16(2) (2018) 1–16. DOI:10.3390/md16020065

Bayuo J., Abukari, M.A., Pelig-Ba, K.B. Desorption of chromium (VI) and lead (II) ions and regeneration of the exhausted adsorbent, *Applied Water Science* 10 (171) (2020) 2735.

Bosecker K. Bioleaching: metal solubilization by microorganisms, *FEMS Microbiology Reviews* 20 (3–4) (1997) 591–604. DOI: 10.1111/j.1574-6976.1997.tb00340.x

Brady D., Duncan J.R. Bioaccumulation of metal cations by Saccharomyces cerevisiae, *Applied Microbiology and Biotechnology* 41 (1993) 149–154.

Brady D., Rose P.D., Duncan, J.R. The use of hollow fiber cross-flow microfiltration in bioaccumulation and continuous removal of heavy metals from solution by Saccharomyces cerevisiae, *Biotechnology and Bioengineering* 44 (1994)1362–1366.

Brodskiy V.A., Gaydukova A.M., Kolesnikov V.A. Influence of pH of the medium on the physicochemical characteristics and efficiency of electroflotation recovery of poorly soluble cerium(III, IV) compounds from aqueous solutions, *Russian Journal of Applied Chemistry* 88 (2015) 1446–1450.

Carlos A.M.H., Sergio F., Electrochemical oxidation of organic pollutants for the wastewater treatment: direct and indirect processes, *Chemical Society Reviews* 35 (2006) 1324–1340.

Chen G. Electrochemical technologies in wastewater treatment, *Separation and Purification Technology* 38 (2004) 11–41. DOI: 10.1016/j.seppur.2003.10.006

Chen J.P., Lim L.L. Recovery of precious metals by an electrochemical deposition method, *Chemosphere* 60(10) (2005) 1384–1392. DOI: 10.1016/j.chemosphere.2005.02.001

Cuhorka J., Wallace E., Mikulášek, P. Removal of micropollutants from water by commercially available nanofiltration membranes, *Science of The Total Environment* 720 (2020) 137474.

Das T.K., Poater A. Review on the use of heavy metal deposits from water treatment waste towards catalytic chemical syntheses, *International Journal of Molecular Science* 22(24) (2021) 13383–133409. DOI: 10.3390/ijms222413383

Devda V., Chaudhary K., Varjani S., Pathak B., Patel A.K., Singhania R.R., Taherzadeh M.J., Ngo H.H., Wong J.W.C., Guo W., Chaturvedi P. Recovery of resources from industrial wastewater employing electrochemical technologies: status, advancements and perspectives, *Bioengineered* 12(1) (2021) 4697–4718. DOI: 10.1080/21655979.2021.1946631

Dhankhar, R., Hooda, A., Fungal biosorption-an alternative to meet the challenges of heavy metal pollution in aqueous solutions, *Environmental Technology* 32(5) (2011) 467–491.

Dydo P., Nemś I., Turek, M. Boron removal and its concentration by reverse osmosis in the presence of polyol compounds, *Separation and Purification Technology* 89 (2012) 171–180.

Ebbers B., Ottosen L.M., Jensen P.E. Electrodialytic treatment of municipal wastewater and sludge for the removal of heavy metals and recovery of phosphorus, *Electrochimica Acta* 181 (2015) 90–99. DOI: 10.1016/j.electacta.2015.04.0973

Es-sahbany H., Berradi M., Nkhili S., Hsissou R., Allaoui M., Loutfi M., Bassir D., Belfaquir M., El Youbi M.S. Removal of heavy metals (nickel) contained in wastewater-models by the adsorption technique on natural clay, *Materials Today: Proceedings* 13 (2019) 866–875.

Fang, D., Liu, X., Zhang, R., Deng, W., Zhou, L. Removal of contaminating metals from soil by sulfur-based bioleaching and biogenic sulfide-based precipitation. *Geomicrobiology Journa*l 30 (2013) 473–478.

Farnane M., Tounsadi H., Elmoubarki R., Mahjoubi F.Z., Elhalil A., Saqrane S., Abdennouri M., Qourzal S., Barka N. Alkaline treated carob shells as sustainable biosorbent for clean recovery of heavy metals: kinetics, equilibrium, ions interference and process optimisation, *Ecological Engineering* 101 (2017) 9–20.

Ferella F., Prisciandaro M., De Michelis, I., Veglio' F. Removal of heavy metals by surfactant-enhanced ultrafiltration from wastewaters, *Desalination* 207 (2007) 125–133.

Folens K., Huysman, S. Van Hulle S., Du Laing G. Chemical and economic optimization of the coagulation-flocculation process for silver removal and recovery from industrial wastewater, *Separation and Purification Technology* 179 (2017) 145–151.

Gadd G.M., Rome L. Biosorption of copper by fungal melanin, Biosorption of copper by fungal melanin, *Applied Microbiology and Biotechnology* 29 (1988) 610–617. DOI: 10.1007/BF00260993

Ghosh, S., Bhattacharya, J., Nitnavare, R., Webster, T.J. Heavy metal removal by *Bacillus* for sustainable agriculture. In *Bacilli in Agrobiotechnology: Plant stress tolerance, bioremediation, and bioprospecting* (Eds: M. T. Islam, M. Rahman, and P. Pandey), Springer International Publishing, Switzerland (2022) 1–30.

Guan W., Tian S., Cao D., Chen Y., Zhao X. Electrooxidation of nickel-ammonia complexes and simultaneous electrodeposition recovery of nickel from practical nickel-electroplating rinse wastewater, *Electrochimica Acta* 246 (2017) 1230–1236.

Guo Z.-R., Zhang G., Fang, J., Dou, X. Enhanced chromium recovery from tanning wastewater, *Journal of Cleaner Production* 14 (2006) 75–79.

He J., Chen J.P. A comprehensive review on biosorption of heavy metals by algal biomass: materials, performances, chemistry, and modeling simulation tools, *Bioresource Technology* 160 (2014) 67–78. DOI: 10.1016/j.biortech.2014.01.068

Hoseinian F.S., Rezai B., Kowsari, E. Optimization and separation mechanism of Ni(II) removal from synthetic wastewater using response surface method, *International Journal of Environmental Science and Technology* 16 (2019) 4915–4924.

Hossain S. M.Z., Mansour N. Biosensors for on-line water quality monitoring – a review, *Arab Journal of Basic and Applied Science* 26 (2019) 502–518. DOI: 10.1080/25765299.2019.1691434

Huang C.-H., Shen S.-Y., Dong C.-D., Kumar M., Chang, J.-H. Removal mechanism and effective current of electrocoagulation for treating wastewater containing Ni(II), Cu(II), and Cr(VI), *Water* 12(9) (2020) 2614. DOI: 10.3390/w12092614

Huang J., Yuan F., Zeng G., Li X., Gu Y., Shi L., Liu W., Shi Y. Influence of pH on heavy metal speciation and removal from wastewater using micellar-enhanced ultrafiltration, *Chemosphere* 173 (2017) 199–206.

Jadhav U.U., Hocheng H. A review of recovery of metals from industrial waste, *Journal of achievements in materials and manufacturing* 54 (2012) 156–167.

Jayakrishnan, D.S. Electrodeposition: the versatile technique for nanomaterials. In *Corrosion Protection and Control Using Nanomaterials* (Eds: V.S. Saji and R. Cook) Woodhead Publishing, Cambridge (2012) 86–125.

Jiang S., Fu F., Qu J., Xiong, Y. A simple method for removing chelated copper from wastewaters: $Ca(OH)_2$-based replacement-precipitation, *Chemosphere* 73 (2008) 785–790.

Jin W., Zhang Y. Sustainable electrochemical extraction of metal resources from waste streams: from removal to recovery, *ACS Sustainable Chemical Engineering*, 8(12) (2020) 4693–4707.

Jin W., Fu Y., Hu M., Wang S., Liu Z. Highly efficient SnS-decorated Bi2O3 nanosheets for simultaneous electrochemical detection and removal of Cd(II) and Pb(II), *Journal of Electroanalytical Chemistry* 856 (2020) 113744.

Kanamrlapudi S.L.R.K.K., Chintalpudi V.K., Muddada S. Application of biosorption for removal of heavy metals from wastewater. In *Biosorption* (Eds: J. Basile and B. Vrana), Intech, London (2018). DOI: 10.5772/intechopen.77315

Kang J., Hu Y., Sun W., Liu R., Gao Z., Guan Q., Tang H., Yin Z. Utilisation of FGD gypsum for silicate removal from scheelite flotation wastewater, *Chemical Engineering Journal* 341 (2018) 272–279.

Khosla N.K., Venkatachalam S., Somasundaran P. Pulsed electrogeneration of bubbles for electroflotation, *Journal of Applied Electrochemistry* 21 (1991) 986–990.

Krishnan S., Zulkapli N.S., Kamyab H., Taib S.M., Din M.F.B.M., Maji Z.A., Chaiprapat S., Kenzo I., Ichikawa Y., Nasrullah M., Chelliapan S., Othman N. Current technologies for recovery of metals from industrial wastes: an overview, *Environmental Technology & Innovation* 22 (2021) 101525. DOI: 10.1016/j.eti.2021.101525

Koren J.P.F., Syversen U. State-of-the-art electroflocculation, *Filtration & Separation* 32(2) (1995) 153–156. DOI: 10.1016/S0015-1882(97)84039-6

Kumar R.N., Nagendran R. Influence of initial pH on bioleaching of heavy metals from contaminated soil employing indigenous Acidithiobacillus thiooxidans. *Chemosphere* 66 (2017) 1775–1781

López-Maldonado E.A., Oropeza-Guzman M.T., Jurado-Baizaval J.L., Ochoa-Terán A. Coagulation-flocculation mechanisms in wastewater treatment plants through zeta potential measurements, *Journal of Hazardous Materials* 279 (2014) 1–10.

Liu W., Sun B., Zhang D., Chen L., Yang, T. Effect of pH on the selective separation of metals from acidic wastewater by controlling potential, *Separation and Purification Technology* 205 (2018) 223–230.

Mani D., Kumar C. Biotechnological advances in bioremediation of heavy metals contaminated ecosystems: an overview with special reference to phytoremediation, *International Journal of Environmental Science and Technology* 11 (2014) 843–872.

Manikandan S., Saraswathi R. Electrocoagulation technique for removing organic and inorganic pollutants (COD) from the various industrial effluents: an overview, *Environmental Engineering Research* 28(4) (2023) 220231. DOI: 10.4491/eer.2022.231

Mansour L.B., Chalbi S., Kesentini I. Experimental study of hydrodynamic and bubble size distributions in electroflotation process. *Indian Journal of Chemical Technology* 14 (2007) 253–257.

Manzoor, K., Ahmad, M., Ahmad, S., Ikram, S. Removal of Pb(ii) and Cd(ii) from wastewater using arginine cross-linked chitosan-carboxymethyl cellulose beads as green adsorbent, *RSC Advances* 9 (2019) 7890–7902.

Martin E.T., McGuire C.M., Mubarak M.S., Peters D.G. electroreductive remediation of halogenated environmental pollutants, *Chemical Reviews* 116(24) (2016) 15198–15234. DOI: 10.1021/acs.chemrev.6b00531

Mickova I. Advanced electrochemical technologies in wastewater treatment. Part II: electro-flocculation and electroflotation, *American Scientific Research Journal for Engineering, Technology, and Sciences (ASRJETS)* 14(2) (2015) 273–294.

Mohsen-Nia M., Montazeri P., Modarress H. Removal of Cu^{2+} and Ni^{2+} from wastewater with a chelating agent and reverse osmosis processes, *Desalination* 217 (2007) 276–281.

Mota I.O., Castro J.A., Casqueira R.G., Junior A.G.O. Study of electroflotation method for treatment of wastewater from washing soil contaminated by heavy metals, *Journal of Materials Research and Technology* 4(2) (2015) 109–113.

Ntagia E., Fiset E., Lima L. D.S., Pikaar I., Zhang X., Jeremiasse A.W., Prevoteau A., Rabaey K. Anode materials for sulfide oxidation in alkaline wastewater: an activity and stability performance comparison. *Water Resources* 149 (2019) 111–119.

Nuhoglu Y., Malkoc E. Investigations of nickel (II) removal from aqueous solutions using tea factory waste, *Journal of Hazardous Materials* 127 (2005) 120. DOI: 10.1016/j.jhazmat.2005.06.030

Ojedokun A.T., Bello O.S. Sequestering heavy metals from wastewater using cow dung, *Water Resources in India* 13 (2016) 7–13.

Olaniran A.O., Balgobind A., Pillay B. Bioavailability of heavy metals in soil: impact on microbial biodegradation of organic compounds and possible improvement strategies, *International Journal of Molecular Sciences* 14 (2013) 10197–10228.

Oliveira E.M.S., Silva F.R., Morais C.C.O., Mielle T., Oliveira B.F., Martínez-Huitle C.A., Motheo A.J., Albuquerque C.C., Castro S.S.L. Performance of (in) active anodic materials for the electrooxidation of phenolic wastewaters from cashew-nut processing industry, *Chemosphere* 201 (2018) 740–748. DOI: 10.1016/j.chemosphere.2018.02.037

Oyedepo T.A. Biosorption of lead (II) and copper (II) metal ions on Calotropis procera (Ait.), *Science Journal of Pure and Applied Chemistry* 1 (2011) 1–7.

Oztekin Y., Yazicigil Z. Recovery of metals from complexed solutions by electrodeposition, *Desalination* 190 (1–3) (2006) 79–88. DOI: 10.1016/j.desal.2005.07.017

Pangeni B., Paudyal H., Inoue K., Ohto K., Kawakita H., Alam S. Preparation of natural cation exchanger from persimmon waste and its application for the removal of cesium from water, *Chemical Engineering Journal* 242 (2014) 109–116.

Parga J.R., Valenzuela J.L., Munive G.T., Vazquez V.M., Rodriguez M. Thermodynamic study for arsenic removal from freshwater by using electrocoagulation process, *Advances in Chemical Engineering and Science* 04 (2014) 548–556.

Qianru Z., Xiaoqing G., Junqiang Y., Peng Z., Geng C., Yaming L., Keliang S., Wangsuo W. Investigation on the thermal activation of montmorillonite and its application for the removal of U(VI) in aqueous solution, *Journal of the Taiwan Institute of Chemical Engineers* 80 (2017) 754–760.

Qu M., Chen J., Huang Q., Chen J., Xu Y., Luo J., Wang K., Gao W., Zheng Y. Bioremediation of hexavalent chromium contaminated soil by a bioleaching system with weak magnetic fields, *International Biodeterioration and Biodegradation* 128 (2018) 41–47.

Raptopoulou C., Palasantza P.-A., Mitrakas M., Kalaitzidou K., Tolkou A., Zouboulis A. Statistical variation of nutrient concentrations and biological removal efficiency of a wastewater treatment plant, *Water Utility Journal* 14 (2016) 5–17.

Razzak S.A., Faruque M.O., Alsheikh Z., Alsheikhmohamad L., Alkuroud D., Alfayez A., Hossain S.M.Z., Hossain M.M. A comprehensive review on conventional and biological-driven heavy metals removal from industrial wastewater, *Environmental Advances* 7 (2022) 100168. DOI: 10.1016/j.envadv.2022.100168

Redha A.A. Removal of heavy metals from aqueous media by biosorption, *Arab Journal of Basic and Applied Sciences* 27(1) (2020) 183–193. DOI: 10.1080/25765299.2020.1756177

Sadegh H., Ali G.A.M., Makhlouf A.S.H., Chong K.F., Alharbi N.S., Agarwal S., Gupta V.K. MWCNTs-Fe3O4 nanocomposite for Hg(II) high adsorption efficiency, *Journal of Molecular Liquids* 258 (2018) 345–353.

Sao, K. A review on heavy metals uptake by plants through biosorption, *International Proceedings of Economics Development and Research* 75(17) (2014) 78–83. DOI: 10.7763/IPEDR

Sarkodie E.K., Jiang L., Li K., Yang J., Guo Z., Shi J., Deng Y., Liu H., Jiang H., Liang Y., Yin H., Liu X. A review on the bioleaching of toxic metal(loid)s from contaminated soil: insight into the mechanism of action and the role of influencing factors, *Front Microbiology* 13 (2022) 1049277.

Scott K., Paton E.M. An analysis of metal recovery by electrodeposition from mixed metal ion solutions – part I. Theoretical behavior of batch recycle operation, *Electrochimica Acta* 38(15) (1993) 2181–2189. DOI: 10.1016/0013-4686(93)80096-I

Selvaraj R., Santhanam M., Selvamani V., Sundaramoorthy S., Sundaram M. A membrane electroflotation process for recovery of recyclable chromium(III) from tannery spent liquor effluent, *Journal of Hazardous Materials* 346 (2018) 133–139.

Shahriari T., Bidhendi G.N., Mehrdadi N., Torabian A. Removal of chromium (III) from wastewater by electrocoagulation method, *KSCE Journal of Civil Engineering* 18 (2014) 949–955.

Silverman M.P. Mechanism of bacterial pyrite oxidation, *Journal of Bacteriology* 94 (1967) 1046–1051.

Soya K., Mihara N., Kuchar D., Kubota M., Matsuda H., Fukuta T. Selective sulfidation of copper, zinc and nickel in plating wastewater using calcium sulphide, *International Journal of Civil and Environmental Engineering* 2 (2010) 93–97.

Srivastava S., Agrawal S.B., Mondal, M.K. Synthesis, characterization and application of Lagerstroemia speciosa embedded magnetic nanoparticle for Cr(VI) adsorption from aqueous solution, *Journal of Environmental Sciences (China)* 55 (2017) 283–293.

Taffarel S.R., Rubio J. Removal of Mn^{2+} from aqueous solution by manganese oxide coated zeolite, *Minerals Engineering* 23 (2010) 1131–1138.

Tonini G.A., Ruotolo, L.A.M. Heavy metal removal from simulated wastewater using electrochemical technology: optimization of copper electrodeposition in a membraneless fluidized bed electrode, *Clean Technologies and Environmental* Policy 19 (2017) 403–415.

Torma A.E. The role of *Thiobacillus ferrooxidans* in hydrometallurgical processes, *Advances in Biochemical Engineering* 6 (1977) 1–37.

Veeken A.H.M., Akoto L., Hulshoff Pol L.W., Weijma J. Control of the sulfide (S2–) concentration for optimal zinc removal by sulfide precipitation in a continuously stirred tank reactor, *Water Research* 37 (2003) 3709–3717.

Vidu R., Matei E., Predescu A.M., Alhalaili B., Pantilimon C., Tarcea C., Predescu C. Removal of heavy metals from wastewaters: a challenge from current treatment methods to nanotechnology applications, *Toxics* 8(4) (2020) 101. DOI: 10.3390/toxics8040101

Wang J., Chen C. Biosorbents for heavy metals removal and their future, *Biotechnology Advances* 27 (2009) 195–226. DOI: 10.1016/j.biotechadv.2008.11.002

Wang L.K., Chen J.P., Hung Y.-T., Shammas N.K. Membrane and desalination technologies, *Membrane and Desalination Technologies* 13 (2011) 7645.

Wu D., Tan Z., Yu H., Li Q., Thé J., Feng X. Use of nanofiltration to reject cobalt (II) from ammoniacal solutions involved in absorption of SO_2/NO_x, *Chemical Engineering Science* 145 (2016) 97–107.

Xing M., Wang J. Nanoscaled zero valent iron/graphene composite as an efficient adsorbent for Co(II) removal from aqueous solution, *Journal of Colloid and Interface Science* 474 (2016) 119–128.

Yang L., Hu W., Chang Z., Liu T., Fang D., Shao P., Shi H., Luo X. Electrochemical recovery and high value-added reutilization of heavy metal ions from wastewater: recent advances and future trends, *Environment International* 152 (2021) 106512. DOI: 10.1016/j.envint.2021.106512

Yuan S., Zhang P., Yang Z., Lv L., Tang S., Liang B. Successive grafting of poly(hydroxyethyl methacrylate) brushes and melamine onto chitosan microspheres for effective Cu(II) uptake, *International Journal of Biological Macromolecules* 109 (2018) 287–302.

Zhang M., Ma X., Li J., Huang R., Guo L., Zhang X., Fan Y., Xie X., Zeng G. Enhanced removal of As(III) and As(V) from aqueous solution using ionic liquid-modified magnetic graphene oxide, *Chemosphere* 234 (2019) 196–203.

Zhang T., Wen T., Zhao Y., Hu H., Xiong B., Zhang Q. Antibacterial activity of the sediment of copper removal from wastewater by using mechanically activated calcium carbonate, *Journal of Cleaner Production* 203 (2018) 1019–1027.

Zhang S., Zhang J., Wang W., Li F., Cheng X. Removal of phosphate from landscape water using an electrocoagulation process powered directly by photovoltaic solar modules, *Solar Energy Materials and Solar Cells* 117 (2013) 73–80.

Zouboulis A.L., Matis K.A., Hancock I.C. Biosorption of metals from dilute aqueous solutions, *Separation and Purification Methods* 26 (1997) 255–295. DOI: 10.1080/03602549708014160

Zyoud A., Alkowni R., Yousef O., Salman M., Hamdan S., Helal M.H., Jaber S.F., Hilal H.S. Solar light-driven complete mineralization of aqueous gram-positive and gram-negative bacteria with ZnO photocatalyst, *Solar Energy* 180 (2019) 351–359. DOI: 10.1016/j.solener.2019.01.034

3 Bioleaching
An Advanced Technique for the Renewal of Essential Metals

Angel Mathew, Anna Nova, and Neetha John

3.1 INTRODUCTION TO BIOLEACHING

Our current consuming habits are depleting the planet's resources while also causing pollution and endangering the survival of our species. Change is required immediately. The waste management paradigm aims to effectively recover resources (energy, metals, and nutrients) from waste streams, improving the effectiveness of industrial and urban operations. Recent UN Global Resources Outlook data indicates that the extractive industries are responsible for half of the world's carbon emissions [1]. Recent data from the UN Global Resources Outlook reveals alarming statistics: the extractive industries account for half of global carbon emissions, with 90% of biodiversity loss and water stress attributed to resource extraction and processing. This unchecked resource exploitation imposes severe and escalating pressures on our climate and the ecosystems that sustain life. Despite a twofold increase in population since 1970, resource extraction has surged threefold, underscoring the urgency of decoupling economic growth from resource consumption and environmental degradation. In response to these challenges, the United Nations' Sustainable Development Goals (SDGs) aim to achieve sustainable management and efficient utilization of natural resources by 2030. This entails enhancing resource efficiency, reducing reliance on raw materials, and boosting recycling to alleviate environmental strain. The concepts of "circular economy" and "zero waste" have gained traction as potential solutions, although achieving these objectives remains formidable [2].

Microbial leaching techniques are gaining traction as effective means to extract metals from low-grade ores and concentrates, which are often uneconomical to process using traditional methods. Due to the advancements in technology and industry, numerous industrial locations are now contaminated with heavy metals and organic pollutants. These pollutants, largely of human origin, pose significant risks to various organisms, including humans. Consequently, both industry and government entities are increasingly mandated to enforce stringent environmental management systems. Environmental biotechnology has emerged as a promising solution, leveraging microorganisms to mitigate the environmental impact of these toxic substances, thereby improving both cost-effectiveness and efficiency. This approach falls under

 DOI: 10.1201/9781003415541-3

the umbrella of biotechnology. For about four decades, the integration of biotechnology into mining has been the subject of extensive research and development across numerous countries. Today, a range of technologies have been commercialized, operating within well-designed and mechanized systems. These technologies collectively fall under the term "biohydrometallurgy," which encompasses various disciplines categorized by the interactions between metals and microbes. Key bioprocessing techniques include bioremediation, biosorption, bioaccumulation, and bioleaching. Bioremediation involves treating acidic mine runoff or tailings to reduce the concentration of contaminants affecting natural ecosystems. In contrast, biosorption and bioaccumulation focus on the sequestration and detoxification of heavy metals in environments where they accumulate. The application of microbial technology in mining, particularly bioleaching, has witnessed significant advancements in metal extraction throughout the 20th century. Notably, research into the microbial consortia thriving in highly acidic, metal-rich environments has laid the groundwork for the development of mineral processing technologies [3].

Interestingly, these biotechnological approaches might have roots dating back to prehistoric times. Historical evidence suggests that civilizations like the Greeks and Romans might have utilized mine water to extract copper over 2,000 years ago. Despite these ancient practices, it's only in the last half century that we've come to understand the pivotal role bacteria play in enriching metals in water from ore deposits and mines. This discovery has revolutionized our approach to metal extraction, offering sustainable and efficient alternatives to conventional mining methods [4].

Metals are indispensable to the global economy, underpinning various products ranging from electric vehicles and low-carbon energy systems to technological and healthcare devices. The production and storage of renewable energy, catalytic processes, digital communication, and green technologies require substantial quantities of critical and scarce metals, including platinum group metals (PGM), rare earth elements (REE), cobalt, vanadium, selenium, and tellurium. These metals are classified as critical due to their high economic value coupled with supply risks arising from geopolitical instability, limited material replacement capacity, and low recycling rates [5].

Given the low concentrations of these metals in natural ores, the adoption of cost-effective and environmentally friendly technologies like bioleaching is imperative for economic viability. Bioleaching is employed commercially to recover metals from low-grade and waste ores, especially copper ores that would otherwise be unprofitable to process using conventional methods. In bioleaching, naturally occurring microorganisms are harnessed to produce mineral or organic acids, enhancing metal solubility through enzymatic processes [6].

Bioleaching is a microorganism-based method of extracting metals from ores or other solid materials. Bacteria or fungi are used in the process to break down the mineral matrix and release the target metal into a solution. Metals such as copper, gold, silver, and uranium are extracted using this method. Biohydrometallurgy is a branch of metallurgy that combines biology and metallurgy. Biology is very important in the natural environment. Because of the variety of biological reactions and their interactions with metals in mineral and dissolved forms, microorganisms are now used in a wide range of metal extraction, metal recovery, and water treatment applications. Bioleaching is a simple and effective method for extracting metal from low-grade ores or mineral concentrates that cannot be processed economically using traditional methods. Bioleaching is the process of solubilization that occurs in nature wherever suitable conditions for the growth of ubiquitous bioleaching organisms exist [7,8].

This procedure consists of several steps. The solid material is first crushed and mixed with water to form a slurry. Microorganisms such as *Acidithiobacillus ferrooxidans* and *Acidithiobacillus thiooxidans*, as well as *Aspergillus niger* and *Penicillium simplicissimum*, are then added to the slurry. These microorganisms thrive in acidic environments and can degrade the mineral matrix of the solid material, releasing the target metal into solution. After that, the metal-rich solution is collected and processed to recover the metal. Precipitation, solvent extraction, or electrowinning are all possibilities. The remainder of the solid material is usually discarded as waste. Bioleaching has a number of advantages over conventional mining methods. It is a less harmful process because it employs natural microorganisms rather than toxic chemicals [9]. It can also be used to extract metals from low-grade ores where traditional methods would not be economically viable. The biorecovery of metal ions through bioleaching is of paramount importance in bridging the resource supply gap and mitigating the environmental impact of conventional mining. By harnessing the metabolic capabilities of microorganisms, bioleaching offers a sustainable and efficient pathway for metal extraction, contributing to the transition toward a more circular and resource-efficient economy. However, bioleaching can be a time-consuming process that takes a long time to extract the desired metal [10,11].

3.2 HISTORY OF BIOLEACHING

The application of biotechnology in mining, particularly through bioleaching, has been a subject of exploration and study across diverse research institutions and industries globally. To appreciate the advancements in biohydrometallurgy, it is crucial to delve into the historical evolution of bioleaching.

The recognition of microbial-mediated mineral dissolution can be traced back to the mid-19th century, although the natural environmental leaching of metals from rocks was documented much earlier, around 20–70 AD. The pioneering work of Georgius Agricola (1494–1555) stands as an early testament to this field, as he discussed the leaching of copper from ores and mine leachates.

Surprisingly, the history of bioleaching extends much further than conventional wisdom suggests. Evidence of the natural recovery of base metals, including copper and zinc, from rock solutions has been documented in ancient civilizations such as Spain, China, and India almost 2,000 years ago. The Rio Tinto region in Spain, for instance, derives its name from the reddish-brown waters indicative of elevated ferric ion concentrations, later attributed to the natural dissolution of iron and copper minerals facilitated by native microorganisms.

A significant breakthrough occurred in 1947, when scientists elucidated the role of reduced sulfur and ferrous ion-oxidizing bacteria in producing ferric-ion-containing sulfuric acid. This acid has the capability to dissolve a wide range of sulfide minerals, laying the foundation for modern bioleaching techniques. Subsequent decades witnessed a surge in research focused on *Acidithiobacillus* bacteria, leading to the commercial application of bioleaching, particularly in the heap and dump leaching of low-grade copper and uranium ores.

During the late 1950s and early 1960s, innovative mining practices were introduced at the Kennecott Copper Company's Bingham Canyon. Run-of-mine low-grade copper ores were stacked into towering waste dumps exceeding 100 m in height and subjected to acidic leaching solutions to economically extract copper. This period also saw the development of heap and "*in situ*" mining techniques leveraging native microorganisms to recover copper and uranium from challenging, low-grade ores [12].

Uranium heap bioleaching marked its inception in Canada during the 1960s, signaling the beginning of a new era in biohydrometallurgy. However, it was not until the 1980s that large-scale commercial bioleaching operations gained momentum, particularly with the heap bioleaching of secondary copper sulfides and oxidized ores. Chile emerged as a leading player in this domain, with the installation of numerous copper heap bioleaching systems since the 1980s.

Subsequent advancements in bioleaching technology saw the development of piloting and prototype stirred tank reactor systems tailored for the thermophilic bioleaching of base metal concentrates, including copper, zinc, and nickel. A landmark achievement in commercial cobalt bioleaching was realized in Kasese, Uganda, in 1999, marking the onset of bioleaching applications for cobalt extraction from pyritic concentrates.

The 2000s witnessed significant strides in the direct bioleaching of base metal concentrates in stirred tank bioreactors, reflecting ongoing innovation and refinement in bioleaching methodologies. To address the challenges posed by refractory chalcopyrite ores and concentrates, novel heap bioleaching processes have been developed, including Geocoat, Geoleach, Aster, reductive, and hybrid bioleaching techniques. These innovative approaches have expanded the scope and applicability of bioleaching, contributing to its growing recognition as a sustainable and economically viable alternative to conventional mineral processing methods [13].

Over the past 25 years, the bioleaching industry has witnessed several transformative milestones that have shaped its evolution and expanded its application. One of the pivotal breakthroughs occurred in 1986 when an agitated tank bioleaching process for sulfide concentrates was developed and commercialized at the Fairview Gold Mine in South Africa. This landmark achievement firmly established bioleaching as a viable and competitive metallurgical method. The engineering intricacies of this process laid the groundwork for the commercialization of agitated tank bioleaching technology, characterized by large-scale reactors necessitating high levels of agitation, aeration, and heat exchange.

In response to the growing demand for larger-scale bioleaching operations, significant innovations in three-phase (gas-liquid-solid) mixing technology have emerged over the past quarter century. These advancements have been instrumental in facilitating the efficient operation of increasingly larger agitated tank bioleaching processes. Another noteworthy development in bioleaching technology was the introduction of "thin layer" leaching at the Lo Aguirre Copper Mine in Chile in 1980, marking the inception of heap bioleaching. This innovative approach involves stacking crushed and acid-cured ore 2–3 m high and subsequently rinsing it.

In 1937, the Girilambone Copper Mine in Australia achieved another breakthrough by incorporating forced aeration into the heap bioleaching process for secondary copper sulfide ores. While heap bioleaching of secondary copper sulfides had been practiced since 1980, the Girilambone operation was the first to integrate forced aeration into its plant design, setting a precedent for subsequent operations.

The discovery and utilization of thermophilic microorganisms capable of thriving at higher temperatures marked a significant advancement in bioleaching technology. This breakthrough has broadened the commercial applicability of bioleaching processes, enabling the more efficient extraction of metals from refractory gold concentrates and chalcopyrite ores. The BioCOP process exemplifies this advancement, employing agitated tank bioleaching with thermophiles to produce 20,000 t/a of copper, although the plant has since been decommissioned for commercial reasons.

Recent years have seen the advent of modern microbiological techniques that have facilitated the identification and understanding of a diverse array of bioleaching microorganisms capable of functioning under varying conditions. This burgeoning knowledge base continues to drive research and innovation in the field, enhancing our ability to leverage the microbiological diversity inherent in bioleaching processes.

Thermophilic heap bioleaching represents the next frontier in bioleaching technology, often referred to as the "Holy Grail" due to its potential to unlock significant value from large, low-grade chalcopyrite ore deposits, which constitute a substantial portion of the world's untapped copper reserves. Ongoing research and development efforts, spearheaded by organizations like Mintek, are making remarkable strides in this domain [10].

3.3 BIOLEACHING OF SOLID WASTE

Bioleaching is used to extract metals from waste, reducing the amount of hazardous material in the waste while also creating a potential source of valuable metals. The process of bioleaching solid waste is similar to mining. To begin, the waste is crushed and mixed with water to form a slurry. The slurry is then treated with microorganisms, which can include bacteria and fungi capable of breaking down waste and releasing metals into a solution. After that, the metal-rich solution is collected and processed to recover the metals. Precipitation, solvent extraction, or electrowinning may be used depending on the metals involved and their concentration in the solution.

Bioleaching of solid waste can be an environmentally friendly and cost-effective method of hazardous waste remediation. It could also be a source of valuable metals that can be recycled and reused. However, the type of waste involved and the availability of suitable microorganisms for bioleaching may limit the process [14–16].

Solid waste streams from the mining and metallurgy, energy generation, and recycling industries may contain relatively high levels of metals that are hazardous if released into the environment. These waste streams have the potential to be valuable metal sources [14]. Some solid wastes, including used refinery catalysts, electronic waste, metal-containing sludge, slag, and fly ash, contain toxic metals like Mo, V, Ni, Cu, Co, and Pb. Few of them are toxic to biotic communities, including humans, and treating these wastes is necessary. The metal content of such wastes would leach into land or water bodies if disposed of directly [10]. Metals have traditionally been extracted from solid waste through chemical leaching with strong acids. However, these methods are only advantageous when recoverable metals are present in relatively high concentrations. Bioleaching may be an alternative treatment method for solid waste materials with low levels of valuable metals or that are otherwise difficult to handle or treat [9].

Here's a step-by-step process outlining how bioleaching of solid waste typically occurs:

Step 1: Waste Preparation

- **Material segregation:** The incoming solid waste undergoes a meticulous sorting process to isolate the desired materials, such as metal-rich waste.
- **Particle size reduction:** The segregated waste is mechanically shredded or crushed to enhance its surface area, facilitating better microbial access and action.

Step 2: Microbial Inoculation

- **Selection of microorganisms:** Acidophilic bacteria or archaea that can thrive in acidic conditions are selected.
- **Inoculation:** The selected microorganisms are introduced into the waste material to initiate the bioleaching process.

Step 3: Fermentation

- **Acid production:** The introduced microorganisms metabolize and produce acids (e.g., sulfuric acid) that help in solubilizing the metals.
- **pH control:** Monitoring and controlling the pH level to maintain an acidic environment optimal for microbial activity.

Step 4: Leaching

- **Metal solubilization:** The acids produced by the microorganisms solubilize the metals present in the waste material, releasing them into the solution.
- **Oxidation-reduction reactions:** Microorganisms may also facilitate redox reactions that aid in metal solubilization.

Step 5: Metal Recovery

- **Solid-liquid separation:** The leached solution is separated from the solid waste using techniques like filtration or sedimentation.
- **Metal precipitation:** Metals are recovered from the solution through precipitation methods or electrolysis.

Step 6: Waste Treatment

- **Residue management:** The remaining waste after metal extraction, now devoid of valuable metals, is subjected to appropriate treatment methods, including composting, landfilling, or further recycling.
- **Water treatment:** Prior to disposal or recycling, the leachate undergoes treatment to eliminate contaminants and adjust its composition to meet regulatory standards.

Step 7: Monitoring and Optimization

- **Process monitoring:** Regular monitoring of microbial activity, pH, and metal concentrations to optimize the bioleaching process.
- **Process optimization:** Based on monitoring data, process conditions like temperature, oxygen availability, and nutrient supplementation are fine-tuned to enhance metal recovery rates and minimize process duration.

Step 8: Quality Control and Analysis

- **Metal quality:** Analyzing the recovered metals for purity and quality.
- **Environmental impact assessment:** A comprehensive assessment of the bioleaching process's environmental footprint is conducted to ensure adherence to environmental regulations and to identify areas for improvement.

By following these steps systematically, bioleaching can be an effective and environmentally friendly method for recovering valuable metals from solid waste while minimizing the environmental impact associated with traditional mining and waste disposal methods [17].

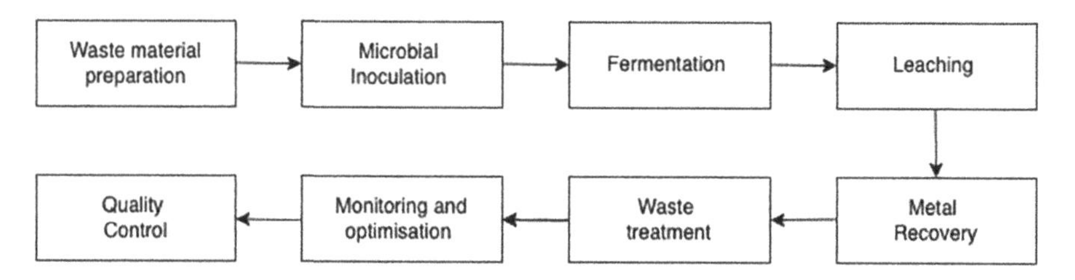

3.4 MICROORGANISMS

Microorganisms, also known as microbes, are tiny living organisms that can only be seen under a microscope. They include bacteria, archaea, fungi, protozoa, and viruses. Despite their small size, microorganisms are incredibly diverse and exist in almost every environment on Earth, from deep-sea vents to the human gut. Microorganisms play a crucial role in various biological processes, including bioleaching. The mineral industry's use of microbial techniques, known as biohydrometallurgy, predates knowledge of the function of microbes in metal extraction. The microorganisms involved in bioleaching are typically acidophilic bacteria that can survive in highly acidic environments, such as *A. ferrooxidans* and *A. thiooxidans*. Other microorganisms that can be involved in bioleaching include archaea, fungi, and algae. For example, some archaea have been found to be involved in the bioleaching of copper, while some fungi are capable of extracting gold from ores. Overall, microorganisms play a crucial role in bioleaching by catalyzing the reactions that release metal ions from ores and making the process more efficient and environmentally friendly [18].

Some of the most used microorganisms in bioleaching include: (Table 3.1)

TABLE 3.1
Microorganisms for Bioleaching

Microorganism	Function	Extract
Acidithiobacillus ferrooxidans	Oxidize Fe^{2+} to Fe^{3+}	Dissolve and move metals like copper, gold, and uranium from ores
Acidithiobacillus thiooxidans	Oxidize elemental sulfur, sulfur compounds, and thiosulfate to sulfate ions	Release valuable metals like copper, zinc, and lead into the solution
Leptospirillum ferrooxidans	Oxidize Fe^{2+} to Fe^{3+}	Release important metals like copper, gold, and nickel into the leaching solution
Thermophilic bacteria: (*Acidithiobacillus caldus, Sulfobacillus thermosulfidooxidans, Thermoacidophilic archaea, Sulfobacillus*)	Oxidize sulfur and ferrous iron	Extract copper, gold, silver, zinc, and iron

1. ***A. ferrooxidans*:** This bacterium oxidizes ferrous iron to ferric iron and produces sulfuric acid as a by-product. It is commonly used in the bioleaching of copper and other metals.
2. ***A. thiooxidans*:** This bacterium oxidizes sulfur compounds, such as elemental sulfur and sulfides, and produces sulfuric acid as a by-product. It is commonly used in the bioleaching of sulfide ores.
3. ***Leptospirillum ferrooxidans*:** This bacterium is known for its ability to oxidize iron and produce sulfuric acid. It is commonly used in the bioleaching of gold and other metals.
4. **Thermophilic bacteria:** Th-bacteria are *thiobacillus*-like bacteria that grow at temperatures around 500°C on pyrite, pentlandite, and chalcopyrite [11].

In addition to their role in metal extraction, microorganisms can also play a role in environmental remediation by breaking down or immobilizing harmful pollutants. For example, some bacteria can oxidize sulfur compounds that produce acid mine drainage, which can be highly toxic to aquatic life [19].

3.4.1 Chemolithoautotrophic Bacteria

Life on Earth predominantly relies on sunlight and photosynthesis to synthesize organic carbon and cellular energy. However, light-independent life forms also thrive, with microbial chemosynthesis playing a pivotal role in sustaining diverse and intricate ecosystems. Chemolithoautotrophic bacteria are a prime example of such microorganisms, utilizing chemical compounds from the bedrock as an energy source for synthesizing their own nutrients. In the caves under study, these chemolithoautotrophic bacteria harness sulfur from the bedrock to sustain their life cycle, potentially supporting higher trophic levels, including crustaceans like crayfish, isopods, and amphipods.

Although chemosynthetic microorganisms inhabit various environments across the globe, they flourish particularly in environments devoid of light and competition from photosynthetic organisms. Such habitats include hydrothermal vents in the deep ocean and subterranean caves. In these nutrient-deficient ecosystems, chemosynthetic bacteria play a crucial role in biomass production. They fix carbon dioxide by oxidizing reduced inorganic compounds like iron, sulfur, and manganese, thereby generating organic matter that serves as an energy source for higher trophic levels.

Deep-sea hydrothermal vents harbor thriving populations of chemolithotrophic bacteria, often engaging in symbiotic relationships with other marine invertebrates. For instance, tube worms (*Riftia pachyptila*) and clams (Calyptogena magnifica) that inhabit these environments possess a unique form of hemoglobin. This specialized hemoglobin not only supplies the bacteria with essential oxygen and hydrogen sulfide for chemoautotrophic metabolism but also facilitates nutrition exchange because these organisms lack conventional digestive systems. It is estimated that these sulfide-rich ecosystems, known as black smokers, support a diverse array of marine life, with over 500 different species thriving in these extreme conditions [20].

Bioleaching typically employs chemolithoautotrophic bacteria from the *Acidithiobacillus*, *Leptospirillum*, and *Sulfobacillus* genera. These bacteria can use the energy released during the oxidation of inorganic compounds like ferrous iron or sulfur to power the oxidation of sulfide minerals. The use of chemolithoautotrophic bacteria in bioleaching has several advantages over traditional metal extraction methods. For example, the process is frequently more efficient and can extract metals from low-grade ores that would not be economically viable to extract using other methods. Furthermore, the process can be carried out in relatively mild conditions, reducing the process's environmental impact. However, there are some disadvantages to using chemolithoautotrophic bacteria for bioleaching. One difficulty is that the bacteria require specific conditions to be effective, such as a low pH and high temperature, which can be difficult to maintain. Furthermore, the process can be slow, which limits its use in some applications. Nonetheless, research is ongoing. Nonetheless, research is ongoing to improve the use of chemolithoautotrophic bacteria in bioleaching and develop new metal extraction techniques. Overall, both heterotrophic and chemolithoautotrophic bacteria are important in bioleaching and are required for efficient and sustainable metal extraction from ores [21].

Chemolithoautotrophic bacteria produce energy by oxidizing inorganic compounds rather than organic compounds. The energy generated by these reactions is used to fix carbon dioxide into organic compounds. This metabolic pathway differs from that used by heterotrophic bacteria, which obtain energy by decomposing organic compounds. They can generate energy from a variety of inorganic compounds, including sulfur, iron, and nitrogen compounds. Some chemolithoautotrophic bacteria, for example, can produce energy by oxidizing sulfur compounds such as hydrogen sulfide, thiosulfate, and elemental sulfur. Others can generate energy by oxidizing iron compounds such as ferrous iron (Fe^{2+}) or iron sulfide (FeS).

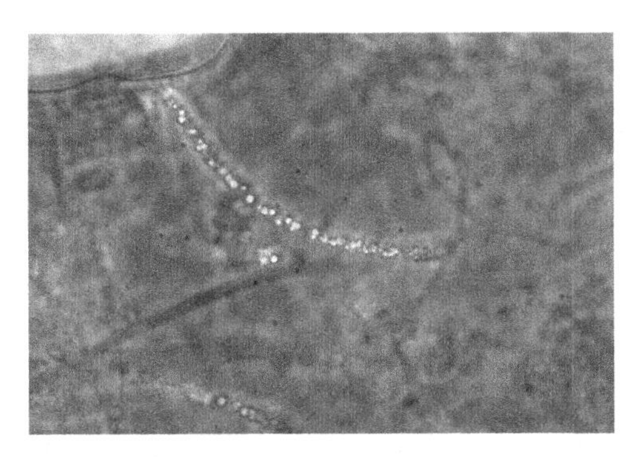

FIGURE 3.1 Elemental sulfur granules present in the tissues of chemolithoautotrophic bacteria from a submerged cave in central Florida [20].

These bacteria play an important role in many ecosystems, especially in harsh environments where organic matter is scarce. Figure 3.1 shows the presence of chemolithoautotrophic bacteria in a submerged cave in central Florida. They can be found in hydrothermal vents, deep-sea sediments, and acidic mine drainage, for example. Chemolithoautotrophic bacteria play an important role in biogeochemical cycles in the environment, such as sulfur and nitrogen cycling. Aside from their ecological significance, chemolithoautotrophic bacteria have a wide range of practical applications. They can be used in bioremediation, which is the process of breaking down pollutants in the environment using living organisms. They can also be used in industrial processes such as sulfuric acid production and wastewater treatment. To produce organic compounds from inorganic sources, these bacteria use a process known as chemosynthesis.

The oxidation of inorganic compounds yields energy in chemosynthesis. To produce organic compounds from inorganic sources, these bacteria use a process known as chemosynthesis. The oxidation of inorganic compounds produces energy, which is used to power the carbon fixation process, in which carbon dioxide is converted into organic molecules. This process is similar to photosynthesis, except that instead of using light energy, chemolithoautotrophic bacteria use inorganic compounds to generate energy. Chemolithoautotrophic bacteria are important in biogeochemical cycles, particularly in sulfur and nitrogen cycling. They are also used in bioremediation to remove pollutants from polluted sites. Chemolithoautotrophic bacteria are also important in bioleaching, particularly in the bioleaching of sulfide minerals. These bacteria can oxidize sulfide minerals like pyrite, chalcopyrite, and sphalerite, releasing metal ions into the solution. This is referred to as indirect bioleaching [22].

3.4.2 Heterotrophic Bacteria

Heterotrophic bacteria comprise a diverse group of microorganisms that rely on organic compounds as their primary source of energy and carbon. Unlike autotrophic bacteria, which can synthesize their own food through processes like photosynthesis, heterotrophic bacteria are dependent on external sources of organic matter, such as decaying organic material or other living organisms, for sustenance. These versatile bacteria inhabit a wide range of environments, including soil, water, and various living organisms, playing a crucial role in ecological processes such as nutrient cycling and decomposition.

In addition to their ecological significance, heterotrophic bacteria hold considerable industrial and biotechnological importance. They are instrumental in the production of fermented foods like cheese, yogurt, and sauerkraut, contributing to the culinary diversity and nutritional value of these food products. Moreover, certain pathogenic heterotrophic bacteria, such as *Streptococcus*, *Staphylococcus*, and *Escherichia coli*, have the potential to cause diseases ranging from minor infections to severe, life-threatening conditions in humans and other animals.

Beyond food production and health, heterotrophic bacteria are extensively utilized in various biotechnological applications, including the production of antibiotics, enzymes, and other biochemicals. They also play a crucial role in environmental remediation processes, such as wastewater treatment and pollutant degradation, showcasing their versatility and adaptability.

One of the most notable applications of heterotrophic bacteria is in the field of bioleaching, a process that harnesses microbial metabolism to extract metals from ores. In bioleaching, certain heterotrophic bacteria, predominantly from the genus *Acidithiobacillus*, *Acidiphilium*, and *Leptospirillum*, oxidize sulfide minerals present in the ore, thereby releasing metal ions into solution. The organic compounds required for the bacterial oxidation process can either be supplied by the ore itself or sourced externally, such as from molasses or sugars.

Heterotrophic bacteria play a pivotal role in the bioleaching process, a microbial-driven method employed for the extraction of metals from ores. Many recent studies are still based on heterotrophic bacteria. Figure 3.2 shows a recent discovery of a new strand of heterotrophic bacteria for the bioleaching of copper [24].

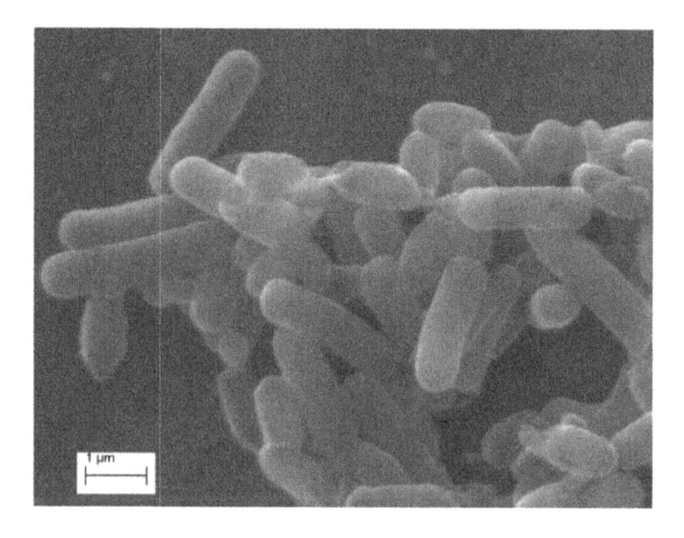

FIGURE 3.2 A new heterotrophic strain for bioleaching of low-grade complex copper ore [23].

The unique metabolic capabilities of these bacteria enable them to oxidize sulfide minerals present in the ore, facilitating the release of metal ions into the solution. Below is a detailed explanation of how heterotrophic bacteria contribute to the bioleaching process:

1. **Energy generation through oxidation:** Heterotrophic bacteria derive their energy from oxidizing organic compounds, which serve as their primary source of metabolic energy. In the context of bioleaching, these bacteria utilize organic compounds, either present in the ore matrix or externally supplied (e.g., molasses or sugars), as an energy source for the oxidation process.
2. **Sulfide mineral oxidation:** The primary role of heterotrophic bacteria in bioleaching is to oxidize sulfide minerals (e.g., iron sulfides) present in the ore. During this oxidative process, the bacteria catalyze the conversion of sulfide minerals into soluble metal ions and sulfuric acid. This transformation renders the metals more accessible and amenable to subsequent extraction processes.
3. **Acid generation:** As a by-product of sulfide mineral oxidation, heterotrophic bacteria produce sulfuric acid, which further enhances the solubilization of metals in the ore. The generated acid lowers the pH of the leaching environment, creating an acidic milieu conducive to metal extraction.
4. **Metal solubilization:** The oxidation of sulfide minerals and the acidification of the leaching environment collectively contribute to the solubilization of metal ions, making them available for subsequent recovery. The solubilized metal ions form metal-rich solutions, which can be processed using techniques such as precipitation, solvent extraction, or electrowinning to recover the desired metals.
5. **Versatility and adaptability:** Heterotrophic bacteria, particularly those from the genera *Acidithiobacillus*, *Acidiphilium*, and *Leptospirillum*, exhibit remarkable adaptability to diverse environmental conditions, including varying pH levels, temperatures, and substrate availability. This adaptability enables these bacteria to thrive in the challenging conditions encountered during bioleaching processes, thereby contributing to the stability and efficiency of the metal extraction process.
6. **Environmental and economic advantages:** Utilizing heterotrophic bacteria in bioleaching offers several environmental and economic benefits. Compared to traditional metallurgical methods; bioleaching is often more environmentally friendly, as it reduces the emission of harmful pollutants and minimizes the ecological footprint associated with metal extraction.

Heterotrophic bacteria play a critical role in bioleaching by facilitating the oxidative dissolution of sulfide minerals, generating acid, and enhancing the solubilization of metals from ores. Their metabolic activities not only enable efficient metal extraction but also contribute to the environmental sustainability and economic viability of the bioleaching process. Ongoing research and technological advancements continue to optimize the use of heterotrophic bacteria in bioleaching and expand the range of metals amenable to this innovative metal extraction [25,26].

3.4.3 Heterotrophic Fungi

Heterotrophic fungi are fungi that get their food and energy from other organisms' organic compounds. They lack photosynthesis and must rely on other organisms for nutrients, such as dead plants and animal matter. They are important in ecosystems because they decompose dead organic matter and recycle nutrients back into the soil. They are also used to make a variety of foods, including bread, cheese, and beer.

Molds, yeasts, and mushrooms are some examples of heterotrophic fungi. Molds are a common type that grows on decaying organic matter and are commonly found in damp conditions. Yeasts are a type of heterotrophic fungus that is widely used in the production of bread, beer, and wine. Mushrooms are heterotrophic fungi that grow on decaying organic matter and are commonly consumed as food. They obtain nutrients by secreting enzymes, which break down complex organic compounds into simpler molecules that the fungus can absorb. The fungus' mycelium then grows toward the source of nutrients, which it can then use to grow and reproduce. Heterotrophic fungi are not commonly used in bioleaching because the process relies on microorganisms' ability to oxidize metals in ores. However, research has shown that certain fungi, such as *A. niger* and *P. simplicissimum*, can indirectly contribute to bioleaching by producing organic acids that improve metal solubilization in ores. These fungi can secrete organic acids like citric, oxalic, and malic acid, which can raise the pH of the solution and dissolve metal ions from the ore. This is referred to as indirect bioleaching.

Heterotrophic fungi, like bacteria, are essential microorganisms that can play a significant role in the bioleaching of metals. While bacteria are often the primary focus in bioleaching processes, fungi can also contribute to metal extraction, especially in environments where both fungi and bacteria coexist. Here's an exploration of the role that heterotrophic fungi play in the bioleaching of metals:

1. **Organic matter decomposition:** Heterotrophic fungi are adept at breaking down complex organic compounds, such as cellulose and lignin, into simpler forms. In the context of bioleaching, fungi can degrade organic matter present in the ore matrix, releasing organic acids and other metabolites that can aid in the solubilization of metals.
2. **Acid production:** Like bacteria, certain fungi can produce organic acids through metabolic processes. These organic acids, including citric, oxalic, and gluconic acids, can contribute to the acidification of the leaching environment, enhancing the solubilization of metal ions from the ore.
3. **Biofilm formation:** Fungi could form complex biofilms on mineral surfaces. These biofilms can act as a protective barrier for the microorganisms, shielding them from adverse environmental conditions and facilitating the efficient colonization of the ore surface. The biofilms can also enhance the attachment of microorganisms to mineral particles, thereby promoting the bioleaching process.

4. **Metal binding and accumulation:** Some fungi possess the capacity to bind and accumulate metals within their biomass. This metal-binding capability can be exploited in bioleaching processes to concentrate and recover metals from dilute solutions, thereby aiding in the concentration and purification of metal ions.
5. **Synergistic interactions with bacteria:** In natural environments, fungi often coexist with bacteria, forming complex microbial communities. These synergistic interactions between fungi and bacteria can enhance the overall efficiency and effectiveness of the bioleaching process. For example, fungi may create microenvironments within biofilms that are conducive to bacterial growth and activity, or they may produce metabolites that stimulate bacterial metabolism.
6. **Adaptability to extreme conditions:** Heterotrophic fungi are known for their ability to thrive in diverse and often extreme environments, including acidic and nutrient-poor conditions commonly encountered in bioleaching operations. Their adaptability to such challenging conditions can contribute to the resilience and stability of the bioleaching process.
7. **Environmental and economic benefits:** Incorporating heterotrophic fungi into bioleaching processes can offer environmental and economic advantages similar to those provided by bacteria. Fungi-mediated bioleaching can be more environmentally sustainable, reducing the use of harmful chemicals and minimizing the ecological impact associated with metal extraction. Additionally, fungi can potentially enhance the recovery of metals from low-grade ores, thereby increasing the economic viability of bioleaching operations.

While heterotrophic fungi are not commonly used in bioleaching, they have been studied for their potential to improve process efficiency. One study discovered that introducing *A. niger* into a bioleaching culture increased the rate of copper extraction from chalcopyrite ores. However, there are some drawbacks to using fungi in bioleaching. Fungal growth can be slow, and the conditions for optimal fungal growth may be different from those for optimal metal extraction. Furthermore, some fungi can produce organic compounds that inhibit the growth of other microorganisms, lowering the overall efficiency of the bioleaching process. Overall, while heterotrophic fungi have shown some promise for improving bioleaching processes, more research is needed to determine their efficacy and optimize their use in this application [26,27].

3.5 MECHANISM OF BIOLEACHING

Processes for extracting metals through microorganisms are typically more environmentally friendly than physicochemical ones. As opposed to roasting and smelting, they do not emit sulfur dioxide or other hazardous gases and do not require a significant amount of energy. As a result, it may be said that this procedure complies with antipollution rules.

Bioleaching utilizes the natural processes of sulfur and iron oxidation by microbes, commonly found in the acidic waters of abandoned mine sites. Iron- and sulfur-oxidizing bacteria play a crucial role in breaking down sulfide minerals by producing sulfuric acid and Fe^{3+} ions. These substances act on the mineral, disrupting the bonds between the sulfide anions and the metals they are attached to, thereby releasing the metals. The oxidation of sulfides to sulfates produces sulfuric acid, and the nutrient-poor conditions of mineral substrates make acidophilic chemolithoautotrophs the ideal organisms for bioleaching. These are microorganisms that thrive in acidic environments, utilizing inorganic sources like sulfur and iron for energy and obtaining carbon from CO_2 fixation. While heterotrophs are present, they mainly support mineral dissolution by breaking down organic compounds that might otherwise inhibit the activity of chemolithotrophs. Various acidophilic species have been effectively employed in bioleaching processes, and the microbial diversity in these systems has been extensively studied and reviewed [28].

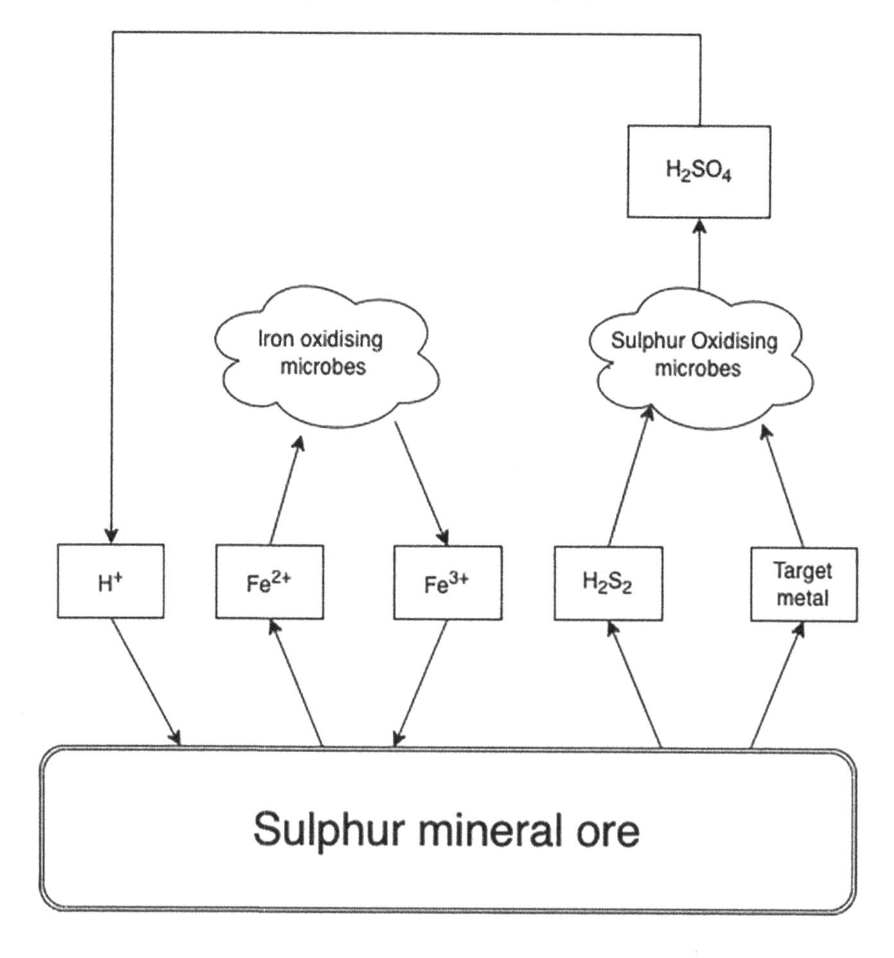

Most naturally existing bacteria and fungi carry out a variety of physiologically significant processes that allow them to expand and procreate. Acidolysis, complexolysis, and redoxolysis are the three major principles on which the mineralytic effects of bacteria and fungi on minerals are based. The oxygen atoms that are on the surface of the metallic complex undergo protonation during acidolysis. Acidolysis is possible with the protons of heterotrophic organic acids (such as malic, oxalic, gluconic, acetic, citric, succinic, pyruvic, and formic acids) as well as bacterial inorganic acids (such as H_2SO_4).

Redoxolysis is a process that uses oxidation-reduction processes to dissolve metals. Through redoxolysis, the electron transfer required for microbial growth takes place. In acidophiles' redoxolysis processes, ferric ions are reduced enzymatically in anaerobic environments with sulfur or hydrogen serving as the electron donor. Another method used by fungi, complexolysis, is important for cyanogenic bacteria to obtain precious metals. In the late stationary phase of microbial growth, glycine is decarboxylated to generate cyanide. Several cyanide-producing microorganisms have the ability of cyanoalanine synthesis to detoxify cyanide to cyanoalanine. This makes the biocyanidation method appealing when there is less dangerous cyanide present in the wastewater streams [19]. Metals can be mobilized by microorganisms using the following procedures:

1. Acid creation (proton formation) in both organic and inorganic substances.
2. Complexing agent excretion (ligand formation).
3. Reduction and oxidation processes: The bioleaching mechanisms involve several steps, including the attachment of microorganisms to the mineral surface, oxidation of metal sulfides, and metal solubilization.

Each metal has different types of mechanisms for extraction. The microbial metal solubilization of sulfide minerals was first explained by a concept comprising two pathways. Through a direct method whereby electrons are acquired directly from the reduced minerals, microorganisms can oxidize metal sulfides. In this situation, tight contact is required to bind the cells to the mineral surface. Within a few minutes or hours, cells will bind to mineral particles suspended in solution.

Another method, known as the "indirect" mechanism, uses the ferric (III) ion to mediate the oxidation of reduced metals. The ferric (III) ion is produced by microbial oxidation of ferrous iron found in rocks. Ferrous iron can be reduced to ferric iron, which can then be oxidized by microorganisms. Ferric iron acts as an oxidant and can oxidize metal sulfides. In this case, iron acts as an electron carrier. It was proposed that no direct physical contact is needed for the oxidation of iron. It was found that the "direct" mechanism frequently prevails over the "indirect" one, primarily because the direct mechanism entails direct physical contact between the bacteria and the mineral surfaces. *Thiobacillus ferrooxidans* has been shown to bind easily to the surface of metal sulfides. In a direct attack, Fe(II) is bound in the cell membrane, and the exopolymer is involved in the electron transfer. It has been noted that there aren't many places for bacteria to adhere and assault metal sulfide particles. Therefore, after maximal attachment has been reached, additional multiplication of attached cells should cause one of each bacterium's two offspring cells to be displaced into the bulk phase. However, the presence of microbes on surfaces does not prove that a direct process exists. The relevance of bacterial adhesion to mineral surfaces is denoted by the phrase contact leaching [18].

Direct

$$2FeS_2 + 3.5O_2 + H_2O \rightarrow Fe_2 + 2H^+ + 2SO_4{}^{2-}$$

$$2Fe_2^+ + 1/2O_2 + 2H^+ \rightarrow 2Fe_3 + H_2O$$

Indirect

$$FeS_2 + 14\ Fe^{+3} + 8H_2O \rightarrow 15Fe^{+2} + 16H^+ + 2SO_4^{-2}$$

$$MS + 2Fe^{+3} \rightarrow M^{+2} + S^0 + 2Fe^{2+}$$

$$S^0 + 1.5O_2 + H_2O \rightarrow 2H^+ + SO_4{}^{2-20}$$

The first step in bioleaching is the attachment of microorganisms, such as chemolithoautotrophic bacteria and heterotrophic fungi, to the mineral surface. This attachment is aided by the production of extracellular polymers, which aid in the adhesion of microorganisms to the mineral surface. Once attached, the microorganisms start oxidizing the metal sulfides in the ore. Chemolithoautotrophic bacteria oxidize metal sulfides by using inorganic compounds, such as iron, sulfur, and hydrogen, as electron donors. Metal ions and sulfuric acid are produced during this oxidation reaction. Solubilization: In the sulfuric acid solution, the metal ions produced by the oxidation reaction are solubilized. Heterotrophic fungi also contribute to metal solubilization by producing organic acids such as oxalic acid and citric acid, which aid in the dissolution of metals in ore.

Metal ions that have been solubilized in the sulfuric acid solution can be recovered using a variety of techniques, including precipitation and solvent extraction. The remaining solids, known as tailings, can be processed further to recover any remaining metals. Overall, bioleaching mechanisms include microorganism attachment to mineral surfaces, oxidation of metal sulfides, metal solubilization in sulfuric acid solution, and metal ion recovery. These steps can be optimized to improve the efficiency and effectiveness of bioleaching for metal extraction. In chemical oxidation, some microorganisms, such as chemolithoautotrophic bacteria, can oxidize metal sulfides in ore. This converts the metal sulfides into soluble metal ions that can be extracted more easily from the ore. As a by-product of their metabolism, heterotrophic microorganisms such as fungi produce organic acids. These organic acids can aid in the dissolution of metals in ore, making them more accessible for extraction. Microorganisms can attach to the surface of ore particles, facilitating metal extraction. It can form a biofilm on ore particles, increasing the surface area available for metal extraction. The microorganisms in the bioleaching process can change the system's redox potential, which can promote metal sulfide oxidation and metal solubilization in the ore. Some microorganisms can produce leaching agents, such as siderophores, which aid in the extraction of metals from ore. Chelating agents, which are molecules with the ability to bind to metal ions and improve their solubility in water, can also be used during the bioleaching process. These chelating compounds, which some microbes can make, can improve the bioleaching procedure. Following their discharge into solution, metal ions can be recovered using a number of techniques, including solvent extraction and precipitation [29].

The essential elements of the leaching have been combined, and a mechanism has been created that has the following characteristics:

Cells must be physically in contact with the surface and connected to the minerals in order to generate and excrete exopolymers. These exopolymeric cell envelopes also contain ferric iron compounds that are complexed with glucuronic residues.

These are thought to be a component of the initial attack.

Sulfur or polythionate granules are created in the periplasm or in the cell envelope during the oxidation of sulfur compounds, and thiosulfate is produced as an intermediary during this process [18,29].

3.5.1 Redoxolysis

Redoxolysis, also known as redox reactions or oxidative-reductive reactions, plays a crucial role in the bioleaching process, particularly in the extraction of metals from metal sulfides and highly oxidized manganese ores. This mechanism involves the transfer of electrons from one molecule to another, leading to the oxidation of one substance and the reduction of another. In the context of bioleaching, redoxolysis facilitates the conversion of insoluble metal compounds into soluble metal ions, thereby making them accessible for extraction.

Mechanism of Redoxolysis: In bioleaching, redoxolysis is primarily mediated by the metabolic activities of microorganisms, such as bacteria and fungi, involved in the process. These microorganisms catalyze the transfer of electrons from the metal sulfides or manganese ores to electron acceptors, typically oxygen or ferric ions (Fe^{3+}). This electron transfer leads to the oxidation of the metal sulfides and the reduction of the electron acceptors.

This procedure is divided into two stages: Microorganisms such as *A. ferrooxidans* or *L. ferrooxidans* oxidize metal sulfides such as pyrite or chalcopyrite to produce metal ions and sulfuric acid in the first stage.

$$FeS_2 + 7/2O_2 + H_2O \rightarrow Fe^{2+} + 2SO_4^- + 2H^+$$

Other microorganisms, such as *A. thiooxidans* or *Acidiphilium acidophilum*, reduce metal ions, such as Fe^{2+} or Cu^{2+}, to produce elemental metals in the second stage. This reaction can be expressed as follows:

$$2Fe^{2+} + S^0 + 2H^+ \rightarrow 2FeS + H_2O$$

The net result of these reactions is the conversion of metal sulfides into soluble metal ions and elemental metals that can be extracted more easily from ore.

In the case of manganese fungal leaching, solubilization occurs through enzymatic reduction of highly oxidized manganese minerals.

$$Mn^{2} + 2e^{-} + 4H^{+} \rightarrow Mn^{2+} + 2H_2O$$

Redoxolysis also increases the acidity of the solution, which can aid in the dissolution of metals in the ore. Overall, redoxolysis is an important mechanism in bioleaching, allowing efficient metal extraction from ores and other mineral sources [15,30].

As a result of redoxolysis, the mobility of metals increases as the insoluble metal sulfides and manganese ores are transformed into soluble metal ions. This enhanced mobility facilitates the leaching and extraction of metals from the ore matrix, making them more accessible for subsequent recovery processes. The activity of specific microorganisms, such as *A. ferrooxidans, A. thiooxidans*, and certain fungal species, is instrumental in accelerating the redoxolysis process in bioleaching. These microorganisms possess enzymes and metabolic pathways that facilitate the oxidation of sulfide minerals and the reduction of electron acceptors, thereby promoting the efficiency of metal extraction.

Various environmental factors, including pH, temperature, and the availability of nutrients and electron acceptors, can influence the rate and efficiency of redoxolysis in bioleaching. Optimal conditions that favor microbial activity and electron transfer are essential for maximizing the efficacy of the redoxolysis-driven bioleaching process.

Ongoing research and technological advancements continue to refine and optimize the redoxolysis-based bioleaching process. Recent studies have focused on identifying novel microorganisms with enhanced redox capabilities, developing bioreactor designs that facilitate efficient electron transfer, and optimizing operational parameters to enhance the overall efficiency and cost-effectiveness of bioleaching processes.

Despite the significant advantages offered by redoxolysis-driven bioleaching, certain challenges, such as the maintenance of optimal environmental conditions, microbial contamination, and the management of by-products and waste, need to be addressed. Innovative solutions, including the use of genetically modified microorganisms with enhanced redox capabilities, the development of sustainable bioreactor designs, and the implementation of integrated waste management strategies, are being explored to overcome these challenges and further enhance the sustainability and efficiency of bioleaching processes.

3.5.2 Acidolysis

Acidolysis is a bioleaching mechanism that uses acid to dissolve metals from ores or other mineral sources. *A. niger* is a fungus that has been found to create organic acids such as citric, oxalic, and gluconic acids during bioleaching. This process is known as acidolysis. Anions and protons are made available for metal leaching by acidolysis and complexolysis by the formation of organic acids. In acidolysis, organic or inorganic acid produced by microorganisms is used to dissolve the metal. Acidolysis is encouraged by low pH, and free metal cations are released and made more mobile by protonation. When pH is less than 7, acidolysis is the predominant bioleaching mechanism. Metal complexes are created when anions combine with metal cations. By weakening links and removing metal ions from the ore surface, protons connect to the metal ore surface and react to it. Metal oxide separates from the solid metal's surface. Oxalic, gluconic, malonic, lactic, acetic, succinic, pyruvic, malonic, isocitric, and formic acids are among the organic acids released by fungi. These are crucial for preserving the acidic pH needed for a better bioleaching process.

Additionally, through the plasma membrane-located proton translocating ATPase, the acids serve as the primary suppliers of protons. This enzyme's function is to restrict anions' access to the cations in a process involving metal compounds, which raises the solubility of the metal ions. In addition, the acids' hydrogen ions (H+) stabilize the metal while facilitating the mobilization of metal chelates. Total production of metal oxides is equal to the maximum number of protons that organic acids can secrete. Acidolysis is described as the most rapid and prevalent leaching mechanism for fungi and other heterotrophic organisms [31].

$$MeO + 2H + Me^{2+} + H_2O$$

MeO is a type of metal oxide that includes Nickel(II) oxide (NiO), ferric oxide (Fe_2O_3), and calcium carbonate ($CaCO_3$), among others.

Acidolysis is commonly used in bioleaching in conjunction with other mechanisms such as oxidation and reduction to extract metals from metal sulfides. Microorganisms produce acid as a by-product of their metabolism in acidolysis. This acid can aid in the dissolution of metals in ore, making them more accessible for extraction. Sulfuric acid, which is produced by the oxidation of sulfur or sulfur-containing compounds, is the most common acid produced in bioleaching. The acid produced during acidolysis can also aid in the solubilization of metal ions in the ore. This can happen as a result of protonation, a process in which the acid reacts with metal ions to form metal complexes that are more soluble in water. Acidolysis, for example, can produce copper sulfate complexes, which are more soluble than copper sulfide in copper sulfide ores.

Factors affecting acidolysis are as follows:

- **pH:** Maintaining an optimal pH that fosters the dissolution of metal compounds without compromising microbial activity is crucial for effective acidolysis.
- **Temperature:** Temperature within the leaching environment plays a significant role in determining the pace of acidolysis. While higher temperatures can expedite the dissolution of metal compounds, excessively high temperatures can hamper microbial activity and compromise the stability of the leaching solution.
- **Concentration:** The concentration of acid employed in the leaching process is a vital determinant of acidolysis efficiency. Elevated acid concentrations can boost the rate of metal dissolution, but they may also escalate operational expenses and pose environmental challenges.

The ore's mineralogical composition, especially the type and prevalence of metal sulfides, can influence the ore's responsiveness to acidolysis. Some minerals might necessitate more rigorous leaching conditions to achieve optimal metal extraction.

Although acidolysis can be executed purely through chemical processes, the involvement of acidophilic microorganisms, like those from the *Acidithiobacillus* genus, can substantially augment the acidolysis process's efficiency in bioleaching. These microorganisms can enhance the leaching environment's acidity and expedite metal compound dissolution through their metabolic processes.

Current research and technological innovations in acidolysis-driven bioleaching are directed toward refining leaching conditions, cultivating new acidophilic microorganisms with superior metal solubilization capacities, and devising inventive leaching approaches to enhance acidolysis efficiency and sustainability. Moreover, advancements in bioreactor designs, integrated control systems, and environmentally friendly acid regeneration methods are under exploration to curtail operational costs and environmental repercussions.

Despite its efficacy in metal recovery, acidolysis-based bioleaching encounters specific challenges, including the production of acidic waste streams, the potential for acid spillages and leaks, and the environmental ramifications of acid consumption. To overcome these challenges and bolster the sustainability of acidolysis-driven bioleaching processes, innovative solutions such as closed-loop acid regeneration systems, the adoption of biocompatible acids and chelating agents, and the deployment of comprehensive waste management approaches are being actively pursued.

Overall, acidolysis is an important mechanism in bioleaching that aids in the solubilization of metals and their accessibility for extraction. The acid produced by microorganisms can aid in the dissolution of metal sulfides, the solubilization of metal ions, and the reduction of solution pH, all of which are important factors in the bioleaching process [32].

3.5.3 Complexolysis

Complexolysis is a bioleaching mechanism that involves the breakdown of metal complexes in the ore to release metal ions that can be extracted more easily. Complexolysis can occur naturally in bioleaching as a result of the acidic environment created by the microorganisms involved in the process.

Many metals found in ores exist as metal complexes, which are molecules composed of a metal ion surrounded by other atoms or molecules known as ligands. Because they are frequently insoluble or less reactive than the metal ion itself, these complexes can make the metal less accessible for extraction.

Through the chelation mechanism, complexolysis is carried out, bringing the metal ions out of the solution and balancing them through acidolysis. The reaction is stabilized by the complexation of the metal ion(s) and organic acid(s). Amino acids, which are also necessary for complexolysis, are released by fungi. Even though filamentous fungus only excretes a very small number of amino acids, the synthesis of metal-complexing ligands such as organic acids and amino acids by actinomycetes, fungi, and other heterotroph species increases metal solubility. The organic acids have great potential to start a metallic complex with the metal ions and are natural chelating agents. The ability of molecules to generate complex metal chelating agents serves as the basis for the solubilization of metal ions. However, compared to the lattice bonds created between solid particles and metal ions, the bonds formed between metal ions and ligands are stronger, which improves the bioleaching process of solid particles. By stabilizing the metal ions generated during acidolysis, the acidolysis mechanism enhances complexolysis. The harmful effect of metal ions on the fungi is reduced by the stability of metal complexes in the solutions. Like citric acid and magnesium, oxalic acid and iron, amino acids with metal ions, and phenol derivatives, the organic ligands form stable complexes with the metal ions. Furthermore, the solubility and mobilization of the metal to be leached are greatly influenced by the natural organic ligands such as gluconate, acetate, maleate, formate, succinate, and oxalate. The acid produced by microorganisms in bioleaching can aid in the breakdown of these metal complexes via protonation or other chemical reactions. This can result in the release of metal ions that can be extracted more easily from the ore. In the case of copper sulfide ores, for example, the acid produced by bioleaching can help to break down the copper sulfide complex, releasing copper ions that can be extracted more easily.

Complexolysis is frequently used in conjunction with other bioleaching mechanisms such as oxidation, reduction, and acidolysis. These mechanisms work together to break down the metal sulfides in the ore and liberate the metal ions for extraction.

Factors Influencing Complexolysis with Microorganisms

- **Microbial diversity:** The type and diversity of microorganisms present in the leaching environment can significantly impact complexolysis efficiency. Some microorganisms produce more effective ligands or thrive under specific conditions, influencing the stability and formation of metal-ligand complexes.
- **Nutrient availability:** Microorganisms require specific nutrients to produce organic ligands that are essential for complexolysis. The availability and type of nutrients, such as carbon, nitrogen, and sulfur sources, can influence the metabolic activities of microorganisms and, consequently, the efficiency of complexolysis.
- **pH levels:** The pH of the leaching solution is crucial as it affects both microbial activity and the stability of metal-ligand complexes. Maintaining an optimal pH level that promotes microbial growth and stable complex formation is essential for efficient complexolysis.
- **Temperature:** The temperature of the leaching environment influences microbial activity and metabolic rates. While elevated temperatures can accelerate microbial metabolism and ligand production, extreme temperatures can inhibit microbial activity, affecting complexolysis efficiency.
- **Oxygen availability:** Oxygen is vital for the aerobic metabolism of many microorganisms involved in complexolysis. Adequate oxygen availability can enhance microbial growth and ligand production, thereby promoting complexolysis efficiency.

The latest updates and advances in microbial complexolysis are advanced molecular techniques, such as metagenomics and metatranscriptomics, which are being employed to analyze and characterize the microbial communities involved in complexolysis. This helps in understanding the complex interactions between different microorganisms and their role in enhancing complexolysis efficiency. Ongoing research focuses on optimizing leaching conditions, such as pH, temperature, and nutrient availability, to maximize microbial activity and ligand production, thereby improving complexolysis efficiency. Biotechnological approaches are being used to develop genetically engineered microbial strains with enhanced ligand-producing capabilities. These novel strains aim to improve the efficiency and sustainability of microbial-mediated complexolysis in bioleaching processes. Innovations in bioreactor design, such as the development of biofilm reactors and membrane bioreactors, are being explored to enhance microbial growth, ligand production, and overall complexolysis efficiency.

Overall, complexolysis is an important bioleaching mechanism that aids in the release of metal ions from metal complexes in the ore. This process can significantly improve the efficiency of metal extraction from ores and other mineral sources [31].

3.6 BIOACCUMULATION

Bioaccumulation is the process by which living organisms accumulate substances in their tissues over time, such as pollutants or nutrients. The aggregation of pollutants in living organisms results in bioaccumulation. When an organism consumes a substance at a rate that exceeds its ability to eliminate it, this process occurs. The substance then accumulates in the organism's tissues, leading to an increase in concentration over time. Bioaccumulation can occur in a variety of organisms, including plants, animals, and microorganisms. Depending on the substance accumulated and the concentration at which it occurs, it can have both positive and negative effects. Bioaccumulation of nutrients can be beneficial in some cases. Certain organisms, for example, may accumulate essential nutrients in their tissues, such as iron or calcium, to support growth and development. The gradual accumulation of substances such as pollutants or nutrients in the tissues of living organisms over time is known as bioaccumulation. Bioaccumulation in the context of bioleaching can refer to the accumulation of metal ions or other substances in the cells of the microorganisms involved in the process. It is an important consideration in environmental science, as it can have significant impacts on ecosystems and human health. For example, bioaccumulation of toxic substances in fish or other aquatic organisms can lead to the consumption of contaminated fish by humans, resulting in health problems [26].

The accumulation of solid particles or precipitation in vacuoles happens when soluble metal ions are transferred between the cell membrane and the cell (metabolism-dependent intracellular uptake). The fungal mycelium, which has different functional groups (sulfate, hydroxyl, carboxyl, phosphate, and amine), is assumed to be responsible for bioaccumulation. It has been demonstrated that these metal ions operate as a cation exchanger by attaching to these functional groups on the cell walls. the method of absorbing metals whereby bioaccumulation takes longer than biosorption and does not need the creation of metabolites. The mycelia of fungi use active metabolic reactions and passive adsorption to solubilize metal ions through bioaccumulation [31].

Microorganisms such as *Acidithiobacillus, ferrooxidans*, or *A. thiooxidans* use various mechanisms to extract metals from metal sulfides in the ore during bioleaching. As they carry out these processes, these microorganisms can accumulate metal ions in their cells. Metal ion concentrations in these microorganisms' cells can rise over time, leading to bioaccumulation. Bioaccumulation has both positive and negative consequences. Bioaccumulation of metal ions can be beneficial in the case of bioleaching because it can help to increase the efficiency of metal extraction. Some microorganisms, for example, have been engineered to over-express metal transporters or other proteins that improve their ability to accumulate metals. This can aid in increasing the concentration of metal ions in these microorganisms' cells, resulting in more efficient metal extraction. However, in some cases, bioaccumulation can be harmful. Pollutants or toxic metals in an organism's tissues can cause negative health effects or even death if the concentration becomes too high. Furthermore, if the microorganisms involved in bioleaching are not properly contained, they have the potential to spread and cause environmental contamination. Bioaccumulation is an important factor to consider in bioleaching and other biotechnological processes. Researchers can optimize these processes to achieve the desired outcomes while minimizing any negative impacts by understanding the mechanisms of bioaccumulation and its potential effects [31,33].

3.7 ADVANTAGES OF BIOLEACHING

Bioleaching has a number of advantages over conventional mining and extraction methods. It is a simple technique that helps to recover the metals from low-grade ores. Bioleaching is a more environmentally friendly method of metal extraction than traditional mining methods, which often involve the use of toxic chemicals and heavy machinery. Bioleaching extracts metals from ores using natural microorganisms and relatively safe chemicals, reducing the environmental impact of metal extraction. It requires less energy to extract metals from ores; bioleaching is a more energy-efficient method of metal extraction than traditional mining methods. It helps to extract the metals that are not possible to extract from other techniques.

The microorganisms involved in bioleaching extract metals at low temperatures and pressures while using little energy. The microorganisms involved in the process can target specific metals in the ore, bioleaching is a selective method of metal extraction. This can lead to a higher concentration of the desired metal in the final product, reducing the need for further processing. It often requires less capital investment than traditional mining methods, bioleaching can be a cost-effective method of metal extraction. Furthermore, when compared to traditional methods, the use of natural microorganisms and relatively safe chemicals can reduce operating costs. Bioleaching can extract metals from low-grade ores that would be too expensive to extract using traditional mining methods. It also shows as the most efficient method in terms of heavy metal solubilization and has a higher potential to remove heavy metals. It also does not require a specific raw material composition. It can be intensively applied at an industrial scale for low-grade ores containing heavy metals at a concentration of less than 0.5% wt. It is also applicable to highly contaminated raw materials [34].

Overall, bioleaching offers several advantages over traditional mining and extraction methods, including its environmentally friendly nature, energy efficiency, selectivity, low cost, and ability to extract metals from low-grade ores. These benefits have increased interest in bioleaching as a viable method of metal extraction in a variety of industries [30,35].

3.8 FACTORS AFFECTING BIOLEACHING

Bioleaching is a complicated process that is influenced by several factors, including:

3.8.1 Microorganisms

The type and abundance of microorganisms present in the bioleaching system can significantly affect the process's efficiency and effectiveness. Different microorganisms extract metals from ores in different ways, and some may be better suited to specific environmental conditions.

3.8.2 pH

The pH of the bioleaching system can affect the activity of the microorganisms involved, which can have a significant impact on the process. Acidophilic microorganisms, which are commonly used in bioleaching, prefer low-pH environments and may be less effective when pH levels are raised.

The pH of the bioleaching process is an important parameter that can affect the development and activity of the microorganisms, as well as the solubility of the metal ions being leached. The oxidation of metal sulfides by microorganisms that produce sulfuric acid as a by-product is typical of bioleaching. As the sulfuric acid combines with the surrounding water, it produces hydronium ions (H_3O^+), enhancing the solution's acidity. The majority of bioleaching operations are carried out in acidic conditions, with a pH range of 1.5–3.5. This acidic environment is required to enhance sulfide mineral oxidation and metal ion dissolution. On the other hand, maintaining the right pH level is crucial, as excessively low or high pH levels can inhibit microbial activity and create hazardous by-products such as hydrogen sulfide or ferric hydroxide.

A variety of approaches can be used to adjust the pH level throughout the bioleaching process. One popular method is to add an acid or a base to the nutrient culture medium to lower or increase the pH. Sulfuric acid is a common acid, and bases such as lime or sodium hydroxide can be employed to elevate the pH level. It should be noted that the acid or base used, as well as the amount used, can have a considerable impact on the bioleaching process. If the pH falls too low, the acidic circumstances can become poisonous to microorganisms, inhibiting their development and activity. Furthermore, too low-pH values might cause the formation of metal sulfides, reducing metal solubility and decreasing process efficiency. On the other hand, if the pH level becomes too high, it might cause the precipitation of metal hydroxides, which can limit metal solubility and process efficiency. Moreover, high pH levels might increase the growth of undesirable microbes, which can compete with the appropriate microorganisms and degrade process efficiency even further. Over-acidification or over-alkalization of the nutrient culture medium can impede microbial growth and activity.

As a result, careful monitoring and management of the pH level are required to ensure optimal conditions for microbes and efficient metal ion leaching. pH control options such as the use of pH buffers or the use of pH sensors and controllers may be used to maintain a constant and appropriate pH level during the bioleaching process [36,37].

3.8.3 Temperature

The activity of microorganisms involved in bioleaching can be influenced by temperature. The optimal growth temperature of different microorganisms varies, and some may be more active at higher or lower temperatures.

Temperature is an important factor in bioleaching, as it affects the growth rate and activity of the microorganisms involved in the process. Different microorganisms have different optimal temperature ranges, and the temperature range also depends on the type of element being leached.

The ideal temperature for bioleaching is determined by the microorganisms used and the type of material being treated. Temperatures in the 25°C–45°C range are commonly employed for bioleaching of important elements such as copper, zinc, and nickel. This is because the majority of the microorganisms utilized in bioleaching are mesophilic, which means they grow best at moderate temperatures. However, thermophilic bacteria can thrive at greater temperatures and can be employed for bioleaching at high temperatures. For example, in the bioleaching of copper, the optimal temperature range is typically between 30°C and 45°C, with some strains of bacteria able to tolerate temperatures up to 60°C. Some mesophilic bacteria, such as *A. ferrooxidans*, have an optimal temperature range of 25°C–35°C for the leaching of copper and zinc, while other thermophilic bacteria, such as Sulfolobusmetallicus, have an optimal temperature range of 60°C–80°C for the leaching of gold.

The temperature also affects the rate of chemical reactions involved in the bioleaching process. Higher temperatures generally increase the reaction rates and thus the rate of metal extraction, but can also lead to decreased microbial activity due to the denaturation of enzymes and other biological structures. Too high temperatures can denature the enzymes produced by the microorganisms, resulting in a decrease in the activity and effectiveness of the process. Temperatures that are excessively low, on the other hand, can slow down the growth and metabolism of microorganisms, resulting in slower and less efficient bioleaching. To ensure optimal bioleaching performance, it is critical to maintaining the appropriate temperature range. The optimal temperature for bioleaching of essential elements depends on the specific microorganisms and the type of metal being extracted, and balancing the effects of temperature on microbial activity and chemical reaction rates is important for maximizing the efficiency of the bioleaching process [36,38].

3.8.4 Oxygen Concentration

Because it is required for the oxidation reactions that break down metal sulfides in the ore, oxygen concentration is critical in bioleaching. The concentration of oxygen in the bioleaching system can therefore impact the efficiency of the process.

Oxygen and carbon dioxide are significant gases in bioleaching because they are required for the growth and metabolism of the microorganisms involved.

The availability and management of oxygen and carbon dioxide are critical aspects to consider in bioleaching because they influence the metabolic activity and efficiency of the microorganisms involved.

The metabolism of the microorganisms engaged in bioleaching requires oxygen. These microbes need oxygen to respire, which is how they turn organic compounds into energy. Aerating the nutrient culture medium with air or pure oxygen normally provides the oxygen. The microorganisms oxidize the metal sulfides in the ore during the process, releasing metal ions and sulfuric acid. To create energy for the bacteria, this activity also requires oxygen as an electron acceptor. As a result, enough oxygen supply is critical for bioleaching efficiency. Excessive oxygen, on the other hand, can be harmful since it can result in the formation of reactive oxygen species, which can harm bacteria. Anaerobic bacteria, on the other hand, do not require oxygen to make energy and can grow in the absence of oxygen. Some anaerobic bacteria can also be employed in bioleaching, although their metabolic activities require alternate electron acceptors such as sulfur or iron.

In contrast, carbon dioxide is created as a by-product of the metabolic process in bioleaching. During cellular respiration, the microorganisms involved in bioleaching transform the carbon source contained in the nutritional culture medium, often glucose, into carbon dioxide. Carbon dioxide build-up can cause a drop in the pH of the medium, which can impair the activity of the microorganisms. Carbon dioxide is often removed from the system by aeration or by sparging the system with air or oxygen to maintain an appropriate pH.

In addition to their involvement in respiration, oxygen, and carbon dioxide can influence the solubility of metals in leached ore. Oxygen can assist in oxidizing the metal ions in the ore, making them more soluble and easier to remove by the microorganisms. Carbonates, on the other hand, can occur when carbon dioxide reacts with metal ions, reducing the efficacy of the bioleaching process.

Overall, the quantities of oxygen and carbon dioxide in the nutritional culture medium must be carefully monitored and managed to maintain optimal microorganism growth and metabolism and maximize bioleaching efficiency [36,39].

3.8.5 Nutrient Availability

Microorganisms involved in bioleaching require certain nutrients to carry out their metabolic processes, such as nitrogen and sulfur. The availability of these nutrients can affect the activity of the microorganisms and, as a result, the process's efficiency.

Nutrient culture media are used to provide the nutrients required for the microorganisms involved in bioleaching to grow and function. The composition of the nutrient culture media is determined by the microbial species used and the type of ore processed.

Carbon, nitrogen, sulfur, and phosphorus are some of the most common nutrients required for microbial growth in bioleaching. Trace elements like iron, magnesium, calcium, and potassium may also be needed in small amounts.

Some of the common nutrient culture media used in bioleaching include:

Nitrogen supplies in nutritional culture media can include ammonium sulfate, urea, or organic nitrogen compounds, whereas carbon sources can include glucose, sucrose, or other sugars.

Mineral salt media contain inorganic salts such as ammonium sulfate, potassium dihydrogen phosphate, magnesium sulfate, and iron sulfate. These media are frequently used to cultivate acidophilic bacteria widely employed in bioleaching. Organic nutrition media include organic components such as yeast extract, peptone, and tryptone. These media are frequently used to cultivate microorganisms that are not precisely acidophilic but can nevertheless perform bioleaching. Complex media are those that contain both organic and inorganic nutrients. They are frequently used to cultivate microorganisms that are difficult to grow in other mediums.

Sulfur media include sulfur compounds such as elemental sulfur, thiosulfate, and sulfide. These media are frequently used to culture bacteria capable of oxidizing sulfur compounds, a crucial stage in the bioleaching process. The type of nutrient culture media employed is determined by the microorganisms used, the properties of the substance being leached, and the desired outcome of the bioleaching process [40,41].

Nutrient culture media influence the rate of bioleaching in addition to providing nutrients for microorganisms. The pH of the nutritional culture medium, for example, is an important element that influences microbial activity. Some microbes are acidophilic, which means they thrive in acidic surroundings, while others are neutrophilic, which means they thrive in neutral pH settings.

3.8.6 Mineralogy of the Ore

The mineralogy of the ore being leached can also have an impact on the efficiency of the bioleaching process. Ores with complex mineralogy or low metal content may require different conditions or different microorganisms to be effectively leached.

3.8.7 Mineral Composition

The mineral composition of the ore being processed can affect the efficiency of bioleaching. Some minerals may be more resistant to bioleaching than others, and the presence of certain minerals can affect the activity of the microorganisms involved in bioleaching.

3.8.8 Surfactants and Organic Extractants

The leaching bacteria are typically inhibited by surfactants and organic compounds employed in solvent extraction, mostly due to a reduction in surface tension and mass transfer of oxygen. For the concentration and recovery of metals from pregnant solution, solvent extraction is currently favored. Combining solvent extraction with bacterial leaching results in enriched solvents in the aqueous phase, which must be eliminated before the barren solution is recirculated to the leaching operations.

3.8.9 Agitation and Mixing

Agitation and mixing of the ore slurry can improve the efficiency of bioleaching by increasing the contact between the microorganisms and the ore particles, enhancing nutrient availability, and promoting oxygen transfer.

Overall, understanding the various factors that can influence bioleaching is important for optimizing the process and achieving the desired outcomes. Researchers can improve the efficiency and effectiveness of bioleaching and its potential as an alternative to traditional metal extraction methods by controlling and manipulating these factors [7,42].

3.9 INOCULUM

Inoculum is a critical component of the bioleaching process. The term inoculum refers to the microorganisms added to a bioleaching system to initiate or accelerate the metal extraction process. In the bioleaching of essential elements, the inoculum is often used to introduce acidophilic or acid-tolerant microorganisms that can survive and thrive in the acidic environment required for the process.

The selection and preparation of the inoculum are important to the bioleaching process's success. The specific metal ore being treated and the traits of the microorganisms involved determine the sort of inoculum utilized in bioleaching. For instance, acidophilic bacteria like *A. ferrooxidans* and *L. ferrooxidans* are frequently utilized as inoculum for copper bioleaching. The minerals in the ore can be oxidized by these microbes, releasing copper ions that can subsequently be recovered from the solution. The inoculum may occasionally come from a natural source, such as a mine or mineral deposit, where the microorganisms have already been proven to be efficient in bioleaching. In other situations, the inoculum may be derived from a lab culture in which the microorganisms have been chosen and adapted for the particular bioleaching procedure.

The density of the microbial inoculum is a pivotal factor influencing the kinetics of leaching through both direct and indirect pathways. A higher concentration of microorganisms can potentially accelerate the rate and efficiency of bioleaching. However, factors such as oxygen availability and nutrient supply can impose limitations on the maximum inoculum density achievable. Thus, maintaining an optimal inoculum density is crucial for achieving peak bioleaching performance. Typically, the microbial concentration peaks between 10^3 and 10^9 cells/mL in continuous stirred tank reactors. To optimize the leaching kinetics, it becomes imperative to augment the microbial population.

One strategy to increase the bacterial population in the leaching solution involves harvesting the biomass through techniques like centrifugation or membrane filtration and then reintroducing it into a reduced volume of the solution. While effective, these methods can be expensive and require sophisticated equipment. An alternative, more cost-effective approach is the utilization of a biofilm reactor, also referred to as bacterial film oxidation. Many acidophilic microorganisms exhibit a unique affinity for jarosite. Thus, under favorable conditions, the formation of a thin jarosite film can provide these microorganisms with a crucial growth site. Pesic and Kim have proposed a mechanism detailing jarosite formation and its role in microbial growth.

Another innovative method to bolster the microbial population is the application of an electrical potential. The growth of bacterial populations is directly proportional to the amount of ferrous iron oxidized. As a general estimate, producing 1 g of biomass necessitates the oxidation of approximately 100 g of ferrous iron. Ideally, when the rates of iron oxidation and reduction are balanced by the applied potential, it promotes a high bacterial population and oxidation potential. This equilibrium further enhances sulfide dissolution kinetics, optimizing the bioleaching process.

In general, the inoculum should be comprised of microorganisms that are adapted to the specific conditions of the bioleaching environment, including pH, temperature, nutrient availability, their activity and efficiency in metal oxidation, their resistance to the environmental conditions of the bioleaching system, and their availability and cost. The inoculum should also be free of contaminants that could interfere with the bioleaching process or damage the equipment being used. There are several approaches that can be used to prepare the inoculum for bioleaching. One popular method is to utilize a mixed culture of microorganisms separated from the environment where the ore or other item being treated was collected. A pure culture of a certain microbe known to be effective at solubilizing the key components of interest is another option. The size of the inoculum can also play a role in the bioleaching process. Generally, a higher inoculum concentration can result in faster bioleaching rates but can also lead to higher costs and greater risks of contamination. Careful optimization of the inoculum concentration is, therefore, necessary to achieve the most efficient and cost-effective bioleaching process. An inoculum that is too tiny may result in sluggish or incomplete solubilization of the required materials. Optimal inoculum size can vary depending on the specific conditions of the bioleaching system and may require optimization through experimentation [43,44].

3.10 METAL RESISTANCE OF MICROORGANISMS

Microorganisms that are used in bioleaching processes need to have the ability to tolerate and resist the toxic effects of the metals being extracted. This is because, during the bioleaching process, the microorganisms are exposed to high concentrations of heavy metals such as copper, zinc, and iron, which can be toxic to most microorganisms. The ability of microorganisms to withstand and thrive in environments with high metal concentrations is referred to as metal resistance, and it depends on a number of variables, including the type of metal, the concentration, and length of exposure, as well as the unique genetic and physiological traits of the microorganisms in question.

The metal resistance mechanisms of microorganisms in bioleaching can include a range of strategies, such as:

3.10.1 Metal Efflux Pumps

These are proteins that are embedded in the cell membrane of microorganisms and actively pump out metal ions, reducing their intracellular concentration. This may aid in lowering the amount of metals present in the cell and prevent damage. Some microbes may also be able to alter the cell membrane to lessen metal uptake or create alternate metabolic pathways to produce energy when metals are present.

3.10.2 Metal-Binding Proteins

Some microorganisms can produce specific proteins that bind to metal ions, preventing them from interfering with cellular processes. For example, some acidophilic bacteria used in bioleaching are able to produce metallothioneins, which are small proteins that can bind to heavy metals and protect the microorganisms from their toxic effects.

3.10.3 Oxidative Stress Response

When exposed to high concentrations of metals, microorganisms can produce enzymes that help to detoxify reactive oxygen species that can damage cellular components.

3.10.4 Biofilm Formation

Some microorganisms can form biofilms on mineral surfaces, which can protect them from toxic metals and create a more favorable environment for bioleaching.

3.10.5 Genetic Adaptation

Over time, microorganisms can evolve genetic mutations that confer resistance to specific metals, allowing them to survive and thrive in harsh environments.

The success of the bioleaching process depends critically on the microorganisms' capacity to withstand metals. The extraction of necessary metals from ores and waste materials using bioleaching can be made more effective and sustainable by choosing and enhancing microorganisms with the appropriate metal resistance mechanisms. A variety of environmental conditions, such as pH, temperature, and the availability of nutrients, can impact the metal resistance of microorganisms throughout the complicated and dynamic process of bioleaching. For the bioleaching process to be optimized and improved in terms of efficiency and efficacy, it is essential to comprehend the processes of metal resistance in bioleaching microorganisms [45,46].

3.11 SPENT HYDROPROCESSING CATALYST

Spent hydroprocessing catalysts are solid materials that are used in the petrochemical industry to remove impurities from crude oil and other hydrocarbons. These catalysts typically contain metals such as nickel (Ni), molybdenum (Mo), phosphorus (P), and aluminum oxide (Al_2O_3). Traditional disposal of used catalysts in landfills has been replaced by effective recovery of valuable metals from the catalysts because of environmental rules and economic factors. The wasted catalysts are treated for metal extraction by various solubilization processes and reused in a range of applications to reduce contamination in land disposal and minimize landfill area, which can make them attractive for use in bioleaching processes for metal recovery [47].

Bioleaching techniques can be used to recycle and recover used hydroprocessing catalysts. The process typically involves a series of steps, including pre-treatment to remove any organic material or other impurities that may interfere with the bioleaching process, inoculation with acidophilic microorganisms, and monitoring of key process parameters such as pH and temperature. Acidophilic microorganisms like *A. ferrooxidans* and *A. thiooxidans*, which can oxidize and solubilize the metals in the catalysts, are frequently used in the bioleaching of hydroprocessing catalysts. In comparison to conventional mineral ores, the use of hydroprocessing catalysts in bioleaching can provide a number of benefits, including a greater metal concentration, more uniform particle size, and the lack of undesirable minerals or contaminants. By enabling the recycling and reuse of these components, the usage of used catalysts can also aid in lowering waste and environmental effects.

In order to facilitate the bioleaching process, the used catalysts are normally pulverized and combined with a liquid medium containing acidophilic bacteria, nutrients, and other additives. The metals in the catalysts are oxidized by the microorganisms and released into a solution where they can be further treated and recovered. Utilizing bioleaching is a reasonably inexpensive and ecologically benign method for recovering metals from used hydroprocessing catalysts. Bioleaching generates fewer pollutants and waste products than conventional smelting and refining methods since it doesn't use hazardous chemicals, high temperatures, or other harsh procedures. Utilizing bioleaching to recover metals from used hydroprocessing catalysts is not without its difficulties. For instance, the catalysts may have high sulfur content, which can prevent some microbes from growing and functioning normally. The catalysts could also include dangerous or toxic substances that need particular handling and disposal methods [3].

Additionally, depending on the particular application, the makeup and features of the used catalysts can differ significantly, necessitating specialized bioleaching procedures and conditions to maximize the recovery of metals. In general, research and development are ongoing in the use of hydroprocessing catalysts in bioleaching, which has the potential to provide a sustainable and affordable method for metal recovery and waste management in the petrochemical industry [46–48].

3.12 CONCLUSIONS

In conclusion, bioleaching emerges as a cost-effective and simplified alternative to traditional metallurgical processes, offering operational advantages and requiring minimal maintenance. Unlike conventional methods that often necessitate a team of specialized personnel to manage complex chemical plants, bioleaching can be efficiently operated with a smaller workforce.

A notable advantage of bioleaching is its capability to achieve high yields, often exceeding 90%, even when processing ores with low metal concentrations. This efficiency is attributed to the metabolic activities of specialized microorganisms that derive energy from the breakdown of minerals into their elemental components, facilitating the extraction of metals from low-grade ores.

Furthermore, bioleaching obviates the need for energy-intensive processes such as crushing and grinding, which are typically employed in conventional mineral processing methods. By eliminating these stages, bioleaching not only reduces operational costs but also minimizes energy consumption, contributing to a more sustainable and environmentally friendly approach to metal extraction.

In scenarios involving high-grade ores, such as copper, bioleaching proves to be an economically advantageous process, further enhancing its appeal as a viable alternative to traditional extraction methods.

In summary, bioleaching offers a synergistic blend of economic feasibility, operational simplicity, and environmental sustainability, positioning it as a promising technology for the future of metal extraction and resource recovery. As research and development in the field of bioleaching continue to advance, the potential applications and benefits of this innovative approach are expected to expand, further consolidating its role in the transition toward more sustainable and efficient mineral processing practices.

REFERENCES

1. Oberle B, Bringezu S, Hatfield Dodds S, et al. *Global Resources Outlook 2019 Natural Resources for the Future We Want.* A Report of the International Resource Panel. United Nations Environment Programme. Nairobi, Kenya.
2. Velenturf APM, Archer SA, Gomes HI, Christgen B, Lag-Brotons AJ, Purnell P. Circular economy and the matter of integrated resources. *Sci Total Environ.* 2019;689:963–969. doi:10.1016/j.scitotenv.2019.06.449
3. Mishra D, Kim DJ, Ahn JG, Rhee YH. Bioleaching: a microbial process of metal recovery; a review. *Met Mater Inter.* 2005;11(3):249–256. doi:10.1007/BF03027450
4. Bosecker K. Bioleaching: metal solubilization by microorganisms. *FEMS Microbiol Rev.* 1997;20(3–4):591–604. doi:10.1111/j.1574-6976.1997.tb00340.x
5. Sajjad W, Zheng G, Din G, Ma X, Rafiq M, Xu W. Metals extraction from sulfide ores with microorganisms: the bioleaching technology and recent developments. *Trans Indian Inst Metals.* 2019;72(3):559–579. doi:10.1007/s12666-018-1516-4
6. Hofmann M, Hofmann H, Hagelüken C, Hool A. Critical raw materials: a perspective from the materials science community. *Sustain Mater Technol.* 2018;17: e00074. doi:10.1016/j.susmat.2018.e00074
7. Bosecker K. Bioleaching: metal solubilization by microorganisms. *FEMS Microbiol Rev.* 1997;20(3–4):591–604. doi:10.1111/j.1574-6976.1997.tb00340.x
8. Sajjad W, Zheng G, Din G, Ma X, Rafiq M, Xu W. Metals extraction from sulfide ores with microorganisms: the bioleaching technology and recent developments. *Trans Indian Inst Metals.* 2019;72(3):559–579. doi:10.1007/s12666-018-1516-4
9. Dopson M, Holmes DS. Metal resistance in acidophilic microorganisms and its significance for biotechnologies. *Appl Microbiol Biotechnol.* 2014;98(19):8133–8144. doi:10.1007/s00253-014-5982-2

10. William J, Mintek N. A mintek perspective of the past 25 years in minerals bioleaching agitated bioleaching of base metal concentrates view project from lab to heap. J *South Afr Inst Min Metall.* 2009;70:567–585. https://www.researchgate.net/publication/230787177
11. Olson GJ, Brierley JA, Brierley CL. Bioleaching review part B: progress in bioleaching: applications of microbial processes by the minerals industries. *Appl Microbiol Biotechnol.* 2003;63(3):249–257. doi:10.1007/s00253-003-1404-6
12. Ehrlich HL. Beginnings of rational bioleaching and highlights in the development of biohydrometallurgy: a brief history. *Eur J Mineral Process Environ Protect.* 2004;4:1303.
13. Natarajan KA. Methods in biohydrometallurgy and developments. *Biotechnol Metals.* 2018;2018:81–106. doi:10.1016/b978-0-12-804022-5.00005-0
14. Vestola EA, Kuusenaho MK, Närhi HM, et al. Acid bioleaching of solid waste materials from copper, steel and recycling industries. *Hydrometallurgy.* 2010;103(1–4):74–79. doi:10.1016/j.hydromet.2010.02.017
15. Srichandan H, Mohapatra RK, Parhi PK, Mishra S. Bioleaching approach for extraction of metal values from secondary solid wastes: a critical review. *Hydrometallurgy.* 2019;189:105122. doi:10.1016/j.hydromet.2019.105122
16. Gomes HI, Funari V, Ferrari R. Bioleaching for resource recovery from low-grade wastes like fly and bottom ashes from municipal incinerators: a SWOT analysis. *Sci Total Environ.* 2020;715:136945. doi:10.1016/j.scitotenv.2020.136945
17. Bosecker K. Bioleaching: metal solubilization by microorganisms. *FEMS Microbiol Rev.* 1997;20(3–4):591–604. doi:10.1111/j.1574-6976.1997.tb00340.x
18. Sajjad W, Zheng G, Din G, Ma X, Rafiq M, Xu W. Metals extraction from sulfide ores with microorganisms: the bioleaching technology and recent developments. *Trans Indian Inst Metals.* 2019;72(3):559–579. doi:10.1007/s12666-018-1516-4
19. Schippers A. Microorganisms involved in bioleaching and nucleic acid-based molecular methods for their identification and quantification. In: Donati, E.R., Sand, W. (eds) *Microbial Processing of Metal Sulfides* (2007). Springer, Dordrecht.
20. Exploration of Deepwater SeamountAssociated Chondrichthyans of the Southern Indian Ocean Using Genetic barcoding, eDNA, and Deepwater Cameras Paul J Clerkin and Jan R McDowell, 16th DSBS Deep-Sea Biology Symposium Brest, 12-17 September 2021
21. Kelly DP. Introduction to the chemolithotrophic bacteria. In: Starr, M.P., Stolp, H., Trüper, H.G., Balows, A., Schlegel, H.G. (eds) *The Prokaryotes* (1981). Springer, Berlin, Heidelberg.
22. Leduc LG, Ferroni GD. Federation of European microbiological societies the chemolithotrophic bacterium ferrooxidans thiobacillus. *FEMS Microbiol Rev.* 1994;14(2):103–119. https://academic.oup.com/femsre/article-abstract/14/2/103/595812
23. Canada É. *Recovery of copper from low-grade ores by Aspergillus niger.* Masters thesis, Concordia University (2001).
24. Hu K, Wu A, Wang H, Wang S. A new heterotrophic strain for bioleaching of low grade complex copper ore. *Minerals.* 2015;6(1):1–11. doi:10.3390/min6010012
25. Rezza I, Salinas E, Elorza M, Sanz de Tosetti M, Donati E. Mechanisms involved in bioleaching of an aluminosilicate by heterotrophic microorganisms. *Process Biochem.* 2001;36(6):495–500. doi:10.1016/S0032-9592(00)00164-3
26. Ilyas S, Anwar MA, Niazi SB, Afzal Ghauri M. Bioleaching of metals from electronic scrap by moderately thermophilic acidophilic bacteria. *Hydrometallurgy.* 2007;88(1–4):180–188. doi:10.1016/j.hydromet.2007.04.007
27. Johnson DB, Roberto FF. Heterotrophic acidophiles and their roles in the bioleaching of sulfide minerals. In: Rawlings, D.E. (eds) *Biomining. Biotechnology Intelligence Unit* (1997). Springer, Berlin, Heidelberg.

28. Jones S, Santini JM. Mechanisms of bioleaching: iron and sulfur oxidation by acidophilic microorganisms. *Essays Biochem.* 2023;67(4):685–699. doi:10.1042/EBC20220257
29. Rohwerder T, Gehrke T, Kinzler K, Sand W. Bioleaching review part A: progress in bioleaching: fundamentals and mechanisms of bacterial metal sulfide oxidation. *Appl Microbiol Biotechnol.* 2003;63(3):239–248. doi:10.1007/s00253-003-1448-7
30. Asghari I, Mousavi SM, Amiri F, Tavassoli S. Bioleaching of spent refinery catalysts: a review. *J Ind Eng Chem.* 2013;19(4):1069–1081. doi:10.1016/j.jiec.2012.12.005
31. Dusengemungu L, Kasali G, Gwanama C, Mubemba B. Overview of fungal bioleaching of metals. *Environ Adv.* 2021;5:100083. doi:10.1016/j.envadv.2021.100083
32. Xu TJ, Ting YP. Fungal bioleaching of incineration fly ash: metal extraction and modeling growth kinetics. *Enzyme Microb Technol.* 2009;44(5):323–328. doi:10.1016/j.enzmictec.2009.01.006
33. Beek B, Böhling S, Bruckmann U, Franke C, Jöhncke U, Studinger G. The assessment of bioaccumulation. In: Beek, B. (eds) *Bioaccumulation – New Aspects and Developments. The Handbook of Environmental Chemistry*, vol 2J (2000). Springer, Berlin, Heidelberg.
34. Baniasadi M, Vakilchap F, Bahaloo-Horeh N, Mousavi SM, Farnaud S. Advances in bioleaching as a sustainable method for metal recovery from e-waste: a review. *J Ind Eng Chem.* 2019;76:75–90. doi:10.1016/j.jiec.2019.03.047
35. Rouchalova D, Rouchalova K, Janakova I, Cablik V, Janstova S. Bioleaching of iron, copper, lead, and zinc from the sludge mining sediment at different particle sizes, pH, and pulp density using acidithiobacillus ferrooxidans. *Minerals.* 2020;10(11):1–28. doi:10.3390/min10111013
36. Deveci H, Akcil A, Alp I. Parameters for control and optimization of bioleaching of sulfide minerals. *Mater Sci Technol.* 2003;22:77–90.
37. Plumb JJ, Muddle R, Franzmann PD. Effect of pH on rates of iron and sulfur oxidation by bioleaching organisms. *Miner Eng.* 2008;21(1):76–82. doi:10.1016/j.mineng.2007.08.018
38. Rasoulnia P, Barthen R, Lakaniemi AM. A critical review of bioleaching of rare earth elements: the mechanisms and effect of process parameters. *Crit Rev Environ Sci Technol.* 2024;51(4):378–427.
39. Petersen J, Minnaar S, Plessis CD. Carbon dioxide and oxygen consumption during the bioleaching of a copper ore in a large isothermal column. *Hydrometallurgy.* 2010;104(3):356–362.
40. Van Hille RP, Bromfield LV, Botha SS, Jones G, Van Zyl AW, Harrison STL. The effect of nutrient supplementation on growth and leaching performance of bioleaching bacteria. *Adv Mater Res.* 2009;71:413–416.
41. Egli T, Zinn M. The concept of multiple-nutrient-limited growth of microorganisms and its application in biotechnological processes. *Biotechnol Adv.* 2003;22:35–43. doi:10.1016/j.biotechadv.2003.08.006
42. Das T, Ayyappan S, Roy Chaudhury G. Factors affecting bioleaching kinetics of sulfide ores using acidophilic microorganisms. *BioMetals.* 1999;12(1):1–10. doi:10.1023/A:1009228210654
43. Ma L, Huang S, Wu P, et al. The interaction of acidophiles driving community functional responses to the re-inoculated chalcopyrite bioleaching process. *Sci Total Environ.* 2021;798:149186. doi:10.1016/j.scitotenv.2021.149186
44. Zhang R, Wei X, Hao Q, Si R. Bioleaching of heavy metals from municipal solid waste incineration fly ash: availability of recoverable sulfur prills and form transformation of heavy metals. *Metals (Basel).* 2020;10(6):1–14. doi:10.3390/met10060815
45. Navarro CA, von Bernath D, Jerez CA. Heavy metal resistance strategies of acidophilic bacteria and their acquisition: importance for biomining and bioremediation. *Biol Res.* 2013;46(4):363–371.

46. Dopson M, Holmes DS. Metal resistance in acidophilic microorganisms and its significance for biotechnologies. *Appl Microbiol Biotechnol.* 2014;98(19):8133–8144. doi:10.1007/s00253-014-5982-2
47. Mishra D, Kim DJ, Ralph DE, Ahn JG, Rhee YH. Bioleaching of spent hydro-processing catalyst using acidophilic bacteria and its kinetics aspect. *J Hazard Mater.* 2008;152(3):1082–1091. doi:10.1016/j.jhazmat.2007.07.083
48. Qu Y, Lian B, Mo B, Liu C. Bioleaching of heavy metals from red mud using Aspergillus niger. *Hydrometallurgy.* 2013;136:71–77. doi:10.1016/j.hydromet.2013.03.006

4 Bioleaching of Lithium Ions from Li-Ion Battery Waste

Kalaiselvam S, Lakshmi Kanthan Bharathi A, Mercy Jacquline B, Harish Kumar U, and Bhuvanesh M

4.1 INTRODUCTION

The concept of lithium-ion battery (LIB) was first developed by Stanley Whittingham in the 1970s and then was commercialized by Sony and Asaki Kasei in 1991 (Yang et al., 2022) for a variety of applications due to minimum self-discharge rate, high energy density, extended charge storage life, lightweight, and safe handling ability properties (Alipanah et al., 2023; Miao et al., 2022). The huge development of transport and information technology industries has increased the usage of electronic gadgets such as mobile phones, laptops, wireless headphones, backup power systems, electric toys, handheld power tools, renewable energy storage devices, medical devices, and electric vehicles (Gholam-Abbas Nazri, 2009). The advance of these devices has led to an increase in the great necessity of electrical energy storage materials and systems in recent years. The electrical energy storage materials mostly used in the above devices to store electricity are lithium metal as a lithium-ion battery, and so the need for lithium in abundance as lithium-ion batteries has rapidly raised (Golmohammadzadeh et al., 2022). China accounts for 77% of the world's total lithium-ion battery output and it is predicted that 700,000 metric tons of lithium-ion batteries will come out as waste at the end of 2025 considering their life span as ten years (Alipanah et al., 2021). As per the research by the Lithium-ion Battery Market, the global demand for lithium-ion batteries is anticipated to increase at 15.2% CAGR by 2032, and by 2050 the lithium is expected to grow by at least 25 times larger than the level of 2022 (Luiz Henrique Ferreira, 2023) which has to be only supplied by the primary sources of lithium.

The primary lithium sources are spodumene and brine, with brine occupying half of the account (Li et al., 2019). However, continuous mining gradually exhausts these main resources and therefore there is a necessity for the investigation of numerous new strategies for lithium extraction from brine resources (Flexer et al., 2018). Conversely, the extended usage of lithium ore for manufacturing lithium-ion batteries increases the demand, which may eventually double the depletion of primary sources available in the future. On the other hand, the waste lithium-ion batteries and other electronic waste that contain traces of lithium are the secondary resources of lithium that serve as a substitute for the primary resources (Srichandan et al., 2019).

DOI: 10.1201/9781003415541-4

The need for recovering lithium from secondary sources like waste lithium-ion batteries is necessary to meet the future demand, as the primary sources may not accomplish the future demand and as the recovery methods for secondary sources require less operating energy and contribute less to environmental pollution compared to extraction from primary sources (Roy et al., 2021). The metal recovery from the secondary sources is usually by the traditional methods such as hydrometallurgy, pyrometallurgy, and bio-hydrometallurgy, where hydrometallurgy uses aqueous chemistry, pyrometallurgy uses heat, and bio-hydrometallurgy merges biology and hydrometallurgy (Jha et al., 2013). The process involved during recycling of waste Li-ion batteries is shown in Figure 4.1 (Biswal & Balasubramanian, 2023).

The various parts of waste lithium-ion batteries include a cathode, anode, electrolyte, battery casing, plastic materials, and miscellaneous parts and their overall weight percentage are 35%, 15%–18%, 11%–12%, 15%–18%, 25%–30%, 5%–8%, and 4% respectively (Heydarian et al., 2018; Horeh et al., 2016). Lithium cobalt oxide ($LiCoO_2$) is the most commonly used cathode material in lithium-ion batteries in almost all electronic gadgets as it has excellent energy density and resilience (Zhang et al., 2022). Lithium and cobalt are the two major materials found in the cathode of waste lithium-ion batteries in weight percentage of 10.3% and 30.4% respectively (Heydarian et al., 2018), along with the traces of Al, Cu, Mn, Ni, and Fe. The concentrations of Mn, Cu, and Ni in ranges of 5%–11%, 6%–12%, and 5%–10%, respectively, were detected in waste lithium-ion batteries during extraction (Roy et al., 2021; Venkata Ratnam et al., 2022). The separation of Li from used lithium-ion batteries has been commercially implemented by the use of hydrometallurgical methods (Zheng et al., 2018). However, due to its inherent disadvantages, which include high energy consumption, difficult conditions for operation, and demanding machinery needs, it is critical to explore for alternate techniques like bio-hydrometallurgy or bioleaching (Xu & Ting, 2009).

Bioleaching or bio-mining is a metal recovery technique of bio-hydrometallurgy that employs tiny living organisms like fungi, chemolithotrophic, and acidophilic bacteria to extract precious metals from secondary sources like low-grade ore/spent LIBs (Biswal & Balasubramanian, 2023). The study conducted with varying gravity conditions on the International Space Station (ISS) by researchers in 2020, demonstrated that microorganisms might be used to bioleach valuable materials from basaltic deposits in space (Cockell et al., 2020). Valuable metals may be economically recovered through the bioleaching process from a variety of electronic wastes. This process is also used commercially nowadays to extract valuable metals like Li, Co, Cu, Zn, Pb, Ni, Mo, Au, and Ag (Zhang et al., 2023). The bioleaching of lithium-ion batteries not only recovers precious metals but also stops contamination of the environment (Zhang et al., 2023). The bioleaching technique has many advantages such as less usage of chemicals, low operation cost, complete metal retrieval, less toxicity to the people and environment, and a less energy-demanding process without the need for sophisticated industrial equipment and elevated process conditions (Moazzam et al., 2021). The schematic for bioleaching process is shown in Figure 4.2. This technique is regarded as a commercially viable solution that offers numerous benefits over traditional approaches in lithium extraction from used LIBs for sustainable governing of recycling industries globally.

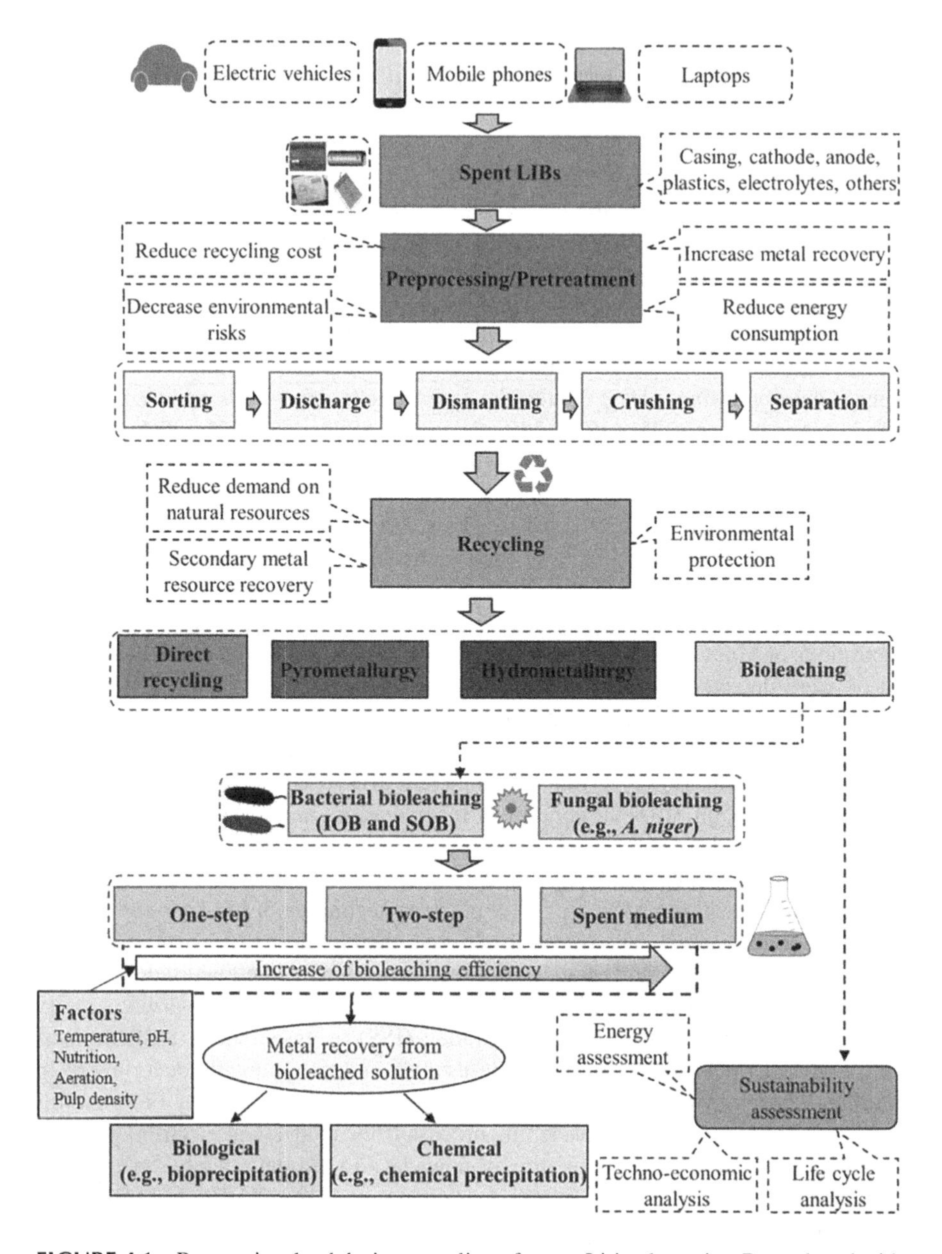

FIGURE 4.1 Process involved during recycling of waste Li-ion batteries. Reproduced with copyright permission from Biswal and Balasubramanian (2023).

Bioleaching of Li-ions is very much important as the lithium-ion battery trash level rise in the future and also the available sources get depleted. To overcome this issue and to meet the future lithium demands without harming the environment and increasing the energy cost for extraction, bioleaching of Li-ion battery is necessary both now and in the future. Analyzing the works of literature, the number of articles

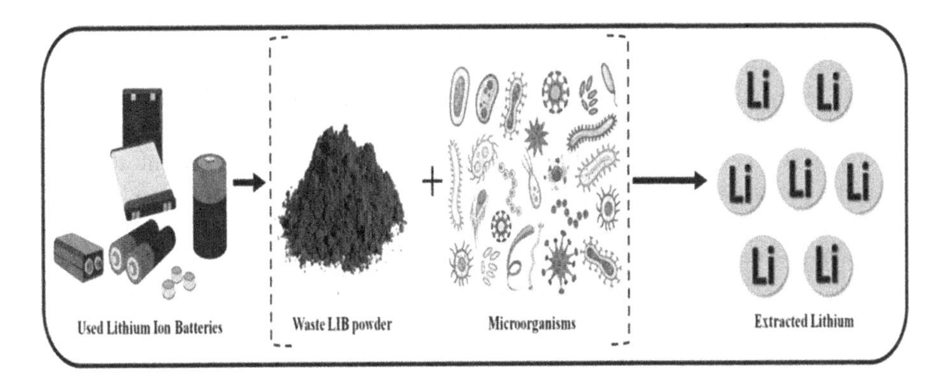

FIGURE 4.2 Schematic of bioleaching of lithium from Li-ion battery waste.

on the bioleaching of lithium ions from waste lithium-ion batteries has increased in recent years, this shows that the bioleaching technique is gaining more interest among researchers as an upcoming eco-friendly waste metal recovery method. On the other hand, there is not much comparative assessment on the effectiveness of fungal and bacterial bioleaching for the separation of valuable elements like Co and Li from waste lithium-ion batteries. The primary aim of this chapter is to give a thorough analysis of the latest advancements in the field of bioleaching and the utilization of microbiological agents, specifically bacteria and fungus, in the recovery of precious lithium metal from spent lithium-ion batteries. The chapter broadly discusses sources of lithium, bioleaching of lithium, microorganisms used in bioleaching, and different methods and mechanisms involved in bioleaching. This chapter also deals with the parameters influencing and affecting bioleaching such as pH, medium composition, temperature, pulp density, cell nutrients, redox potential, substrate concentration, types of microorganisms, interactions between microbes and metals, along with hybrid bioleaching. In conclusion, the chapter contracts with a significant discussion on bioleaching sustainability, advantages, and upcoming opportunities of bioleaching to give an insight to the researchers into the use of the bioleaching technique for lithium extraction from the waste lithium-ion batteries.

4.2 RESOURCES OF LITHIUM

Solids and liquids are the two classifications of lithium resources. The minerals ores, electronic waste, and waste lithium-ion batteries from electronic gadgets are categorized under solid lithium resources, whereas the seawater, geo-thermal brines, and salt-lake brines are categorized under liquid lithium resources (Li et al., 2019; Loganathan et al., 2017). The solid and liquid lithium resources are further classified into primary and secondary resources.

4.2.1 Primary Resources

Minerals such as spodumene, petalite, lepidolite, and amblygonite and salar brines are categorized under the primary resources of lithium which are non-renewable.

Spodumene mineral is found all over the world and the biggest spodumene mine is in Australia. Mining process of minerals for extracting lithium is difficult, expensive, and has many environmental risks. Since 59% of the global reserves of lithium are found in salar brines, there has been a lot of interest in the recovery of lithium from aqueous resources (Li et al., 2019). However, because of the high concentrations of competing ions, including calcium, magnesium, sodium, and potassium, and the low fraction of lithium, extraction from the brine is challenging and requires preliminary treatment procedures (Moazzam et al., 2021).

4.2.2 Secondary Resources

Extraction of lithium from secondary resources like waste lithium-ion batteries and other e-waste has become an economical and environmentally beneficial alternative for primary resources. The secondary resource has advantage of reducing the hazardous waste thrown into the environment that causes harm to environment and people health. Li-ion batteries has aluminum cathode coated with lithium cobalt oxide ($LiCoO_2$) from which Li can be extracted (Zhang et al., 2023).

4.3 MICROORGANISMS USED IN LI BIOLEACHING

Bioleaching method was adapted for the extraction of Li-ions from waste lithium-ion batteries (LIB) as it is an eco-friendly method and involves microbial participation for leaching. The microorganisms used in Li bioleaching play a major role in leaching process and are classified based on microorganisms' thrive temperature, nutrition and energy sources, and pH.

4.3.1 Classification of Microorganisms According to Thrive Temperature

The microorganisms based on the temperature at which their metabolism occurs are classified as mesophilic organisms (below 40°C) (Gu et al., 2018), moderate thermophilic organisms (45°C–70°C) (Kaksonen et al., 2017), thermophilic organism (70°C or above), and heterotrophic fungi (Rasoulnia et al., 2016).

4.3.1.1 Mesophile Bacteria (S/Fe-Oxidizing Bacteria)

The metallic components in LIB can be solubilized by mesophilic microbes, which are also referred as S/Fe oxidizing bacteria. In LIB bioleaching, *Acidithiobacillus (Aci.)* type bacteria have better tolerance for metal toxicity and use elemental sulfur (S_8) or ferrous ion (Fe^{2+}) as its main source of energy and atmospheric carbon dioxide as a source of carbon. The dissolution of metal happens by biogenic sulfuric acid (H_2SO_4) and ferric ion (Fe^{3+}) production, facilitating the oxidation of metal (Wu et al., 2020). The majority of metal ion remains in the leaching solution during bacterial bioleaching carried out in an acidic pH range of 1–3. *Aci. ferrooxidans, Aci. thiooxidans,* and *Leptospirillum (Lep.) ferrooxidans* are some of the extensive mesophilic organisms used for the bioleaching of LIBs. However, further types of *Acidithiobacillus* bacteria are not acidic enough to contribute to bioleaching (Hoque & Philip, 2011). The use of ferrous ion (Fe^{2+}) as a donor for electron is not seen in *Aci. thiooxidans* and *Aci. caldus* types bacteria (Gu et al., 2018).

4.3.1.2 Moderate Thermophilic and Thermophilic S/Fe-Oxidizing Bacteria

The higher temperatures during the leaching process encourage process kinetics and the microorganisms/bacteria that proliferate at higher temperatures are advantageous (Kaksonen et al., 2017). The introduction of thermophilic bacteria increases the speed of the reaction and also increases the metal recovery rate (Gu et al., 2018). *Thermoplasma acidophilum, Sulfobacillus acidophilus,* and *Sulfobacillus thermosulfidooxidans* are the commonly used microorganisms that can grow up to 50°C and are used in the bioleaching of LIB and electronic waste (Ilyas et al., 2018). Few moderate thermophilic organisms are sulfur-oxidizing and iron-oxidizing like *Aci. caldus* and *Lep. ferriphilum,* respectively. *Sulfolobus acidocaldarius, Sulfolobu solfataricus, Sulfolobus brierley,* and *Sulfolobus ambioalous* are examples of extreme thermophiles that thrive at 75°C–80°C, at pH 1–3. Due to their tendency for growth at high temperatures, thermophiles exhibit higher rates of metal bioleaching than moderate and mesophilic thermophiles.

4.3.1.3 Heterotrophic Fungi

Heterotrophic fungi, such as *Penicillium (Pen.) simplicissimum, Penicillium (Pen.) chrysogenum,* and *Aspergillus (Asp.) niger,* have been used in the bioleaching of e-waste (Rasoulnia et al., 2016). *Aspergillus niger* is chosen over other fungi as it is easy to handle, harvest, ferment variety of organic acids, and has high yields. The fungal bioleaching of LIB has the following merits than bacterial bioleaching like the occurrence of bioleaching at neutral pH since LIB are alkaline. In order to make organic acids like citric acid, gluconic acid, lactic acid, malic acid, oxalic acid, etc., that helps heterotrophic fungi with carbon supplement during metal leaching (Vakilchap et al., 2016), the leftover sugars and agricultural waste are used as main sources of carbon nutrition. The process of carbon supply makes the bioleaching process costly and also metal recovery by fungal bioleaching is low compared to acidophilic bacteria.

4.3.2 Classification of Microorganisms Based on Energy Source and Nutrition

Microorganisms are categorized under various types according to the energy source and the nutrition as follows: chemolithotrophic organisms, chemoorganotrophic organisms, microbial consortium, and mixed cultures. Chemolithotrophs proliferate greatly in an acidic pH level and use the inorganic substances as a source of energy, whereas chemoorganotrophs use carbon obtained from organic materials as a source of energy.

4.3.2.1 Chemolithotrophic Bioleaching

Chemolithotrophic organisms are the most common microbes used in acidic conditions with the pH range of 2 and below. Mesophiles, moderate thermophiles, and thermophiles come under this category. They get energy by oxidizing ferrous ion (Fe^{2+}) to their respective ferric ion (Fe^{3+}) and reducing sulfur compounds like thiosulfate ($S_2O_3^{2-}$) and octasulfur (S_8) to ferric ion (Fe^{3+}) and sulfuric acid (H_2SO_4), respectively (Srichandan et al., 2019). Sulfate formed causes the reaction environment's pH to shift toward a highly acidic state, creating the ideal conditions for the solubilization of metals. However, ferric ion promotes the formation of sulfuric acid and soluble metal in the process in a microbe-free manner by oxidizing metal sulfides on the surface of the solid metal powder (Gu et al., 2018). *Acidithiobacillus* and *Leptospirillum genera* are the very frequently used chemolithotrophs that include *Aci. ferrooxidans, Aci. thiooxidans,* and *Lep. ferrooxidans.*

Acidithiobacillus thiooxidans was employed to bioleach waste coin cells and 20%, 60%, and 99% of Mn, Co, and Li were, respectively, extracted (Naseri et al., 2019a). Sulfur-oxidizing bacteria separated from the mine pit was used to extract lithium from LIB and obtained an extraction efficiency of 93.64% by combining the selective adsorption technique with granular-activated carbon particles (Huang et al., 2019).

4.3.2.2 Chemoorganotrophic Bioleaching

Chemoorganotrophic microorganisms, otherwise referred as heterotroph microorganisms, acquire their necessary energy from the organic compounds. Heterotrophic microorganisms that are mostly active are used for bioleaching of LIB and electronic waste that includes filamentous fungi, such as species of *Aspergillus (Asp.), Penicillium,* and cyanogenic bacteria. Since the oxides in the industrial wastes are high, the shortage of energy sources occurs due to the reduction of organic acid formation from chemolithotrophic bacteria; so, heterotrophs, particularly fungi, are preferred for lithium bioleaching and for bioleaching industrial waste, as fungi are not much sensitive to highly concentrated heavy metals. Fungi engage in metal ion mobilization with two primary mechanisms that include the secretion of metabolites and the accumulation or buildup of metal ions. Lactic, formic, citric, pyruvic, succinic, oxalic, malic, and tartaric acids are the major organic-type metabolites secreted during leaching by the heterotroph fungus. The secreted acids are mild, so fungus typically bioleach metals over a wider range of pH from 3 to 7. The fungi employ three ways for metal mobilization: acidolysis, redoxolysis, and complexolysis. The acidolysis is a process that displaces protons from the source surface as a result of oxygen atom protonation by fungi (Xu & Ting, 2009). Redoxolysis is the process of oxidation and/or reduction of metals by bacteria and fungi to produce where the insoluble metals are transformed during leaching to their own soluble state and ferric (Fe) ions. Further, complexolysis is the development of complex among organic acids and metal ions, involving the chelation of metals with the ligands (Dutta et al., 2023; Sethurajan & Gaydardzhiev, 2021). The species of bacteria, such as Bacillus, Escherichia, and Pseudomonas, and fungi species like clitocybe and polysporus, are regarded as the primary cyanogenic bioleachers and exhibit cyanogenic leaching conditions at a pH range from 7 to 11 and at the temperature range of 25°C–35°C. For leaching Li and Co from rechargeable waste LIB, the three strains of fungi, namely *Penicillium chrysogenum, Penicillium simplicissimum,* and *Aspergillus niger,* were discovered, and on comparing all three, *Asp. niger* medium showed the maximum concentration of the oxalic acid secreted.

4.3.2.3 Microbial Consortium and Mixed Cultures

Microbial consortium is referred to when two or more number of microbe species coexist symbiotically in the medium or culture. A mixed culture is an association that is not based on function, but a consortium illustrates an association of long-term groups of microorganisms which altogether carry out similar activity. Utilizing microbial consortiums has become more common in industries owing to their advanced functional stability and larger efficiency in comparison with single strains (Kang et al., 2020). Heydarian et al. used an optimal inoculum ratio forming *Aci. ferrooxidans* and *Aci. thiooxidans* mixtures to bioleach spent LIB with the Li recovery of 99.2% efficiency (Heydarian et al., 2018). Some examples of mixed culture and consortiums are *Lep. ferriphilum* dominating consortium and mixed culture of sulfur-oxidizing bacteria *Aci. ferrooxidans* and heterotrophic yeast *Rhodotorula rubra*. Liao et al. (2022) inoculated a mixed culture of *Aci. caldus* and *Sulfobacillus thermosulfidooxidans* to increase metal leaching rates by 6%–20%. When LIBs are recovered in mixed cultures of autotrophic and heterotrophic microorganisms, a small amount (0.2 g/L) of yeast extract might prevent a decrease in microbial activity due to Fe2+ deficit (Wu et al., 2020).

4.3.3 Classification of Microorganisms Based on pH

Acidophiles belong to the bacterial variety/kind that attains growth in the optimal pH range from 1.8 to 2.5. Acidophiles are a bacterial group that requires an acidic environment with a pH in the range of 2.0–2.5 for proliferation and working during bioleaching. *Aci. ferrooxidans* and *Aci. thiooxidans* are some of the examples of acidophiles. Fungi, such as Penicillium, come under neutrophils that have an optimum pH range from 2.0 to 8.0 for the bioleaching process to take place and grow.

4.4 DIFFERENT METHODS INVOLVED IN LI BIOLEACHING

Bioleaching procedures are classified into two types namely: (1) contact bioleaching (one- and two-step process) and (2) non-contact bioleaching (spent-medium process) (Moazzam et al., 2021). The schematic of different methods used for bioleaching is depicted in Figure 4.3.

4.4.1 One-Step Bioleaching Method

One-step method is a simple method of leaching where bioleaching is carried out during the culture of microorganisms which are inoculated in the medium along with spent lithium-ion battery (LIB) powder simultaneously (Heydarian et al., 2018).

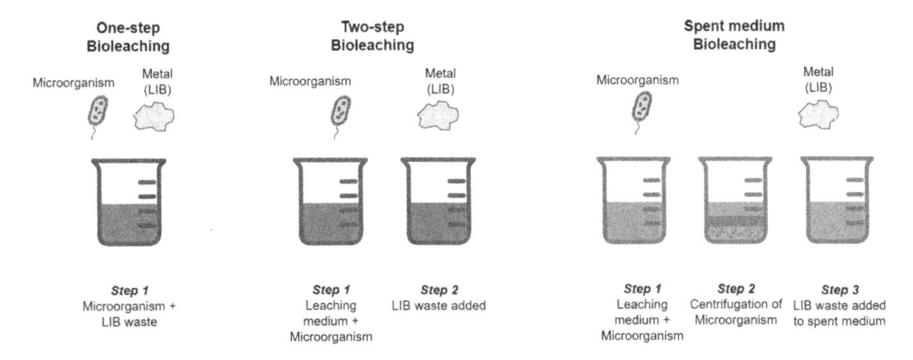

FIGURE 4.3 Schematic of different bioleaching methods of lithium. Reproduced with copyright permission from Moazzam et al. (2021).

The operational costs decrease (Bahaloo-Horeh et al., 2018) when employing this method, as both the procedures of bioleaching and fermentation are performed as a single stage. This method can be used for spent LIBs containing fewer toxic components. Sedlakova-Kadukova et al. (2016) used the mixed culture of *Aci. ferrooxidans, Aci. Thiooxidans*, and achieved a recovery yield of 80% Li with 1% pulp density, 1.5 pH. Hartono et al. (2017) used one-step bioleaching with locally isolated bacterial strains and achieved a yield of 62.8% Li with a pulp density of 2 mg/mL, pH 7, and a temperature of 30°C. Naseri et al. (2019a) achieved a leaching of 100% lithium with *Asp. niger (Aspergillus niger)* using one-step method.

4.4.2 Two-Step Bioleaching Method

In the presence of solid waste, microbial growth cannot be preferred in some circumstances because of the harmful nature of e-waste and LIB. Hence, in some cases, two-step approach is preferred to increase the microbial growth rate. As the name implies, two steps are involved where in the first step, the growth of microorganisms is carried out without the presence of lithium-ion battery (LIB) waste until it reaches the logarithmic growth state. In the next step, the solid LIB waste is added to the medium of the first step with microorganisms to instigate the bioleaching process. Microorganisms entering the exponential phase can be seen by the sudden drop in pH value originating from organic acid secretion. In the fungus germination bioleaching method, the LIB powder's hindrance to fungus growth is restricted by producing metal extraction and organic acid. It is used in spent LIBs containing more toxic components (Biswal & Balasubramanian, 2023). This method is more suitable for industry applications as the efficiency of bioleaching gets enhanced with microbial production of acidic solution prior to the addition of solid LIB waste into the medium. Xin et al. (2016) used a mixed culture of *Aci. thiooxidans* and *Lep. ferriphilum* and achieved a recovery yield of 98% Li with two-step bioleaching. *Aci. thiooxidans*is used as a microbial agent by (Biswal et al., 2018) and observed higher recovery of lithium (66%) than cobalt using this bioleaching method. Similarly, it was used by (Naseri et al., 2019a) for extraction from spent LIBs and observed a higher recovery of lithium (99%).

4.4.3 Spent-Medium Bioleaching

Although the yield of two-step approach of bioleaching is more than one-step approach (Rasoulnia et al., 2021), the combination of solid waste and spent-medium after the separation of microorganisms from the culture results in increased efficiency. In this approach, the microbes are cultured initially in the growth medium without the presence of solid LIB waste until the production of metabolites reaches a stationary state. To carry out further process of bioleaching, the cultured medium with microorganisms is centrifuged and filtered to get the cell-free medium (Biswal et al., 2018). Finally, the bioleaching process gets started with the addition of LIB powder to the cell-free medium.

The spent-medium approach allows the optimization of biological and chemical processes independently. Due to toxicity from the batteries, there is no damage in the cell wall and high pulp density usage is not shortened. On comparing with one- and two-step methods, the processing time of this method is less (Bahaloo-Horeh et al., 2018).

However, the spent-medium method requires high operating costs because of the extra bioreactor tanks used and metabolites produced at two separate states. Among the above bioleaching methods, spent-medium method was found to have the maximum performance (Naseri et al., 2019a). On comparing the three bioleaching methods by using the microorganism, *Asp. niger*, the one-step, two-step, and spent-medium methods take 30, 27, and 16 days, respectively, for incubation and complete leaching of lithium-ion battery powder. Among this, the highest recovery (lithium 95%, cobalt 45%, copper 100%, and nickel 38%) was achieved with spent-medium method. Bahaloo-Horeh and Mousavi ((2017) used spent-medium method and achieved a yield of 100% Li with *Asp. niger*. The parameters exhibited by the microorganism for its leaching efficiency will be quite different with different methods of bioleaching (Faraji et al., 2018).

4.5 MECHANISMS INVOLVED IN LITHIUM BIOLEACHING

The mechanism of bioleaching is classified into direct (contact) and indirect (non-contact) bioleaching based on microorganisms' interaction with the metal surface, as shown in Figure 4.4 (Tao & Dongwei, 2014).

4.5.1 CONTACT OR DIRECT BIOLEACHING

In contact or direct bioleaching, the mineral particles under suspension absorb the cells within a period of time. This mechanism occurs by electron transfer from the cells attached to the mineral surface as the reduced metals get oxidized by microorganisms. During biofilm formation, extra-cellular polymeric substances (EPS) that contain lipids, polysaccharides, and proteins are considered as main factors. In this mechanism, the reactions of bioleaching occur by the adhesion of cells onto metal surfaces. The main interactions responsible for the above cell adhesion are hydrophobic and electrostatic forces (Wang et al., 2018). The chelated Fe^{3+} in EPS creates positive charges on the cell surface, which are responsible for the electrostatic interactions. Furthermore, few protein molecules stay within EPS and accumulate the

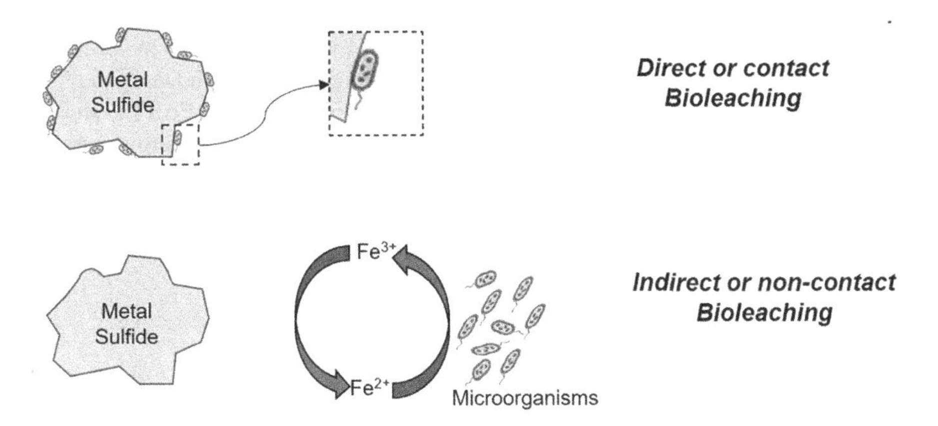

FIGURE 4.4 Bioleaching mechanisms based on microorganism interaction with metal particles. Reproduced with copyright permission from Tao and Dongwei (2014).

sulfur for successive reduction reactions to produce sulfuric acid. Eqs. (4.1) and (4.2) show the direct leaching mechanism (Sand et al., 2001)

$$FeS_2 + H_2O + 3.5O_2 \rightarrow 2H^+ + 2SO_4^{2-} + Fe^{2+} \quad (4.1)$$

$$Fe^{2+} + 2H^+ + 0.5O_2 \rightarrow 2Fe^{3+} + H_2O \quad (4.2)$$

4.5.2 Non-Contact or Indirect Bioleaching

In non-contact or indirect bioleaching, the microorganisms do not come in direct contact with the metal particles during the process. The process is mediated by acids like organic and inorganic acids produced by microorganisms. Contact between microorganisms and solid substrate is not necessary, as the metal dissolution process is mainly carried by the leaching agents produced by the microbial action (Horeh et al., 2016).

Inorganic acids like sulfuric acid (Gu et al., 2018) and organic acids like citric, gluconic, malic, and oxalic acids are mostly employed during non-contact bioleaching (Horeh et al., 2016). Fe^{3+} is reduced by Fe^{2+} oxidation and is known to be a strong oxidizer. The complexing agent affects the metal sulfide dissolution leading to stable complex formation after the metal ions' chelation. The mechanisms involved in this type of bioleaching are shown in Eqs. (4.3)–(4.5) (Sand et al., 2001), where bivalent metal is denoted by M

$$FeS_2 + 14Fe^{3+} + 8H_2O \rightarrow 15Fe^{2+} + 16H^+ + 2SO_4^{2-} \quad (4.3)$$

$$MS + 2Fe^{3+} \rightarrow M^{2+} + S^0 + 2Fe^{2+} \quad (4.4)$$

$$S^0 + 1.5O_2 + H_2O \rightarrow 2H^+ + 2SO_4^{2-} \quad (4.5)$$

The direct method has more efficiency than indirect method, as the microorganisms are in physical contact with solid surface. However, indirect bioleaching is more suitable than the direct method because of its flexibility to optimize the process. The mechanism for bioleaching of metals, such as Co, Ni, and Mn, in LIBs differs from that of lithium bioleaching (Wang et al., 2018). Bioleaching of lithium proceeds with H_2SO_4 (biogenic acid) production through bio-oxidation of sulfur, which does not depend on the addition of EPS. The extraction of Li and Co through bioleaching process on LIBs by iron-oxidizing and sulfur-oxidizing bacteria is shown in the following equations, where Eq. (4.6) represents $LiCoO_2$ dissolution and Eq. (4.7) shows the release of Li by the effect of products from bacterial activity and energy sources

$$4LiCoO_2 + 12H^+ \rightarrow 4Li^+ + 4Co^{2+} + 6H_2O + O_2 \quad (4.6)$$

$$2FeSO_4 + 2LiCoO_2 + 4H_2SO_4 \rightarrow Fe_2(SO_4)_3 + 2CoSO_4^{2+} + Li_2SO_4 + 4H_2O \quad (4.7)$$

4.6 THERMODYNAMIC EVALUATION

The evaluation of the thermodynamic process during the bioleaching of lithium-ion batteries gives insight into energy balance and potential reactions that take place during catalytic reactions.

4.6.1 Catalyst Selection for Bioleaching

The bioleaching process has a slow reaction rate for metal recovery from spent LIB which is a demerit. The inclusion of catalysts, surfactants, and cultures with a modified medium may help to increase the rate of bioleaching. Catalysts, such as activated carbon, Ag^+, Bi^{3+}, Sn^{2+}, Cu^{2+}, Co^{2+}, Ni^{2+}, and the illumination of light are used to increase the reaction rate (Niu et al., 2014). Increasing the standard potential difference of the catalyst and waste LIB has led to higher recovery of metal and faster reaction. Gibbs-free batteries, the descending order of catalyst efficacy is Bi^{3+}, Cu^{2+}, Sn^{2+}, Ni^{2+}, Co^{2+}, and Ag^+. The catalytic role of Cu^{2+} and Ag^+ in leaching lithium is shown in Eqs. (4.8) and (4.9), and the catalysts appear to boost the liberation of Li^+ from the Co compound

$$Cu^{2+} + 2LiCoO_2 \rightarrow CuCo_2O_4 + 2Li^+ \tag{4.8}$$

$$Ag^+ + LiCoO_2 \rightarrow AgCoO_2 + 2Li^+ \tag{4.9}$$

4.6.2 Potential Reactions in Lithium-Ion Batteries

$CuCo_2O_4$ and $AgCoO_2$ were determined to have Gibbs-free energy of 578.0 and 245.0 kJ/mol, respectively. The reactions in Eqs. (4.10) and (4.12) are thermodynamically favorable

$$G_r^0 = G_f^0(\text{products}) - G_f^0(\text{reactants}) \tag{4.10}$$

$$CuCo_2O_4 + 2Fe^{2+} + 8H^+ \rightarrow 2Fe^{3+} + Cu^{2+} + 2Co^{2+} + 4H_2O \tag{4.11}$$

$$AgCoO_2 + Fe^{2+} + 4H^+ \rightarrow Fe^{3+} + Ag^+ + Co^{2+} + 2H_2O \tag{4.12}$$

In these reactions, cobalt is a trivalent used in the compounds and reasonable for Co^{3+} reduced to Co^{2+} with Fe^{3+}. The oxidation and reduction potential increases with Fe^{3+} involving microbial oxidation of Fe^{2+} and coupling of Fe^{2+} and Co^{3+}.

4.7 FACTORS AFFECTING LITHIUM BIOLEACHING

The bioleaching efficiency solely relies on the mineral and chemical composition of the source to be leached, as well as the efficiency of the microbes used. The attainment of maximum yield for extraction of metal is contingent upon the leaching factors matching the ideal growth conditions of the microorganisms. A summary of lithium recovery from Li-ion batteries by bioleaching is listed in Table 4.1. To improve the bioleaching efficiency, the influencing factors need to be strictly controlled and optimized, and the relevant contents have been well summarized (Sethurajan & Gaydardzhiev, 2021).

4.7.1 pH at Initial Stage

The pH of the solution is critical in improving metal leaching efficiency. Maintaining the pH of the bioleaching process is essential because of its impact on microbes activity and reproduction (Rasoulnia et al., 2021). The operating system and the

TABLE 4.1
Review Summary of Bioleaching Recovery of Lithium from Li-Ion Batteries

Microorganisms	Metal Sources	Nutrient Media	Temperature	Pulp Density w/v	pH	Method	Lithium Bioleaching Efficiency %	Other Metals Efficiency %	References
Acidithiobacillus ferrooxidans (PTCC 1647)	Spent LIBs	–	30°C	10 g/L		–	67	Co 19, Mn 50, Ni 34	Nazerian et al. (2023)
Acidithiobacillus caldus and *Sulfobacillus thermosulfdooxidans*	Spent $LiCoO_2$ batteries powders	Modified 9 K+S powder 10 + 6 g/L Fe^{2+}	30°C	20 g/L	2.5	Two-step	100	99 Co	Liao et al. (2022)
Penicillium citrinum	Spent coin cells (SCCs) powder	3.9% (w/v) PDA medium	30°C	20 g/L	6	One step	70	Mn 53	Naseri et al. (2022)
Acidithiobacillus ferrooxidans (DSMZ 1927)	Spent LIBs	–	30°C	100 g/L	–	–	89.9	Co 90.4 Mn 91.8, Ni 85.5	Do et al. (2022)
Acidithiobacillus ferrooxidans (isolated)	Spent LIBs	–	30°C	Inoculum concentration: 20% (v/v)	2.5	–	–	Co 57.8	Hariyadi, Masago et al. (2022)
Acidithiobacillus ferrooxidans (isolated)	Spent LIBs	–	30°C	10 g/L inoculum concentration: 20% (v/v)	2–4	–		Co 73.95	Putra et al. (2022)
Acidithiobacillus ferrooxidans and *Acidithiobacillus. thiooxidans*	Spent LIBs	–	30°C	1% (w/v)	2.0	–	–	Co 99.95, Ni 99.95	Noruzi et al. (2022)
Aspergillus niger (isolated)	Spent LIBs	–	30°C	Carbon sources: glucose, incubation time: 21 days	–	–	72	Co 57	Hariyadi et al. (2022)

(Continued)

TABLE 4.1 (*Continued*)
Review Summary of Bioleaching Recovery of Lithium from Li-Ion Batteries

Microorganisms	Metal Sources	Nutrient Media	Temperature	Pulp Density w/v	pH	Method	Lithium Bioleaching Efficiency %	Other Metals Efficiency %	References
Acidithiobacillus ferrooxidans (DSMZ 1927)	Electrodes powder	Modifed 9 K medium +$FeSO_4 \cdot 7H_2O$ 150 g/L	30°C	1% (w/v)	2.0	Two-step	82	Li 89, Mn 92, Ni 90	Roy et al. (2021)
Aspergillus niger and *Aspergillus tubingensis*	Spent LIBs	–	30°C	1% (w/v)	–	–	95	Co 60, Mn ~98, Ni ~80, Al ~82	Alavi et al. (2021)
Acidophilic Microbial Consortia	$LiCoO_2$ powder		42°C	5% (w/v)	1.25	Two-step	98.1	Co 96.3	Liu et al. (2020)
Acidithiobacillus thiooxidans, *Leptospirillum ferriphilum*, and *Acidithiobacillus ferrooxidans*	Cathodes $LiNi_xCo_yMn_{1-x-y}O_2$ powder	A mineral salt medium with sulfur (1.0% w/v) +$FeSO_4 \cdot 7H_2O$ 20 g/L	30°C	4% (w/v)	1.0	Two-step	100	Ni 42, Co 40, Mn 40	Wang, Kijkla, et al. (2022); Wang, Yang, et al. (2022)
Aspergillus niger (PTCC 5010)	Spent LIBs	–	30°C	10% (w/v) carbon source: glucose	4.5	–	73.3	–	Kazemian et al. (2020)

(*Continued*)

TABLE 4.1 (*Continued*)
Review Summary of Bioleaching Recovery of Lithium from Li-Ion Batteries

Microorganisms	Metal Sources	Nutrient Media	Temperature	Pulp Density w/v	pH	Method	Lithium Bioleaching Efficiency %	Other Metals Efficiency %	References
Penicillium chrysogenum PTCC 5037	Spent LIBs	–	30°C	10% (w/v) carbon source: glucose	4.5	–	54.6	–	Kazemian et al. (2020)
Acidithiobacillus caldus, *Leptospirillum. ferriphilum*, *Sulfobacillus* spp., and *Ferroplasma* spp.	Spent LIBs	–	30°C	10 g/L	1.8	–	84	Co 99.90	Ghassa et al. (2020)
Leptospirillum ferriphilum and *Sulfobacillus thermosulfidooxidans* *Leptospirillum ferriphilum*	Spent LIBs	–		15 g/L	1.2	–	98.1	Co 99.3	Liu et al. (2020)
Acidothiobacillus thiooxidan	Spent coin cells powder	9 K medium $+FeSO_4{\cdot}7H_2O$ (44.2 g/L)	30°C	30 (g/L)	2.0	Two-step	99	–	Naseri et al. (2019a)
Acidothiobacillus ferrooxidans (PTCC1647)	Anodes and cathodes powder	9 K medium $+FeSO_4{\cdot}7H_2O$ 44.2 g/L	30°C	1.0–10% (w/v)	2.0	Two-step	100	Co 88, Mn 20 (4 S/L)	Naseri et al. (2019b)
Acidothiobacillus thiooxidans (PTCC1717)	Anodes and cathodes powder	9 k Medium +S powder 5 g/L	30°C	1.0–5.0% (w/v)	2.0	Two-step	99	Co 60, Ni 20 (3 w/v)	Naseri et al. (2019b)

(Continued)

TABLE 4.1 (*Continued*)
Review Summary of Bioleaching Recovery of Lithium from Li-Ion Batteries

Microorganisms	Metal Sources	Nutrient Media	Temperature	Pulp Density w/v	pH	Method	Lithium Bioleaching Efficiency %	Other Metals Efficiency %	References
Acidothiobacillus ferrooxidans (PTCC1647) and *Acidothiobacillus thiooxidans* (PTCC1717)	Electrodes	Modifed 9k medium + S powder 5 g/L +$FeSO_4 \cdot 7H_2O$ 36.7 g/L	32°C	4.0% (w/v)	1.5	Two-step	99.2	Co 50.4, Ni 89.4	Heydarian et al. (2018)
Acidothiobacillus ferrooxidans and *Acidothiobacillus thiooxidan*	Cathode powder from laptop LIB	Basal salts medium (pH 1.8) with both 110 g/L soluble $FeSO_4 \cdot 7H_2O$ and 5 g/L sterile S powder	22°C	10% (w/v)	1.8	Spent-medium	60	Co 53.2, Ni 48.7, Mn 81.8, Cu 74.4	Boxall et al. (2018)
Acidothiobacillus thiooxidans (80,191)	Spent Li-ion Battery Powder	Basel 317 +S power 1%	30°C	0.25% (w/v)	3.3 and 2.4	One step	22	Co 66	Biswal et al. (2018)
Aspergillus niger (isolated)	Spent Li-ion Battery Powder	Sucrose medium	30°C	0.25% (w/v)	2.4	One step	100	Co 82	Biswal et al. (2018)
Aspergillus niger (PTCC 5010)	Cathode and anode powder	Sucrose medium	30°C	1.0 (w/v)	2.5	One step	100	Co 38, Cu 94, Mn 72, Ni 45	Bahaloo-Horeh et al. (2018)
Locally isolated bacterial strains	cathode ($LiCoO_2$) powder	3 g/L of meat extract, 5 g/L of peptone and 5 g/L of NaCl	30°C	2 (mg/mL)	7	One step	62.83	–	Hartono et al. (2017)

(*Continued*)

TABLE 4.1 (*Continued*)
Review Summary of Bioleaching Recovery of Lithium from Li-Ion Batteries

Microorganisms	Metal Sources	Nutrient Media	Temperature	Pulp Density w/v	pH	Method	Lithium Bioleaching Efficiency %	Other Metals Efficiency %	References
Aspergillus niger	Cathodes and anodes powder	Sucrose medium	30°C	2% (w/v)	5.44	Spent-medium	100	Cu 100, Mn 77, Al 75	Bahaloo-Horeh and Mousavi (2017)
Aspergillus niger (PTCC 5210)	Cathodes and anodes powder	Sucrose medium	30°C	1% (w/v)	5.44	Spent-medium	–	Co 64, Ni 54	Bahaloo-Horeh and Mousavi, (2017)
Acidothiobacillus ferrooxidans (isolated) and *Acidothiobacillusthiooxidans* (isolated)	$LiNi_xCo_yMn_{1-x-y}O_2$, $LiMn_2O_4$ and $LiFePO_4$	Basic medium +S powder 16 g/L +pyrite 16 g/L (1:1 ratio)	30°C	1.0% (w/v)	1.5	One step	98	Ni 97, Co 96, Mn 90	Xin et al. (2016)
Acidothiobacillus ferrooxidans and *Acidothiobacillus thiooxidans*	Cathode powder	K_2HPO_4 0.1 g/L, $(NH_4)_2SO_4$ 2.0 g/L, KCl 0.1 g/L, $MgSO_4{\cdot}7H_2O$ 4.0 g/L, $FeSO_4{\cdot}7H_2O$ 44.2 g/L, S powder 4 g/L	30°C	1% (w/v)	1.5	One step	80	Co 67	Sedlakova-Kadukova et al. (2016)
Acidothiobacillus thiooxidans (isolated)	$LiNi_xCo_yMn_{1-x-y}O_2$	Basic medium +S powder 16 g/L +pyrite 16 g/L	30°C	1% (w/v)	1.0	One step	85	Mn 19, Co 10, Ni 10	Xin et al. (2016)

(*Continued*)

TABLE 4.1 (*Continued*)
Review Summary of Bioleaching Recovery of Lithium from Li-Ion Batteries

Microorganisms	Metal Sources	Nutrient Media	Temperature	Pulp Density w/v	pH	Method	Lithium Bioleaching Efficiency %	Other Metals Efficiency %	References
Acidothiobacillus ferrooxidans (isolated)	$LiNi_xCo_yMn_{1-x-y}O_2$	Single basic Medium +S powder 16 g/L +pyrite 16 g/L	30°C	1% (w/v)	1.0	One step	31	Mn 42,Co 23, Ni 23	Xin et al. (2016)
Aspergillus niger (PTCC 5210)	Cathodes and anodes powder	Sucrose medium	30°C	1.0 (w/v)	6	Spent-medium	95	Co 45, Cu 100, Mn 70, Ni 38	Horeh et al. (2016)
Acidothiobacillus ferrooxidans and *Acidothiobacillus thiooxidans*	spent LIBs	–	30°C	10 g/L	1.5	–	80	Co 67	Sedlakova-Kadukova et al. (2016)
Alicyclobacillus spp. and *Sulfobacillus* spp.	Electrodes powder	Basic medium +S powder 16 g/L +pyrite 16 g/L	35°C	2% (w/v)	1.0	Two-step	89	Co 72	Niu et al. (2014)
Acidothiobacillus ferrooxidans	Cathode material powder	9 K medium $+FeSO4{\cdot}7H_2O$ 444.8 g/L	35°C	1% (w/v)	2.0	Two-step	–	98.4 Co	Zeng et al. (2013)
Acidothiobacillus ferrooxidans (isolated)	Spent LIBs	–	30°C	1% (s/v) bacteria inoculation: 5% (v/v)	1.5	–	–	Co 47.60	Li et al. (2013)

(*Continued*)

TABLE 4.1 (*Continued*)
Review Summary of Bioleaching Recovery of Lithium from Li-Ion Batteries

Microorganisms	Metal Sources	Nutrient Media	Temperature	Pulp Density w/v	pH	Method	Lithium Bioleaching Efficiency %	Other Metals Efficiency %	References
Acidothiobacillus ferrooxidans	$LiCoO_2$ powder	Modifed 9 K medium +$FeSO4 \cdot 7H_2O$ 44.8 g/L +Copper 0.75 g/L	35°C	1% (w/v)	2.0	One step	–	99.9 Co after 6 days	Zeng et al. (2012)
Acidothiobacillus ferrooxidans (isolated)	Spent LIBs	–	30°C	1% (s/v)		–	–	Co 99.90	Zeng et al. (2012)
Mixed culture of sulfur-oxidizing and iron-oxidizing bacteria	Electrodes powder	9 K medium +S+FeS2	30°C	1% (w/v)	1.0	Two-step	80	Co 90	Xin et al. (2009)
Acidothiobacillus ferrooxidans (ATCC19859)	Cathode waste materials of LIBs	9 k medium +S power +$Fe2^+$ ion 3 g/L	30°C	0.5% (w/v)	2.5	One step	0.5	Co 65	Mishra et al. (2008)

microorganism being used determine the pH maintained during leaching (Patel et al., 2015). The acidophiles like *Aci. thiooxidans* and *Aci. ferrooxidans* require acidic environment for development and bioleaching process with a pH value range of 2.0–2.5, whereas fungi like Penicillium requires a pH range of 2.0–8.0 (Gu et al., 2018). Wang et al. (2014) investigated the progress of a combination of cultures on *Aci. thiooxidans* and *Aci. ferrooxidans* at multiple pH levels and found that the ideal range for pH is between 1.8 and 2.5, with the highest concentration of microbial cells attained at pH 2. Fungi, on the other hand, have a significantly wider optimal range of pH for proliferation and biological leaching activity, including Penicillium, with an ideal pH level of 2–8 (Srichandan et al., 2019). Park et al. (2014) examined the influence of pH on biological leaching by conducting arsenic bioleaching by *Aci. ferrooxidans* at the following pH: 2.2, 2, and 1.8, keeping all other parameters like pulp density constant throughout the experiments. It was shown that arsenic leaching at pH of 2.2 remained substantially lower than recovery at other pH values. Further investigation revealed the production of jarosite at pH 2.2, resulting in ineffective arsenic leaching. Horeh et al. (2016) looked at pH variations caused by *Asp. niger* biological leaching of LIBs. In the procedure, the pH decreased considerably, suggesting that the fungus had entered the stage of lag and were producing acidic compounds. After a while, the acids and H ions were consumed because of metal dissolution and metal complex formation, and the pH increased (Bahaloo-Horeh & Mousavi, 2017). LIBs, the most prevalent second-generation source of lithium ions, are highly acidic, with a pH of 9.6. As a result, lower pH media are recommended for lithium biological leaching from LIBs. In this context, Boxall et al. (2018) performed indirect biological leaching for lithium extraction from LIBs, at a relatively low pH of 1.8. To conclude, pH impacts both bacterial growth and metal breakup, selecting the optimum pH for a biological leaching procedure is essential. As a result, in addition to finding the appropriate pH for effective metal extraction process, a researcher ought to investigate the ideal pH for the development and function of the employed microorganisms. Furthermore, in biological leaching tests, tracking pH fluctuations during the entire procedure is equally critical as regulating the pH.

4.7.2 Temperature

Temperature plays a primary role in the bioleaching process, as it influences growth, metabolism of the microorganisms, and bioleaching reaction (Rasoulnia et al., 2021). The performance of lithium bioleaching from LIBs at different temperatures was evaluated to know the effect of temperature (Niu et al., 2014). Wang et al. (2014) investigated the impact of temperature on the development of a hybrid culture of mesophiles such as *Aci. thiooxidans* and *Aci. ferrooxidans* in the range of 25°C–45°C. It has been demonstrated that raising the ambient temperature from 25°C to 30°C accelerated the development of the germs. Nonetheless, there was a significant drop in cell numbers at elevated temperatures. On the contrary, thermophilic bacteria have an ideal temperature range of 40°C–45°C. This makes them perfect for bioleaching methods that include elevated temperatures and therefore better reaction efficiency and recovery of metal yield (Bahaloo-Horeh et al., 2019; Priya & Hait, 2017).

The ideal range of temperatures for the development and functioning of a fungus is typically between 20°C and 40°C (Habibi et al., 2020). Enzymes and proteins participating in the microbial processes begin to get damaged at temperatures above the ideal one (Aghababaie et al., 2016, 2020). The most frequent temperature in the biological leaching research on lithium extraction using LIBs was 30°C, which offered the best circumstances for microbes to develop and become active. Lithium was recovered from LIBs using a variety of bioleaching microorganisms at 30°C, including the fungus *Asp. niger* (Bahaloo-Horeh & Mousavi, 2017), *Aci. thiooxidans* (Naseri et al., 2019a), a combination culture of iron- and sulfur-oxidizing bacteria (Xin et al., 2009), and *Lep. ferriphilum* (Xin et al., 2016). Another study showed that a better leaching effect was achieved with cell-free media after separating microorganisms in the logarithmic growth phase from the system. In this way, both *Asp. niger* and *Aci. thiooxidans* resulted in an increase in lithium leaching rate of close to 10% (Bahaloo-Horeh & Mousavi, 2017; Biswal et al., 2018). Choosing the ideal temperature in each research is ultimately a trade-off, since it affects both the total cost of the system and the efficacy of the microbes in the recovery of metals (Rasoulnia et al., 2021). This means the primary factors that one should consider about are the various types of microbes and the range of temperatures at which they thrive.

4.7.3 Pulp Density

The crucial factor considered in the bioleaching process is pulp density (ratio of solid mass to liquid volume). It has impact on leaching process parameters, such as industrial reactor design, composition, process expense, and subsequent separation procedures (Petrus et al., 2018). When the pulp density increased from 1% to 4%, more harmful and hazardous chemicals were present in the medium, which inhibited the development of microorganisms (Niu et al., 2014). The employed pulp density must be equal to the maximum quantity feasible to improve the biological leaching process's profitability. Thus, it is crucial to investigate how higher pulp densities affect the microbial communities (Gu et al., 2018). Wang et al. (2014) examined the impact of varying pulp densities on a mixed culture of mesophiles' ability to extract copper sulfide and chalcopyrite from ore. The highest recovering rate for copper sulfide was obtained at an average pulp density equal to 5%, with a significant decrease occurring at greater pulp densities. Nevertheless, the maximum yield of extraction was identified at a pulp density lower than 10%, since the population of bacteria could not get sufficient energy from tiny amounts of chalcopyrite. The pre-adapted bacteria were employed in a related investigation to assess the impact of varying pulp densities on copper biological leaching from chalcopyrite (Nguyen et al., 2018). A combined culture of bacteria that was previously subjected to the sludge particles for six months was employed by Zhang et al. (2020). To determine the optimum pulp density, five distinct pulp densities that ranged from 5% to 20% were incorporated in the bioleaching process. Ultimately, more than 90% of the metals were retrieved with an enhanced pulp density of 15%, which was thought to be the ideal pulp density, under the ideal pH and temperature conditions (Zhang et al., 2020). The growth and efficiency of microorganisms are adversely affected by high pulp densities. The efficiency of bioleaching is typically reduced by increased pulp density as a result of increase in the toxicity due to the presence of heavy metals from sources. Pulp densities 1%, 2%, and 4% on lithium and cobalt recovery from lithium-ion batteries, with a combined culture of sulfur- and iron-oxidizing bacteria, show a significant decrease in the leaching efficiencies of both metals (Niu et al., 2014). In a separate investigation on Li biological leaching from LIB, Naseri et al. (2019a) conducted the recovery procedure at different pulp densities ranging from 10 to 50 g/L. The ideal pulp density was found to be 30 g/L, at which point the recovery yield of lithium was 99%.

4.7.4 Medium Composition

According to Bahaloo-Horeh et al. (2019), the development and activity of the biological leaching microbes are significantly influenced by the culture medium and cell nutrients. The medium composition mainly depends on the leaching mechanism and the type of used microorganisms (Potysz et al., 2018). For microorganisms to grow, proliferate, and function properly, a culture medium must have important chemical components. Some of the elements required in adequate media include nitrogen, phosphorus, and magnesium. The 9K medium containing phosphate, potassium, and ammonium is one of the most often used media in the bioleaching processes. Due to the low levels of iron and sulfur and a shortage carbon-based components, bioleaching of e-waste like LIBs can be hard (Bahaloo-Horeh et al., 2019; Niu et al., 2014). As an electron donor, S^0 is essential for sulfur metabolism because it provides the bioleaching microorganisms with energy. The activity and efficiency of solubilizing metal gets improved at high levels of S^0. However, surplus amount can harm the bioleaching process (Gu et al., 2018).

4.7.5 Cell Nutrients

The bioleaching of e-waste such as LIBs can be difficult due to lower levels of iron and sulfur and the absence of carbon-based materials. The nutritional medium in the presence of soluble S^0, Fe^{2+} along with additional nutrients becomes necessary for the development and bioleaching activity of used microorganisms. Bioleaching of Li with a rich nutrient medium and low nutrient medium was conducted, obtaining a recovery yield of 80% and a bioleaching efficiency of 35% for Li, which is very less for the low nutrient medium (Sedlakova-Kadukova et al., 2016).

4.7.6 Aeration

The selection of aeration between aerobic and anaerobic condition plays a significant role in the bioleaching process. In general, aerobic conditions are preferred by most bioleaching microorganisms for their growth and activity. In addition to supplying the microbes with sufficient carbon source, aeration provides them with an adequate electron acceptor (Brunnström et al., 2017). As the solubility of O_2 and CO_2 in water is limited, high-rate aeration is necessary to provide adequate levels of the gases for microbial activity and metabolism. Some kinds of bioleaching bacteria such as *Aci. ferrooxidans* exhibit growth and activity under anaerobic environments (Zhao et al., 2019). They generate energy by oxidizing Fe^{2+} or sulfur compounds, but they can also activate anaerobic metabolism and change Fe^{3+} ion to Fe^{2+} ion.

4.7.7 Types of Microorganisms

The bioleaching process involves employing microorganisms to remove the metal from the ore using various strains of fungal and bacteria species. To recover lithium from lepidolite, the bioleaching processes of *Rho. mucilaginosa, Asp. niger, Aci. Thiooxidans,* and *Aci. ferrooxidans* were compared. The results showed that bacteria extract the greatest amount of lithium, but the process is time-consuming and there is no evidence of lithium in the abiotic control from lepidolite (Sedlakova-Kadukova et al., 2020). Reichel et al. (2017) found 11% lithium recovery using a consortium of sulfur-oxidizing bacteria, but they were unable to find a plausible explanation for bioleaching efficiency to be higher than chemical method. Lithium bioleaching from spodumene was carried out using *Asp. niger,* with a maximum Li recovery of 0.75 mg L^{-1} with no bioaccumulation (Rezza et al., 1997). Hartono et al. (2017) found a prospective bacterium strain that recovers Li up to 63% of its original state after 15 days. A benefit of employing fungi and yeast for recovering Li via leachate is their strong inclination to gather Li (Sedlakova-Kadukova et al., 2020).

4.7.8 Redox Potential

The capability of a soil solution to absorb or release electrons is measured by its redox potential. Numerous metal ions in the soil have various redox states. Mobility and solubility of metals are significantly influenced by their oxidation states. Redox potential is a vital factor in the process of bioleaching, since it largely influences the types of metabolic bacterial communities. Fe^{2+} is bio-oxidized to Fe^{3+}, thereby increasing the medium's redox potential. Higher amounts of Fe imply high oxidation–reduction potentials, which are beneficial for metal leaching (Billy et al., 2018). High redox potential is encouraged for the efficient utilization of sulfur throughout the bioleaching process. Restricting the oxygen supply or introducing chemical reductants, however, can manage the redox potential. In this regard, pyrite will bioleach better in a high redox potential environment and chalcopyrite will bioleach better in a low redox potential environment (Gericke et al., 2010). A direct oxidation of the metal substrate itself or an indirect oxidation mediated by a redox pair, like Fe^{3+} ions and the element iron found in the metallic substrate, are the two processes that have been postulated for the metal bio-oxidation conducted by the self-sustaining bacteria (Mishra et al., 2008). The process demonstrates that the metal dissolves by an integrated acidolysis/redoxolysis mechanism; its oxidation rate, pH, and ferric iron content all affect the kinetics (Niu et al., 2014). The reaction suggests that a combined acidolysis/redoxolysis mechanism dissolves the metal; its kinetics are influenced by the pH, ferrous iron content, and rate of oxidation (Xin et al., 2016). There was no mention of metal interaction throughout the disintegration method using heterotrophic microbes and fungus-motivated biological leaching. During their processes of metabolism, these microorganisms release protons and organic acids, which aid in oxidizing metals from LIB with the assistance of the O_2/H_2O redox combination. The protonated metals then form complexes with the naturally occurring acids (Biswal et al., 2018).

4.7.9 Surfactants and Natural Extracts

Surfactants have drawn attention for their capacity to remove contaminants from a variety of mediums due to their contaminant removal efficiency, eco-friendliness, versatility, and green chemistry background (Kumar et al., 2021). *Bacillus cereus* and *Bacillus subtilis* are examples of heterotrophic bacteria that may create surfactants, which have been demonstrated to extract metals (such as Cr), from used electronics boards(Karwowska et al., 2014). The surfactants' unique structure demonstrates contaminants with hydrophilic and hydrophobic compounds (He et al., 2016). The reduction in oxygen transfer and surface tension due to surfactants can accelerate the leaching rate (Aioub et al., 2019). The outer membrane of bacteria and mineral surface properties can be altered to improve the interfacial action between minerals and microorganisms as the process depends on interface interactions at the multiphase level (Fang et al., 2014). Surfactants have significant solubilization activity on particular pollutants and become harmful to the microbe community, even though surfactant adsorption is said to have limits (Kumar et al., 2021). This aids in harmful metal removal through complexation and ion exchange processes linked to surfactants. Biosurfactants promote the pollutants' microbial breakdown, which speeds up the removal procedure in addition to the beneficial effects of solubilizing and desorbing contaminants. Additional to the surfactants, chelating agents, ligand ions, and organic solvents are used to improve pollutant removal. One study showed that 0.5% (w/v) of sulfur is ideal for both metallic dissolution and sediment destabilization in the biological leaching procedure. Higher sulfur concentrations (more than 0.5% w/v) prohibit elemental sulfur from oxidizing (Chen & Lin, 2001).

4.7.10 Substrate Concentration

A quick and low-cost way for metal extraction of industrial waste is bioleaching, which can treat or remove heavy metals from contaminated sediments. In sludge with a sucrose level of 10 g/L, heavy metals are not eliminated. Due to a lack of sugar content, the fungus (*Aspergillus niger*) had inadequate growth during the process. It was found that Mn > Zn > Ni > Pb was the order of efficiency in removing metal from sludge under bioleaching conditions using fungi. The experiment's optimal sucrose concentration was stated to be 100 g L^{-1} (the highest). The highest metal removal rates for Ni, Mn, and Zn during the process were 24, 75, and 48%, respectively. The effectiveness of metal elimination gets increased with higher sucrose concentration (Chen & Wang, 2019). The leaching efficiency of zinc and nickel was found to be 95% and 94%, respectively, with a concentration of citric acid as 5,340 mg/L (Dacera & Babel, 2008). In addition, it was found that using 400 g/L sucrose, Zn and Pb were removed with an efficiency of 88% and 82%, respectively (Xu & Feng, 2016).

4.8 HYBRID BIOLEACHING

Bioleaching is a less expensive process but needs more process time because of complicated anabolic and catabolic reactions in the cells of microbes. Though chemical leaching is fast and efficient for lithium recovery, it causes environmental issues due to the harmful chemical agents used during extraction (Pant et al., 2012, 2018). To overcome these disadvantages of both the methods, a new technique was proposed as hybrid bioleaching that incorporates chemical and biological leaching methods together. In hybrid bioleaching, the combination of ligand with microorganism is used for lithium extraction from the lithium-ion batteries (LIB). The ligand and microorganism can be recovered and reused further after the hybrid bioleaching process. Hybrid bioleaching increases the rate of metal recovery and yield in an ecological and sustainable way. The schematic of hybrid bioleaching technique is shown in Figure 4.5. Certain ligands are used in hybrid leaching procedures in conjunction with bacteria or fungi to extract metals. Using the compound ethylene diamine tetraacetic acid and the bacteria *Serratia plymuthica*, metals were hybrid recovered from waste cathode ray tubes (Pant et al., 2018). Dolker and Pant (2019) used hybrid method that combine chemical and biological procedure for the extraction of lithium from waste LIB by integrating *Lysinibacillus sphaericus* and citric acid and that offered an important prospect for metal recovery. Initially, the treatment with citric acid separated copper (Cu) and aluminum (Al) from the Li-ion battery powder while parting lithium (Li) and cobalt (Co) in the remaining concentrate. The quantity of lithium (Li) leached with *Lysinibacillus sphaericus* and separate citric acid was 201.2 and 56.2 ppm, respectively, and significantly enhanced throughout the hybrid process. The hybrid technique increased Li leaching by 25%. Based on the comparison studies and results, the biological metal extraction or bioleaching process was proposed as the most inexpensive methods for lithium metal extraction from LIB; however the selection of an appropriate method depends on the metal recovery efficiency, ecological issues, and the potential profits. In several disciplines, including bio-extraction and

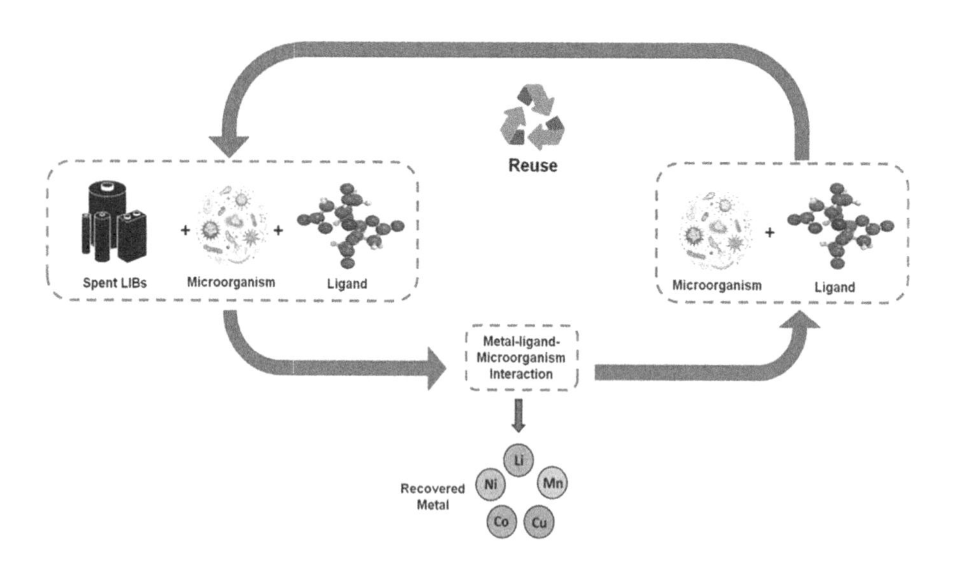

FIGURE 4.5 Schematic diagram of hybrid bioleaching process. Reproduced with copyright permission from Pant et al. (2012).

bio-sensing, the incorporation of nanomaterials into traditional systems has led to improved performances. In order to improve the Li extraction rate from waste LIB, hybridizing of nanomaterials with bioleaching microorganisms appears to be an intriguing field of study that requires investigation by researchers.

4.9 MATERIAL YIELD ENHANCEMENT AND INDUSTRIAL BIOLEACHING OF LI-ION

In the bioleaching process, yield refers to the efficiency of the process for recovering lithium. The efficiency can be formulated differently for different recovery methods using the same initial theory. The yield percentage for leaching by hydrometallurgy can be calculated using Eq. (4.13) (Chen & Ho, 2018)

$$Y_p = \frac{m_1}{m_2} \times 100\% \tag{4.13}$$

where Y_p – the yield percentage, m_1 – the quantity of metal leached, and m_2 – the quantity of raw resource metal. Several researches indicated that the yield for bioleaching, as in Eq. (4.14) (Bahaloo-Horeh & Mousavi, 2017),

$$\text{Recovery of material} = \frac{v_s \times X_s}{X_f \times m_f} \times 100\% \tag{4.14}$$

where v_s – the volume of the leach solution used for bioleaching (l), X_s – the concentration of metal in leached liquid ($mg \bullet L^{-1}$), X_f – the metal content of the resource powder ($mg \bullet g^{-1}$), and m_f – the total mass of the resource powder (kg). Additionally, the research done so far has optimized, the parameters discussed above such as pH, type of microorganisms, pulp density, medium composition, and methods of bioleaching, which has no direct influence on the material percentage of recovery.

Separation of lithium from lithium-ion batteries (LIB) by industries is observed in many countries like France, Germany, Canada, USA, China, and Switzerland (Swain, 2017). The method used for recycling includes conventional methods such as hydrometallurgy, pyrometallurgy, and other chemical techniques, and its combination excludes the bioleaching process. Currently, the industries use the above techniques and recycle only 3% of lithium-ion batteries (Wang et al., 2014). The metal recovery industry in the USA uses cryogenic treatment to reduce the reactivity of LIBs, as they are reactive to atmospheric conditions. Then the batteries are processed and the powder obtained is added to the high pH medium along with lithium hydroxide. This method gives 97% of lithium recovery and is one of the best methods for LIB recycling (Anderson, 2018).

In recent discovery, the traces of lithium-rich bauxite were found in Jammu and Kashmir, which is the primary resource. However, the excavation and extraction are tedious processes resulting in environmental impacts. Even the lithium extraction process requires a large land area, resulting in the elimination of native vegetation from the Himalayan regions. Converting waste to useful resource is one of the best ways to reduce world trash and to reach a sustainable goal. So, recovering lithium from the secondary resources like waste lithium-ion battery is a significantly important.

Numerous research and development work have been done to separate lithium from LIB by applying the process of hydrometallurgy, which undergoes leaching with acid followed by lithium carbonate precipitation. Though, bioleaching has numerous benefits (Bahaloo-Horeh et al., 2018; Borja et al., 2016), there are also few studies on the applications of bioleaching in industry for recovery from waste LIBs. The expanding of bioleaching method for metal recovery from LIBs and other electronic waste is lowered due to the factors such as slow growth of microorganisms, which consumes much time, density of waste, danger to ecosystem, and small treatment capacity. But still a lot of research is needed on the extraction of lithium by bioleaching from lithium-ion battery waste to reduce the landfill use and depletion of primary sources.

4.10 CONCLUSIONS

The extraction of lithium metal from waste lithium-ion batteries (LIB) by bioleaching method is a sustainable process that gives cent percent recovery. Due to the extensive rise in the exploitation of lithium-ion batteries (LIB) for e-vehicles, energy storage devices, mobile phones, and hand--operated electronic gadgets, the need for lithium metal also rises. To meet the demand, depending only on the primary sources will not be a better option, as many primary sources are getting depleted and some countries have no resources. Therefore, the recovery of lithium by bioleaching processes from used LIBs, which is the secondary source, is very essential. Though there are many other chemical and thermal methods for metal recovery, applying the bioleaching technique will lead to environmental benefits by conserving natural resources, avoiding harmful chemicals in the atmosphere, and developing mining strategies for better recovery. Since the bioleaching of LIBs is time-consuming, which is the only disadvantage; optimizing factors such as temperature, type of microorganisms, pH and nutrition during the process is essential and yet to be explored more in future. Integrating and remodeling the leaching methods and processes involved need to be focused by researchers and scientists to improve the recovery rate of lithium from LIBs, which may pave the way for industries to adapt to the bioleaching process on a large scale. Adapting the bioleaching technique by industries will reduce energy consumption, import cost of lithium, and e-waste landfill. To maximize the bioleaching process in industries, the separation of electrolyte, binders, and anode from waste LIBs needs to be concentrated on more during recycling. There are many scopes in the future for the collection, storage, dismantling, segregation, processing, bioleaching, manufacturing, and export of noble metals from LIBs for the upcoming generation.

REFERENCES

Aghababaie, M., Beheshti, M., Razmjou, A., & Bordbar, A.-K. (2016). Covalent immobilization of Candida rugosa lipase on a novel functionalized $Fe_3O_4@SiO_2$ dip-coated nanocomposite membrane. *Food and Bioproducts Processing*, *100*, 351–360. https://doi.org/10.1016/j.fbp.2016.07.016

Aghababaie, M., Beheshti, M., Razmjou, A., & Bordbar, A.-K. (2020). Enzymatic biodiesel production from crude *Eruca sativa* oil using *Candida rugosa* lipase in a solvent-free system using response surface methodology. *Biofuels*, *11*(1), 93–99. https://doi.org/10.1080/17597269.2017.1345359

Aioub, A. A. A., Li, Y., Qie, X., Zhang, X., & Hu, Z. (2019). Reduction of soil contamination by cypermethrin residues using phytoremediation with Plantago major and some surfactants. *Environmental Sciences Europe*, *31*(1), 26. https://doi.org/10.1186/s12302-019-0210-4

Alavi, N., Partovi, K., Majlessi, M., Rashidi, M., & Alimohammadi, M. (2021). Bioleaching of metals from cellphones batteries by a co-fungus medium in presence of carbon materials. *Bioresource Technology Reports*, *15*, 100768. https://doi.org/10.1016/j.biteb.2021.100768

Alipanah, M., Reed, D., Thompson, V., Fujita, Y., & Jin, H. (2023). Sustainable bioleaching of lithium-ion batteries for critical materials recovery. *Journal of Cleaner Production*, *382*, 135274. https://doi.org/10.1016/j.jclepro.2022.135274

Alipanah, M., Saha, A. K., Vahidi, E., & Jin, H. (2021). Value recovery from spent lithium-ion batteries: a review on technologies, environmental impacts, economics, and supply chain. *Clean Technologies and Recycling*, *1*(2), 152–184. https://doi.org/10.3934/ctr.2021008

Anderson, C. G. (2018). The production of critical materials as by products. *Aspects in Mining & Mineral Science*, *2*(2), 532. https://doi.org/10.31031/AMMS.2018.02.000532

Bahaloo-Horeh, N., & Mousavi, S. M. (2017). Enhanced recovery of valuable metals from spent lithium-ion batteries through optimization of organic acids produced by Aspergillus niger. *Waste Management*, *60*, 666–679. https://doi.org/10.1016/j.wasman.2016.10.034

Bahaloo-Horeh, N., Mousavi, S. M., & Baniasadi, M. (2018). Use of adapted metal tolerant Aspergillus niger to enhance bioleaching efficiency of valuable metals from spent lithium-ion mobile phone batteries. *Journal of Cleaner Production*, *197*, 1546–1557. https://doi.org/10.1016/j.jclepro.2018.06.299

Bahaloo-Horeh, N., Vakilchap, F., & Mousavi, S. M. (2019). Bio-hydrometallurgical methods for recycling spent lithium-ion batteries. *Recycling of Spent Lithium-Ion Batteries*, *7*, 161–197. https://doi.org/10.1007/978-3-030-31834-5_7

Billy, E., Joulié, M., Laucournet, R., Boulineau, A., Vito, E. De, & Meyer, D. (2018). Dissolution mechanisms of $LiNi_{1/3}Mn_{1/3}Co_{1/3}O_2$ positive electrode material from lithium-ion batteries in acid solution. *ACS Applied Materials & Interfaces*, *10*(19), 16424–16435. https://doi.org/10.1021/acsami.8b01352

Biswal, B. K., & Balasubramanian, R. (2023). Recovery of valuable metals from spent lithium-ion batteries using microbial agents for bioleaching: a review. *Frontiers in Microbiology*, *14*, 1197081. https://doi.org/10.3389/fmicb.2023.1197081

Biswal, B. K., Jadhav, U. U., Madhaiyan, M., Ji, L., Yang, E.-H., & Cao, B. (2018). Biological leaching and chemical precipitation methods for recovery of Co and Li from spent lithium-ion batteries. *ACS Sustainable Chemistry & Engineering*, *6*(9), 12343–12352. https://doi.org/10.1021/acssuschemeng.8b02810

Borja, D., Nguyen, K., Silva, R., Park, J., Gupta, V., Han, Y., Lee, Y., & Kim, H. (2016). Experiences and future challenges of bioleaching research in South Korea. *Minerals*, *6*(4), 128. https://doi.org/10.3390/min6040128

Boxall, N. J., Cheng, K. Y., Bruckard, W., & Kaksonen, A. H. (2018). Application of indirect non-contact bioleaching for extracting metals from waste lithium-ion batteries. *Journal of Hazardous Materials*, *360*, 504–511. https://doi.org/10.1016/j.jhazmat.2018.08.024

Brunnström, H., Johansson, A., Westbom-Fremer, S., Backman, M., Djureinovic, D., Patthey, A., Isaksson-Mettävainio, M., Gulyas, M., & Micke, P. (2017). PD-L1 immunohistochemistry in clinical diagnostics of lung cancer: inter-pathologist variability is higher than assay variability. *Modern Pathology*, *30*(10), 1411–1421. https://doi.org/10.1038/modpathol.2017.59

Chen, S.-Y., & Lin, J.-G. (2001). Effect of substrate concentration on bioleaching of metal-contaminated sediment. *Journal of Hazardous Materials*, *82*(1), 77–89. https://doi.org/10.1016/S0304-3894(00)00357-5

Chen, S.-Y., & Wang, S.-Y. (2019). Effects of solid content and substrate concentration on bioleaching of heavy metals from sewage sludge using aspergillus niger. *Metals*, *9*(9), 994. https://doi.org/10.3390/met9090994

Chen, W.-S., & Ho, H.-J. (2018). Recovery of valuable metals from lithium-ion batteries NMC cathode waste materials by hydrometallurgical methods. *Metals*, *8*(5), 321. https://doi.org/10.3390/met8050321

Cockell, C. S., Santomartino, R., Finster, K., Waajen, A. C., Eades, L. J., Moeller, R., Rettberg, P., Fuchs, F. M., Van Houdt, R., Leys, N., Coninx, I., Hatton, J., Parmitano, L., Krause, J., Koehler, A., Caplin, N., Zuijderduijn, L., Mariani, A., Pellari, S. S., & Demets, R. (2020). Space station biomining experiment demonstrates rare earth element extraction in microgravity and Mars gravity. *Nature Communications*, *11*(1), 5523. https://doi.org/10.1038/s41467-020-19276-w

Dacera, D. D. M., & Babel, S. (2008). Removal of heavy metals from contaminated sewage sludge using Aspergillus niger fermented raw liquid from pineapple wastes. *Bioresource Technology*, *99*(6), 1682–1689. https://doi.org/10.1016/j.biortech.2007.04.002

Do, M. P., Jegan Roy, J., Cao, B., & Srinivasan, M. (2022). Green closed-loop cathode regeneration from spent NMC-based lithium-ion batteries through bioleaching. *ACS Sustainable Chemistry & Engineering*, *10*(8), 2634–2644. https://doi.org/10.1021/acssuschemeng.1c06885

Dolker, T., & Pant, D. (2019). Chemical-biological hybrid systems for the metal recovery from waste lithium ion battery. *Journal of Environmental Management*, *248*, 109270. https://doi.org/10.1016/j.jenvman.2019.109270

Dutta, D., Rautela, R., Gujjala, L. K. S., Kundu, D., Sharma, P., Tembhare, M., & Kumar, S. (2023). A review on recovery processes of metals from E-waste: a green perspective. *Science of The Total Environment*, *859*, 160391. https://doi.org/10.1016/j.scitotenv.2022.160391

Fang, F., Zhong, H., Jiang, F., Li, Z., Chen, Y., & Zhan, X. (2014). Influence of surfactants on bioleaching of arsenic-containing gold concentrate. *Journal of Central South University*, *21*(10), 3963–3969. https://doi.org/10.1007/s11771-014-2384-7

Faraji, F., Golmohammadzadeh, R., Rashchi, F., & Alimardani, N. (2018). Fungal bioleaching of WPCBs using Aspergillus niger: observation, optimization and kinetics. *Journal of Environmental Management*, *217*, 775–787. https://doi.org/10.1016/j.jenvman.2018.04.043

Flexer, V., Baspineiro, C. F., & Galli, C. I. (2018). Lithium recovery from brines: a vital raw material for green energies with a potential environmental impact in its mining and processing. *Science of The Total Environment*, *639*, 1188–1204. https://doi.org/10.1016/j.scitotenv.2018.05.223

Gericke, M., Govender, Y., & Pinches, A. (2010). Tank bioleaching of low-grade chalcopyrite concentrates using redox control. *Hydrometallurgy*, *104*(3–4), 414–419. https://doi.org/10.1016/j.hydromet.2010.02.024

Ghassa, S., Farzanegan, A., Gharabaghi, M., & Abdollahi, H. (2020). Novel bioleaching of waste lithium ion batteries by mixed moderate thermophilic microorganisms, using iron scrap as energy source and reducing agent. *Hydrometallurgy*, *197*, 105465. https://doi.org/10.1016/j.hydromet.2020.105465

Gholam-Abbas Nazri, G. P. (2009). *Lithium Batteries* (G. A. Nazri & G. Pistoia, Eds.). Springer, Berlin, Germany. https://doi.org/10.1007/978-0-387-92675-9

Golmohammadzadeh, R., Faraji, F., Jong, B., Pozo-Gonzalo, C., & Banerjee, P. C. (2022). Current challenges and future opportunities toward recycling of spent lithium-ion batteries. *Renewable and Sustainable Energy Reviews*, *159*, 112202. https://doi.org/10.1016/j.rser.2022.112202

Gu, T., Rastegar, S. O., Mousavi, S. M., Li, M., & Zhou, M. (2018). Advances in bioleaching for recovery of metals and bioremediation of fuel ash and sewage sludge. *Bioresource Technology*, *261*, 428–440. https://doi.org/10.1016/j.biortech.2018.04.033

Habibi, A., Shamshiri Kourdestani, S., & Hadadi, M. (2020). Biohydrometallurgy as an environmentally friendly approach in metals recovery from electrical waste: a review. *Waste Management & Research*, *38*(3), 232–244. https://doi.org/10.1177/0734242X19895321

Hariyadi, A., Masago, A. R., Febrianur, R., & Rahmawati, D. (2022). Optimization fungal leaching of cobalt and lithium from spent Li-ion batteries using waste spices candlenut. *Key Engineering Materials*, *938*, 177–182. https://doi.org/10.4028/p-lkr100

Hariyadi, A., Sholikah, U., Gotama, B., & Ghony, M. A. (2022). Biohydrometallurgy for cobalt recovery from spent Li-ion batteries using acidophilic bacteria isolated from acid mine drainage. *CHEMICA: Jurnal Teknik Kimia*, *9*(2), 88. https://doi.org/10.26555/chemica.v9i2.22328

Hartono, M., Astrayudha, M. A., Petrus, H. T. B. M., Budhijanto, W., & Sulistyo, H. (2017). Lithium recovery of spent lithium-ion battery using bioleaching from local sources microorganism. *Rasayan Journal of Chemistry*, *10*(3), 897–903. https://doi.org/10.7324/RJC.2017.1031767

He, Z., Liu, G., Yang, X., & Liu, W. (2016). A novel surfactant, N, N-diethyl-N′-cyclohexylthiourea: synthesis, flotation and adsorption on chalcopyrite. *Journal of Industrial and Engineering Chemistry*, *37*, 107–114. https://doi.org/10.1016/j.jiec.2016.03.013

Heydarian, A., Mousavi, S. M., Vakilchap, F., & Baniasadi, M. (2018). Application of a mixed culture of adapted acidophilic bacteria in two-step bioleaching of spent lithium-ion laptop batteries. *Journal of Power Sources*, *378*, 19–30. https://doi.org/10.1016/j.jpowsour.2017.12.009

Hoque, M. E., & Philip, O. J. (2011). Biotechnological recovery of heavy metals from secondary sources – an overview. *Materials Science and Engineering: C*, *31*(2), 57–66. https://doi.org/10.1016/j.msec.2010.09.019

Horeh, N. B., Mousavi, S. M., & Shojaosadati, S. A. (2016). Bioleaching of valuable metals from spent lithium-ion mobile phone batteries using Aspergillus niger. *Journal of Power Sources*, *320*, 257–266. https://doi.org/10.1016/j.jpowsour.2016.04.104

Huang, T., Liu, L., & Zhang, S. (2019). Recovery of cobalt, lithium, and manganese from the cathode active materials of spent lithium-ion batteries in a bio-electro-hydrometallurgical process. *Hydrometallurgy*, *188*, 101–111. https://doi.org/10.1016/j.hydromet.2019.06.011

Ilyas, S., Kim, M., & Lee, J. (2018). Integration of microbial and chemical processing for a sustainable metallurgy. *Journal of Chemical Technology & Biotechnology*, *93*(2), 320–332. https://doi.org/10.1002/jctb.5402

Jha, M. K., Kumari, A., Jha, A. K., Kumar, V., Hait, J., & Pandey, B. D. (2013). Recovery of lithium and cobalt from waste lithium ion batteries of mobile phone. *Waste Management*, *33*(9), 1890–1897. https://doi.org/10.1016/j.wasman.2013.05.008

Kaksonen, A. H., Boxall, N. J., Usher, K. M., Ucar, D., & Sahinkaya, E. (2017). *Biosolubilisation of Metals and Metalloids*. Springer, Switzerland (pp. 233–283). https://doi.org/10.1007/978-3-319-58622-9_8

Kang, D., Jacquiod, S., Herschend, J., Wei, S., Nesme, J., & Sørensen, S. J. (2020). Construction of simplified microbial consortia to degrade recalcitrant materials based on enrichment and dilution-to-extinction cultures. *Frontiers in Microbiology*, 10, 3010. https://doi.org/10.3389/fmicb.2019.03010

Karwowska, E., Andrzejewska-Morzuch, D., Łebkowska, M., Tabernacka, A., Wojtkowska, M., Telepko, A., & Konarzewska, A. (2014). Bioleaching of metals from printed circuit boards supported with surfactant-producing bacteria. *Journal of Hazardous Materials*, *264*, 203–210. https://doi.org/10.1016/j.jhazmat.2013.11.018

Kazemian, Z., Larypoor, M., & Marandi, R. (2020). Evaluation of myco-leaching potential of valuable metals from spent lithium battery by *Penicillium chrysogenum* and *Aspergillus niger*. *International Journal of Environmental Analytical Chemistry*, *103*(3), 514–527. https://doi.org/10.1080/03067319.2020.1861605

Kumar, M., Bolan, N. S., Hoang, S. A., Sawarkar, A. D., Jasemizad, T., Gao, B., Keerthanan, S., Padhye, L. P., Singh, L., Kumar, S., Vithanage, M., Li, Y., Zhang, M., Kirkham, M. B., Vinu, A., & Rinklebe, J. (2021). Remediation of soils and sediments polluted with polycyclic aromatic hydrocarbons: to immobilize, mobilize, or degrade? *Journal of Hazardous Materials*, *420*, 126534. https://doi.org/10.1016/j.jhazmat.2021.126534

Li, L., Zeng, G. S., Luo, S. L., Deng, X. R., & Xie, Q. J. (2013). Influences of solution pH and redox potential on the bioleaching of LiCoO2 from spent lithium-ion batteries. *Journal of the Korean Society for Applied Biological Chemistry*, *56*, 187–192.

Li, X., Mo, Y., Qing, W., Shao, S., Tang, C. Y., & Li, J. (2019). Membrane-based technologies for lithium recovery from water lithium resources: a review. *Journal of Membrane Science*, *591*, 117317. https://doi.org/10.1016/j.memsci.2019.117317

Liao, X., Ye, M., Liang, J., Guan, Z., Li, S., Deng, Y., Gan, Q., Liu, Z., Fang, X., & Sun, S. (2022). Feasibility of reduced iron species for promoting Li and Co recovery from spent LiCoO2 batteries using a mixed-culture bioleaching process. *Science of The Total Environment*, *830*, 154577. https://doi.org/10.1016/j.scitotenv.2022.154577

Liu, X., Liu, H., Wu, W., Zhang, X., Gu, T., Zhu, M., & Tan, W. (2020). Oxidative stress induced by metal ions in bioleaching of LiCoO2 by an acidophilic microbial consortium. *Frontiers in Microbiology*, *10*, 3058. https://doi.org/10.3389/fmicb.2019.03058

Loganathan, P., Naidu, G., & Vigneswaran, S. (2017). Mining valuable minerals from seawater: a critical review. *Environmental Science: Water Research & Technology*, *3*(1), 37–53. https://doi.org/10.1039/C6EW00268D

Luiz Henrique, F. (2023), Lithium consumption projections until 2050: power generation in focus - an in-depth analysis, optimized and sustainable lubrication, pizzani lubrificantes. https://www.linkedin.com/newsletters/ecoinspire-your-newsletter-7047977900976816128/

Miao, Y., Liu, L., Zhang, Y., Tan, Q., & Li, J. (2022). An overview of global power lithium-ion batteries and associated critical metal recycling. *Journal of Hazardous Materials*, *425*, 127900. https://doi.org/10.1016/j.jhazmat.2021.127900

Mishra, D., Kim, D.-J., Ralph, D. E., Ahn, J.-G., & Rhee, Y.-H. (2008). Bioleaching of metals from spent lithium ion secondary batteries using Acidithiobacillus ferrooxidans. *Waste Management*, *28*(2), 333–338. https://doi.org/10.1016/j.wasman.2007.01.010

Moazzam, P., Boroumand, Y., Rabiei, P., Baghbaderani, S. S., Mokarian, P., Mohagheghian, F., Mohammed, L. J., & Razmjou, A. (2021). Lithium bioleaching: an emerging approach for the recovery of Li from spent lithium ion batteries. *Chemosphere*, *277*, 130196. https://doi.org/10.1016/j.chemosphere.2021.130196

Naseri, T., Bahaloo-Horeh, N., & Mousavi, S. M. (2019a). Bacterial leaching as a green approach for typical metals recovery from end-of-life coin cells batteries. *Journal of Cleaner Production*, *220*, 483–492. https://doi.org/10.1016/j.jclepro.2019.02.177

Naseri, T., Bahaloo-Horeh, N., & Mousavi, S. M. (2019b). Environmentally friendly recovery of valuable metals from spent coin cells through two-step bioleaching using Acidithiobacillus thiooxidans. *Journal of Environmental Management*, *235*, 357–367. https://doi.org/10.1016/j.jenvman.2019.01.086

Naseri, T., Pourhossein, F., Mousavi, S. M., Kaksonen, A. H., & Kuchta, K. (2022). Manganese bioleaching: an emerging approach for manganese recovery from spent batteries. *Reviews in Environmental Science and Bio/Technology*, *21*(2), 447–468. https://doi.org/10.1007/s11157-022-09620-5

Nazerian, M., Bahaloo-Horeh, N., & Mousavi, S. M. (2023). Enhanced bioleaching of valuable metals from spent lithium-ion batteries using ultrasonic treatment. *Korean Journal of Chemical Engineering*, *40*(3), 584–593. https://doi.org/10.1007/s11814-022-1257-2

Nguyen, K. A., Borja, D., You, J., Hong, G., Jung, H., & Kim, H. (2018). Chalcopyrite bioleaching using adapted mesophilic microorganisms: effects of temperature, pulp density, and initial ferrous concentrations. *Materials Transactions*, *59*(11), 1860–1866. https://doi.org/10.2320/matertrans.M2018247

Niu, Z., Zou, Y., Xin, B., Chen, S., Liu, C., & Li, Y. (2014). Process controls for improving bioleaching performance of both Li and Co from spent lithium ion batteries at high pulp density and its thermodynamics and kinetics exploration. *Chemosphere*, *109*, 92–98. https://doi.org/10.1016/j.chemosphere.2014.02.059

Noruzi, F., Nasirpour, N., Vakilchap, F., & Mousavi, S. M. (2022). Complete bioleaching of Co and Ni from spent batteries by a novel silver ion catalyzed process. *Applied Microbiology and Biotechnology*, *106*(13–16), 5301–5316. https://doi.org/10.1007/s00253-022-12056-0

Pant, D., Giri, A., & Dhiman, V. (2018). Bioremediation techniques for E-waste management. *Waste Bioremediation*, *2018*, 105–125. https://doi.org/10.1007/978-981-10-7413-4_5

Pant, D., Joshi, D., Upreti, M. K., & Kotnala, R. K. (2012). Chemical and biological extraction of metals present in E waste: a hybrid technology. *Waste Management*, *32*(5), 979–990. https://doi.org/10.1016/j.wasman.2011.12.002

Park, J., Han, Y., Lee, E., Choi, U., Yoo, K., Song, Y., & Kim, H. (2014). Bioleaching of highly concentrated arsenic mine tailings by Acidithiobacillus ferrooxidans. *Separation and Purification Technology*, *133*, 291–296. https://doi.org/10.1016/j.seppur.2014.06.054

Patel, B., Tipre, D., & Dave, S. (2015). Biomining of base metals from sulphide minerals. In *Microbiology for Minerals, Metals, Materials and the Environment* (pp. 35–58). CRC Press, Boca Raton, FL. https://doi.org/10.1201/b18124-3

Petrus, H. B. T. M., Wanta, K. C., Setiawan, H., Perdana, I., & Astuti, W. (2018). Effect of pulp density and particle size on indirect bioleaching of Pomalaa nickel laterite using metabolic citric acid. *IOP Conference Series: Materials Science and Engineering*, *285*, 012004. https://doi.org/10.1088/1757-899X/285/1/012004

Potysz, A., van Hullebusch, E. D., & Kierczak, J. (2018). Perspectives regarding the use of metallurgical slags as secondary metal resources – a review of bioleaching approaches. *Journal of Environmental Management*, *219*, 138–152. https://doi.org/10.1016/j.jenvman.2018.04.083

Priya, A., & Hait, S. (2017). Comparative assessment of metallurgical recovery of metals from electronic waste with special emphasis on bioleaching. *Environmental Science and Pollution Research*, *24*(8), 6989–7008. https://doi.org/10.1007/s11356-016-8313-6

Putra, R. A., Al Fajri, I., & Hariyadi, A. (2022). Metal bioleaching of used lithium-ion battery using *Acidophilic ferrooxidans* isolated from acid mine drainage. *Key Engineering Materials*, *937*, 193–200. https://doi.org/10.4028/p-sd859o

Rasoulnia, P., Barthen, R., & Lakaniemi, A.-M. (2021). A critical review of bioleaching of rare earth elements: the mechanisms and effect of process parameters. *Critical Reviews in Environmental Science and Technology*, *51*(4), 378–427. https://doi.org/10.1080/10643389.2020.1727718

Rasoulnia, P., Mousavi, S. M., Rastegar, S. O., & Azargoshasb, H. (2016). Fungal leaching of valuable metals from a power plant residual ash using Penicillium simplicissimum : evaluation of thermal pretreatment and different bioleaching methods. *Waste Management*, *52*, 309–317. https://doi.org/10.1016/j.wasman.2016.04.004

Reichel, S., Aubel, T., Patzig, A., Janneck, E., & Martin, M. (2017). Lithium recovery from lithium-containing micas using sulfur oxidizing microorganisms. *Minerals Engineering*, *106*, 18–21. https://doi.org/10.1016/j.mineng.2017.02.012

Rezza, I., Salinas, E., Calvente, V., Benuzzi, D., & de Tosetti, M. I. S. (1997). Extraction of lithium from spodumene by bioleaching. *Letters in Applied Microbiology*, *25*(3), 172–176. https://doi.org/10.1046/j.1472-765X.1997.00199.x

Roy, J. J., Cao, B., & Madhavi, S. (2021). A review on the recycling of spent lithium-ion batteries (LIBs) by the bioleaching approach. *Chemosphere*, *282*, 130944. https://doi.org/10.1016/j.chemosphere.2021.130944

Roy, J. J., Srinivasan, M., & Cao, B. (2021). Bioleaching as an eco-friendly approach for metal recovery from spent NMC-based lithium-ion batteries at a high pulp density. *ACS Sustainable Chemistry & Engineering*, *9*(8), 3060–3069. https://doi.org/10.1021/acssuschemeng.0c06573

Sand, W., Gehrke, T., Jozsa, P.-G., & Schippers, A. (2001). (Bio)chemistry of bacterial leaching—direct vs. indirect bioleaching. *Hydrometallurgy*, *59*(2–3), 159–175. https://doi.org/10.1016/S0304-386X(00)00180-8

Sedlakova-Kadukova, J., Marcincakova, R., Luptakova, A., Vojtko, M., Fujda, M., & Pristas, P. (2020). Comparison of three different bioleaching systems for Li recovery from lepidolite. *Scientific Reports*, *10*(1), 14594. https://doi.org/10.1038/s41598-020-71596-5

Sedlakova-Kadukova, J., Mražíková, A., Velgosová, O., & Luptakova, A. (2016). Metal bioleaching from spent lithium-ion batteries using acidophilic bacterial strains. *Journal of the Polish Mineral Engineering Society*, *17*(1), 117–120. https://www.researchgate.net/publication/311433998

Sethurajan, M., & Gaydardzhiev, S. (2021). Bioprocessing of spent lithium ion batteries for critical metals recovery – a review. *Resources, Conservation and Recycling*, *165*, 105225. https://doi.org/10.1016/j.resconrec.2020.105225

Srichandan, H., Mohapatra, R. K., Parhi, P. K., & Mishra, S. (2019). Bioleaching approach for extraction of metal values from secondary solid wastes: a critical review. *Hydrometallurgy*, *189*, 105122. https://doi.org/10.1016/j.hydromet.2019.105122

Swain, B. (2017). Recovery and recycling of lithium: A review. In *Separation and Purification Technology* (Vol. 172, pp. 388–403). Elsevier B.V. https://doi.org/10.1016/j.seppur.2016.08.031**

Tao, H., & Dongwei, L. (2014). Presentation on mechanisms and applications of chalcopyrite and pyrite bioleaching in biohydrometallurgy – a presentation. *Biotechnology Reports*, *4*, 107–119. https://doi.org/10.1016/j.btre.2014.09.003

Vakilchap, F., Mousavi, S. M., & Shojaosadati, S. A. (2016). Role of Aspergillus niger in recovery enhancement of valuable metals from produced red mud in Bayer process. *Bioresource Technology*, *218*, 991–998. https://doi.org/10.1016/j.biortech.2016.07.059

Venkata Ratnam, M., Senthil Kumar, K., Samraj, S., Abdulkadir, M., & Nagamalleswara Rao, K. (2022). Effective leaching strategies for a closed-loop spent lithium-ion battery recycling process. *Journal of Hazardous, Toxic, and Radioactive Waste*, *26*(2), 671. https://doi.org/10.1061/(ASCE)HZ.2153-5515.0000671

Wang, D., Kijkla, P., Saleh, M. A., Kumseranee, S., Punpruk, S., & Gu, T. (2022). Tafel scan schemes for microbiologically influenced corrosion of carbon steel and stainless steel. *Journal of Materials Science & Technology*, *130*, 193–197. https://doi.org/10.1016/j.jmst.2022.05.018

Wang, D., Yang, C., Saleh, M. A., Alotaibi, M. D., Mohamed, M. E., Xu, D., & Gu, T. (2022). Conductive magnetite nanoparticles considerably accelerated carbon steel corrosion by electroactive Desulfovibrio vulgaris biofilm. *Corrosion Science*, *205*, 110440. https://doi.org/10.1016/j.corsci.2022.110440

Wang, J., Tian, B., Bao, Y., Qian, C., Yang, Y., Niu, T., & Xin, B. (2018). Functional exploration of extracellular polymeric substances (EPS) in the bioleaching of obsolete electric vehicle $LiNi_xCoyMn_{1-x-y}O_2$ Li-ion batteries. *Journal of Hazardous Materials*, *354*, 250–257. https://doi.org/10.1016/j.jhazmat.2018.05.009

Wang, J., Zhu, S., Zhang, Y., Zhao, H., Hu, M., Yang, C., Qin, W., & Qiu, G. (2014). Bioleaching of low-grade copper sulfide ores by Acidithiobacillus ferrooxidans and Acidithiobacillus thiooxidans. *Journal of Central South University*, *21*(2), 728–734. https://doi.org/10.1007/s11771-014-1995-3

Wang, Y., Song, Z., Everaert, N., De Ketelaere, B., Willemsen, H., Decuypere, E., & Buyse, J. (2014). The anorectic effects of alpha-lipoicacid are mediated by central AMPK and are not due to taste aversion in chicken (Gallus gallus). *Physiology & Behavior*, *132*, 66–72. https://doi.org/10.1016/j.physbeh.2014.04.047

Wu, W., Li, X., Zhang, X., Gu, T., Qiu, Y., Zhu, M., & Tan, W. (2020). Characteristics of oxidative stress and antioxidant defenses by a mixed culture of acidophilic bacteria in response to Co^{2+} exposure. *Extremophiles*, *24*(4), 485–499. https://doi.org/10.1007/s00792-020-01170-4

Xin, B., Zhang, D., Zhang, X., Xia, Y., Wu, F., Chen, S., & Li, L. (2009). Bioleaching mechanism of Co and Li from spent lithium-ion battery by the mixed culture of acidophilic sulfur-oxidizing and iron-oxidizing bacteria. *Bioresource Technology*, *100*(24), 6163–6169. https://doi.org/10.1016/j.biortech.2009.06.086

Xin, Y., Guo, X., Chen, S., Wang, J., Wu, F., & Xin, B. (2016). Bioleaching of valuable metals Li, Co, Ni and Mn from spent electric vehicle Li-ion batteries for the purpose of recovery. *Journal of Cleaner Production*, *116*, 249–258. https://doi.org/10.1016/j.jclepro.2016.01.001

Xu, T.-J., & Ting, Y.-P. (2009). Fungal bioleaching of incineration fly ash: metal extraction and modeling growth kinetics. *Enzyme and Microbial Technology*, *44*(5), 323–328. https://doi.org/10.1016/j.enzmictec.2009.01.006

Xu, Y., & Feng, Y.-Y. (2016). Feasibility of sewage sludge leached by Aspergillus Niger in land utilization. *Polish Journal of Environmental Studies*, *25*(1), 405–412. https://doi.org/10.15244/pjoes/60861

Yang, J., Zhao, X., Ma, M., Liu, Y., Zhang, J., & Wu, X. (2022). Progress and prospect on the recycling of spent lithium-ion batteries: ending is beginning. *Carbon Neutralization*, *1*(3), 247–266. https://doi.org/10.1002/cnl2.31

Zeng, G., Deng, X., Luo, S., Luo, X., & Zou, J. (2012). A copper-catalyzed bioleaching process for enhancement of cobalt dissolution from spent lithium-ion batteries. *Journal of Hazardous Materials*, *199–200*, 164–169. https://doi.org/10.1016/j.jhazmat.2011.10.063

Zeng, G., Luo, S., Deng, X., Li, L., & Au, C. (2013). Influence of silver ions on bioleaching of cobalt from spent lithium batteries. *Minerals Engineering*, *49*, 40–44. https://doi.org/10.1016/j.mineng.2013.04.021

Zhang, L., Zhou, W., Liu, Y., Jia, H., Zhou, J., Wei, P., & Zhou, H. (2020). Bioleaching of dewatered electroplating sludge for the extraction of base metals using an adapted microbial consortium: process optimization and kinetics. *Hydrometallurgy*, *191*, 105227. https://doi.org/10.1016/j.hydromet.2019.105227

Zhang, S., Qi, M., Guo, S., Sun, Y., Tan, X., Ma, P., Li, J., Yuan, R., Cao, A., & Wan, L. (2022). Advancing to 4.6V review and prospect in developing high-energy-density LiCoO2 cathode for lithium-ion batteries. *Small Methods*, *6*(5), 202200148. https://doi.org/10.1002/smtd.202200148

Zhang, X., Shi, H., Tan, N., Zhu, M., Tan, W., Daramola, D., & Gu, T. (2023). Advances in bioleaching of waste lithium batteries under metal ion stress. *Bioresources and Bioprocessing*, *10*(1), 19. https://doi.org/10.1186/s40643-023-00636-5

Zhao, Q., Yang, H., & Tong, L. (2019). Effect of carbonaceous matter on copper behavior in bioleaching from waste printed circuit boards. *Journal of Physics: Conference Series*, *1347*(1), 12079. https://doi.org/10.1088/1742-6596/1347/1/012079

Zheng, X., Zhu, Z., Lin, X., Zhang, Y., He, Y., Cao, H., & Sun, Z. (2018). A mini-review on metal recycling from spent lithium ion batteries. *Engineering*, *4*(3), 361–370.

5 Biological Recovery of Gold from e-Wastes

Pandya, Shivamkumar N.

5.1 INTRODUCTION: SYNOPSIS OF E-WASTE AND RECOVERY OF GOLD

The disposal of electronic devices, such as televisions, computers, and mobile phones, is known as electronic waste, or e-waste, and it is an issue that is expanding quickly. In addition to potentially dangerous chemicals and materials, these devices frequently contain valuable materials like gold, silver, and copper (Kiddee et al., 2013; Liu et al., 2021).

An estimated 50 million tons of e-waste are produced annually, according to the World Health Organization, posing serious environmental risks. For environmentally safe and responsible disposal, the field of e-waste management is therefore essential (Tiwari & Dhawan, 2014; Wang et al., 2019). To recover valuable materials and lessen environmental pollution, this involves using sophisticated recycling techniques (Cui & Zhang, 2008; Choi et al., 2021).

The importance of recovering gold from e-waste has increased because of its limited supply and economic value. Conventionally, a variety of techniques like chemical leaching and smelting are employed (Zhang & Xu, 2016. However, a substantial amount of the environmental harm caused by e-waste is due to the extraction of gold, which makes the need for alternative techniques like bioleaching even more pressing (Kaya, 2016; Kim et al., 2020). A sustainable and environmentally beneficial alternative for recovering gold and other metals is bioleaching and other biological techniques (Reith et al., 2009; Ilyas et al., 2010). These techniques, which use biomass from plants or microorganisms like bacteria and fungi, can extract and concentrate metals while using less energy and having a smaller negative environmental impact (Brandl et al., 2001; Ahluwalia & Goyal, 2007).

5.2 THE IMPORTANCE OF MANAGING E-WASTE AND GOLD RECOVERY

5.2.1 Environmental Defense

Improper e-waste disposal can result in the release of hazardous chemicals into the environment, causing pollution and environmental damage. e-Waste accounts for roughly 70% of toxic waste in landfills, highlighting the importance of effective management (Li et al., 2021). The goal of environmental stewardship in e-waste

DOI: 10.1201/9781003415541-5

management is to mitigate these effects by ensuring safe electronic device disposal and recycling (Kiddee et al., 2013).

5.2.2 Conservation of Natural Resources

e-Waste frequently contains precious metals such as gold, silver, and copper. One ton of mobile phones, for example, can yield up to 350 g of gold. Gold recovery from e-waste not only conserves these valuable resources but also reduces the need for new mining activities, which have their own environmental costs (Zhang & Xu, 2016).

5.2.3 Opportunities for Employment

e-Waste management that is effective creates job opportunities in the recycling, refurbishment, and disposal industries. According to proper e-waste management could generate over 30,000 jobs. These opportunities have the potential to stimulate economic growth while also benefiting local communities (Wilson et al., 2015).

5.2.4 Reducing Conflict Minerals

It is possible to lessen the demand for conflict minerals by recovering metals from e-waste. War-torn areas are frequently the source of conflict minerals, which are then sold to fund armed conflicts. Thus, managing e-waste may also have geopolitical advantages.

5.2.5 Upholding Human Rights and Labor Standards

Poor e-waste practices may result in labor exploitation and hazardous working conditions. It is essential for e-waste management to adhere to human rights and labor standards, thereby ensuring worker safety and dignity (Figure 5.1).

5.3 E-WASTE AND GOLD CONTENT

See Table 5.1.

5.4 CONVENTIONAL GOLD RECOVERY TECHNIQUES

5.4.1 Pyrometallurgy

This high-temperature process can reach up to 1,000°C to melt e-waste for metal separation. It can yield gold recovery rates of around 95%–98% but is highly energy intensive, consuming up to 300 kWh per ton of e-waste processed. However, it can result in harmful emissions if not well managed (Zhang & Xu, 2016).

5.4.2 Hydrometallurgy

In this chemical-based process, agents like cyanide dissolve metals from pulverized e-waste. Although this process requires less energy, it consumes about 5,000 L of

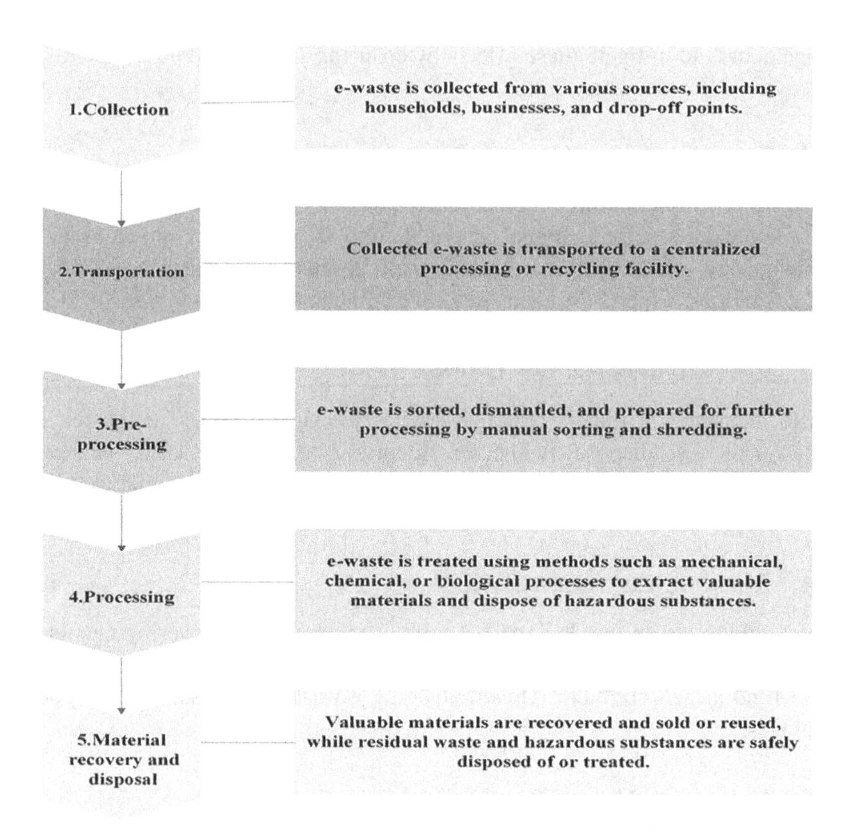

FIGURE 5.1 An overview of e-waste management procedures.

TABLE 5.1
Devices and their Approximate Gold Content per Unit

Device	Approximate Gold Content (grams/unit)		Reference
Computer motherboard	Article I.	0.1–0.2	Hagelüken (2006)
Cell phone	Article II.	0.024	Kiddee et al. (2013)
Television	Article III.	0.01–0.02	Hagelüken (2006)
Laptop	Article IV.	0.1–0.3	Hagelüken (2006)
Tablet	Article V.	0.01–0.03	Kiddee et al. (2013)

water per ton of e-waste. Effective for low gold concentrations, it also poses environmental and health risks due to chemical usage.

5.4.3 Gravity Separation

This method leverages density differences for separation, achieving up to 80% gold recovery when e-waste is abundant in other metals. It is a more cost-effective option but not suitable for all e-waste types.

5.4.4 Flotation

Introducing air bubbles facilitates metal separation, achieving approximately 85% gold recovery efficiency (Wang et al., 2019). Commonly used for circuit boards with low gold concentrations, this method is relatively eco-friendly but may require specific equipment.

5.4.5 Cyanide Leaching

This method can achieve gold recovery rates of up to 92% but requires the use of large amounts of cyanide, a hazardous substance. Therefore, it poses significant environmental and health risks if not managed properly (Akcil et al., 2018) (Table 5.2).

5.5 BENEFITS ECONOMICALLY FROM BIO-BASED METAL RECOVERY

5.5.1 Eco-Pure Nature

Because they avoid using dangerous chemicals, like cyanide or mercury, which are commonly used in conventional methodologies, biological techniques, like bioleaching, have a lower environmental impact and promote a more environmentally friendly method of recovering metals (Brierley, 2008; Ilyas et al., 2007).

5.5.2 Financial Sustainability

Biological methods are less dependent on energy and chemicals, which makes them a more economical option than conventional metallurgical processes.

TABLE 5.2
Conventional Gold Recovery Methods: Advantages, Drawbacks, and Key References

Method	Advantages	Drawbacks	References
Pyrometallurgy	High recovery rates; well-established technique	Energy intensive; toxic emissions	Kaya (2016), Zhang and Xu (2016)
Hydrometallurgy	Less energy intensive; suitable for low-concentration sources	Requires large amounts of water and chemicals; potential environmental and health hazards	
Gravity separation	Cost-effective; minimal chemicals and energy input	Less effective for fine particles or low-grade sources	Wills and Napier-Munn (2006)
Flotation	Suitable for low-concentration sources; selective recovery	Less efficient for very fine particles; waste slurry disposal	
Cyanide leaching	High recovery rates for low-concentration sources	Requires large amounts of toxic cyanide; environmental and health risks	Akcil and Mudder (2003)

5.5.3 Resource Performance

According to Dorado et al. and Chakraborty et al., bioleaching represents an effective method of recovering metals from low-grade ores and e-waste, materials that may be too costly to process using traditional methods. This helps to advance a paradigm that is more resource efficient.

5.5.4 Improved Extraction from Complicated Matrix

Biological techniques are superior to conventional methods in the extraction of metals because they can easily navigate the intricately embedded metals in the complex matrices of e-waste (Cui & Zhang, 2008).

5.5.5 Reduced Energy Consumption

The low energy requirements of biological processes, which work well at lower temperatures, represent significant energy savings when compared to the high temperature and high energy requirements of conventional processes (Ilyas et al., 2007).

5.6 BIOLOGICAL METHODS FOR RECOVERY OF GOLD FROM E-WASTE

5.6.1 Biosorption

Biosorption is the process by which metal ions from aqueous solutions passively cling to biological materials, such as microbial cell walls. Ionic-exchange mechanisms, which mostly involve hydrogen ions, enhance the sorption of gold ions onto the surface of cells. Gold ions may participate in complexation reactions or precipitation events after this initial binding event, depending on the presence and kind of other species in the aqueous matrix. As per Eisenler (2003), certain bacteria possess the ability to expel extracellular polymeric substances (EPS) that are enhanced with several functional moieties, such as carboxyl, hydroxyl, amine, and sulfhydryl groups. However, the process is more nuanced than mere adherence; it involves complex mechanisms like ion exchange, complexation, and micro-precipitation. "*Cupriavidus metallidurans* CH34 (oxic)" stands as a testament to the potential of this method, demonstrating a gold removal efficiency of 92.8%. In controlled settings, it managed to recover gold concentrations of 26.78 mg at a pH of 5 and temperatures oscillating between 15°C and 22°C (Romero, 2018). Another organism of note is "*Fucus vesiculosus*," which displayed a gold removal efficiency of 90%–100% at a pH of 7.

5.6.2 Bioleaching

Utilizing microbial metabolic processes to remove metals from ores is the core of bioleaching; it goes beyond simple microbial metal extraction. Metal ions can be released into a solution when certain microorganisms oxidize metal sulfides. The metal ions can be released into the solution by the bacteria as they break down the metal sulfides.

After that, the solution can be processed to extract the desired metals. An economical, environmentally benign, and effective way to recover metal is through bioleaching. *Chromobacterium violaceum* oxidizes gold sulfide in a process known as bioleaching. This resultant solution holds the potential for further processing to recover the desired metals. "*Chromobacterium violaceum*" has been studied extensively for its prowess in this domain. It showcases an efficiency rate of 83% and can recover gold at concentrations of 0.34 mg/L. Operating optimally at a pH of 8.5 and a temperature of 6.5°C, this microorganism emphasizes the potential of bioleaching (Reith et al., 2010).

5.6.3 Bioelectrosorption

In the field of biomining, bioelectrosorption is a cutting-edge method that effectively recovers gold ions by combining the complementary effects of electric fields and microbial metabolism. Bioelectrosorption, in contrast to conventional bioleaching or biosorption techniques, uses the electroactive properties of particular microbial strains to enable a specific, metabolically active binding of gold ions onto the cellular matrix. An exemplary example is the microbe *Shewanella oneidensis* MR-1 (anoxic), which has been shown to possess an extraordinary 88.5% gold removal efficiency. This equates to impressive quantities of gold per gram of dry cell weight, ranging from 26.05 to 70 mg of trivalent gold (Au[III]). Empirical studies have optimized the conditions under which this microbial strain operates, pinpointing a neutral pH level of 7 as ideal for maximum gold recovery. Furthermore, it has been observed that the process efficacy is temperature dependent; it exhibits optimal activity within a 5°C–10°C range in the absence of hydrogen gas (H_2) and significantly broadens to a 30°C–100°C range in the presence of H_2. These parameters, delineated in studies by Nakajima in 2003 and De Corte et al. in 2018, underscore the unprecedented potential and adaptability of bioelectrosorption in sustainable gold recovery (Nakajima et al., 2003).

5.6.4 Biosolubilization

Biosolubilization emphasizes the microbial dissolving of metals in a unique way, albeit it is related to bioleaching. Here, bacteria create organic acids that bind and solubilize metals instead of immediately converting metal sulfides. As a result, metals are extracted and placed in a state suitable for additional processing. The chelation of trivalent gold ions by citric acid, a metabolic byproduct of *Pseudomonas aeruginosa*, is a classic illustration of the chemical processes behind biosolubilization. This reaction's stoichiometry is as follows: $Au^{3+} + C_6H_8O_7 \rightarrow Au(C_6H_5O_7)^{3+} + 3H^+$. "*Bacillus subtilis*" and "*Pseudomonas aeruginosa*" are prime representatives of this method. The latter, in particular, boasts a gold removal efficiency of 89% and manages to recover gold concentrations as high as 500 mg Au/L. The process reaches its zenith of efficiency at a pH of 6.5 and a temperature of 25°C.

5.6.5 Bioaccumulation

Bioaccumulation is the biological process by which organisms progressively absorb and accumulate metals from their environment. This is not about a fleeting interaction;

rather, it has to do with the long-term storage of metals within an organism. Microorganisms like "*Plectonema boryanum*" and "*Arthrobacter* spp. 61B" have shown remarkable abilities in this field. The resultant gold nanoparticles are usually trapped in the microbial cell wall or the cytoplasm after the reaction. The morphological and physiological traits of the microbial species in question may influence the geometric and dimensional properties of these nanoparticles. The latter, for example, can recover gold particles with sizes ranging from 30 to 80 nm and showed a gold removal efficiency of 7%–12%. This process has been studied and optimized for conditions like a pH of 7.4 at room temperature (Kalbaghishvili et al., 2012).

5.6.6 Biomineralization

Biomineralization is a novel, environmentally friendly method for extracting Au from e-waste. After the pretreatment phase, specific strains of biomineralizing bacteria are introduced into the pre-treated matrix. These bacteria enzymatically solubilize gold ions, leading to the precipitation of gold nanoparticles. For optimizing bacterial enzymatic activity and gold ion solubility, operational parameters such as a pH level of approximately 2.5 and a thermodynamic condition of 25°C are critical. Biomineralization is distinguished from biosorption by its ability to bio-integrate metal ions into the bacterial cellular or extracellular matrix. It has been noted that the effectiveness of bacterial strains in the removal and recovery of gold ions varies and that the medium's nutrient availability also plays a role. Notably, under controlled environmental conditions, some heterotrophic bacteria have shown the ability to recover gold at high concentrations, specifically 395 mg/L or 1,134 ng/g, and have demonstrated gold ion removal efficiencies of up to 60%. These features highlight biomineralization as a promising, environmentally sound method of recovering gold ions from e-waste that has demonstrable benefits for the environment (Reith & McPhail, 2006) (Table 5.3).

5.7 CONCLUSION

This comprehensive examination elucidates the burgeoning importance of bio-based techniques in the recovery of gold from e-waste, drawing particular attention to advancements in bioelectrosorption, biosorption, bioleaching, biomineralization, and bioaccumulation. The exploration of these technologies reveals promising data, such as gold recovery efficiencies of up to 100% in bioaccumulation and 92.8% in biosorption, emphasizing their potential both in terms of environmental sustainability and economic viability. These methods not only operate at lower thermodynamic and energy thresholds but also significantly mitigate the environmental footprint through the minimization of hazardous reagents, making them ideal for low-grade ores and complex e-waste matrices.

In summation, the biotechnological approaches scrutinized herein present themselves as not just environmentally benign but also economically advantageous avenues for gold recovery. These advancements underscore the urgent need for further focused research aimed at optimization, scalability, and comprehensive life-cycle assessments. The transition to these greener methods holds the potential to

TABLE 5.3
Biological Gold Extraction: Comparative Evaluation of Methods, Advantages, Disadvantages and Parameters, and Gold Recover Efficiency

Method	Advantages	Disadvantages	Parameters	Gold Recover Efficiency	References
Bioelectrosorption	Efficient in anoxic conditions; Low cost and energy consumption	May require specific microbial strains; Limited by the surface area of the electrode	pH: 1, Temp.: 37°C	88.5%	Nakajima et al. (2003), De Corte et al. (2011)
Biosorption	Wide range of applicability; High selectivity and capacity	May need pretreatment of biomass; Possible interference from other metals	pH: 5, Temp.: 28°C	92.8%	Romero (2018)
Bioleaching	Effective for complex ores; Environmentally friendly	Can be slower; May produce toxic by-products	pH: 8.5, Temp.: 25°C	83%	Reith et al. (2010), Das et al. (2017)
Biomineralization	Stable and long-lasting gold recovery; Potential for nanotechnology applications	May depend on the availability of gold complexing ligands	pH: 2.5, Temp.: 25°C	60%	Reith and McPhail (2006)
Bioaccumulation	Effective for certain metal concentrations; Can be enhanced by genetic engineering	Specific to organisms; May pose ethical and biosafety issues	pH: 7–12, Temp.: 20°C–30°C	Varies (e.g., 90%–100% for *Rhizoclonium hieroglyphicum*)	Kalbegishvili et al. (2012)

substantially mitigate the e-waste crisis and foster a more sustainable management of precious metallic resources.

REFERENCES

Ahluwalia, S. S., & Goyal, D. (2007). Microbial and plant derived biomass for removal of heavy metals from wastewater. *Bioresource Technology*, 98(12), 2243–2257.

Akcil, A., & Mudder, T. (2003). Microbial destruction of cyanide wastes in gold mining: process review. *Biotechnology Letters*, 25(6), 445–450.

Brandl, H., Bosshard, R., & Wegmann, M. (2001). Computer-munching microbes: metal leaching from electronic scrap by bacteria and fungi. *Hydrometallurgy*, 59(2–3), 319–326.

Brierley, J. A. (2008). A perspective on developments in biohydrometallurgy. Hydrometallurgy, 94(1-4), 2–7.

Choi, J.-W., Bediako, J. K., Kang, J.-H., Lim, C.-R., Dangi, Y. R., Kim, H.-J., Cho, C.-W., & Yun, Y.-S. (2021). In-situ microwave-assisted leaching and selective separation of Au (III) from waste printed circuit boards in biphasic aqua regia-ionic liquid systems. Separation and Purification Technology, 255, 117649.

Choi, J.-W., Bediako, J. K., Kang, J.-H., Lim, C.-R., Dangi, Y. R., Kim, H.-J., Cho, C.-W., & Yun, Y.-S. (2021). In-situ microwave-assisted leaching and selective separation of Au (III) from waste printed circuit boards in biphasic aqua regia-ionic liquid systems. Separation and Purification Technology, 255, 117649.

Cui, J., & Zhang, L. (2008). Metallurgical recovery of metals from electronic waste: a review. *Journal of Hazardous Materials*, 158(2–3), 228–256.

Das, S., Natarajan, G., & Ting, Y.P. (2017). Bio-extraction of precious metals from urban solid waste. *AIP Conference Proceedings*, 1805, 020004.

De Corte, A. et al. (2011). Development and validation of a terrestrial biotic ligand model predicting the effect of cobalt on root growth of barley (Hordeum vulgare). *Environmental Pollution*, 159, 148–154.

Eisler, R. (2003). Biorecovery of gold. *Indian Journal of Experimental Biology*, 41(9), 967–971.

Hagelüken, C. (2006). Recycling of electronic scrap at Umicore Precious Metals Refining. *Acta Metallurgica Slovaca*, 12, 111–120.

Ilyas, S., Anwar, M. A., Niazi, S. B., & Ghauri, M. A. (2007). Bioleaching of metals from electronic scrap by moderately thermophilic acidophilic bacteria. Hydrometallurgy, 88(1-4), 180–188.

Ilyas, S., Ruan, C., Bhatti, H.N., Ghauri, M.A., & Anwar, M.A. (2010). Column bioleaching of metals from electronic scrap. *Hydrometallurgy*, 101(3–4), 164–173.

Johnston, C.W., Wyatt, M.A., Li, X., Ibrahim, A., Shuster, J., Southam, G., & Magarvey, N.A. (2013). Gold biomineralization by a metallophore from a gold-associated microbe. *Nature Chemical Biology*, 9(4), 241–243.

Kalabegishvili, T. L., Kirkesali, E. I., Rcheulishvili, A. N., Ginturi, E. N., Murusidze, I. G., Pataraya, D. T., Gurielidze, M. A., Tsertsvadze, G. I., Gabunia, V. N., Lomidze, L. G., Gvarjaladze, D. N., Frontasyeva, M. V., Pavlov, S. S., Zinicovscaia, I. I., Raven, M. J., Francinah, S. N., & Arnaud, F. (2012). Synthesis of gold nanoparticles by some strains of Arthrobacter genera. Journal of Materials Science and Engineering A: Structural Materials: Properties, Microstructure and Processing, 2(2), 164–173.

Kaya, M. (2016). Recovery of metals and nonmetals from electronic waste by physical and chemical recycling processes. *Waste Management*, 57, 64–90.

Kiddee, P., Naidu, R., & Wong, M.H. (2013). Electronic waste management approaches: an overview. *Waste Management*, 33(5), 1237–1250.

Kim, K. R., Choi, S., Yavuz, C. T., & Nam, Y. S. (2020). Direct Z-Scheme Tannin–TiO_2 heterostructure for photocatalytic gold ion recovery from electronic waste. ACS Sustainable Chemistry & Engineering, 8(19), 7359–7370.

Liu, J., Deng, Z., Yu, H., & Wang, L. (2021). Ferrocene-based metal-organic framework for highly efficient recovery of gold from WEEE. Chemical Engineering Journal, 410, 128360.

Nakajima, A. et al. (2003). Strain profiling of HfO_2/Si(001) interface with high-resolution Rutherford backscattering spectroscopy. *Applied Physics Letters*, 83, 296–298.

Reith F. & McPhail D.C. (2006). Effect of resident microbiota on the solubilization of gold in soil from the Tomakin Park gold mine, New South Wales, Australia. *Geochimica et Cosmochimica Acta*, 70, 1421–1438.

Reith, F. et al. (2010). Biomineralization of gold: biofilms on bacterioform gold. *Science*, 330, 66–69.

Reith, F., Etschmann, B., Grosse, C., et al. (2009). Mechanisms of gold biomineralization in the bacterium *Cupriavidus metallidurans. Proceedings of the National Academy of Sciences*, 106(42), 17757–17762.

Romero Mosquera, S.S.M. (2018). Microbial-electrochemical systems for metal recovery. Ghent University Doctoral Dissertation.

Tiwari, D., & Dhawan, N.G. (2014). E-waste management: an emerging challenge to manage and recover valuable resources. *International Journal of Environmental Research and Development*, 4(3), 253–260.

Wills, B.A., & Napier-Munn, T. (2006). *Wills' Mineral Processing Technology: An Introduction to the Practical Aspects of Ore Treatment and Mineral Recovery* (7th ed.). Elsevier Science & Technology, Amsterdam, The Netherlands.

Zhang, L., & Xu, Z. (2016). A review of current progress of recycling technologies for metals from waste electrical and electronic equipment. *Journal of Cleaner Production*, 127, 19–36.

Wilson, M. L., Renne, E., Roncoli, C., Agyei-Baffour, P., & Tenkorang, E. Y. (2015). Integrated assessment of artisanal and small-scale gold mining in Ghana—Part 3: Social sciences and economics. International Journal of Environmental Research and Public Health, 12(7), 8133–8156

6 Bioleaching of e-Wastes
Copper Extraction

G. Damaru Mani, Javvadi K J N S Thanishka, Ananya Namala, Reddem Poojitha Reddy, and Jayato Nayak

6.1 INTRODUCTION

In current times, the use of electrical and electronic machinery has increased incrementally, this also applies to e-waste yield. Annually, e-waste measured in millions is brought out globally. For the most part, e-waste is a term referring to electrical and electronic equipment (EEE) that is discarded, unregenerated, and deteriorated. The stream of waste is accumulated into a legislative act by the European Union (EU)—Waste Electrical and Electronic Equipment (WEEE). It is currently part of the most rapidly expanding categories of waste (Sharifidarabad et al., 2024). This stream covers a wide range of products, including cell phones, computers, TVs, fridges, and other kitchen appliances, lighting, healthcare supplies, and solar energy sources. The longest possible permitted period of storage for electronic waste (e-waste) is 180 days (about 6 months). So, hence these must be utilized again shortly (Mim et al., 2023).

Printed circuit boards, or PCBs, are essential elements of electronic gadgets that offer kinetic and electronic interfaces for diverse components. These are mostly recycled components. PCBs tend to be made up of 30% metal and 70% nonmetallic elements. PCBs are said to range from 20% to 35%, especially in metallic copper formulations. Lead (10%–20%), nickel (1%–2%), copper (28%), and other materials (plastic products, the element bromine, transparent material, and porcelain) have been documented (Murugesan et al., 2020). Improper handling of copper discovered in e-wastes can lead to health risks. Prolonged exposure can cause lung ailments and skin disorders, while brief exposure might irritate the digestive and respiratory systems.

Bioleaching is our primary technique employed to extract copper. Biohydrometallurgy, additionally referred to as the bioleaching process, is a natural and ecological method that uses microorganisms (molds, bacterial cells, and phage) to dissolve minerals from poor-quality ores, mining accumulates, or e-wastes (Zhao et al., 2020). Stated differently, the process involves the effect of microorganisms to transform insoluble ferrous elements found in a rigid matrix to become soluble (Samarasekere et al., 2024). Additionally, biological leaching generates minimal secondary waste and allows for easier metal recovery during the process. However, two significant obstacles to the widespread use of bioleaching are the extended leaching period and low metal yields. Certain metabolites secreted by microbes help to mobilize and solubilize insoluble metals from e-waste (Pang et al., 2023).

DOI: 10.1201/9781003415541-6

Bioleaching for the extraction of copper has substantial financial, ecological, and operational advantages. It cuts expenditures by 20%–50%, lowering costs to \$1,000–\$2,400 for each ton versus \$2,000–\$3,000 with standard approaches. Energy usage is reduced by up to 50%, lowering CO_2 emissions from 2–3 to 1–1.5 tons per ton of copper. Bioleaching produces 20%–50% fewer waste products and requires 30%–40% less water. It uses bacteria such as *Acidithiobacillus* to effectively treat low-grade ores (0.3%–1% copper) with recovery rates of 60%–90%, despite the longer time required for extraction. Bioleaching can stand as an extremely sustainable technology for the recovery of copper as shown by, where 96% of copper was recovered using *Acidithiobacillus thiooxidans*.

Bioleaching uses microorganisms such as *Acithiobacillus ferroxidase* for the extraction of copper. It oxidizes mineral sulfides, delivering metal ions in the leaching stream. *A. ferroxidase*, an extremophilic, chemo lithotrophic bacteria, oxidizes iron (Fe^{2+}) and sulfur (S^0) to produce ferrous ions (Fe^{3+}) and H^+, which target chalcopyrite ($CuFeS_2$). The involved biochemical reaction scheme is shown as follows:

$$CuFeS_2 + 4H^+ + O_2 \rightarrow Cu_2{}^+ + Fe^{2+} + 2S^0 + 2H_2O$$

$$4Fe^{2+} + 4H^+ + O_2 \rightarrow 4Fe^{3+} + 2H_2O$$

$$2S^0 + 3O_2 + 2H_2O \rightarrow 2SO_4{}^{2-} + 4H^+$$

$$CuFeS_2 + 4Fe^{3+} \rightarrow Cu_2{}^+ + 2S^0 + 5Fe^{2+}$$

6.2 RESOURCES OF COPPER

Copper is most abundantly available on Earth's crust, with an average content of about 0.25%. However, the concentration of copper in most deposits is quite low, typically ranging from trace amounts to a few percent (Zulkernain et al., 2023). Only in few copper deposits, such as porphyry copper deposits, copper content can be found at higher concentrations, reaching up to 3%–5%.

Figure 6.1 shows worldwide copper production from mining resources. Copper in nature often exists in the form of compounds, primarily as sulfide minerals like chalcopyrite, bornite, enargite, and chalcocite, as well as oxide minerals like cuprite and malachite (Sulaiman Zangina et al., 2023). These minerals are the primary sources of copper extraction and among these resources, porphyry copper deposits are the most important, which is nearly 76% of the total explored copper. Magma seeping into the Earth's crust creates massive low-grade deposits known as porphyry copper deposits (Rao et al., 2021).

Overall, while copper is widespread in the Earth's crust, its economic extraction is primarily dependent on the concentration and accessibility of high-grade deposits, with porphyry copper deposits being the dominant source of explored copper resources.

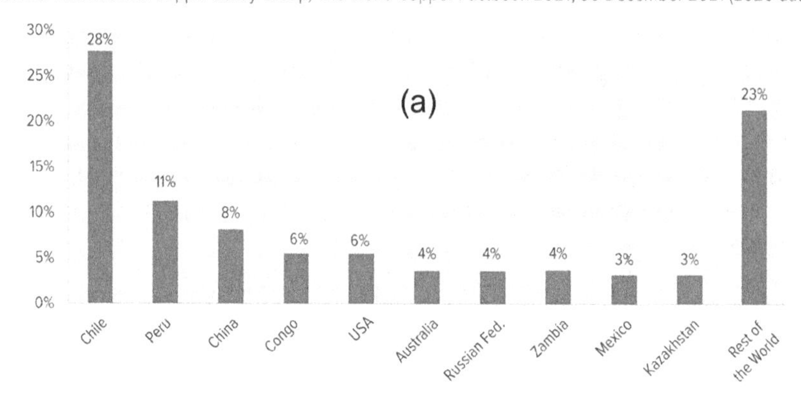

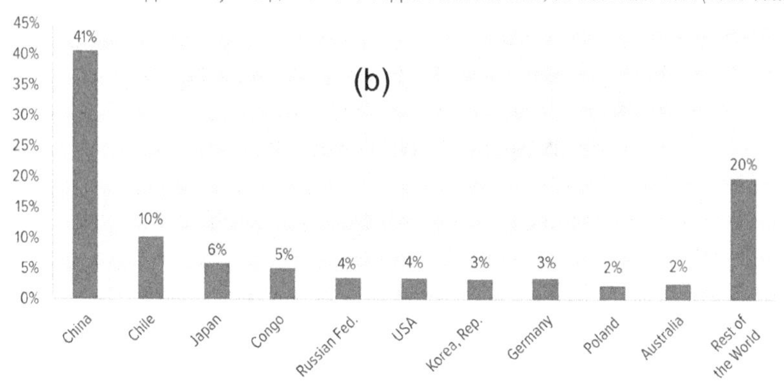

FIGURE 6.1 Global copper production (a) world copper mine production and (b) global copper refined production. Reproduced from Global X Research Team, Copper, Explained, Nov, 22, 2022, Last Viewed June 19th, 2024, URL: https://www.globalxetfs.com/copper-explained/#:~:text=invest%20in%20copper%3F-,How%20is%20Copper%20Produced%3F,pit%20or%20underground%20mining%20techniques

6.2.1 Resources of Copper for Extraction

According to data from the United States Geological Survey (USGS), global copper mine production was approximately 1.1 million tons in 2023, which is a 11% decrease from that in 2022. This production includes copper extracted from both primary sources (ores) and secondary sources (recycled scrap).

6.2.1.1 Primary Resources

Primary resources of copper refer to the naturally occurring sources from which copper is extracted through mining and processing. These primary resources include:

- Mining
- Crushing and grinding
- Concentration
- Smelting
- Refining
- Casting and forming (Royaei et al., 2019)

6.2.1.2 Secondary Resources

Secondary resources of copper refer to recycled copper obtained from various sources after it has served its initial purpose. This recycled copper can be recovered from materials or products that contain copper components or copper alloys which include:

- Scrap metal
- Industrial waste
- Obsolete products (outdated electronics and machinery equipment)
- Post-consumer waste (electrical wiring and cables) (Zulkernain et al., 2023).

6.3 CONVENTIONAL METHODOLOGIES OF COPPER EXTRACTION

6.3.1 Hydrometallurgical Copper Extraction

Sulfuric acid is used in hydrometallurgical processes for removing metals from WEEE, and in PCBs that are high in metal. The major goal is to look into the leaching of metals including iron (Fe), copper (Cu), zinc (Zn), and aluminum (Al) using hydrogen peroxide, a strong oxidant, in a solution of sulfuric acid (Dusengemungu et al., 2021). This examines the effects of temperature and the rate at which oxidants—specifically, hydrogen peroxide—are added to the extraction efficiency of copper and other precious metals from waste PCBs (WPCBs). Because of its extensive use and strong oxidation–reduction potential, hydrogen peroxide is used as the oxidizing agent and is being studied in relation to the oxidation processes of sulfides. WPCB samples are taken from different electric and electronic appliances like radios, computers, printers, mobile phones, laptops, and calculators. Hydrogen peroxide (30%) and sulfuric acid (95%–97%) are used (Mathew et al., 2021).

The PCBs are broken down, crushed, and ground into small pieces to release copper and other materials. The material is then coarsely ground using a ring mill. Samples are sieved after being ground. X-ray fluorescence (XRF), scanning electron microscopy (SEM), and Inductively coupled plasma mass spectrometry (ICP-MS) are used to characterize the samples (Ishak et al., 2022).

6.3.2 Extraction of Copper through Induced Morphological Changes Using Supercritical CO_2

Researchers are looking into supercritical fluids more and more because they offer an intriguing, environmentally friendly substitute for acid leaching. It has been demonstrated that supercritical H_2O processes are useful for getting rid of BFRs (brominated flame retardants) (Zulkernain et al., 2024). Even the combination of acid leaching and supercritical water has been used to recover metals from PCBs. Supercritical water oxidation (SCWO) and depolymerization (SCWD), two pre-treatment techniques, together with HCl leaching, showed efficacy in recovering 99.8% of the Cu that was recovered (Benzal et al., 2020a,b). Using supercritical CO_2 and aqueous acid, a unique method can be created to separate the leaching streams of copper and other metals from WPCB fragments. A high-pressure and high-temperature reactor was used to study the effects of $scCO_2$ with different co-solvents by synthesizing metal and polymer laminates to simplify PCB samples (Boopathi et al., 2024). The fundamental study findings indicate that $scCO_2$'s function is to physically alter the polymer rather than to directly extract metal from PCBs (Hubau et al., 2024). This method, which pretreats e-waste with supercritical CO_2 and acid, shows promise for extracting metal from it with enhanced sustainability of the environment. Nowadays, most reported hydrometallurgical methods make use of finely crushed PCB as the beginning feedstock and $scCO_2$'s function in treating e-waste. Considering the technology suggested in this research, intensive mechanical pre-processing could be reduced, and lowering the acidity level to efficiently extract Cu from e-waste (Hsu et al., 2021).

6.3.3 Pyrometallurgy

The process of refining non-ferrous metals from metallurgical materials at high temperatures is known as pyrometallurgy. Removal of important metals from garbage involves procedures including incinerator, melting, drossing, smelting, and roasting (Abbadi et al., 2021). The variates of e-waste and the specification of the melting process determine which pyrometallurgy technologies to be used. With this method, many metals, such as Cu, Au, Ni, Pd, Ag, Se, and Pb, can be recovered from various e-waste sources (Jiankang et al., 2024). The procedure uses a lot of energy and needs to be done at temperatures between 300°C and 900°C. To maintain high temperatures, managing the heat and the materials to be treated is required which is one of the most difficult tasks to perform. And produce hazardous gases that include corrosive acids like H_2SO_4, HCl, and HNO_3 (Ahmadinia et al., 2024). Furthermore, the copper in PCBs catalyzes the combustion process which results in the creation of dioxin. The majority of pyrometallurgical methods lose Al and Fe to the slag and only produce substantial Cu recovery (Zulkernain et al., 2023).

Although pyrometallurgy has shown to be efficient and profitable, its uses are restricted by the possibility of hazardous gas leakage, the production of dioxin, and the waste that results from the development of slag. Given the limitations of pyrometallurgy and the rise in e-waste creation, emphasis must be given to the treatment of hazardous gases generated during sustainable metal recovery (Baniasadi et al., 2021).

6.3.4 Nitric Acid and Ammonia-Ammonium Salt

At room temperature, diluted nitric acid can oxidize copper to copper nitrate without the need for another oxidizing agent. Acid gas is produced during leaching.

High-purity copper cannot be electroplated directly using copper nitrate. Before electroplating, it needs to be extracted and back extracted into a pure copper sulfate solution (Sharifidarabad, 2024). In the presence of an oxidizing agent, copper reacts with ammonia or ammonium salt and generates a copper–ammonia complex, which dissolves the copper. Both chemical equilibrium and the copper leaching rate are impacted by the pH of the leaching solution. The resulting copper–ammonia complex solution needs to be transformed into pure copper sulfate solution before being utilized for electroplating (Carmen Falagán et al., 2024). Figure 6.2 depicts a schematic flowchart of e-waste recycling methods, especially based on copper extractions.

6.4 BIOLEACHING, THE NEED OF THE HOUR

Microorganisms are used in the biohydrometallurgy process that dissolves components from solid materials and these can be recovered using separation procedures (Huang et al., 2020). Bioleaching, also referred to as the biohydrometallurgical process, is a simple approach that has several benefits over the traditional hydrometallurgical process, including increased safety and efficiency, reduced energy and operating costs, simpler management, normal operating conditions at room temperature and atmospheric pressure, environmental friendliness, and fewer industrial steps without the need for skilled labor (Zulkernain et al., 2023).

There are two methods of operation for it: single-stage and two-stage. The leaching of biohydrometallurgy involves using a variety of organisms, such as *Acidobacillus ferrooxidans, Acidobacillus thiooxidans, Sulfolobus* sp., *Pseudomonas* sp., *Bacillus* sp., *Aspergillus* sp., and *Penicillium* sp. Biohydrometallurgy is a successful process that has gained immense attention (Sharifidarabad, 2024). The exploitation of different microorganisms for bioleaching of copper, growing on different nutrient media and metal resources, are tabulated in Table 6.1.

Bioleaching involves acidophilic bacteria and archaea which metabolize iron and sulfur compounds to dissolve minerals. They thrive at pH below 3 and across various temperatures. At moderate temperatures (<40°C), mesophilic genera like *Acidithiobacillus* and *Acidiphilium* are dominant. As temperatures rise (40°C–60°C), thermophilic bacteria such as *Alicyclobacillus* and *Sulfobacillus*, alongside *Acidithiobacillus caldus* and thermophilic *archaea* of the *Thermoplasmatales*, become prevalent. Above 60°C, *Sulfolobales* archaea like *Sulfolobus* and *Metallosphaera* dominate, known for their extreme thermophilic and mineral oxidizing capabilities. Factors like pH, temperature, presence of oxygen, and carbon sources influence microbial diversity in these processes (Vera et al., 2022).

6.5 RECENT DEVELOPMENTS IN BIOEXTRACTION OF COPPER

6.5.1 Extraction of Copper Using Acidithiobacillus Ferroxidase in WPCB

Acidithiobacillus ferroxidase, a chemo lithotrophic bacteria that is aerobic, generates energy by deteriorating ferrous ions (Fe^{2+}) to ferric ions (Fe^{3+}) and using molecule-level oxygen (O_2) as an ending electron receiver. In the current investigation,

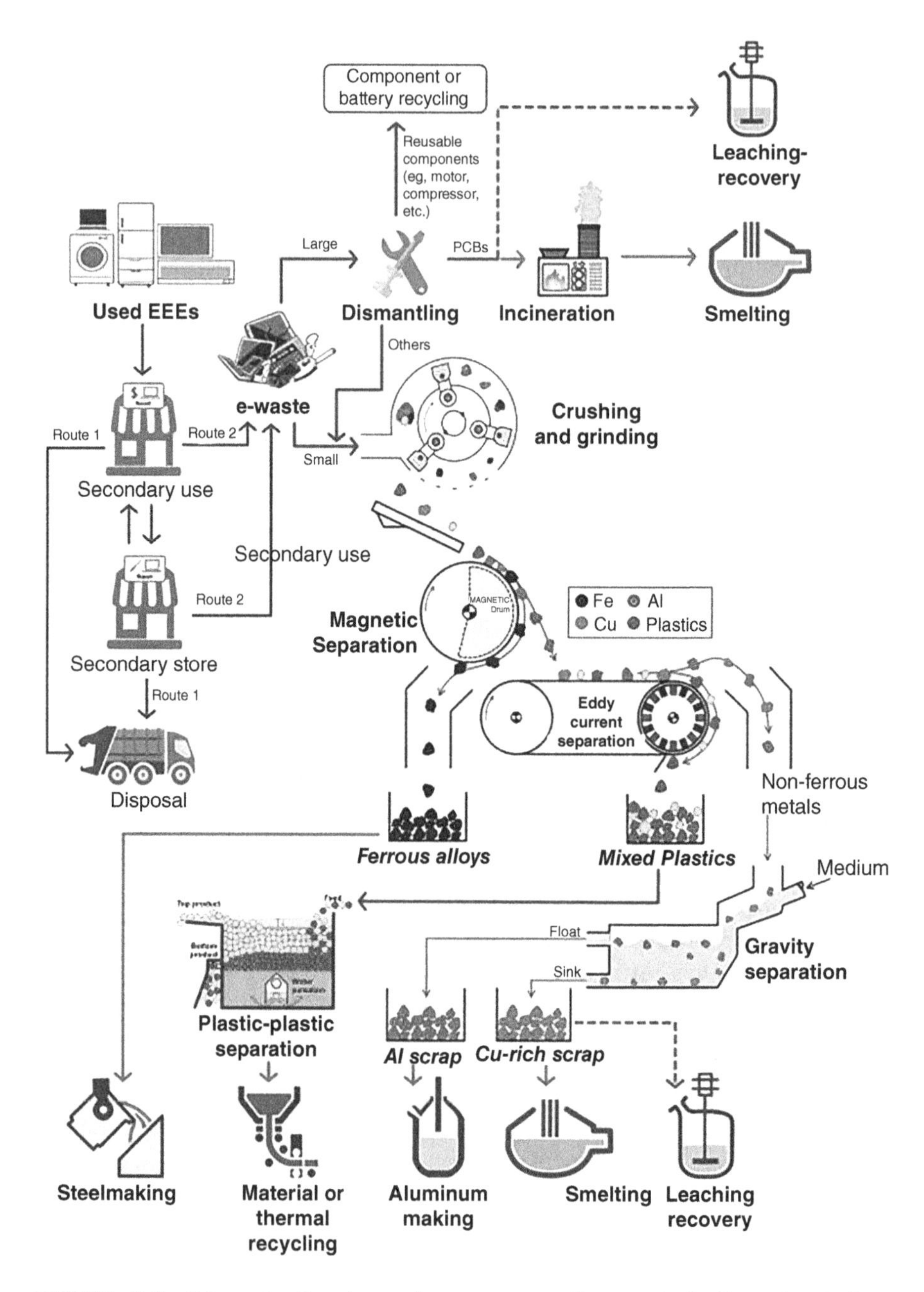

FIGURE 6.2 Schematic flowchart of e-waste recycling method. Reproduced from Sharifidarabad (2024).

the impact of dissolved oxygen, or DO, concentrations in the cultivation medium on the development of cells and the extraction of copper from discarded printed circuit boards (PCBs) were examined in *A. ferroxidase* (Sakthivel et al., 2022). The whole

TABLE 6.1
Review summary of bioleaching recovery of copper by using different microorganisms

Microorganism	Metal source	Nutrient media	Temperature	pH	Copper bioleaching efficiency	Other metals	Method	References
Acidithiobacillus thiooxidans and Leptospirillum ferriphilum	Powdered spent batteries	Basal salts medium	22°C	1.8	Cu (96%)	Mn, Zn	Spent medium	Xia et al. (2018)
Chromobacterium violaceum	SIM card	Luria-Bertani (LB) medium.	20°C	6.5	Cu (13.7%)	Au, Ag	Double step bioleaching	Sahni et al. (2016)
Acidithiobacillus sp. and *Leptospirillum* sp.	PCBs	ATCC 164 medium	25°C	1.5	Cu	Ni	NM[a]	Dey et al. (2023)
Aspergillus niger	Cathod and snodes powder	Sucrose medium	30°C	5.4	Cu (94%)	Mn, Al	One step	Bahaloo Horeh and Mousavi (2017)
Aspergillus niger (PTCC 5010)	Cathod and snodes powder	Sucrose medium	30°C	2.5	Cu (94%)	Mn, Ni	One step	Bahaloo Horeh and Mousavi (2017)
Pseudomonas chlororaphis	PCBs	Luria-Bertani (LB) medium	25°C	5.0	Cu (52.3%)	Au, Ag	NM[a]	Ruan et al. (2014)
Sulfobacillus thermosulfidooxidans	wires, and connectors	DSMZ 866M	40°C	3.5	Cu (95%)	Al, Zn, Ni	NM[a]	Ilyas and Lee (2014)
Acidiphilium acidophilum (ATCC 27807)	PCBs	Modified Brock medium	25°C	2.0	Cu (3.6%)	Ni	NM[a]	Hudec et al. (2005)

(*Continued*)

TABLE 6.1 (*Continued*)
Review summary of bioleaching recovery of copper by using different microorganisms

Microorganism	Metal source	Nutrient media	Temperature	pH	Copper bioleaching efficiency	Other metals	Method	References
Acidaianus brierleyi	PCBs	Modified DSMZ 88 media	70°C	2.0	Cu (81.4%)	NM[a]	NM[a]	Dey et al. (2023)
C. Violaceum, P. fluorescens, P. aeruginosa	PCBs	Luria-Bertani (LB) medium	25°C	6.0	Cu (83%)	Ag, Au, Zn	NM[a]	Pradhan (2012)
C. Violaceum	PCBs	Luria-Bertani (LB) medium	25°C	6.0	Cu (83%)	Au	NM[a]	Dey et al. (2023)
A. thiooxidans	PCBs	Modified 9K medium	25°C	1.5	Cu (98%)	NM	NM[a]	Hong and Valix (2014)
A. ferrooxidans	PCBs	Modified 9K medium	25°C	1.5	Cu (96.8%)	AL, Zn	NM[a]	Yang et al. (2014)
Acidithiobacillus, Gallionella	PCBs	NM[a]	20°C	6.5	Cu (95%)	NM	NM	Xiang et al. (2010)
Acidithiobacillus, Gallionella	wires, and connectors	NM	20°C	6.5	Cu (96.8%)	Zn, Al	NM[a]	Dey et al. (2023)
A ferrivorans, A thiooxidans, Pseudomonas putida, Pseudomonas fluorescens	PCBs	Luria-Bertani (LB) medium	25°C	6.5	Cu (98.4%)	Au	NM[a]	Isildar et al. (2015)

(*Continued*)

TABLE 6.1 (*Continued*)
Review summary of bioleaching recovery of copper by using different microorganisms

Microorganism	Metal source	Nutrient media	Temperature	pH	Copper bioleaching efficiency	Other metals	Method	References
Thermoplasma acidophilim Sulfobacillus, thermosulfidooxidans	wires, and connectors	NM[a]	NM[a]	NM	Cu (86%)	Al, Zn,Ni	NM[a]	Ilyas et al. (2024)
P. putida	PCBs	Luria-Bertani (LB) medium	25°C	6.5	Cu (98%)	Au	NM[a]	Isildar et al. (2016)
Bacillus megaterium	PCBs	Nutrient Agar or Nutrient Broth	25°C	6.0	Cu (72%)	Au	NM[a]	Arshadi et al. (2016)

growing phase was separated into two phases, which were cell development and copper recovery. In the previous step, an extremely low DO level was used to support bacterial development while minimizing significant Fe^{2+} oxidation. In the more recent stages, higher DO was utilized to boost copper recovery (Wang et al., 2021). Furthermore, the change in DO between lesser to higher values was calculated by modeling the Gauss function. For a concentration of 18 g/L PCBs while maintaining DO at 10% for the first 64 h (about 2.5 days) and increasing to 20% later, resulted in an overall copper recovery rate of 94.1%. More critically, copper leaching times were reduced from 108 to 60 h (about 2.5 days). It was claimed that the use of a DO-shifted method to improve the extraction of copper from PCBs with shorter leaching times is conceivable (Liang et al., 2021).

Typically, copper recovery rates are approximately 90%. The ideal parameters for bioleaching are in the range of temperature of 30°C–35°C with a pH of 2.0–2.5 (Yang et al., 2020). Under these circumstances, the bacteria efficiently oxidize sulfide minerals, facilitating copper solubility. The result is a copper-rich effluent solution, which may be further refined (Pirsaheb et al., 2021).

6.5.2 Fungal Bioleaching

The advances in metal bioleaching studies are reported on the use of filamentous fungus. This review focuses on fungal bioleaching from diverse metal sources, such as scrap from mines, electronic debris, solid mining wastes, wasted catalysts, and poor-quality ores. Filamentous fungi are employed in bioleaching because they emit naturally occurring acids and help metal ions dissolve in the water-based state (Zeng et al., 2020). Fungal bioleaching uses fungi's oxidative ability to remove metals from mining dumps and electronic trash. Mining waste containing minerals like Ni, Fe, Cu, and Pb can be successfully treated by fungus including Aspergillus, Fusarium micro and Hypo Crea. These species can survive high levels of metal concentrations and can even dissolve refractory metal molecules (Amber Trivedi et al., 2023). Fungi may recover elements, such as aluminum, copper, zinc, and gold, from electronic trash by creating acid compounds that aid in the extraction of metals. According to few studies, fungal strains, notably *Aspergillus niger*, are known to enhance metal in the phytoextraction process and can be utilized as a replacement to biological leaching agents, emphasizing the need for more study into optimizing fungal and herb mixtures for enhanced metal extraction (Jaiswal et al., 2024). Fungal bioleaching also uses fungus to remove metals from a variety of solid materials, including incinerator waste, ash, polluted soil, and used catalysts. Fungi such as *Aspergillus niger* and *Penicillium* have shown to absorb metals including Cu, Ni, Co, Fe, Cd, and Pb from ashes and polluted soils (Nili et al., 2022).

Filamentous fungi demand both sugar and oxygen, and they compete with indigenous microflora for nutrition. Despite fungi's superior resistance to metals that are dissolved, their value in soil and waste remediation remains untapped. According to the latest research, fungal bioleaching is a more economical and ecologically benign approach to recovering metals from wasted accelerators than standard chemical-based methods (Dusengemungu et al., 2021). *Aspergillus niger* and *Penicillium simplicissimum* have metal extraction rates of 70%–90% at a temperature range of

25°C to 30°C along with the pH ranging between 3.0 to 6.0. Further essential parameters are the content of glucose or sucrose as a carbon source, which is generally 50–100 g/L, and the pulp's density, which is 1%–10% (w/v) (Chaerun et al., 2017). Following the process, the end product will be a metal-rich drainage solution that may be used to recover and purify metal copper, in addition, the procedure produces natural acids such as citric and oxalic acids as byproducts. This environmentally conscious process enables the effective and consistent recovery of precious metals from diverse ores (Xia et al., 2018) (Figure 6.3).

6.5.3 In Situ Bioleaching

When copper is bioleached in situ, it is done utilizing a biologically generated ferric iron solution. Once the valuable metals have been recovered, the solution is recycled back into the in situ reactor and reoxidized by iron-oxidizing bacteria (IOB) (Bakhshoude et al., 2023). When compared to traditional excavation-intensive

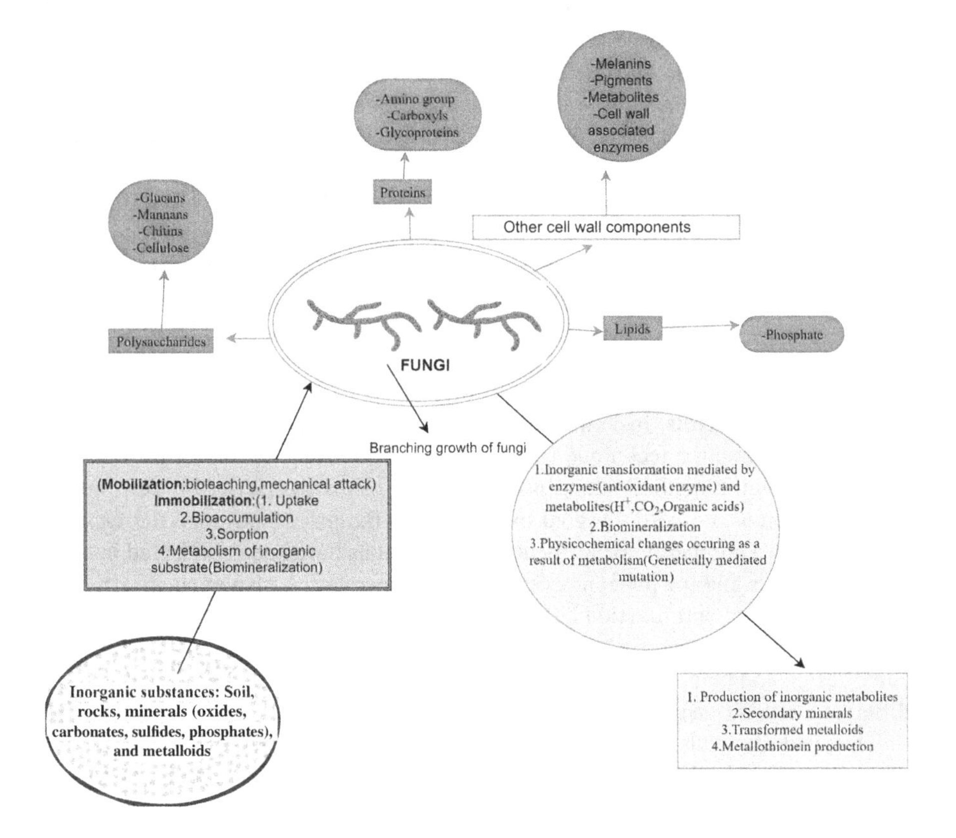

FIGURE 6.3 Simplified filamentous fungi interaction with inorganic substances activated bioprocesses such as mobilization: (bioleaching and mechanical attack) and immobilization (bioaccumulation, biosorption, and biomineralization). The cell wall structure, composition, and filamentous branching growth increase fungi's comprehensive metabolic capabilities. Reproduced with copyright permission from Dusengemungu et al. (2021).

mining operations, the idea of in situ biomining is enticing (Ranjbar et al., 2020). Nevertheless, there are still potential environmental issues with this technology, like leaching agents contaminating groundwater and leftover IOB from the onsite e-wastes causing uncontrolled leaching and iron oxidation (e.g., Seredkin et al., 2016. Therefore, at the conclusion of the procedure, the IOB must be rendered inactive (Bomberg et al., 2023). If the deposit of ore is depleted, the in situ leaching process must be properly stopped to prevent the uncontrollably high production of acidic mine drainage (AMD). AMD is caused by the accelerated oxidation of sulfidic ores in the presence of oxygen, either water or another oxidant, like ferric iron (Jaiswal et al., 2024). As previously verified in simulation studies, it is crucial to make sure that there is no leakage of acidic and metal-rich lixiviant fluid from the leaching location to the surroundings by carefully stopping the process at the end of operations (Samarasekere et al., 2024).

In situ biological leaching includes Temperature: The ideal temperature for bioleaching is normally between 30°C and 50°C, based on the microorganisms utilized. pH values: The optimal pH range for bioleaching procedure is 1.5–3.0 (Jiankang et al., 2024). Strong acidophiles flourish in these circumstances, allowing for quick metal solubilization. In situ bioleaching normally produces a solution rich in the required metal ions, which are subsequently recovered using different processes such as precipitation, solvent extraction, or electro-winning (Acosta Hernández et al., 2023). The concentration in order of Cu in the output of the stone reactors throughout the pre-water cleanse and water cleanse stages was poor, at most 4 and 0.4 mg/L. However, at the moments of microbiological collection, the Cu concentration in order in the pre-water cleanse phase was smaller. Total iron concentration in order was 0.24 mg/L (Malin Bomberg et al., 2023).

6.5.4 Heap Bioleaching

For low-grade minerals, including copper, gold, silver, and uranium, heap leaching is a popular extraction technique (Pinho et al., 2023). There are novel uses for this technique in soil remediation and nonmetallic minerals like saltpeter. Few studies have been conducted with the goal of optimizing the process, despite the fact that modeling and a number of experimental investigations have been conducted in order to better understand the phenomenon and how it operates (Arslan et al., 2021). Most of the research on optimization has been done from a technical standpoint, whether it is using models or experiments. Creating a methodology for the heap leaching circuit's design and planning is the goal of this endeavor. A superstructure with several different circuits is suggested.

The yield of metals extracted by bacterial leaching is affected by environmental, natural, and Physico-chemical variables, just like any other procedure involving live organisms. For the microbes engaged in the process to develop, ideal humidity, pH, temperature, energy sources, and nutrition levels must exist, as well as the lack of any potential inhibitors (Dusengemungu et al., 2021). These circumstances are necessary for the development of microbes.

There are several benefits associated with heaps plus dumps, including easy-to-use equipment and functioning, minimal startup and running expenses, and respectable yields. However, one must acknowledge that the procedure is severely limited. To start, the heaped material is diverse, making it nearly impossible to exercise close process oversight. The only things that can be done are to change the pH and supply some nutrients. Heap reactors are more challenging to effectively aerate than tank reactors (Samarasekere et al., 2024).

Heap bioleaching produces a leachate solution improved with dispersed metal ions, usually copper (Määttä et al., 2022). The ideal pH range for heap bioleaching is between 1.5 and 3.0, which allows acidophilic microorganisms to thrive and facilitate the leaching process. Temperature: Optimal temperatures for heap bioleaching generally vary from 30°C to 60°C, with thermophilic microbes operating well at the upper end of this temperature range (Pang et al., 2023). As the last part of heap bioleaching, pyrite oxidation reduces pH, particularly up to 3, where Fe^{3+} is an oxidant that acts up on sulfide minerals. Acidophiles replenish Fe^{3+}, which promotes copper oxidation and produces sulfuric acid. Throughout the year, the pH decreases to roughly around a value of 10 (Jia et al., 2021).

6.6 CONCLUSIONS

Electronics and electrical sectors have grown due to population growth and higher living standards. e-Waste volumes are increasing dramatically, posing serious environmental hazards. e-Waste contains rare earth elements and critical metals, making it economically useful. Bioleaching optimizes microbial activity using numerous physicochemical variables, making it an ecofriendly and sustainable metal recovery approach. Bioleaching recovers more metal from e-waste than standard procedures. Innovative and sustainable metal bioleaching methods are essential for e-waste management. Future studies should improve biorecovery kinetics by enhancing metal–microbe interactions. Genetically engineered microorganisms, thermophiles, fungi, bacteria, and archaea should be tested under varied settings to improve metal recovery. Microbial culture in bioleaching mediums with alternate carbon sources like agro-industrial waste can boost metabolite secretion and metal recovery. Pre-biorecovery e-waste must be detoxicated by inventing and executing methods to efficiently separate dangerous elements from waste.

REFERENCES

Abbadi, A., Rácz, Á., Bokányi, L. Exploring the comminution process of waste printed circuit boards in recycling: a review. *Journal of Material Cycles and Waste Management*, 26, 1326–1348, 2024. https://doi.org/10.1007/s10163-024-01945-3

Acosta Hernández, I., Muñoz Morales, M., Fernandez Morales, F., Rodríguez, L., Camacho, J. Removal of heavy metals from mine tailings by in-situ bioleaching coupled to electrokinetics. *Environmental Research*, 238, 117183, 2023. https://doi.org/10.1016/j.envres.2023.117183.

Ahmadinia, M., Setchi, R., Evans, S.L. et al. Transforming E-waste into value: a circular economy approach to pcb recycling. In: Scholz, S.G., Howlett, R.J., Setchi, R. (eds) *Sustainable Design and Manufacturing 2023. SDM 2023. Smart Innovation, Systems and Technologies*, vol 377. Springer, Singapore, 2024. https://doi.org/10.1007/978-981-99-8159-5_24

Arshadi, M., Mousavi, S.M., Rasoulnia, P. Enhancement of simultaneous gold and copper recovery from discarded mobile phone PCBs using Bacillus megaterium: RSM based optimization of effective factors and evaluation of their interactions. *Waste Management*, 57, 158–167, 2016. https://doi.org/10.1016/j.wasman.2016.05.012.

Arslan, V. Bacterial leaching of copper, zinc, nickel and aluminum from discarded printed circuit boards using acidophilic bacteria. *Journal of Material Cycles and Waste Management*, 23, 2005–2015, 2021. https://doi.org/10.1007/s10163-021-01274-9

Bahaloo-Horeh, N., Mousavi, S. M. Enhanced recovery of valuable metals from spent lithium-ion batteries through optimization of organic acids produced by *Aspergillus niger*. *Waste Management*, 60, 666–679, 2017. https://doi.org/10.1016/j.wasman.2016.10.034.

Bakhshoude, M., Darezereshki, E., Bakhtiari, F. Thermoacidophilic bioleaching of copper sulfide concentrate in the presence of chloride ions. *Journal of Central South University*, 30, 749–762, 2023. https://doi.org/10.1007/s11771-023-5276-x

Baniasadi, M., Graves, J.E., Ray, D.A. et al. Closed-loop recycling of copper from waste printed circuit boards using bioleaching and electrowinning processes. *Waste and Biomass Valorization*, 12, 3125–3136, 2021. https://doi.org/10.1007/s12649-020-01128-9

Benzal, E., Cano, A., Solé, M. et al. Copper recovery from PCBs by Acidithiobacillus ferrooxidans: toxicity of bioleached metals on biological activity. *Waste and Biomass Valorization*, 11, 5483–5492, 2020a. https://doi.org/10.1007/s12649-020-01036-y

Benzal, E., Solé, M., Lao, C. et al. Elemental copper recovery from E-wastes mediated with a two-step bioleaching process. *Waste and Biomass Valorization*, 11, 5457–5465, 2020b. https://doi.org/10.1007/s12649-020-01040-2

Bomberg, M., Miettinen, H., Hajdu-Rahkama, R., Lakaniemi, A-M., Anacki, W., Witecki, K., Puhakka, J. A., Ineich, T., Slabbert, W., Kinnunen, P. Indirect in situ bioleaching is an emerging tool for accessing deeply buried metal reserves, but can the process be managed? – A case study of copper leaching at 1 km depth. *Environmental Technology & Innovation*, 32, 103375, 2023. https://doi.org/10.1016/j.eti.2023.103375.

Boopathi, S. Experimental study on mechanical strength and durability of E-waste mixed concrete. In: Sahoo, S., Yedla, N. (eds) *Recent Advances in Mechanical Engineering. ICRAMERD 2023. Lecture Notes in Mechanical Engineering*. Springer, Singapore, 2024. https://doi.org/10.1007/978-981-97-1080-5_3

Chaerun, S. K., Sulistyo, R. S., Minwal, W. P., Mubarok, M. Z. Indirect bioleaching of low-grade nickel limonite and saprolite ores using fungal metabolic organic acids generated by Aspergillus niger. *Hydrometallurgy*, 174, 29–37, 2017. https://doi.org/10.1016/j.hydromet.2017.08.006.

Dey, S., Shekhawat, M. S., Pandey, D. K., Ghorai, M., Anand, U., Hoda, M., Bhattacharya, S., Bhattacharjee, R., Ghosh, A., Nongdam, P., Kumar, V., Dey, V. Chapter 10 - Microbial community and their role in bioremediation of polluted e-waste sites. In: Kumar, V., Bilal, M., Shahi, S. K., Garg, V. K. (eds), *Developments in Applied Microbiology and Biotechnology, Metagenomics to Bioremediation*, 261–283. Academic Press, 2023. https://doi.org/10.1016/B978-0-323-96113-4.00006-8.

Dusengemungu, L., Kasali, G., Gwanama, C., Mubemba, B. Overview of fungal bioleaching of metals. *Environmental Advances*, 5, 100083, 2021. https://doi.org/10.1016/j.envadv.2021.100083

Falagán, C., Sbaffi, T., Williams, G.B., Bargiela, R., Dew, D. W., Hudson-Edwards, K. A. Nutrient optimization in bioleaching: are we overdosing? *Frontiers in Microbiology*, 15:1359991, 2024. doi: 10.3389/fmicb.2024.1359991

Hong, Y., Valix, M. Bioleaching of electronic waste using acidophilic sulfur oxidising bacteria. *Journal of Cleaner Production*, 65, 465–472, 2014. https://doi.org/10.1016/j.jclepro.2013.08.043.

Hsu, E., Durning, C.J., West, A.C., Park, A.H.A. Enhanced extraction of copper from electronic waste via induced morphological changes using supercritical CO_2. *Resources, Conservation & Recycling*, 168, 2021, 105296. https://doi.org/10.1016/j.resconrec.2020.105296

Huang, M.Q., Wu, A.X. Numerical analysis of aerated heap bioleaching with variable irrigation and aeration combinations. *Journal of Central South University*, 27, 1432–1442, 2020. https://doi.org/10.1007/s11771-020-4379-x

Hubau, A., Pino-Herrera, D.O., Falagán, C. et al. Influence of the nutrient medium composition during the bioleaching of polymetallic sulfidic mining residues. *Waste and Biomass Valorization*, 15, 561–575, 2024. https://doi.org/10.1007/s12649-023-02090-y

Hudec, M. A. R., Sodhi, M., Goglia-Arora, D. Biorecovery of metals from electronic waste. In *7th Latin American and Caribben Conference for Engineering and Technology*, SAN Cristobal, Venezuela, 2009.

Ilyas, S., Lee, J-c. Bioleaching of metals from electronic scrap in a stirred tank reactor. *Hydrometallurgy*, 149, 50–62, 2014. https://doi.org/10.1016/j.hydromet.2014.07.004.

Ilyas, S., Srivastava, R.R., Kim, H. Bioleaching of post-consumer $LiCoO_2$ batteries using aspergillus niger. In: Forsberg, K., et al. (eds) *Rare Metal Technology 2024. TMS 2024. The Minerals, Metals & Materials Series*. Springer, Singapore, 2024. https://doi.org/10.1007/978-3-031-50236-1_18

Ishak, K.E.H.K., Ismail, S., Razak, M.I.B.A. Recovery of copper and valuable metals from E-waste via hydrometallurgical method. *Materials Today: Proceedings*, 66, 3077–3081, 2022. https://doi.org/10.1016/j.matpr.2022.07.395

Işıldar, A., van de Vossenberg, J., Rene, E. R., van Hullebusch, E. D., Lens, P. N. L. Two-step bioleaching of copper and gold from discarded printed circuit boards (PCB). *Waste Management*, 57, 149–157, 2016. https://doi.org/10.1016/j.wasman.2015.11.033.

Jaiswal, S.K., Mukti, S.K., Barriers affecting formal recycling of e-waste in Indian Context. In: Bhardwaj, A., Pandey, P.M., Misra, A. (eds) *Optimization of Production and Industrial Systems. CPIE 2023. Lecture Notes in Mechanical Engineering*. Springer, Singapore, 2024. https://doi.org/10.1007/978-981-99-8343-8_26

Jia, Y., Sun, H., Tan, Q., Xu, J., Feng, X., Ruan, R. Industrial heap bioleaching of copper sulfide ore started with only water irrigation. *Minerals*, 11, 1299, 2021. https://doi.org/10.3390/min11111299

Jiankang, W., Bowei, C., Kuangdi, X., Bioleaching. In: Xu, K. (eds) *The ECPH Encyclopedia of Mining and Metallurgy*. Springer, Singapore, 2024. https://doi.org/10.1007/978-981-19-0740-1_1391-1

Johnson, M., Fitzpatrick, C., Wagner, M., Huisman, J. Modelling the levels of historic waste electrical and electronic equipment in Ireland. *Resources, Conservation and Recycling*, 131, 1–16, 2018. https://doi.org/10.1016/j.resconrec.2017.11.029.

Liang, G., Li, P., Liu, W. et al. Enhanced bioleaching efficiency of copper from waste printed circuit boards (PCBs) by dissolved oxygen-shifted strategy in *Acidithiobacillus ferrooxidans*. *Journal of Material Cycles and Waste Management*, 18, 742–751, 2016. https://doi.org/10.1007/s10163-015-0375-x

Määttä, L., Hajdu-Rahkama, R., Oinonen, C., Puhakka, J. A. Effects of metal extraction liquors from electric vehicle battery materials production on iron and sulfur oxidation by heap bioleaching microorganisms. *Minerals Engineering*, 178, 107409, 2022. https://doi.org/10.1016/j.mineng.2022.107409.

Mathew, A.A., Parthasarathy, P., Vivekanandan, S. Development of copper nanoparticles from E-waste for biomedical applications. In: Komanapalli, V.L.N., Sivakumaran, N., Hampannavar, S. (eds) *Advances in Automation, Signal Processing, Instrumentation, and Control. I-CASIC 2020. Lecture Notes in Electrical Engineering*, vol 700. Springer, Singapore, 2021. https://doi.org/10.1007/978-981-15-8221-9_64

Mim, S.H., Tabassum, F., Ripa, T.A. et al. Applying a privacy policy for E-waste management in Bangladesh. In: Bhattacharyya, S., Banerjee, J.S., Köppen, M. (eds) *Human-Centric Smart Computing. ICHCSC 2023. Smart Innovation, Systems and Technologies*, vol 376. Springer, Singapore, 2023. https://doi.org/10.1007/978-981-99-7711-6_26

Murugesan, M.P., Kannan, K., Selvaganapathy, T. Bioleaching recovery of copper from printed circuit boards and optimization of various parameters using response surface methodology (RSM). *Materials Today: Proceedings*, 26, 2720–2728, 2020. https://doi.org/10.1016/j.matpr.2020.02.571

Nili, S., Arshadi, M., Yaghmaei, S. Fungal bioleaching of e-waste utilizing molasses as the carbon source in a bubble column bioreactor. *Journal of Environmental Management*, 307, 114524, 2022, https://doi.org/10.1016/j.jenvman.2022.114524.

Pang, C., Lin, M., Wu, Y., Ruan, J. Enhanced Au bioleaching from waste central processing unit (CPU) slots in a bioreactor with 10 mA direct current. *Resources, Conservation & Recycling*, 198, 107171, 2023. https://doi.org/10.1016/j.resconrec.2023.107171

Pinho, S.C., Ferraz, C.A., Almeida, M.F. Copper recovery from printed circuit boards using ammonia–ammonium sulphate system: a sustainable approach. *Waste and Biomass Valorization*, 14, 1683–1691, 2023. https://doi.org/10.1007/s12649-022-01953-0

Pirsaheb, M., Zadsar, S., Rastegar, S. O., Gu, T., Hossini, H., Bioleaching and ecological toxicity assessment of carbide slag waste using Acidithiobacillus bacteria, *Environmental Technology & Innovation*, 22, 101480, 2021, https://doi.org/10.1016/j.eti.2021.101480.

Pradhan, J. Metals bioleaching from electronic waste by Chromobacterium violaceum and Pseudomonads sp. *Waste Management & Research: the Journal of the International Solid Wastes and Public Cleansing Association, ISWA*, 30, 2012. doi: 10.1177/0734242X12437565

Ranjbar, M., Ranjbar Hamghavandi, M., Fazaelipoor, M.H. et al. Development of a kinetic model of the bacterial dissolution of copper concentrate. *Mining, Metallurgy & Exploration*, 37, 345–353, 2020. https://doi.org/10.1007/s42461-019-00114-7

Rao, M.D., Singh, K.K., Morrison, C.A., Love, J.B. Recycling copper and gold from E-waste by a two-stage leaching and solvent extraction process. *Separation and Purification Technology*, 263, 118400, 2021. https://doi.org/10.1016/j.seppur.2021.118400

Royaei, M.M., Pourbabaee, A., Noaparast, M. Copper recovery from chalcopyrite flotation concentrate using mixed mesophilic strains subjected to microwave and ultrasound irradiation. *Iranian Journal of Science and Technology, Transaction A, Science*, 43, 335–343, 2019. https://doi.org/10.1007/s40995-017-0426-3

Ruan, J., Zhu, X., Qian, Y., Hu, J. A new strain for recovering precious metals from waste printed circuit boards. *Waste Management*, 34(5), 901–907, 2014. https://doi.org/10.1016/j.wasman.2014.02.014.

Sakthivel, U., Swaminathan, G., Anis, J.J.J. Strategies for quantifying metal recovery from waste electrical and electronic equipment (WEEE/E-waste) using mathematical approach. *Process Integration and Optimization for Sustainability*, 6, 781–790, 2022. https://doi.org/10.1007/s41660-022-00250-6

Samarasekere, P.W. Microbial remediation technologies for mining waste management. In: Bala, K., Ghosh, T., Kumar, V., Sangwan, P. (eds) *Harnessing Microbial Potential for Multifarious Applications. Energy, Environment, and Sustainability*. Springer, Singapore, 2024. https://doi.org/10.1007/978-981-97-1152-9_3

Seredkin, M., Zabolotsky, A., Jeffress, G. In situ recovery, an alternative to conventional methods of mining: Exploration, resource estimation, environmental issues, project evaluation and economics, *Ore Geology Reviews*, 79, 500–514, 2016. https://doi.org/10.1016/j.oregeorev.2016.06.016.

Sharifidarabad, H. Cu recovery from E-wastes. In: González, D.F. (eds) *Copper Overview – From Historical Aspects to Applications*. Intechopen, London, 2024. https://doi.org/10.5772/.1004994

Sulaiman Zangina, A., Ahmed, A., Nuhu, Z. et al. E-waste quantification and the associated valuable resources in Kano metropolis, Nigeria. *Journal of Material Cycles and Waste Management*, 25, 3664–3673, 2023. https://doi.org/10.1007/s10163-023-01793-7

Trivedi, A., Kanaujia, K., Upvan, K., Hait, S. Chapter 4, Fungal biotechnology for the recovery of critical metals from e-waste current research trends and prospects. In: Debnath, B., Das, A., Chowdary, P. A., Bhattacharyya, S. (eds) *Development in E-waste Management*, 168–183, Boca Raton, CRC Press, 2023. https://doi.org/10.1201/9781003301899

Vera, M., Schippers, A., Hedrich, S., et al. Progress in bioleaching: fundamentals and mechanisms of microbial metal sulfide oxidation – part A. *Applied Microbiology and Biotechnology* 106, 6933–6952, 2022. https://doi.org/10.1007/s00253-022-12168-7

Wang, X., Yang, H.Y., Zhang, Q. et al. Effect of particle size on bioleaching of low-grade nickel ore in a column reactor. *Journal of Central South University*, 28, 1333–1341, 2021. https://doi.org/10.1007/s11771-021-4706-x

Xia, M., Bao, P., Liu, A., Wang, M., Shen, L., Yu, R., Liu, Y., Chen, M., Li, J., Wu, X., Qiu, G., Zeng, W. Bioleaching of low-grade waste printed circuit boards by mixed fungal culture and its community structure analysis. *Resources, Conservation and Recycling*, 136, 267–275, 2018. https://doi.org/10.1016/j.resconrec.2018.05.001.

Xiang, Y., Wu, P., Zhu, N., Zhang, T., Liu, W., Wu, J., Li, P. Bioleaching of copper from waste printed circuit boards by bacterial consortium enriched from acid mine drainage. *Journal of Hazardous Materials*, 184(1–3), 812–818, 2010. https://doi.org/10.1016/j.jhazmat.2010.08.113.

Yang, B., Zhao, C., Luo, W., Liao, R., Gan, M., Wang, J., Liu, X., Qiu, G. Catalytic effect of silver on copper release from chalcopyrite mediated by Acidithiobacillus ferrooxidans. *Journal of Hazardous Materials*, 392, 122290, 2020, https://doi.org/10.1016/j.jhazmat.2020.122290.

Yanga, Y., Chena, S., Li, S., Chena, M., Chena, H., Liub, B. Bioleaching waste printed circuit boards by Acidithiobacillus ferrooxidans and its kinetics aspect. *Journal of Biotechnology*, 173, 24–30, 2014. https://doi.org/10.1016/j.jbiotec.2014.01.008

Zeng, W.M., Cai, Y.X., Hou, C.W. et al. Influence diversity of extracellular DNA on bioleaching chalcopyrite and pyrite by Sulfobacillus thermosulfidooxidans ST. *Journal of Central South University*, 27, 1466–1476, 2020. https://doi.org/10.1007/s11771-020-4382-2

Zhaoa, Q., Tonga, L., Kamalib, A.R., Sande, W., Yang, H. Role of humic acid in bioleaching of copper from waste computer motherboards. *Hydrometallurgy*, 197, 2020, 105437. https://doi.org/10.1016/j.hydromet.2020.105437

Zulkernain, N.H., Basant, N., Ng, C.C. et al. Recovery of precious metals from E-wastes through conventional and phytoremediation treatment methods: a review and prediction. *Journal of Material Cycles and Waste Management*, 25, 2726–2752, 2023. https://doi.org/10.1007/s10163-023-01717-5

7 Bioleaching of Manganese from Industrial Waste Effluents

Anitha Thulasisingh and Trisha G

7.1 INTRODUCTION

7.1.1 Industrial Waste

Industrial waste refers to any waste generated during the production or manufacturing process within industries. Industrial waste comprises a diverse array of materials as shown in Figure 7.1. Generated from manufacturing processes, mining operations, construction activities, and other industrial activities, this waste stream poses significant risks to ecosystems, human health, and the environment (Tong et al., 2018). Improper disposal and inadequate treatment can lead to contamination of air, water bodies, and soil, exacerbating pollution and endangering biodiversity. Industrial waste poses a multifaceted challenge to modern society, encompassing environmental, economic, and health concerns. As industries continue to expand globally, the volume and complexity of waste generated have escalated, necessitating urgent attention and sustainable solutions (Roy et al., 2014).

7.1.2 Overview of Industrial Waste Generation and Environmental Concerns

Industrial waste is a significant by-product of modern industrial processes, encompassing a wide array of materials and substances that are no longer useful for production.

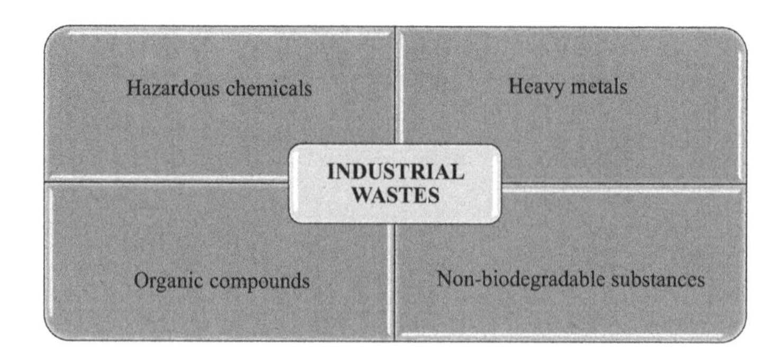

FIGURE 7.1 Components of industrial wastes.

DOI: 10.1201/9781003415541-7

From manufacturing plants to chemical refineries, industrial activities generate copious amounts of waste, ranging from solid debris to hazardous chemicals. This waste poses considerable environmental concerns due to its potential to pollute air, water, and soil, as well as its detrimental effects on ecosystems and human health. One of the primary issues associated with industrial waste is its sheer volume. Industrial processes, particularly in sectors like manufacturing, construction, and mining, produce immense quantities of waste on a daily basis. This accumulation can overwhelm waste management systems and lead to improper disposal practices, such as illegal dumping or inadequate treatment, exacerbating environmental pollution (Pradhan et al., 2015).

Furthermore, industrial waste often contains hazardous substances that pose serious risks to both the environment and public health. Chemical pollutants, heavy metals, and toxic compounds can leach into soil and groundwater, contaminating drinking water sources and rendering land unfit for agriculture or habitation. Airborne emissions from industrial activities, such as particulate matter and volatile organic compounds, contribute to air pollution, respiratory ailments, and even climate change. Improper handling and disposal of industrial waste can also result in ecological damage. Contaminated water bodies can disrupt aquatic ecosystems, leading to fish kills and biodiversity loss. Soil contamination can impair plant growth and soil fertility, affecting agricultural productivity and food security. Moreover, wildlife may suffer adverse effects from exposure to hazardous substances, leading to population declines and ecological imbalances (Chen et al., 2020).

Addressing the environmental concerns associated with industrial waste requires comprehensive waste management strategies. These strategies should prioritize waste reduction and prevention through improved production processes, resource efficiency, and recycling initiatives. Additionally, proper treatment and disposal methods, such as incineration, landfilling, or hazardous waste treatment facilities, must be employed to minimize environmental contamination.

Regulatory frameworks and enforcement mechanisms play a crucial role in ensuring compliance with environmental standards and promoting responsible waste management practices among industries. Government agencies, environmental organizations, and industry stakeholders must collaborate to develop and implement policies that prioritize environmental protection while fostering sustainable industrial development (Ferrier et al., 2022).

7.1.3 Manganese

Manganese, a chemical element represented by the symbol Mn and atomic number 25, holds a significant place in various aspects of human life, ranging from industrial applications to biological functions. Manganese plays a crucial role in various industrial processes due to its diverse properties. One of its primary applications is in steelmaking, where it serves as a crucial alloying element, imparting strength, hardness, and corrosion resistance to steel. Manganese is also utilized in the production of batteries, particularly alkaline batteries and lithium-ion batteries, where it contributes to improved performance and stability. Additionally, manganese compounds find application in the manufacture of pigments, ceramics, and fertilizers, as well as in water treatment and environmental remediation processes (Youcai et al., 2014).

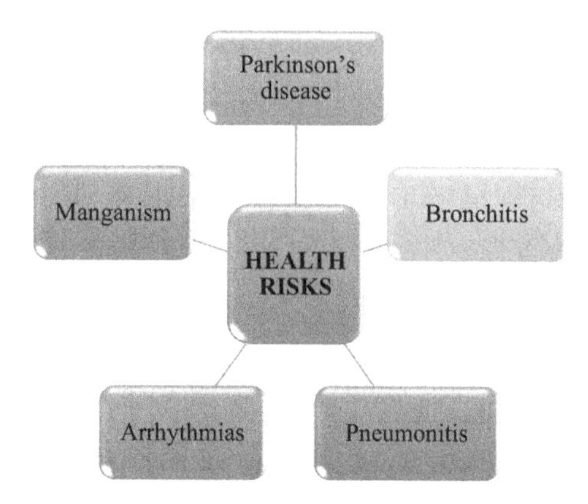

FIGURE 7.2 Health risks associated with manganese inhalation.

Excessive manganese levels in soil and water can lead to environmental contamination and adverse ecological impacts. Industrial activities, mining operations, and agricultural practices are among the primary sources of manganese pollution, which can result in soil degradation, water quality deterioration, and toxicity to aquatic organisms. Excessive exposure to manganese, particularly through occupational settings or environmental pollution, can pose significant health risks. Inhalation of manganese dust or fumes, as seen in occupational settings such as mining or welding, can lead to various health issues as shown in Figure 7.2. Therefore, stringent regulations and occupational safety measures are essential to mitigate manganese-related health hazards and protect human well-being (Pak et al., 2012).

7.1.4 Importance of Manganese as a Strategic Metal

Manganese holds significant importance as a strategic metal in industrial waste due to its diverse properties and potential for recycling and reuse. Manganese is a vital alloying element in steelmaking, where it enhances the strength, hardness, and corrosion resistance of steel. Industrial waste from steel manufacturing processes often contains manganese-rich residues, such as slags and dust. Recycling these wastes allows for the recovery of manganese, reducing the need for primary manganese ore extraction and conserving natural resources. Recycling manganese from spent batteries reduces the demand for materials and minimizes environmental impacts associated with mining and processing (Sahoo et al., 2011).

Manganese compounds serve as catalysts in chemical reactions and industrial processes, such as petroleum refining, water treatment, and synthesis of organic compounds. Industrial waste streams from chemical manufacturing industries may contain manganese-containing catalysts or by-products. Recycling these wastes enables the recovery of valuable manganese compounds, promoting resource efficiency and reducing the need for fresh production. Waste generated from the production and application of manganese-based pigments can be collected and recycled, thus minimizing waste disposal and conserving manganese resources. Manganese compounds are employed in environmental remediation, such as the treatment of contaminated soil and water. Industrial waste containing manganese-based adsorbents or treatment residues may contain recoverable manganese, which can be reused in remediation processes or repurposed for other applications (Contestabile et al., 2011).

7.1.5 Introduction to Bioleaching as a Sustainable Solution for Manganese Recovery

Bioleaching is a microbial-mediated process that utilizes microorganisms to solubilize metals from ores or waste materials. In the case of manganese recovery, certain bacteria and fungi are capable of oxidizing manganese minerals, releasing soluble manganese ions into solution. These microorganisms typically thrive in acidic environments and produce organic acids or other metabolites that facilitate metal solubilization. Bioleaching, as a sustainable solution for manganese recovery, presents an innovative and environmentally friendly approach to extract manganese from ores and industrial waste materials. Unlike traditional mining and metallurgical processes, bioleaching is a low-impact and environmentally benign method that does not involve the use of harsh chemicals or generate toxic by-products. By harnessing the natural metabolic activities of microorganisms, bioleaching offers an eco-friendly alternative to conventional extraction methods, reducing energy consumption and minimizing environmental pollution (Sildar et al., 2019).

One of the key advantages of bioleaching for manganese recovery is its sustainability. Bioleaching can be applied to a wide range of manganese-containing materials, including low-grade ores, mining residues, and industrial waste streams. This versatility makes bioleaching particularly suitable for recovering manganese from sources that are economically or environmentally challenging to process using conventional techniques. Unlike chemical leaching methods, which often result in the dissolution of various metals and impurities, bioleaching can be tailored to target specific metal species, thereby improving the purity and quality of the extracted manganese. This selectivity reduces the need for downstream purification steps and enhances the overall efficiency of the recovery process. Manganese-rich waste materials, such as mine tailings, slag, or industrial residues, can be treated using bioleaching to recover valuable manganese compounds. By converting waste into a valuable resource, bioleaching contributes to waste reduction, resource conservation, and sustainable materials management (He et al., 2021).

7.2 MANGANESE IN INDUSTRIAL WASTE

7.2.1 Sources and Types of Manganese Containing Industrial Wastes

Manganese, a transition metal widely used in industrial processes, plays a crucial role in various sectors such as steel production, batteries, ceramics, and fertilizers. However, its widespread use also leads to the generation of industrial wastes containing manganese. Understanding the sources and types of these wastes is essential for effective waste management and environmental protection (Valenzuela et al., 2019).

7.2.1.1 Sources of Manganese-Containing Industrial Wastes

Metallurgical process: The primary source of manganese-containing industrial wastes stems from metallurgical activities, particularly steelmaking. During the production of steel, manganese is often added as an alloying element to enhance the material's strength and durability. Consequently, the by-products of steel production, such as slag and dust, contain significant amounts of manganese.

Mining operations: Mining activities aimed at extracting manganese ores contribute to the generation of industrial wastes. The processing of manganese-rich ores generates tailings and waste rock, which may contain residual manganese compounds. Additionally, runoff and leaching from mining sites can contaminate surrounding water bodies with manganese.

Chemical manufacturing: Industries involved in the production of chemicals, such as those used in batteries, pigments, and fertilizers, can produce manganese-containing wastes. Chemical synthesis processes may result in the formation of by-products or waste streams containing manganese compounds.

Wastewater treatment: Wastewater from various industrial processes can contain manganese as a contaminant. Industries such as mining, metal finishing, and electroplating discharge wastewater containing manganese ions due to the use of manganese-containing chemicals in their operations.

Waste electrical and electronic equipment (WEEE): Manganese-containing materials are present in electronic devices and electrical equipment. The disposal of obsolete or end-of-life electronics contributes to the accumulation of manganese-containing industrial wastes in landfills or informal recycling operations (Ferraro et al., 2019).

7.2.1.2 Types of Manganese-Containing Industrial Wastes

Metallurgical processes, particularly steelmaking, generate slag and dust as by-products. Mining operations produce tailings and waste rock during the extraction and processing of manganese ores. Chemical manufacturing processes can yield residues or by-products containing manganese compounds. Industrial wastewater contaminated with manganese ions is a significant source of manganese-containing industrial waste. Discarded electronic devices and electrical equipment constitute a growing source of manganese-containing

TABLE 7.1
Components Present in Manganese Industrial Wastes

Types of Manganese-Containing Industrial Wastes	Components Present in the Industrial Wastes
Slag and dust	Manganese oxide
Tailings and waste rocks	Manganese minerals
Process residues	Filter cake
	Sludge
	Spent catalysts
Contaminated wastewater	Manganese ions
Electronic waste	Components such as batteries, circuit boards, and connectors may contain manganese compounds

industrial wastes. These components can lead to the environmental risks if not properly managed. Table 7.1 provides information about the components present in different types of manganese-containing industrial wastes (Brandaleze et al., 2018).

7.2.2 Composition and Characteristics of Manganese-Rich Wastes

Manganese-rich wastes are generated from various industrial processes and activities where manganese compounds are utilized. These wastes exhibit diverse compositions and characteristics depending on their sources and origins. Understanding the composition and characteristics of manganese-rich wastes is essential for effective waste management and environmental protection (Ghosh et al., 2016).

7.2.2.1 Composition of Manganese-Rich Wastes

Manganese oxides and hydroxides: Manganese-rich wastes often contain manganese oxides and hydroxides as primary constituents. These compounds result from the oxidation of manganese during industrial processes such as mining, metallurgy, and chemical manufacturing. Common manganese oxides include pyrolusite (MnO_2), hausmannite (Mn_3O_4), and manganite (MnOOH), while manganese hydroxides include vernadite and birnessite.

Manganese silicates: In mining operations, manganese ores may contain silicate minerals such as rhodonite, rhodochrosite, and braunite. Consequently, manganese-rich wastes from mining activities may also contain manganese silicates as significant components.

Manganese carbonates: Manganese-rich carbonate minerals, such as rhodochrosite ($MnCO_3$), are commonly found in manganese ores. Therefore, waste materials generated from mining and mineral processing may contain manganese carbonates, contributing to the overall composition of manganese-rich wastes.

Other metal contaminants: Manganese-rich wastes may also contain other metal contaminants depending on the specific industrial processes involved. Common contaminants include iron, nickel, cobalt, and chromium, which can co-occur with manganese in ores or as impurities in waste materials.

Organic compounds and organic matter: In some industrial contexts, manganese-rich wastes may contain organic compounds or organic matter originating from chemical processes or organic-rich environments. These organic constituents can influence the physicochemical properties and behavior of manganese-rich wastes (Kozlov et al., 2017).

7.2.2.2 Characteristics of Manganese-Rich Wastes

High manganese content: The primary characteristic of manganese-rich wastes is their high manganese content, which distinguishes them from other industrial wastes. The manganese content can vary significantly depending on the source and processing methods but is typically present in substantial quantities.

Variable oxidation states: Manganese exhibits multiple oxidation states, including +2, +3, +4, +6, and +7, depending on the chemical environment. Consequently, manganese-rich wastes may contain manganese in different oxidation states, influencing their chemical reactivity and environmental behavior.

Acid-base properties: Manganese-rich wastes can exhibit acid-base properties due to the presence of manganese oxides and hydroxides. Manganese oxides, such as MnO_2, can act as strong oxidizing agents, while manganese hydroxides can contribute to pH buffering and adsorption processes in the environment.

Solubility and leachability: The solubility and leachability of manganese-rich wastes depend on factors such as particle size, mineralogy, and pH. Manganese compounds can dissolve in acidic or alkaline environments, leading to the release of manganese ions into soil and water, posing environmental risks.

Potential toxicity and environmental impacts: Manganese-rich wastes may exhibit toxicity to organisms and ecosystems, particularly at high concentrations. Manganese toxicity can affect aquatic organisms, plants, and humans, leading to adverse health effects and ecological disturbances (Lucheta et al., 2019).

7.2.3 Environmental and Economic Implications of Manganese Disposal

7.2.3.1 Environmental Implications of Manganese Disposal

Water contamination: Improper disposal of manganese-containing wastes can lead to water contamination. Manganese leaching from landfills or waste disposal sites can contaminate groundwater and surface water bodies. Elevated levels of manganese in water can have detrimental effects on aquatic ecosystems and human health, causing toxicity and impairing water quality.

Soil degradation: Manganese disposal can result in soil degradation and contamination. Manganese-rich wastes deposited on land can leach into the soil, affecting soil pH and nutrient availability. High concentrations of manganese in soil can inhibit plant growth, disrupt soil microbial communities, and lead to long-term soil degradation.

Air pollution: Certain disposal methods, such as incineration or combustion, can release manganese-containing particulate matter into the air. Inhalation of manganese particles can pose health risks to nearby communities, particularly affecting respiratory health. Airborne manganese emissions may also contribute to environmental pollution and ecosystem damage.

Ecological impacts: Manganese disposal can have adverse effects on terrestrial and aquatic ecosystems. Accumulation of manganese in sediments, soils, and water bodies can disrupt ecosystem dynamics and biodiversity. Manganese toxicity can harm aquatic organisms, plants, and soil fauna, leading to ecological imbalances and ecosystem degradation.

Long-term persistence: Manganese is a persistent environmental contaminant, meaning that it can persist in the environment for extended periods. Improperly disposed of manganese-containing wastes may continue to release manganese into the environment over time, exacerbating environmental contamination and posing ongoing risks to ecosystems and human health (Zhang et al., 2017).

7.2.3.2 Economic Implications of Manganese Disposal

Clean-up costs: Addressing environmental contamination resulting from manganese disposal incurs significant clean-up costs. Remediation of contaminated water bodies, soil, and air requires substantial financial resources and may involve complex engineering and treatment technologies. These clean-up costs place a financial burden on governments, industries, and communities.

Healthcare expenses: Exposure to manganese contamination can result in adverse health effects, leading to increased healthcare expenses. Health problems associated with manganese toxicity, such as neurological disorders and respiratory ailments, require medical treatment and healthcare services, contributing to healthcare expenditures and public health costs.

Resource loss: Manganese-containing wastes represent a loss of valuable resources that could otherwise be recovered and reused. Manganese is a finite resource, and its inefficient disposal represents a missed opportunity for resource conservation and sustainable utilization. Implementing recycling and recovery technologies can help recover valuable manganese resources and reduce the need for primary extraction.

Regulatory compliance: Compliance with environmental regulations and standards regarding manganese disposal entails costs for industries and businesses. Meeting regulatory requirements for waste management, pollution control, and environmental monitoring necessitates investments in infrastructure, technology upgrades, and compliance measures, impacting the operating costs and competitiveness of enterprises.

Social costs: The environmental and health impacts of manganese disposal impose social costs on affected communities. Reduced quality of life, decreased property values, and social stigma associated with contaminated environments can detrimentally affect communities' well-being and livelihoods. Addressing these social costs requires holistic approaches that consider community engagement, empowerment, and social justice (Tian et al., 2010).

7.3 FUNDAMENTALS OF BIOLEACHING

7.3.1 Definition and Principles of Bioleaching

7.3.1.1 Definition of Bioleaching

Bioleaching is a microbial process used to extract metals from ores and concentrates using microorganisms such as bacteria and fungi. In bioleaching, these microorganisms catalyze the oxidation of metal sulfides present in ores, releasing metal ions into solutions. The metal ions can then be recovered through subsequent processes. Bioleaching is an environmentally friendly and cost-effective alternative to conventional methods of metal extraction, as it reduces the need for harsh chemicals and energy-intensive processes. This biotechnological approach is applied in industries such as mining, metallurgy, and environmental remediation to recover valuable metals from low-grade ores, complex mineral deposits, and waste materials (Xia et al., 2011).

7.3.1.2 Principles of Bioleaching

Microbial metabolism: At the core of bioleaching lies the metabolic activities of acidophilic microorganisms, predominantly bacteria and archaea. These microorganisms thrive in acidic environments and possess enzymes capable of oxidizing metal sulfide minerals. Key metabolic pathways involved include iron and sulfur oxidation, which result in the release of metal ions from sulfide ores.

Chemical and biological interactions: Bioleaching involves intricate chemical and biological interactions between microorganisms and mineral surfaces. Microbial attachment to mineral substrates, followed by enzymatic oxidation of metal sulfides, initiates the dissolution of metals into solution. Organic acids and other metabolites produced by microorganisms aid in mineral dissolution and solubilization.

Redox reactions: Bioleaching catalyzes redox reactions that transform insoluble metal sulfides into soluble metal ions. Microorganisms such as *Acidithiobacillus ferrooxidans* and *Acidithiobacillus thiooxidans* oxidize ferrous iron (Fe^{2+}) and reduced sulfur compounds (S, HS^-) to ferric iron (Fe^{3+}) and sulfate (SO_4^{2-}), respectively. These reactions liberate metal ions such as copper, zinc, and nickel from sulfide minerals.

pH regulation: Acidophilic microorganisms play a pivotal role in regulating the pH of bioleaching environments. By metabolizing sulfur and producing sulfuric acid (H_2SO_4), these microorganisms maintain low pH conditions conducive to metal solubilization. Optimal pH ranges vary depending on the specific microbial species and the metal being leached.

Temperature and oxygen requirements: Bioleaching processes are influenced by temperature and oxygen availability. Acidophilic microorganisms thrive in moderate to high temperatures, typically ranging from 25°C to 45°C, depending on the microbial species. Adequate oxygen supply is essential to sustain microbial growth and metabolic activity, particularly in aerobic bioleaching processes (Akcil et al., 2015).

7.3.2 Microbial Mechanisms Involved in Manganese Oxidation and Solubilization

Manganese oxidation and solubilization are essential processes in biogeochemical cycling, influencing the mobility, bioavailability, and speciation of manganese in diverse environments. Microorganisms, including bacteria and fungi, play key roles in mediating these processes through enzymatic reactions and metabolic activities. Understanding the microbial mechanisms involved in manganese oxidation and solubilization is critical for elucidating biogeochemical cycles and developing biotechnological applications (Das et al., 2011).

7.3.2.1 Microbial Manganese Oxidation

Microbial manganese oxidation refers to the enzymatic conversion of soluble manganese ions (Mn^{2+}) into insoluble manganese oxides or hydroxides, such as MnO_2 or $MnO(OH)_2$. Several bacterial and fungal species are known to catalyze manganese oxidation, utilizing various mechanisms:

Bacterial manganese oxidizers: Bacteria belonging to genera such as Leptothrix, Pseudomonas, and Bacillus are capable of oxidizing manganese. These bacteria produce extracellular enzymes, such as multicopper oxidases and manganese peroxidases, which catalyze the oxidation of Mn^{2+} to MnO_2. The formed manganese oxides often precipitate as biogenic minerals, contributing to the formation of manganese-rich microbial mats and biofilms in natural environments.

Fungal manganese oxidizers: Fungi, particularly species belonging to the genera Aspergillus and Penicillium, also possess manganese oxidation capabilities. Fungal hyphae produce oxidative enzymes, including laccases and peroxidases, which facilitate the oxidation of Mn^{2+} ions in the presence of oxygen. Fungal-mediated manganese oxidation contributes to mineral weathering and soil formation processes in terrestrial ecosystems (Furini et al., 2015).

7.3.2.2 Microbial Solubilization

Microbial manganese solubilization involves the release of manganese ions from insoluble manganese minerals, making them available for microbial uptake or environmental cycling. Certain microbial species employ mechanisms to solubilize manganese minerals.

Acidic dissolution: Some bacteria and fungi produce organic acids, such as citric acid, oxalic acid, and gluconic acid, which chelate manganese ions from insoluble minerals, such as MnO_2 or $MnCO_3$, leading to their dissolution. These organic acids lower the pH of the surrounding environment, enhancing manganese solubilization through chemical weathering processes.

Reduction-mediated solubilization: Certain microorganisms possess the ability to reduce insoluble manganese oxides under anaerobic conditions, releasing soluble Mn^{2+} ions. This reduction-mediated solubilization process involves the enzymatic reduction of manganese oxides by microbial electron transfer mechanisms, such as microbial respiration or fermentation (Jana et al., 2015).

7.3.3 Key Factors Influencing Bioleaching Efficiency

pH: pH plays a critical role in bioleaching efficiency as it influences the activity and stability of microbial enzymes involved in metal oxidation and solubilization. Different microorganisms exhibit varying pH optima for metal bioleaching. Acidophilic microorganisms thrive in low pH environments (<3), where they efficiently oxidize sulfide minerals and solubilize metals. Maintaining the appropriate pH range is crucial for optimizing microbial activity and metal dissolution rates in bioleaching operations.

Temperature: Temperature significantly affects the growth, metabolic activity, and efficiency of microbial consortia in bioleaching processes. Elevated temperatures can enhance microbial activity and accelerate metal oxidation and solubilization rates. However, extreme temperatures beyond the thermophilic range (>45°C) may inhibit microbial growth and enzyme activity, leading to reduced bioleaching efficiency. Therefore, controlling and optimizing temperature conditions within the suitable range for microbial consortia is essential for maximizing bioleaching performance.

Microbial consortia: The composition and diversity of microbial consortia are crucial determinants of bioleaching efficiency. Acidophilic bacteria and archaea, such as *Acidithiobacillus ferrooxidans*, *Acidithiobacillus thiooxidans*, and *Leptospirillum sp.*, are commonly utilized in bioleaching operations due to their ability to oxidize sulfide minerals and solubilize metals. Additionally, heterotrophic microorganisms, such as iron- and sulfur-oxidizing bacteria, contribute to the regeneration of oxidizing agents and the maintenance of redox balance in bioleaching heaps. Harnessing synergistic interactions between different microbial species and optimizing their growth conditions can enhance bioleaching efficiency (Vrind et al., 2013).

Oxygen supply: Oxygen availability is crucial for supporting aerobic microbial metabolism and facilitating metal oxidation reactions in bioleaching processes. Efficient oxygen transfer to the microbial consortia is essential for maintaining aerobic conditions and sustaining microbial activity throughout the bioleaching heap. Adequate aeration and agitation of the leaching solution ensure optimal oxygen diffusion and distribution, thereby maximizing metal solubilization rates and bioleaching efficiency.

Nutrient availability: Microbial growth and activity in bioleaching systems depend on the availability of essential nutrients, including carbon, nitrogen, phosphorus, and trace elements. Carbon sources, such as organic compounds or CO_2, serve as energy and carbon substrates for microbial metabolism. Nitrogen and phosphorus are essential for protein synthesis and cellular growth, while trace elements, such as iron, sulfur, and potassium, act as cofactors for enzyme activities. Maintaining balanced nutrient concentrations and providing supplemental nutrients when necessary are critical for sustaining microbial activity and maximizing bioleaching efficiency.

Ore characteristics: The mineralogical composition, particle size, and surface area of the ore strongly influence bioleaching efficiency. Complex ores containing a mixture of sulfide minerals may require longer leaching times and more extensive microbial consortia to achieve complete metal recovery. Fine-grained ores provide a larger surface area for microbial attachment and metal oxidation, leading to enhanced bioleaching efficiency. Understanding the ore characteristics and tailoring bioleaching strategies accordingly is essential for optimizing metal extraction rates and resource utilization (Bertini et al., 2017).

7.4 MICROBIAL CONSORTIA FOR MANGANESE BIOLEACHING

7.4.1 Identification and Characterization of Manganese-Oxidizing Microbes

Several approaches are employed to identify and characterize manganese-oxidizing microorganisms.

7.4.1.1 Identification of Manganese-Oxidizing Microbes

Microbiological cultivation: Isolation and cultivation of manganese-oxidizing microorganisms from environmental samples involve selective media and culture conditions tailored to their physiological requirements. Enrichment cultures and isolation techniques, such as streak plating and serial dilution, are utilized to obtain pure cultures of manganese oxidizers for further characterization.

Molecular techniques: Molecular methods, including 16S rRNA gene sequencing, metagenomic analysis, and fluorescence in situ hybridization (FISH), are employed to identify manganese-oxidizing microorganisms and elucidate their phylogenetic relationships. PCR-based techniques targeting genes encoding manganese oxidation enzymes, such as MnxG and MofA, enable the detection and characterization of manganese oxidizers in complex microbial communities.

Proteomic and metabolomic analyses: Proteomic and metabolomic approaches provide insights into the functional capabilities and metabolic pathways of manganese-oxidizing microorganisms. Proteomic analysis involves the identification and quantification of proteins involved in manganese oxidation, while metabolomic profiling elucidates the metabolic intermediates and by-products associated with manganese oxidation pathways (Palmer et al., 2016).

7.4.1.2 Characterization of Manganese-Oxidizing Microbes

Physiological characteristics: Physiological studies determine the optimal growth conditions, temperature range, pH tolerance, and nutrient requirements of manganese oxidizers. Growth kinetics, substrate utilization patterns, and metabolic activities are evaluated to understand their ecophysiology and environmental niche preferences.

Biochemical properties: Biochemical assays assess the enzymatic activities and functional properties of manganese oxidation enzymes, such as manganese peroxidases, multicopper oxidases, and cytochrome c oxidases. Enzyme assays, protein purification, and kinetic studies elucidate the catalytic mechanisms and substrate specificities of manganese-oxidizing enzymes.

Genomic and genetic analysis: Genomic sequencing and genetic manipulation techniques provide genomic insights into manganese-oxidizing microorganisms. Comparative genomics, gene expression profiling, and gene knockout studies identify key genes and regulatory elements involved in manganese oxidation pathways, facilitating the elucidation of molecular mechanisms and evolutionary relationships (Naseri et al., 2022).

7.4.2 Selection and Optimization of Microbial Consortia for Enhanced Bioleaching Performance

7.4.2.1 Selection of Microbial Consortia

Functional diversity: Selecting microbial consortia with diverse metabolic capabilities ensures the efficient oxidation and solubilization of target metals from ores and concentrates. Acidophilic bacteria, such as *Acidithiobacillus sp.*, and fungi, such as *Aspergillus sp.*, contribute to the oxidation of sulfide minerals and the release of metal ions.

Stability and robustness: Robust microbial consortia with stable and resilient populations are essential for maintaining bioleaching performance under varying environmental conditions. Consortia with broad temperature and pH tolerance, as well as resistance to inhibitory factors, exhibit greater stability and reliability in bioleaching operations.

Synergistic interactions: Selection of microbial consortia with complementary metabolic pathways and cooperative interactions enhances bioleaching efficiency. Synergistic relationships between autotrophic and heterotrophic microorganisms facilitate the regeneration of oxidizing agents, nutrient recycling, and redox balance maintenance in bioleaching heaps (Sethurajan, 2015).

7.4.2.2 Optimization of Microbial Consortia

Strain selection and screening: Screening microbial isolates from diverse environments for their metal oxidation capabilities allows the identification of high-performing strains for bioleaching applications. Screening criteria may include metal tolerance, growth kinetics, and enzyme activities relevant to metal oxidation pathways.

Mixed culture development: Developing mixed cultures by co-culturing multiple microbial strains or species enhances bioleaching efficiency through synergistic interactions and metabolic complementarity. Optimizing the composition and ratio of microbial species in mixed cultures promotes efficient metal oxidation and solubilization.

Genetic engineering: Genetic engineering techniques enable the modification of microbial consortia to enhance their bioleaching capabilities. Engineered strains with improved metal tolerance, enzyme activity, and metabolic efficiency can be developed for targeted bioleaching applications.

Adaptation and evolution: Subjecting microbial consortia to selective pressures and environmental conditions mimicking bioleaching environments promotes adaptation and evolution of consortia with enhanced bioleaching performance. Serial cultivation, adaptive laboratory evolution, and directed evolution strategies facilitate the selection of strains with improved metal oxidation and solubilization abilities (Valix et al., 2011).

7.4.3 Strategies for Improving Microbial Activity and Metal Recovery Rates

Improving microbial activity and metal recovery rates in bioleaching processes requires a multifaceted approach that considers various factors influencing microbial metabolism, metal solubilization, and environmental conditions. Here are several strategies for enhancing microbial activity and metal recovery rates,

Optimization of growth conditions: Maintaining optimal pH conditions for microbial growth and metal solubilization is crucial. Adjusting pH levels within the range suitable for the specific microbial consortia involved in bioleaching can enhance their activity and metabolic efficiency. Controlling temperature within the range conducive to microbial growth and activity is essential. Optimal temperatures promote enzymatic reactions, metabolic pathways, and microbial proliferation, leading to improved metal recovery rates. Ensuring adequate oxygen supply to microbial consortia is critical for aerobic metabolism and metal oxidation processes. Optimizing aeration and agitation in bioleaching reactors facilitates oxygen transfer and diffusion, enhancing microbial activity and metal solubilization.

Selection and engineering of microbial consortia: Choosing microbial strains with robust metal oxidation capabilities and metabolic versatility is key. Screening and selecting high-performing strains from diverse environmental sources can improve bioleaching efficiency and metal recovery rates. Developing mixed microbial cultures with synergistic interactions and complementary metabolic pathways enhances bioleaching performance. Combining autotrophic and heterotrophic microorganisms in mixed cultures can improve metal solubilization rates and redox balance maintenance. Engineering microbial strains to enhance metal tolerance, enzyme activity, and metabolic efficiency can boost bioleaching performance. Genetic modification techniques enable the optimization of microbial consortia for targeted metal extraction and recovery from ores and concentrates (Ismail et al., 2014).

Nutrient and electron donor supplementation: Providing essential nutrients, such as carbon, nitrogen, phosphorus, and trace elements, supports microbial growth and activity. Supplementation with organic substrates or inorganic nutrients can alleviate nutrient limitations and enhance metal oxidation rates in bioleaching processes. Supplementing electron donors, such as ferrous iron (Fe^{2+}), sulfur compounds, or organic compounds, facilitates microbial metabolism and metal oxidation reactions. Electron donor addition can stimulate microbial activity and improve metal solubilization rates in bioleaching systems.

Process optimization and monitoring: Optimizing the composition of leaching solutions, including acid concentration, redox potential, and microbial nutrients, influences microbial activity and metal recovery rates. Adjusting leaching solution parameters based on microbial metabolic requirements and environmental conditions can enhance bioleaching efficiency. Implementing real-time monitoring and control systems allows for continuous assessment of microbial activity and metal recovery rates. Monitoring key parameters, such as pH, temperature, dissolved oxygen, and metal concentrations, enables timely adjustments and optimization of bioleaching operations.

Environmental management and remediation: Managing environmental factors, such as temperature, humidity, and light exposure, can influence microbial activity and metal recovery rates. Providing favorable environmental conditions within bioleaching reactors or heaps promotes microbial growth and enhances metal solubilization efficiency. Proper management of waste materials, by-products, and process residues generated during bioleaching operations is essential. Recycling and reuse of waste streams minimize environmental impact and resource depletion, contributing to sustainable metal extraction practices (Tang et al., 2016).

7.5 PROCESS OPTIMIZATION

7.5.1 Optimization of Environmental Conditions

Acidic conditions: Manganese bioleaching typically occurs under acidic conditions ($pH < 3$), which favor the dissolution of manganese minerals and microbial activity. Adjusting pH levels within the optimal range for manganese-oxidizing microorganisms, such as *Acidithiobacillus sp.* and fungal species, enhances bioleaching efficiency.

Moderate temperature: Maintaining moderate temperatures (typically 25°C–45°C) supports microbial growth and metabolic activity in manganese bioleaching systems. Temperature optimization ensures optimal enzymatic reactions and microbial proliferation, leading to improved manganese solubilization rates.

Aeration and agitation: Ensuring adequate oxygen supply through aeration and agitation of the leaching solution promotes aerobic metabolism and metal oxidation reactions. Optimizing oxygen transfer and distribution within bioleaching reactors or heaps enhances microbial activity and manganese oxidation efficiency.

Carbon and nitrogen sources: Providing carbon and nitrogen sources, such as glucose, sucrose, or ammonium sulfate, supports microbial growth and activity in manganese bioleaching systems. Nutrient supplementation alleviates nutrient limitations and enhances manganese solubilization rates.

Phosphorus and trace elements: Supplementing phosphorus and trace elements, such as potassium, magnesium, and iron, facilitates enzyme synthesis and metabolic pathways involved in manganese oxidation. Ensuring adequate availability of essential nutrients promotes microbial activity and bioleaching efficiency (Varvara, 2016).

Real-time monitoring: Implementing real-time monitoring systems allows for continuous assessment of environmental parameters, microbial activity, and manganese solubilization rates. Monitoring key variables, such as pH, temperature, dissolved oxygen, and manganese concentrations, enables timely adjustments and optimization of bioleaching conditions.

Control strategies: Utilizing feedback control loops and automated systems enables precise control of environmental conditions in manganese bioleaching operations. Adjusting parameters based on real-time data and process optimization algorithms ensures optimal performance and maximizes manganese recovery rates.

Waste management: Proper management of waste materials, process residues, and by-products generated during manganese bioleaching is essential for environmental sustainability. Recycling and reuse of waste streams minimize environmental impact and resource depletion, contributing to sustainable metal extraction practices.

Microbial adaptation: Subjecting microbial consortia to selective pressures and environmental conditions promotes the adaptation and evolution of strains with enhanced manganese oxidation capabilities. Adaptive laboratory evolution strategies facilitate the selection of microbial populations adapted to specific bioleaching conditions (Crundwell et al., 2010).

7.5.2 Particle Size Reduction and Surface Area Enhancement Techniques

Particle size reduction and surface area enhancement techniques in manganese bioleaching aim to increase the accessibility of manganese minerals to microbial activity and chemical reactions, leading to improved metal recovery rates and process efficiency. There are several techniques commonly employed for particle size reduction and surface area enhancement in manganese bioleaching.

Mechanical crushing and grinding: Mechanical crushing of manganese ore or concentrate reduces particle size and increases surface area, facilitating microbial access to manganese minerals. Crushers and jaw crushers are commonly used for primary crushing. Grinding further reduces particle size to micron-scale dimensions, enhancing surface area and promoting efficient mineral dissolution. Ball mills, rod mills, and vertical stirred mills are used for grinding manganese ores to fine particle sizes.

High-pressure grinding rolls (HPGR): High-pressure grinding rolls apply high pressure between two counter-rotating rolls to crush and grind manganese ores. HPGR technology produces finer particle sizes and generates more micro-fractures in the ore matrix, increasing surface area and promoting manganese dissolution.

Chemical treatment: Pre-treating manganese ores with acid solutions, such as sulfuric acid or hydrochloric acid, can partially dissolve carbonate and silicate minerals, resulting in increased porosity and surface area.

Acid leaching exposes more manganese minerals to subsequent bioleaching processes. Alkali leaching with solutions such as sodium hydroxide or ammonium hydroxide can selectively dissolve gangue minerals, leaving behind manganese-rich residues with enhanced surface area. Alkali leaching improves the efficiency of subsequent particle size reduction and bioleaching steps (Chaffron et al., 2010).

Ultrasonic treatment: Ultrasonic treatment involves subjecting manganese ores or concentrates to high-frequency ultrasonic waves, causing cavitation and micro-fracturing of mineral particles. Ultrasonic disintegration enhances particle breakage and surface area, facilitating microbial access to manganese minerals.

Microwave-assisted processing: Microwave-assisted processing applies electromagnetic radiation to manganese ores, inducing thermal and mechanical effects that promote particle size reduction and surface area enhancement. Microwave irradiation can selectively heat and fracture manganese minerals, improving bioleaching efficiency.

Electrochemical fragmentation: Electrochemical fragmentation involves applying electrical pulses to manganese ores immersed in electrolyte solutions, inducing electrochemical reactions and mechanical stresses that break down mineral particles. Electrochemical fragmentation enhances surface area and promotes manganese dissolution in bioleaching processes (Carlos, 2017).

7.5.3 Nutrient Supplementation and Substrate Preparation for Maximizing Bioleaching Efficiency

Nutrient supplementation and substrate preparation are essential aspects of maximizing bioleaching efficiency by providing microbial consortia with the necessary nutrients and substrates to support their growth, metabolism, and metal oxidation activities. Some of the strategies for nutrient supplementation and substrate preparation in the bioleaching process are as follows.

Carbon source selection: Supplementing organic carbon sources, such as glucose, sucrose, or molasses, provides energy and carbon substrates for microbial metabolism. Organic carbon sources support the growth and activity of heterotrophic microorganisms, which play roles in mineral oxidation and redox balance maintenance. Utilizing gaseous carbon dioxide (CO_2) as a carbon source promotes autotrophic growth of acidophilic bacteria, such as *Acidithiobacillus sp.*, which utilize CO_2 as a carbon substrate for biomass synthesis through carbon fixation pathways.

Nitrogen and phosphorus supplementation: Supplementing nitrogen sources, such as ammonium sulfate or urea, supports microbial protein synthesis and cellular growth. Nitrogen availability is crucial for enzyme production and metabolic activities involved in metal oxidation and solubilization. Providing phosphorus sources, such as potassium phosphate or phosphoric acid, enhances microbial energy metabolism and enzyme activity. Phosphorus is essential for ATP production, phosphorylation reactions, and biosynthesis processes in bioleaching microorganisms.

Trace element addition: Supplementing ferrous iron (Fe^{2+}) and elemental sulfur (S^0) provides electron donors for microbial respiration and metal oxidation reactions. Iron and sulfur compounds serve as electron acceptors and donors in the oxidation of sulfide minerals and the regeneration of oxidizing agents. Supplementing trace elements, such as magnesium, potassium, and manganese, as well as micronutrients like zinc, copper, and molybdenum, ensures optimal enzyme function and metabolic activity in bioleaching microorganisms. Trace elements act as cofactors for enzyme activities and play roles in cellular processes essential for metal oxidation (Gibsher et al., 2011).

Substrate preparation: Pre-conditioning manganese ores or concentrates through washing, crushing, and grinding treatments improves their accessibility to microbial activity and chemical reactions. Particle size reduction and surface area enhancement techniques increase the exposed surface area of manganese minerals, facilitating microbial access and metal solubilization. Pre-treating ores with acid solutions, such as sulfuric acid or hydrochloric acid, can dissolve carbonate and silicate minerals, exposing more manganese-rich surfaces for microbial oxidation. Acid treatment enhances the efficiency of subsequent bioleaching processes and metal recovery rates.

Buffering and pH control: Supplementing buffering agents, such as calcium carbonate or sodium bicarbonate, helps maintain stable pH conditions in bioleaching systems. Buffering agents counteract fluctuations in pH caused by microbial metabolism and acid production, ensuring optimal microbial activity and metal solubilization rates. Monitoring and controlling pH levels within the optimal range for microbial activity (typically pH 1.5–3.0) is essential. pH adjustment with sulfuric acid or alkali solutions ensures favorable conditions for manganese oxidation reactions and microbial growth in bioleaching processes.

Process optimization: Monitoring nutrient concentrations, such as carbon, nitrogen, phosphorus, and trace elements, in leaching solutions enables real-time adjustments and optimization of nutrient supplementation strategies. Balancing nutrient ratios and maintaining optimal concentrations promote microbial growth and bioleaching efficiency. Continuous assessment of substrate properties ensures efficient utilization of manganese ores and concentrates in bioleaching processes (Maruthamuthu et al., 2012).

7.6 SCALE-UP AND INDUSTRIAL APPLICATIONS

7.6.1 Challenges and Considerations in Scaling Up Bioleaching Processes

Scaling up bioleaching processes from laboratory or pilot-scale to commercial production poses several challenges and considerations. Some of the key challenges and considerations in scaling up bioleaching processes are as follows.

Reactor design: Designing and constructing large-scale bioleaching reactors that provide efficient mixing, aeration, and heat transfer while minimizing energy consumption and capital costs is challenging. Reactor geometry, material selection, and agitation mechanisms must be optimized to ensure uniform microbial activity and metal solubilization.

Hydrodynamics: Managing fluid dynamics and mass transfer in large-scale bioleaching heaps or tanks to ensure adequate nutrient and oxygen distribution throughout the system is critical. Proper hydrodynamic design prevents nutrient depletion, microbial stratification, and biomass settling, which can impair bioleaching performance.

Temperature control: Maintaining optimal temperature conditions for microbial activity and metal oxidation at scale requires efficient heating or cooling systems. Heat dissipation and thermal gradients within large-scale bioleaching systems must be managed to prevent overheating or temperature fluctuations that can affect microbial viability and process efficiency.

Water usage: Large-scale bioleaching operations consume significant quantities of water for ore processing, leaching, and waste management. Minimizing water usage and implementing water recycling and treatment systems are essential for reducing environmental impact and conserving freshwater resources.

Acid generation: Bioleaching processes produce acidic effluents containing sulfuric acid and metal ions, which can pose environmental risks if not properly managed. Implementing effective acid neutralization, containment, and remediation measures mitigates the potential for acid mine drainage and soil or water contamination.

Microbial contamination: Large-scale bioleaching operations are susceptible to microbial contamination by non-beneficial microorganisms, which can compete with metal-oxidizing microbes or degrade process performance. Implementing stringent biosecurity measures, such as microbial monitoring, sterilization protocols, and aseptic practices, helps control contamination risks (Panchaud et al., 2018).

Capital investment: Scaling up bioleaching processes requires substantial capital investment in infrastructure, equipment, and operational facilities. Assessing the economic feasibility and return on investment (ROI) of large-scale bioleaching projects is essential to secure funding and financial support from stakeholders and investors.

Operating costs: Managing operational costs, including energy consumption, labor, maintenance, and reagent usage, is critical for ensuring the economic viability of large-scale bioleaching operations. Optimizing process efficiency, resource utilization, and supply chain logistics helps minimize operating expenses and maximize profitability.

Environmental regulations: Large-scale bioleaching projects must comply with environmental regulations and permitting requirements governing air emissions, water discharge, waste management, and ecosystem protection. Obtaining permits and approvals from regulatory authorities and implementing environmental monitoring and reporting systems are essential for ensuring regulatory compliance and social license to operate.

Scaling up laboratory results: Translating laboratory-scale findings and pilot-scale trials into scalable and commercially viable bioleaching processes requires technology transfer and knowledge transfer between research institutions, academia, and industry partners. Collaborative efforts and interdisciplinary expertise are essential for overcoming technical challenges and optimizing large-scale bioleaching operations (Sand et al., 2013).

7.6.2 Integration of Bioleaching into Existing Industrial Practices for Manganese Recovery

The integration of bioleaching into existing industrial practices for manganese recovery offers a sustainable, cost-effective, and environmentally friendly approach to metal extraction. By leveraging microbial processes and biotechnological innovations, bioleaching enhances metal recovery rates, reduces environmental impact, and promotes the transition toward a more sustainable mining industry. Table 7.2 provides the information about the advantages of bioleaching upon integration with industrial practices (McEwan, 2019).

7.7 ENVIRONMENTAL AND ECONOMIC BENEFITS

7.7.1 Environmental Advantages of Bioleaching over Conventional Methods

Bioleaching offers several environmental advantages over conventional methods of metal extraction, making it a more sustainable and environmentally friendly option for mining operations. Some of the key environmental advantages of bioleaching are as follows.

TABLE 7.2
Advantages of Integrated Bioleaching

Existing Practices	Advantages
Sustainable metal extraction	Harnesses the metabolic activities of microorganisms to solubilize manganese minerals from ores and concentrates, minimizing environmental impact and reducing the carbon footprint associated with conventional mining and metallurgical processes
Resource utilization	Allows for the utilization of low-grade manganese ores, waste materials, and secondary resources that may not be economically viable with traditional extraction methods. This maximizes resource utilization and extends the lifespan of manganese reserves
Process optimization	Enhances metal extraction rates, improves metal purity, and reduces processing costs
Environmental benefits	Minimizes land disruption, water pollution, and ecosystem degradation

Reduced environmental footprint: Bioleaching reduces the environmental footprint of mining operations by minimizing land disturbance, habitat destruction, and ecosystem disruption associated with conventional mining methods such as open-pit mining and underground mining. Bioleaching processes typically require smaller surface areas and produce less waste rock compared to conventional mining operations.

Lower energy consumption: Bioleaching processes generally require lower energy consumption compared to conventional metallurgical processes such as smelting and roasting. Bioleaching relies on microbial activity to catalyze metal oxidation reactions, which occur at ambient temperatures and pressure, reducing the need for energy-intensive heating and cooling processes.

Decreased greenhouse gas emissions: Bioleaching generates fewer greenhouse gas emissions compared to conventional metallurgical processes that rely on fossil fuel combustion and energy-intensive operations. By reducing the use of fossil fuels and energy consumption, bioleaching helps mitigate climate change and air pollution associated with mining activities.

Minimal acid mine drainage: Bioleaching processes produce acidic leach solutions containing metal ions, but the acidity is typically lower and more manageable compared to the sulfuric acid generated by acid mine drainage from sulfide mineral oxidation in conventional mining. Proper management of bioleaching solutions can minimize the risk of acid mine drainage and associated environmental impacts (Mehta et al., 2010).

Recycling and reuse of process solutions: Bioleaching solutions can often be recycled and reused in the leaching process, reducing water consumption and minimizing the generation of wastewater. Recycling process solutions reduce the need for freshwater intake and wastewater treatment, contributing to water conservation and sustainable resource management.

Recovery of valuable metals from low-grade ores: Bioleaching enables the extraction of valuable metals from low-grade ores, mineral residues, and waste materials that may not be economically viable with conventional extraction methods. By utilizing microbial processes to solubilize metals from refractory ores, bioleaching reduces the need for extensive mining of high-grade ore deposits, preserving natural resources and reducing environmental impact.

Remediation of mining waste: Bioleaching can be employed for the remediation and reclamation of mining waste materials, such as tailings and mine spoils, by extracting residual metals and reducing the environmental hazards associated with abandoned mine sites. Bioleaching technologies offer a sustainable approach to mine site rehabilitation and environmental remediation (Vikentiev, 2015).

7.7.2 Economic Feasibility and Cost-Effectiveness of Manganese Recovery through Bioleaching

The economic feasibility and cost-effectiveness of manganese recovery through bioleaching depend on various factors, including ore grade, mineralogy, processing costs, market prices, and environmental considerations. While bioleaching offers several environmental advantages over conventional methods, its economic viability is influenced by a combination of technical, operational, and financial factors.

Ore grade and mineralogy: The ore grade and mineralogy of manganese deposits significantly impact the economic feasibility of bioleaching. Higher-grade ores with greater manganese content are generally more economically viable for bioleaching, as they yield higher metal recoveries and require less processing. Additionally, the mineralogy of manganese ores, including the presence of sulfides, carbonates, and silicates, affects the ease of manganese dissolution and the efficiency of bioleaching processes.

Processing costs: The processing costs associated with bioleaching include ore preparation, microbial inoculation, leaching, solution recovery, and metal extraction. These costs can vary depending on the scale of operations, equipment requirements, labor expenses, and reagent consumption. Optimizing process parameters, minimizing energy consumption, and maximizing metal recovery rates are essential for controlling processing costs and enhancing the economic feasibility of bioleaching (Mironov et al., 2018).

Market prices and demand: Market prices for manganese products, such as manganese ore, ferroalloys, and electrolytic manganese metal (EMM), directly influence the economic viability of bioleaching projects. Fluctuations in global manganese prices, supply-demand dynamics, and market competition impact the revenue generated from manganese recovery through bioleaching. Understanding market trends and demand forecasts is crucial for assessing the long-term economic feasibility of bioleaching operations.

Capital investment and operating costs: The initial capital investment required to establish bioleaching facilities, including infrastructure, equipment, and microbial technologies, represents a significant financial commitment. Operating costs, such as labor, maintenance, utilities, and environmental compliance, also contribute to the overall cost structure of bioleaching projects. Achieving economies of scale, optimizing resource utilization, and minimizing overhead expenses are key strategies for enhancing cost-effectiveness and profitability (Stankovic et al., 2014).

7.7.3 Contribution to Sustainable Resource Management and Circular Economy Principles

Manganese recovery through bioleaching contributes significantly to sustainable resource management and circular economy principles by promoting resource conservation, minimizing waste generation, and reducing environmental impact.

Utilization of low-grade ores and waste materials: Bioleaching enables the extraction of manganese from low-grade ores, mineral residues, and waste materials that are not economically viable for conventional mining and metallurgical processes. By utilizing microbial processes to solubilize metals from refractory ores and industrial by-products, bioleaching maximizes resource utilization and extends the lifespan of mineral reserves.

Reduction of environmental footprint: Bioleaching reduces the environmental footprint of mining operations by minimizing land disturbance, habitat destruction, and ecosystem disruption associated with conventional mining methods. By employing microbial processes at ambient temperatures and pressure, bioleaching reduces energy consumption, greenhouse gas emissions, and water usage, mitigating environmental impacts and preserving natural resources.

Remediation of mining waste: Bioleaching can be employed for the remediation and reclamation of mining waste materials, such as tailings and mine spoils, by extracting residual metals and reducing environmental hazards associated with abandoned mine sites. Bioleaching technologies offer a sustainable approach to mine site rehabilitation, environmental remediation, and ecosystem restoration, contributing to land conservation and biodiversity preservation (Marakushev et al., 2014).

Recovery of valuable metals: Bioleaching facilitates the recovery of valuable metals, such as manganese, copper, gold, and uranium, from ores, concentrates, and industrial residues. By selectively solubilizing target metals from mineral matrices, bioleaching maximizes metal recovery rates and minimizes the generation of waste materials, supporting the efficient use of resources and promoting a closed-loop approach to metal extraction.

Integration into circular economy practices: Bioleaching integrates into circular economy practices by transforming waste streams and secondary resources into valuable commodities. By converting low-grade ores, mining residues, and metallurgical wastes into marketable products, bioleaching contributes to the circularity of resource flows and reduces reliance on virgin materials, fostering a more sustainable and resilient economy.

Water conservation and recycling: Bioleaching processes typically require less water compared to conventional mining and metallurgical processes, as water usage is minimized through recycling and reuse of process solutions. By implementing water conservation measures and closed-loop water management systems, bioleaching reduces freshwater intake, wastewater discharge, and environmental contamination, promoting sustainable water stewardship.

Social and economic benefits: Bioleaching creates opportunities for job creation, technology innovation, and community development in regions where mining activities are prevalent. By supporting local economies, fostering innovation ecosystems, and promoting social responsibility, bioleaching contributes to inclusive growth, poverty alleviation, and sustainable development goals (Blayda et al., 2018).

7.8 CHALLENGES AND FUTURE DIRECTIONS

7.8.1 Current Challenges and Limitations in Bioleaching Technology

While bioleaching technology offers significant potential for sustainable metal extraction, it also faces several challenges and limitations that need to be addressed for widespread adoption and optimization. Addressing these challenges and limitations in bioleaching technology requires ongoing research, innovation, and collaboration among academia, industry, and regulatory bodies. By overcoming technical barriers, optimizing process efficiency, and addressing environmental and economic considerations, bioleaching technology can emerge as a sustainable and economically viable solution for metal extraction in the mining industry.

Bioleaching—challenges and limitations are (Filippov et al., 2017):

- low metal recovery rates;
- slow reaction rates;
- microbial adaptation and competition;
- nutrient limitations;
- ph control and acid management;
- environmental and regulatory compliance;
- capital and operational costs;
- scale-up and technology transfer.

7.8.2 Emerging Trends and Research Directions for Improving Bioleaching Efficiency

Several emerging trends and research directions are shaping efforts to improve the efficiency and effectiveness of bioleaching technology. These trends focus on enhancing metal recovery rates, optimizing process parameters, and addressing key challenges associated with bioleaching operations. These emerging trends and research directions in bioleaching reflect a growing emphasis on innovation, sustainability, and efficiency in metal extraction processes. By leveraging advances in microbiology, biotechnology, engineering, and environmental science, researchers are driving the development of next-generation bioleaching technologies with enhanced performance and broader applicability in the mining industry (Nozhati et al., 2020). Figure 7.3 suggests different research areas for improving efficiency of bioleaching.

7.8.3 Potential Synergies with Other Biotechnological and Metallurgical Processes

Bioleaching can be integrated with other biotechnological and metallurgical processes to create synergies that enhance overall metal recovery efficiency, reduce environmental impact, and promote sustainable mining practices (Alekseyev et al., 2012). Synergies between bioleaching and other biotechnological and metallurgical processes offer opportunities to improve metal extraction efficiency, optimize resource utilization, and mitigate environmental impact in the mining industry. By integrating diverse technologies and harnessing the complementary strengths of microbial processes, electrochemical reactions, and biochemical conversions, synergistic approaches drive innovation and promote sustainability in metal extraction and resource recovery (Fu et al., 2018). Figure 7.4 suggests bioleaching synergies with other biotechnology and metallurgical process.

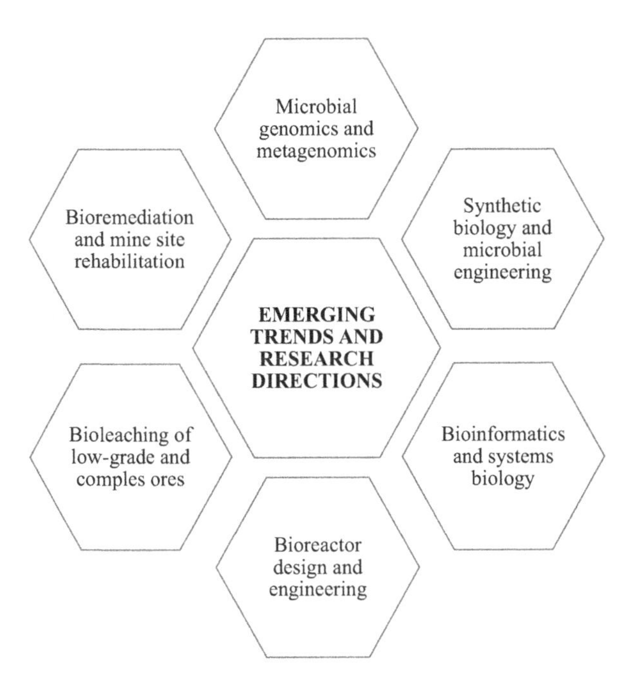

FIGURE 7.3 Prominent emerging trends and research directions in bioleaching.

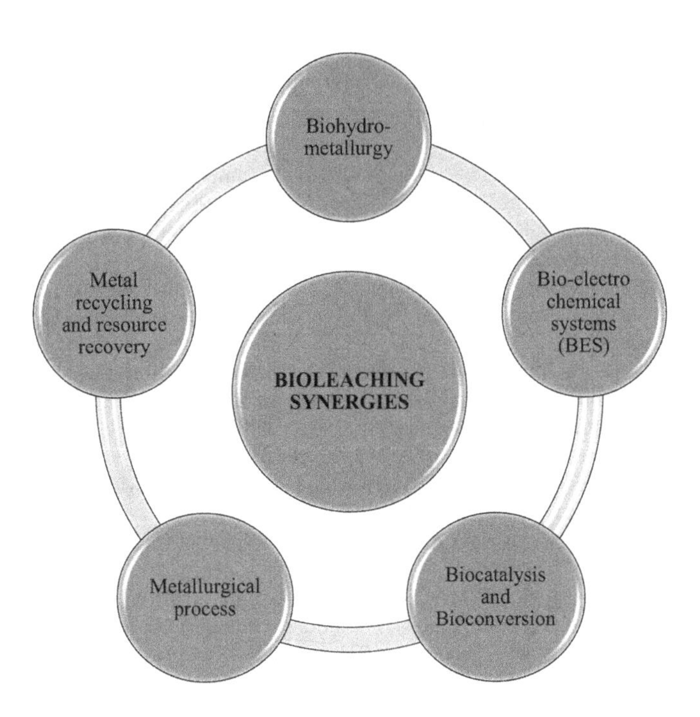

FIGURE 7.4 Bioleaching synergies.

7.9 CONCLUSION

Bioleaching of manganese from industrial wastes presents a promising and environmentally sustainable approach for

metal recovery. By harnessing the metabolic activities of microorganisms to solubilize manganese minerals from ores, concentrates, and waste materials, bioleaching offers several advantages over conventional mining and metallurgical processes. Through optimization of process parameters, microbial consortia selection, and integration with other biotechnological and metallurgical techniques, bioleaching enables efficient metal extraction, resource conservation, and environmental remediation. While challenges such as low metal recovery rates, slow reaction kinetics, and economic feasibility persist, ongoing research and technological advancements continue to expand the applicability and effectiveness of bioleaching in industrial settings. Overall, bioleaching of manganese from industrial wastes represents a sustainable and economically viable solution for metal recovery, contributing to the transition toward a more circular and environmentally responsible mining industry.

REFERENCES

Akcil, A., Veglio, F. and Ferella, F. (2015) 'A review of metal recovery from spent petroleum catalysts and ash', *Waste Manag*, Vol. 45, pp. 420–433.

Alekseyev, V. and Marin, B. (2012) 'Structural-chemical heterogeneity of natural crystals and the microgeochemical direction in ontogeny of minerals', *Geol Ore Deposits*, Vol. 141, pp. 3–21.

Bertini, I. and Cavallaro, G. (2017) 'Metals in the omics world: copper homeostasis and cytochrome c oxidase assembly in a new light', *J Biol Inorg Chem*, Vol. 13, pp. 3–14.

Blayda, T. and Vasyleva, T.V. (2018) 'Biotechnological methods for processing substandard ores and wastes', *Chem Society*, Vol. 2, pp. 6–25.

Brandaleze, E., Benavidez, L. and Santini, S. (2018) 'Treatments and recycling of metallurgical slags', *Recovery Util Metall Solid Waste*, Vol. 8, pp. 90–95.

Carlos, R. (2017) 'Biomining of metals: how to access and exploit natural resource sustainably', *Microb Biotechnol*, Vol. 10, pp. 1191–1193.

Chaffron, S., Rehrauer, H. and Pernthaler, J. (2010) 'A global network of coexisting microbes from environmental and whole-genome sequence data', *Genome Res*, Vol. 20, pp. 947–959.

Chen, S., Zhong, D. and Tang, C. (2020) 'Practical experience in large-scale development of Zijinshan low-grade gold-copper mine', *Mining Metal Explore*, Vol. 37, pp. 1339–1347.

Contestabile, M., Panero, S. and Scrosati, B. (2011) 'A laboratory-scale lithium-ion battery recycling process', *J Power Sources*, Vol. 92, pp. 65–69.

Crundwell, P.R. and Holmes, T.A. (2010) 'How do bacteria interact with minerals', *JSA Inst Mining Metall*, Vol. 11, pp. 299–312.

Das, B., Suklab, N. and Pradhan, S. (2011) 'Manganese bio mining: a review', *Biores Technol*, Vol. 102, pp. 7381–7387.

Ferraro, A., Farina, I. and Race, M. (2019) 'Pre-treatments of MSWI fly-ashes: a comprehensive review to determine optimal conditions for their reuse and/or environmentally sustainable disposal', *Rev Environ Sci Biotechnol*, Vol. 18, pp. 453–471.

Ferrier, J., Csetenyi, L. and Gadd, G.M. (2022) 'Fungal transformation of natural and synthetic cobalt-bearing manganese oxides and implications for cobalt biogeochemistry', *Environ Microbiol*, Vol. 24, pp. 667–677.

Filippov, C., Izart, H., Johnson, R. and Kahnt, H. (2017) 'The BIOMORE project—a new mining concept for extracting metals from deep ore deposits using biotechnology', *Mining Rep*, Vol. 153, pp. 436–445.
Fu, G., Fu, L. and Jiang, X. (2018) 'Extraction of manganese oxide ore by a reduction acidolysis-selective leaching method', *Mining Metall*, Vol. 35, pp. 215–220.
Furini, A. and Manara, G. (2015) 'Environmental phytoremediation: plants and microorganisms at work', *Front Plant Sci*, Vol. 6, pp. 19–27.
Ghosh, S., Mohanty, A. and Sukla, A.P. (2016) 'Greener approach for resource recycling: manganese bioleaching', *Chemosphere*, Vol. 154, pp. 628–639.
Gibsher, A.A., Tomilenko, A.M., Sazonov, M.A. and Ryabukha, A.L. (2011) 'The Gerfed gold deposit: fluids and PT-conditions for quartz vein formation', *Geol Geophys*, Vol. 52, pp. 1851–1867.
He, D., Shu, J. and Wang, R. (2021) 'A critical review on approaches for electrolytic manganese residue treatment and disposal technology: reduction, pretreatment, and reuse', *J Hazard Mater*, Vol. 418, pp. 126–135.
Ismail, A.A., Ali, E.A., Ibrahim, I.A. and Ahmed, M.S. (2014) 'A comparative study on acid leaching of low-grade manganese ore using some industrial wastes as reductants', *Canad J Chem Eng*, Vol. 82, pp. 1296–1300.
Jana, R.K., Singh, D.D.N. and Roy, S.K. (2015) 'Alcohol-modified hydrochloric acid leaching of sea nodules', *Hydrometallurgy*, Vol. 38, pp. 289–298.
Kozlov, A.H. and Kulikova, I.P. (2017) 'Leaching products in bacterial breed-culture system at biochemical degradation by silicate bacteria of diatomite, zeolite and bentonite', *Bull Samara Sci Center Russ Acad Sci*, Vol. 2, pp. 281–288.
Lucheta, M.C., Palmieri, A.L.V., Carmo, P.M.P. and Silva, R.V. (2019) 'Bioleaching for copper extraction of marginal ores from the Brazilian amazon region', *Metals*, Vol. 56, pp. 705–719.
Marakushev, N.A., Paneyakh, S.A. and Marakushev, L. (2014) 'Sulphide ore formation and its hydrocarbon specialization', *Monograph Geos*, Vol. 89, pp. 184–189.
Maruthamuthu, S., Palanichamy, S. and Manickam, S.T. (2012) 'Microfouling of manganese-oxidizing bacteria in Tuticorin harbour waters', *Curr Sci*, Vol. 82, pp. 7–10.
McEwan, A.G. (2019) 'New insights into the protective effect of manganese against oxidative stress', *Mol Microbiol*, Vol. 72, pp. 812–818.
Mehta, K.D., Chitrangada, D. and Pandey, B.D. (2010) 'Leaching of copper, nickel and cobalt from Indian Ocean manganese nodules by Aspergillus niger', *Hydrometallurgy*, Vol. 105, pp. 89–95.
Mironov, A.Y. and Shishkin, A.A. (2018) 'Extraction of cooper from aqueous solutions using iron powder', *Ind Agric Ecol*, Vol. 1, pp. 97–102.
Naseri, T., Pourhossein, F. and Mousavi, S.M. (2022) 'Manganese bioleaching: an emerging approach for manganese recovery from spent batteries', *Rev Environ Sci Biotechnol*, Vol. 21, pp. 447–468.
Nozhati, A. and Azizi, Y. (2020) 'Leaching of copper and zinc from the tailings sample obtained from a porcelain stone mine: feasibility, modeling, and optimization', *Environ Sci Pollut Resour*, Vol. 27, pp. 6239–6252.
Pak, K.R., Lim, O.Y., Lee, H.K. and Choi, S.C. (2012) 'Aerobic reduction of manganese oxide by Salmonella sp. strain MR4', *Biotechnology*, Vol. 24, pp. 1181–1184.
Palmer, F.E., Staley, J.T. and Counsell, J.B.A. (2016) 'Identification of manganese-oxidizing bacteria from desert varnish', *Geomicrobiology*, Vol. 4, pp. 343–360.
Panchaud, A., Affolter, M. and Moreillon, P. (2018) 'Experimental and computational approaches to quantitative proteomics: status quo and outlook', *J Proteomics*, Vol. 71, pp. 19–33.
Pradhan, N., Nathsarma, K.C., Srinivasa, R.K. and Mishra, B.K. (2015) 'Heap bioleaching of chalcopyrite: a review', *Miner Eng*, Vol. 21, pp. 355–365.

Roy, B.P., Paice, M.G., Archibald, F.S. and Misra, S.K. (2014) 'Creation of metal-complexing agents, reduction of manganese dioxide, and promotion of manganese peroxidase-mediated Mn (III) production by cellobiose: quinone oxidoreductase from Trametes versicolor', *J Biol Chem*, Vol. 269, pp. 19745–19750.

Sahoo, R.N., Naik, P.K. and Das, S.C. (2011) 'Leaching of manganese ore using oxalic acid as reductant in sulphuric acid solution', *Hydrometallurgy*, Vol. 62, pp. 157–163.

Sand, T., Gehrke, P.G. and Jozsa, A. (2013) '(Bio) chemistry of bacterial leaching-direct vs. indirect bioleaching', *Hydrometallurgy*, Vol. 59, pp. 159–175.

Sethurajan, S. (2015) 'Metallurgical sludge's, bio/leaching and heavy metals recovery (Zn, Cu)', *Environ Eng*, Vol. 98, pp. 34–38.

Sildar, A., Hullebusch, E.D. and Lenz, M. (2019) 'Biotechnological strategies for the recovery of valuable and critical raw materials from waste electrical and electronic equipment (WEEE)—a review', *J Hazard Mater*, Vol. 362, pp. 467–481.

Stankovic, I., Moric, A., Pavic, S. and Cvetkovic, V. (2014) 'Bioleaching of copper from old flotation tailings samples', *J Serb Chem Soc*, Vol. 79, pp. 1–17.

Tang, J.A. and Valix, M. (2016) 'Leaching of low-grade limonite and nontronite ores by fungi metabolic acids,' *Minerals Eng*, Vol. 19, pp. 1274–1279.

Tian, X.K., Wen, X.X., Yang, C. and Liang, Y.J. (2010) 'Reductive leaching of manganese from low-grade manganese dioxide ores using corncob as reductant in sulfuric acid solution', *Hydrometallurgy*, Vol. 100, pp. 157–160.

Tong, Z., Su, H., Wen, Y. and Sun, Y. (2018) 'Reductive leaching of manganese from low-grade manganese ore in H2SO4 using cane molasses as reductant', *Hydrometallurgy*, Vol. 93, pp. 136–139.

Valenzuela, L., Chi, A., Beard, S., Orell, A. and Jerez, C.A. (2019) 'Metagenomics and proteomics in biomining microorganisms', *Biotechnol Adv*, Vol. 24, pp. 197–211.

Valix, M., Usai, F. and Malik, R. (2011) 'Fungal bioleaching of low-grade laterite ores', *Minerals Eng*, Vol. 14, pp. 197–203.

Varvara, J. (2016) 'Researching the hazardous potential of metallurgical solid wastes', *Polish J Environ Studies*, Vol. 25, pp. 147–152.

Vikentiev, K. (2015) 'Invisible and microscopic gold in pyrite: research methods and new data for Urals sulfide ore', *Geol Ore Deposits*, Vol. 57, pp. 267–298.

Vrind, J.P.M., Boogerd, F.C. and Jong, E.W. (2013) 'Manganese reduction by a marine Bacillus species', *J Bacteriol*, Vol. 167, pp. 30–34.

Xia, J.L., Yang, Y., Nie, Z.Y. and Qiu, G.Z. (2011) 'Sulfur oxidation activities of pure and mixed thermophiles and sulfur speciation in bioleaching of chalcopyrite', *Bioresour Technol*, Vol. 102, pp. 3877–3882.

Youcai, L., Qingquan, L. and Lifeng, F. (2014) 'Study on hydrometallurgical process and kinetics of manganese extraction from low-grade manganese carbonate ores,' *Int J Mining Sci Technol*, Vol. 24, pp. 567–571.

Zhang, R., Xia, Q. and Nie, H. (2017) 'Sulfur activation related extracellular proteins of Acidithiobacillus ferrooxidans,' *Trans Nonferrous Metals Soc*, Vol. 18, pp. 1398–1402.

8 Bio-Recovery of Nickel from Acid Mine Drainage

Shirsendu Banerjee, Suraj K. Tripathy, Sankha Chakrabortty, and Jayato Nayak

8.1 INTRODUCTION

Acid mine drainage (AMD), a toxic byproduct of mining activities, poses a significant environmental threat due to its high acidity, metal content, and detrimental impact on water quality. Nickel (Ni), a valuable heavy metal but toxic in high concentrations, is often present in AMD, necessitating remediation strategies that are not only effective but also sustainable. Conventional methods for Ni removal from AMD often involve chemical precipitation or adsorption, leading to secondary waste generation and potentially significant economic drawbacks (Akcil et al., 2011). Emerging as a promising alternative, bio-recovery harnesses the metabolic capabilities of microbes to remove and potentially even recover Ni from AMD. This bio-based approach offers several advantages, including reduced chemical input: Minimizing the need for harmful chemicals and associated risks; Potential metal recovery: Transforming a waste product into a valuable resource with economic benefits; Improved environmental sustainability: Lower environmental impact compared to conventional methods. Over recent years, considerable attention has been directed toward enhancing the sustainability of the mining sector, as evidenced by numerous studies (Gunkel-Grillon et al., 2014; Hu et al., 2014; Ali et al., 2017). Presently, the mining industry grapples with persistent and mounting pressure from environmental advocates to transition toward more sustainable mining practices (Nagajyoti et al., 2010; Crini and Lichtfouse, 2018). Consequently, stakeholders within the mining sector are compelled to explore novel methodologies that harmonize both waste management efficacy and operational expenses. A significant challenge confronting the mining industry revolves around the substantial volumes of waste generated, which pose significant environmental concerns. Among these waste materials, AMD stands out as particularly noteworthy. AMD arises from the oxidation of sulfide minerals in the presence of oxygen, water, and microorganisms. Although it is a naturally occurring phenomenon, mining activities expedite its formation, leading to profound environmental ramifications, especially on soil quality, water bodies, and aquatic ecosystems (Kefeni et al., 2017). Notable sulfide minerals contributing to AMD include pyrite, marcasite, pyrrhotite, chalcocite, chalcopyrite, molybdenite, millerite, galena, sphalerite, and arsenopyrite. Figure 8.1 provides an illustrative depiction elucidating the process of AMD generation within the mining industry.

DOI: 10.1201/9781003415541-8

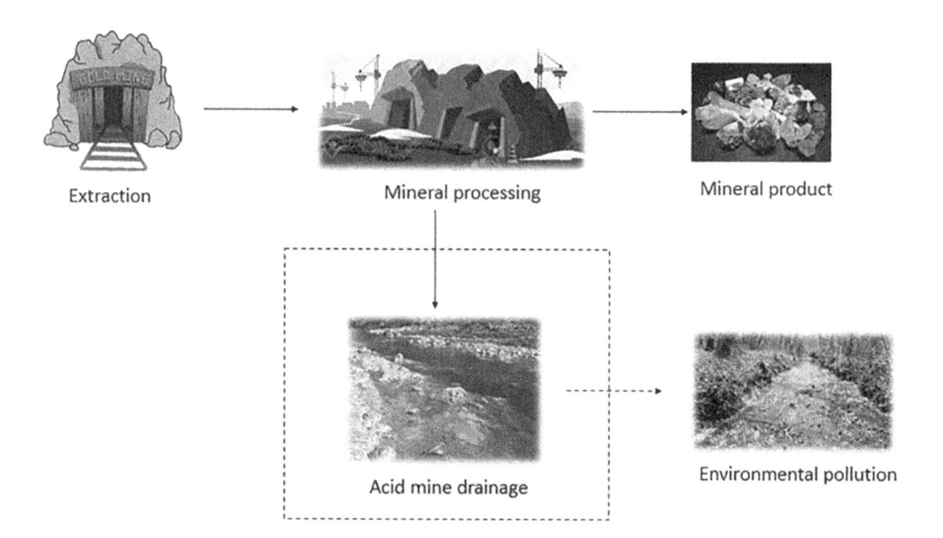

FIGURE 8.1 Acid mine drainage (AMD) generation scenario.

The environmental issues stemming from AMD predominantly arise from the low pH levels of these discharges and the elevated concentrations of dissolved metals, which possess potential toxicity (Simate and Ndlovu, 2014).

Furthermore, heavy metals present in the soil adversely affect plant life, particularly when found in high concentrations or specific combinations, leading to cellular tissue damage and disrupting homeostasis mechanisms. Such disturbances hinder plant growth (Simate and Ndlovu, 2014; Masindi et al., 2018). The availability of essential nutrients crucial for plant development closely correlates with soil pH. A low pH level immobilizes nitrogen, phosphorus, and potassium in the soil, while calcium and magnesium contents may become deficient. Conversely, toxic elements like aluminum, iron, and manganese may leach into the soil, heightening toxicity levels (Simate and Ndlovu, 2014; Masindi et al., 2018). Given the array of environmental challenges posed by AMD, it is imperative to implement appropriate treatment measures in areas where mining operations are prominent. Inadequate management of AMD could precipitate accelerated environmental degradation (Johnson and Hallberg, 2005; Sheoran and Sheoran, 2006; Anawar, 2015; Bejan and Bunce, 2015; Nleya et al., 2016; Ryan et al., 2017). Furthermore, there is potential for recovering valuable solid products from AMD waste, offering not only environmental benefits but also economic advantages. Moreover, certain techniques may yield clean water suitable for reuse, further enhancing sustainability efforts.

8.2 SOURCES OF AMD

AMD emerges predominantly from the oxidation of sulfide minerals in mined rock when they come into contact with air and water, resulting in the formation of sulfuric acid. Scholarly research indicates various sources contributing to AMD.

8.2.1 Abandoned Mines

While the last ore may be extracted, the environmental legacy of abandoned mines can linger for decades, even centuries. One of the most persistent and problematic threats posed by these sites is AMD, a toxic cocktail of acidic water laden with dissolved metals and minerals. Understanding the complex interplay of geochemical and biological processes driving AMD generation in abandoned mines is crucial for mitigating its devastating impacts on surrounding ecosystems and human health.

Chemical Engine of Acidification:

At the heart of AMD generation lies the abiotic and biotic oxidation of sulfide minerals, primarily pyrite (FeS_2) present within ore bodies or surrounding rocks. When exposed to air and water, pyrite undergoes a series of exothermic reactions:

$$4FeS_2 + 8O_2 + 14H_2O \rightarrow 16H^+ + 8SO_4^{2-} + 4Fe^{2+} \quad (8.1)$$

This acidic environment fuels further pyrite oxidation, creating a self-perpetuating cycle:

$$2Fe^{2+} + \tfrac{1}{2}O_2 + 2H^+ \rightarrow 2Fe^{3+} + H_2O \quad (8.2)$$

Ferric iron (Fe^{3+}) generated in reaction (2) can oxidize additional pyrite, amplifying acid production. Additionally, microorganisms known as acidophiles, particularly bacteria and archaea, can accelerate iron oxidation through biological pathways, further contributing to AMD generation (Daraz et al., 2023).

8.2.2 Active Mines

While the core chemical processes responsible for AMD generation remain similar to abandoned mines, active operations create a unique set of dynamics:

Continuous exposure: Mining activities constantly expose fresh sulfide minerals to air and water, fueling rapid oxidation and AMD production.

Active water management: Drainage and pumping systems required for mine operations can mobilize and accelerate the flow of acidic water and contaminants.

Waste rock management: Improper storage and handling of waste rock containing sulfide minerals can significantly contribute to AMD generation.

Geochemical variability: Differences in geology and rock types within the mine site can create varied AMD characteristics and complexities in treatment strategies.

8.2.3 Waste Rock Dumps

The formation of AMD in WRDs involves a chain of interrelated geochemical and microbiological processes:

Sulfide mineral oxidation: The primary culprit is the oxidation of sulfide minerals, commonly pyrite (FeS_2), present in the rock. When exposed to air and water, pyrite undergoes chemical reactions, releasing ferrous iron (Fe^{2+}) and sulfuric acid (H_2SO_4), leading to a decrease in pH and an increase in acidity.

Microbial activity: Acidophilic bacteria and archaea contribute significantly to AMD generation by accelerating the oxidation of ferrous iron to ferric iron (Fe^{3+}). Fe^{3+} further oxidizes pyrite, creating a self-perpetuating cycle of acid production.

Geochemical factors: The mineral composition, particle size distribution, and pH of the WRD influence the rate and severity of AMD generation. Additionally, factors like climate and fluctuating water levels can impact the oxidation processes.

8.2.4 Tailings Dams

The consequences of AMD contamination from tailings dams extend far beyond the dam itself:

Water pollution: AMD acidifies rivers, streams, and groundwater, rendering them unsuitable for drinking, irrigation, and aquatic life. The acidic nature disrupts sensitive ecosystems, harms biodiversity, and disrupts food webs.

Metal mobilization and dispersion: Dissolved metals like iron, copper, and aluminum contaminate soil and water bodies, impacting both terrestrial and aquatic ecosystems. These metals can accumulate in plants and animals, entering the food chain and posing health risks.

Infrastructure damage: The corrosive nature of AMD can damage bridges, pipelines, and other infrastructure, leading to economic losses and impacting vital services.

Soil degradation: Acidic leachate infiltrating the soil alters its pH, impacting nutrient availability and microbial activity, ultimately reducing soil fertility and productivity.

There are various other factors that lead to AMD such as spoil heaps, transportation and storage, underground mines, and so on. Efforts to mitigate AMD encompass various strategies, such as water treatment, proper waste management, reclamation initiatives, and adherence to best practices in mining operations, as advocated by several researchers including Brady et al. and Moran et al..

8.3 AMD TREATMENT TECHNIQUES

Combining multiple treatment techniques in a comprehensive AMD management plan tailored to the specific characteristics of each site is often necessary to achieve effective and sustainable remediation. Regular monitoring of water quality and ecosystem health is also crucial for assessing the long-term success of AMD treatment efforts. Some techniques typically used in the industries are discussed below.

8.3.1 Selective Metal Precipitation

Selective metal precipitation is based on the differences in solubilities of the metal compounds. The most common reactive agents for metal precipitation are sodium hydroxide (alkaline) or sodium sulfide. The main advantages of this method are the volume reduction in the resulting sludge and the possible valorization of metals. During alkaline precipitation, hydrolysis propagates continuously which leads to polymerization of metal oxo/hydroxo series and eventually results in saturation of the solution. Once saturated, metal ions will form clusters of respective metal oxo/hydroxo complexes and precipitate from the solution. Depending on the solution matrix property and chemical characteristics different metal ions may precipitate at different pH ranges (Honaker et al., 2018; Vaziri Hassas et al., 2020; Zhang and Honaker, 2020). Therefore, products with a high percentage of critical elements can be recovered at different pH ranges. However, in sulfide precipitation, a high degree of metal recovery at lower pH is reported. Additionally, similar metal ions such as Cu^{2+}, Co^{2+}, Zn^{2+}, and Mn^{2+} can be recovered by fractional precipitation technique (Liu et al., 2017). Sampaio et al. (2010) selectively removed Zn from an aqueous mixture of Zn and Ni using sodium sulfide as the precipitant.

Many works are reported based on selective precipitation, and the most relevant and recent literature are discussed. Figure 8.2 describes a tentative process for selective metal precipitation. In a recent work by Li et al. (2022) reported a sulfide precipitation technique for recovery of high purity Ni, Co, and Mn from AMD. Initially, the AMD has undergone sulfide precipitation and concentration by increasing the pH to 6.5. Afterward, a preconcentrated slurry having 3,794 mg/L Mn, 59 mg/L Co, and 127 mg/L Ni was obtained by collecting the precipitates formed in the pH range 6.50–10.00. Further, to purify the critical elements, re-dissolution tests were performed on the preconcentrated slurry. It was found that the majority of Co and Ni were dissolved by reducing the pH to 5.00, while more than 50% of Mn remained undissolved, leading to a solid residue containing around 30 wt.% Mn. After successive intermediate treatments, finally, Co and Ni products with almost 94% and 100% purity were recovered by both sulfide and alkaline precipitation, respectively.

Masindi et al. (2017) reported a work on selective precipitation of the metals dissolved in 0.5 L of AMD using calcined cryptocrystalline magnesite, a new material with high neutralization capacity. Fe was recovered at pH 3–3.5, Al at pH $\geq$ 6.5, Mn at pH $\geq$ 9.5, Cu at pH $\geq$ 7, Zn at pH $\geq$ 8, Pb at pH $\geq$ 8, and Ni at pH $\geq$ 9, reaching efficiency greater than 99% for all of them within these pH ranges.

8.3.2 Selective Adsorption

Selective adsorption has evolved as an important technique used for the recovery of various critical elements present in AMD (Mudhoo et al. 2012). Figure 8.3 shows the schematic overview of the process. The adsorbents used are highly porous in nature which provide a large surface for adsorption such as limestone, clay, dolomite, slag, and zeolites (Gu et al., 2018; Saha and Sinha, 2018). However, the presence of inter-particle diffusion may result in a decrease in adsorption rate and available capacity, especially for macromolecules (Hu et al., 2006). To address this, various nano-sized particles with large surface area with small diffusion resistance are used

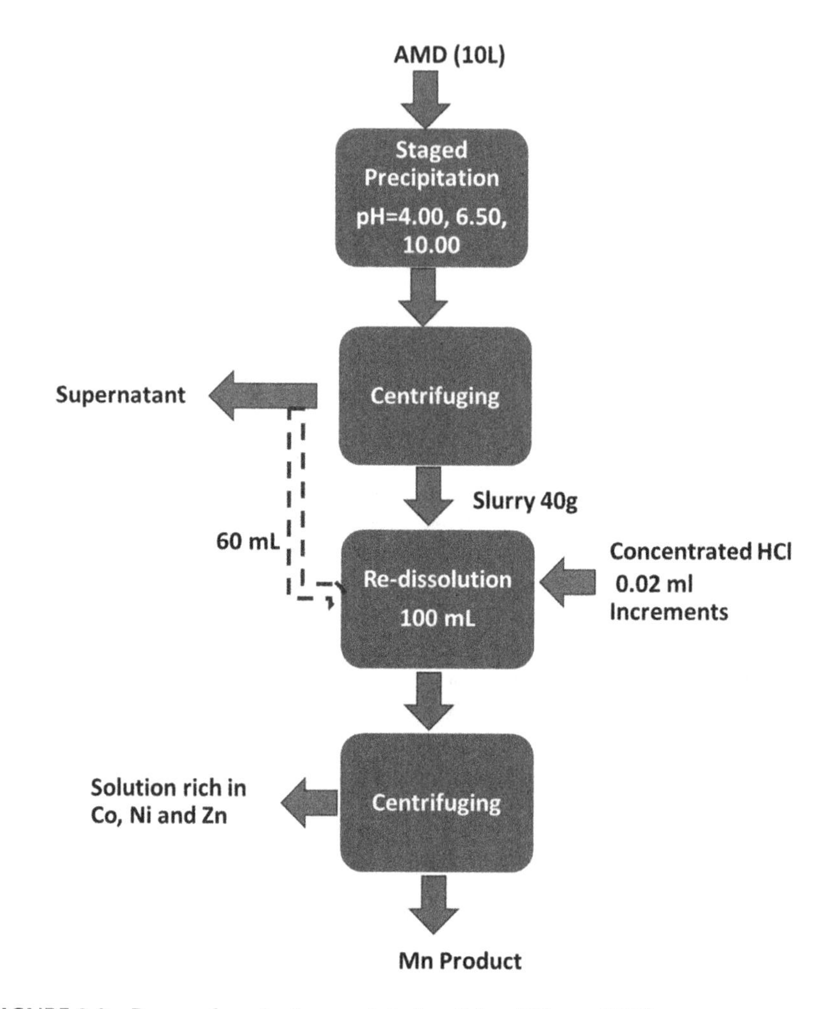

FIGURE 8.2 Process for selective precipitation (Li and Zhang, 2022)

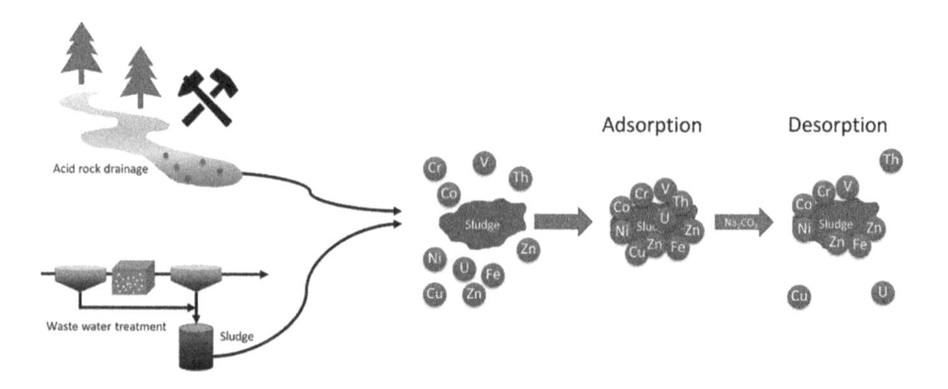

FIGURE 8.3 Overview of adsorption process (Barthen et al., 2022).

for water treatment. Hu et al. (2006) synthesized nano-scale maghemite (g-Fe_2O_3) by sol-gel technique and utilized them for selective removal of Cr (VI), Cu (II), and Ni through adsorption. They have noticed that the removal efficiency was highly pH dependent and the optimal pH for selective removal of Cr, Cu, Ni was found to be 2.5, 6.5, and 8.5, respectively. Regeneration and re-adsorption studies demonstrated that the maghemite nanoparticles could be recovered efficiently.

In a work reported by Iakovleva et al. (2015), limestone was utilized for neutralization and elimination of Cu^{2+}, Zn^{2+}, Fe^{3+}, and Ni^{2+} ions from a real AMD sample. It has been reported that in some cases fly ash can be used as a pretreatment agent as an alternative to limestone and dolomite (Figure 8.4). It can decontaminate the AMD by removing heavy metals such as iron (Fe), aluminum (Al), manganese (Mn), calcium (Ca), magnesium (Mg), zinc (Zn), copper (Cu), cadmium (Cd), nickel (Ni), and cobalt (Co) (Gitari et al., 2008).

In New Zealand, a pilot-scale AMD treatment system was implemented at different locations. A laboratory trial system consisting of a limestone leaching column was also conducted. In these trial experiments for months, it was shown to achieve promising results for Fe and Al recovery (97%), Al (100%), and Ni (66%) as well as reducing the AMD alkalinity (Trumm and Watts, 2010).

In a work reported by Ríos et al. (2008), two variants of zeolites are used to treat an AMD sample from Wales. A batch reactor was employed, where 0.25 and 1 g of sorbents reacted with a volume of 20 mL AMD. The phillipsite-Na showed a lower efficiency compared to the faujasite, whose selectivity for the elimination of the metal ions was Fe > As > Pb > Zn > Cu > Ni > Cr (Ríos et al., 2008).

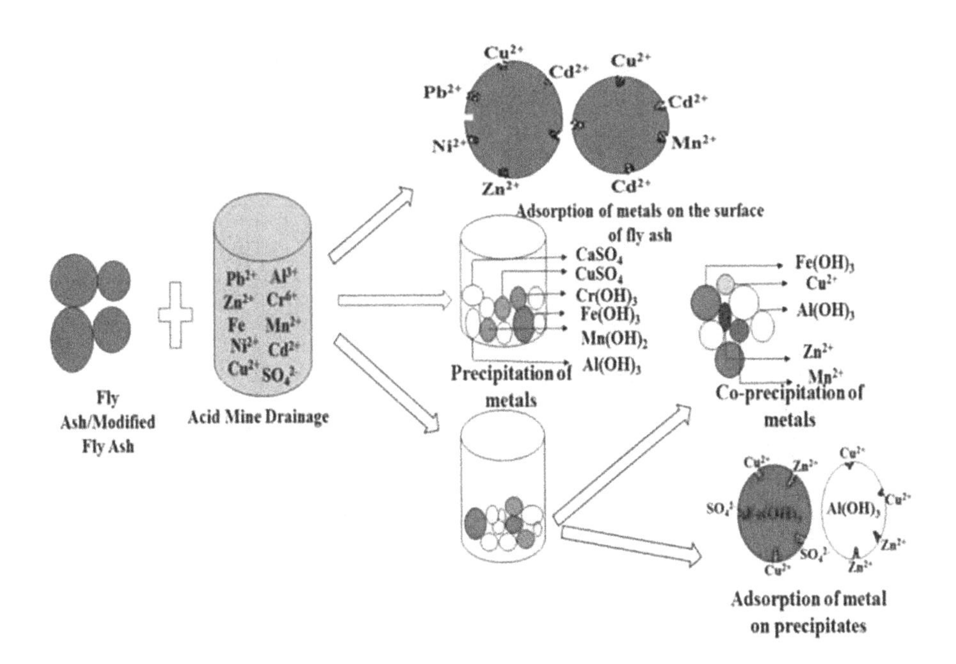

FIGURE 8.4 Removal of heavy metals using fly ash/modified fly ash (Saha and Sinha, 2018).

8.3.3 Electrochemical Processes

It is possible to generate electricity and other metal products simultaneously by electrochemical processes. This is the main advantage of this method compared to other AMD treatment techniques. Microbial fuel cell is one of the promising technologies in this area, which mainly converts lignocellulosic biomass into electricity by the metabolic activity of certain microorganisms such as *Desulfovibrio desulfuricans* or *Escherichia coli* (Simate and Ndlovu, 2014). In a research work by Luo et al. (2014), a dual chamber microbial electrolysis cell was utilized to treat a multicomponent AMD. It resulted successfully in the recovery of Cu^{2+}, Ni^{2+}, and Fe^{2+} while producing H_2 at 0.4–1.1 m^3/day. The cost of energy was offset by the H_2 generation.

Leon-Fernandez et al. (2021) have treated a synthetic AMD containing 500 mg/L copper (Cu^{2+}) and iron (Fe^{+3}), and 50 mg/L nickel (Ni^{2+}) and tin (Sn^{2+}) using a bio-electrochemical system (BES). The presence of electroactive bacteria improved the performance of such reactor configuration, by contrast with systems with abiotic anodes (Figure 8.5).

8.4 BIOLEACHING TECHNIQUES FOR NICKEL RECOVERY FROM AMD

Bioleaching techniques for nickel recovery from AMD typically involve the use of microorganisms to extract nickel from sulfide ores present in the drainage. Some of the most used processes are shown below.

8.4.1 Heap Bioleaching:

This technique involves stacking up crushed ore in heaps and irrigating them with a leaching solution containing microorganisms. As the solution percolates through the heap, the microorganisms catalyze the oxidation of sulfide minerals, releasing nickel ions into the solution. Figure 8.6 shows a typical heap bioleaching process.

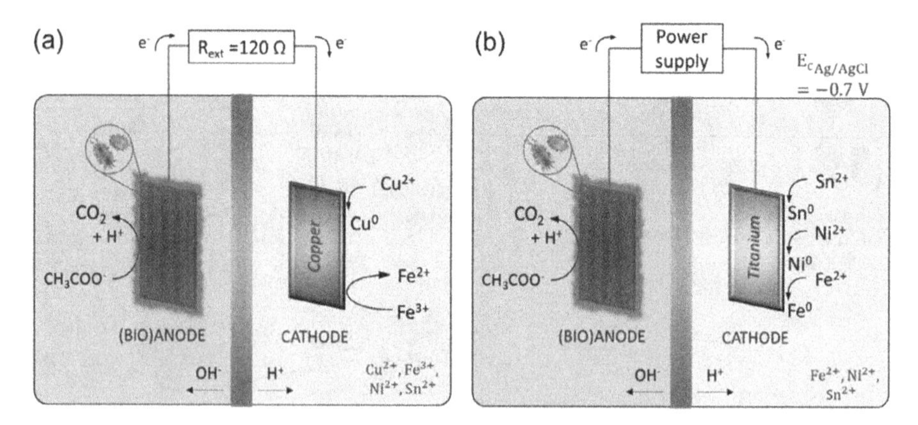

FIGURE 8.5 BES reactor (a) under microbial fuel cell mode, Cu as cathode material, and (b) under microbial electrolysis cell mode, Ti as cathode material (Leon-Fernandez et al., 2021).

FIGURE 8.6 Heap bioleaching.

The process of heap leaching involves the application of a solution containing sulfuric acid, oxidizers (such as oxygen and iron (III) ions), and microorganisms (such as *Thiobacillus ferrooxidans* and *Thiobacillus thiooxidans*) onto the surface or within a heap. This solution is distributed uniformly across the heap through basins, drainage channels, perforated pipes, or pulverization. Channels or pipes then collect the solution enriched with non-ferrous metals, which is subsequently directed for further processing (Svetlov et al., 2015; Khalezov, 2013).

Although both bacterial oxidation and non-bacterial heap leaching are established in the gold and copper industries, their combined use in nickel extraction is unique. The BioHeap process treats crushed ore, so avoiding the need for fine grinding and concentration steps usually needed for most bacterial oxidation techniques. This in turn leads to cost savings which allows the processing of lower-grade nickel sulfide resources, which may currently be uneconomic to exploit.

8.4.2 Tank Bioleaching

In tank bioleaching (Figure 8.7), the ore slurry is contained in tanks where it is agitated and aerated. Microorganisms are introduced into the tanks along with a leaching solution. This method allows for better control of conditions such as temperature, pH, and oxygen levels, leading to higher efficiency.

In the realm of sustainable mining practices, tank bioleaching emerges as a promising solution, offering both environmental benefits and economic advantages in the recovery of nickel from AMD.

Tank bioleaching is a biohydrometallurgical process that utilizes microbial activity to extract metals from ores or waste materials. In the context of nickel recovery from AMD, tank bioleaching involves the cultivation of specific microorganisms capable of oxidizing sulfide minerals containing nickel. These microorganisms, often including *Acidithiobacillus* spp., *Thiobacillus* spp., and *Leptospirillum* spp., facilitate the dissolution of nickel-bearing minerals, releasing the metal into solution.

FIGURE 8.7 Tank bioleaching.

The process of tank bioleaching for nickel recovery from AMD typically involves the following steps:

AMD collection: AMD, characterized by its acidic pH and high metal content, is collected from mining sites or AMD-affected areas.

Tank setup: Tanks or reactors are constructed to provide an optimal environment for microbial activity. Parameters such as pH, temperature, and nutrient availability are carefully controlled to support the growth and activity of bioleaching microorganisms.

Inoculation: The tank is inoculated with a culture of selected microorganisms capable of oxidizing sulfide minerals and solubilizing nickel from the AMD.

Bioleaching process: As the microorganisms metabolize sulfur compounds present in the AMD, they generate sulfuric acid, which aids in the dissolution of nickel-bearing minerals. This process results in the release of nickel ions into the solution.

Metal recovery: The solution containing dissolved nickel is collected from the tank and subjected to further processing steps, such as precipitation or solvent extraction, to recover the nickel in a concentrated form.

Tank bioleaching offers several advantages over traditional metal recovery methods: By harnessing microbial activity, tank bioleaching minimizes the use of harsh chemicals and reduces the environmental impact associated with conventional mining and metallurgical processes. It allows for the selective recovery of nickel from AMD, maximizing resource utilization and minimizing waste generation. Despite the initial investment in tank setup and microbial inoculation, tank bioleaching can offer long-term cost savings through reduced chemical consumption and enhanced metal recovery efficiency. In addition to nickel recovery, tank bioleaching can contribute to the remediation of AMD-affected environments by neutralizing acidity and reducing metal concentrations, thereby mitigating environmental hazards.

8.4.3 Stirred Tank Reactors

Stirred Tank Reactors (STRs) are similar to tank bioleaching but involve continuous agitation to ensure thorough mixing of the ore slurry and the leaching solution as shown in Figure 8.8. This agitation enhances mass transfer and microbial activity, improving nickel recovery rates.

8.4.4 Column Bioleaching

Column bioleaching involves packing columns with ore particles and circulating a leaching solution containing microorganisms through the columns. The solution trickles down through the ore bed, facilitating contact between the microorganisms and the sulfide minerals to release nickel (Figure 8.9). Column bioleaching offers a controlled and efficient approach for metal recovery from AMD. It involves passing the acidic solution containing dissolved metal sulfides through a column packed with a suitable substrate colonized by the bioleaching bacteria. As the solution flows through the column, the microbes facilitate the dissolution and mobilization of nickel. The enriched pregnant leachate, containing the dissolved nickel, can then be separated for further processing and metal recovery.

8.4.5 Biooxidation

Biooxidation is a pretreatment step in which sulfide minerals are oxidized by microbial action before subjecting the ore to conventional leaching processes. This can improve the efficiency of subsequent nickel recovery processes by exposing more nickel-bearing minerals to leaching.

FIGURE 8.8 Stirred tank reactor.

FIGURE 8.9 Column bioreactors.

8.4.6 Reactor Bioleaching

In this technique, bioleaching is carried out in specially designed bioreactors where conditions such as temperature, pH, and nutrient supply can be tightly controlled to optimize microbial activity and nickel recovery.

In situ bioleaching: In situ bioleaching involves stimulating microbial activity within the ore deposit itself by injecting a leaching solution into the ground. This technique is often used for low-grade or deep-seated ore deposits where traditional mining methods are not feasible.

Each of these bioleaching techniques has its advantages and limitations, and the choice of method depends on factors such as the nature of the ore, the scale of operation, and economic considerations. Additionally, optimizing the process conditions and selecting suitable microorganisms are critical for maximizing nickel recovery efficiency. The recovery of valuable metals and other elements such as Cu, Zn, Mn, Ni, and Fe from AMD are broadly classified as active and passive treatment techniques. However, the active methods are mostly explored and are available in literature. The most common active recovery techniques include selective metal precipitation, adsorption, and electrochemical processes (Ayora et al., 2016; Chen et al., 2014; Larsson et al., 2018; Motsi et al., 2009; Oh et al., 2016; Park et al., 2015; Pozo et al., 2017; Seo et al., 2017). In this section, these common techniques are elaborated in detail.

8.5 CONCLUSION

The bio-recovery of nickel from AMD represents a remarkable convergence of environmental stewardship and resource efficiency. Through the ingenious application of microbial activity, bioleaching offers a sustainable solution to the dual challenges of mitigating AMD while simultaneously recovering valuable metals like nickel. By harnessing the metabolic prowess of acidophilic microorganisms such as

Thiobacillus spp., this innovative approach not only minimizes the environmental footprint of mining activities but also presents economic opportunities for resource recovery and recycling. As we continue to explore and refine the bioleaching process, we move closer to a future where the principles of sustainability and responsible resource management guide our interactions with the natural world.

REFERENCES

Akcil, D.A., Kutucuoğlu, K., Ünlü, O. 2011. Nickel recovery from acidic leachates using microorganisms. *Miner. Eng.* 24(8), 806–812. https://doi.org/10.1016/j.mineng.2011.04.010

Ali, S.H., Giurco, D., Arndt, N. et al. 2017. Mineral supply for sustainable development requires resource governance. *Nature* 543, 367–372. https://doi.org/10.1038/nature21359

Anawar, H.M. 2015. Sustainable rehabilitation of mining waste and acid mine drainage using geochemistry, mine type, mineralogy, texture, ore extraction and climate knowledge. *J. Environ. Manag.* 158, 111–121. https://doi.org/10.1016/j.jenvman.2015.04.045

Ayora, C., Macías, F., Torres, E., Lozano, A., Carrero, S., Nieto, J.-M., P´erez-L´opez, R., Fern´andez-Martínez, A., Castillo-Michel, H. 2016. Recovery of rare earth elements and yttrium from passive-remediation systems of acid mine drainage. *Environ. Sci. & Technol.* 50, 8255–8262. https://doi.org/10.1021/acs.est.6b02084

Barthen, R., Sulonen, M.L., Peräniemi, S., Jain, R., Lakaniemi, A.M. 2022. Removal and recovery of metal ions from acidic multi-metal mine water using waste digested activated sludge as biosorbent. *Hydrometallurgy* 207, 105770.

Bejan, D., Bunce, N.J. 2015 Acid mine drainage: electrochemical approaches to prevention and remediation of acidity and toxic metals. *J. Appl. Electrochem.* 45, 1239–1254. https://doi.org/10.1007/s10800-015-0915-z

Chen, T., Yan, B., Lei, C., Xiao, X. 2014. Pollution control and metal resource recovery for acid mine drainage. *Hydrometallurgy* 147, 112–119. https://doi.org/10.1016/j.chemosphere.2022.137089

Crini, G., Lichtfouse, E. 2018. Advantages and disadvantages of techniques used for wastewater treatment. *Environ. Chem. Lett.* 17, 145–155. https://doi.org/10.1007/s10311-018-0785-9

Daraz, U., Li, U., Ahmad, I., Iqbal, R., Ditta, A. 2023. Remediation technologies for acid mine drainage: Recent trends and future perspectives. *Chemosphere* 311(2), 137089.

Gitari, W.M., Petrik, L.F., Etchebers, O., Key, D.L., Okujeni, C. 2008. Utilization of fly ash for treatment of coal mines wastewater: Solubility controls on major inorganic contaminants. *Fuel* 87, 2450–2462.

Gu, S., Kang, X., Wang, L. et al. 2018. Clay mineral adsorbents for heavy metal removal from wastewater: a review. *Environ. Chem. Lett.* 8, 8–9. https://doi.org/10.1007/s10311-018-0813-9

Gunkel-Grillon, P., Laporte-Magoni, C., Lemestre, M., Bazire, N. 2014. Toxic chromium release from nickel mining sediments in surface waters, New Caledonia. *Environ. Chem. Lett.* 12, 511–516. https://doi.org/10.1007/s10311-014-0475-1

Honaker, R.Q., Zhang, W., Yang, X., Rezaee, M. 2018. Conception of an integrated flowsheet for rare earth elements recovery from coal coarse refuse. *Miner. Eng.* 122, 233–240. https://doi.org/10.1016/j.mineng.2018.04.005

Hu, J., Chen, G., Lo, I.M.C. 2006. Selective removal of heavy metals from industrial wastewater using maghemite nanoparticle: performance and mechanisms. *J. Environ. Eng.* 132(7), 709–715. https://doi.org/10.1061/(ASCE)0733–9372(2006)132:7(709)

Hu, X., Yuan, X., Dong, L. 2014. Coal fly ash and straw immobilize Cu, Cd and Zn from mining wasteland. *Environ. Chem. Lett.* 12, 289–295. https://doi.org/10.1007/s10311-013-0441-3

Iakovleva, E., Mäkilä, E., Salonen, J., Sitarz, M., Wang, S., Sillanpää, M. 2015. Acid mine drainage (AMD) treatment: neutralization and toxic elements removal with unmodified and modified limestone. *Ecol. Eng.* 81, 30–40. https://doi.org/10.1016/j.ecoleng.2015.04.046

Johnson, D.B., Hallberg, K.B. 2005. Acid mine drainage remediation options: a review. *Sci. Total Environ.* 338, 3–14. https://doi. org/10.1016/j.scitotenv.2004.09.002

Kefeni, K.K., Msagati, T.A.M., Mamba, B.B. 2017. Acid mine drainage: prevention, treatment options, and resource recovery: a review. *J. Clean Prod.* 151, 475–493. https://doi.org/10.1016/j.jclepro.2017.03.082

Khalezov, B.D. 2013. *Heap Leaching of Copper and Copper-Zinc Ores: (Russian Experiece).* Ekaterinburg: Publ. UrD RAS, 346 p.

Larsson, M., Nosrati, A., Kaur, S., Wagner, J., Baus, U., Nyd´en, M. 2018. Copper removal from acid mine drainage-polluted water using glutaraldehyde-polyethyleneimine modified diatomaceous earth particles. *Heliyon* 4, e00520. https://doi.org/10.1016/j.heliyon.2018.e00520

Leon-Fernandez, L.F., Medina-Díaz, H.L., Pérez, O.G., Romero, L.R., Villaseñor, J., Fernández-Morales, F.J. 2021. Acid mine drainage treatment and sequential metal recovery by means of bioelectrochemical technology. *J. Chem. Technol. Biotechnol.* 96(6), 1543–1552.

Li, Q., Zhang, W. 2022. Process development for recovering critical elements from acid mine drainage. *Resour. Conserv. Recycl.* 180, 106214.

Liu, W., Sun, B., Zhang, D., Chen, L., Yang, T. 2017. Selective separation of similar metals in chloride solution by sulfide precipitation under controlled potential. *Miner. Met. Mater. Soc.* 69, 2358–2363. https://doi.org/10.1007/s11837-017-2526-0

Luo, H., Liu, G., Zhang, R. et al. 2014. Heavy metal recovery combined with H2 production from artificial acid mine drainage using the microbial electrolysis cell. *J. Hazard Mater.* 270, 153–159. https://doi.org/10.1016/j.jhazm at.2014.01.050

Masindi, V., Gitari, M.W., Tutu, H., DeBeer, M. 2017. Synthesis of cryptocrystalline magnesite–bentonite clay composite and its application for neutralization and attenuation of inorganic contaminants in acidic and metalliferous mine drainage. *J. Water Process Eng.* 15, 2–17. https://doi.org/10.1016/j.jwpe.2015.11.007

Motsi, T., Rowson, N.A., Simmons, M.J.H. 2009. Adsorption of heavy metals from acid mine drainage by natural zeolite. *Int. J. Miner. Process.* 92, 42–48. https://doi.org/10.1016/j.minpro.2009.02.005

Mudhoo, A., Garg, V.K., Wang, S. 2012. Removal of heavy metals by biosorption. *Environ. Chem. Lett.* 8, 199–216. https://doi. org/10.1016/j.biortech.2003.10.032

Nagajyoti, P.C., Lee, K.D., Sreekanth, T.V.M. 2010. Heavy metals, occurrence and toxicity for plants: a review. *Environ. Chem. Lett.* 8, 199–216. https://doi.org/10.1007/s10311-010-0297-8

Nleya, Y., Simate, G.S., Ndlovu, S. 2016. Sustainability assessment of the recovery and utilisation of acid from acid mine drainage. *J. Clean Prod.* 113, 17–27. https://doi.org/10.1016/j.jclepro.2015.11.005

Oh, C., Han, Y.-S., Park, J.H., Bok, S., Cheong, Y., Yim, G., Ji, S. 2016. Field application of selective precipitation for recovering Cu and Zn in drainage discharged from an operating mine. *Sci. Total Environ.* 557, 212–220

Park, I., Tabelin, C. B., Jeon, S., Li, X., Seno, K., Ito, M., & Hiroyoshi, N. (2018). A review of recent strategies for acid mine drainage prevention and mine tailings recycling. Chemosphere, 219, 588–606. https://doi.org/10.1016/j.chemosphere.2018.11.053

Park, S.M., Shin, S.Y., Yang, J.S. et al. 2015. Selective recovery of dissolved metals from mine drainage using electrochemical reactions. *Electrochim. Acta.* 181, 248–254. https://doi.org/10.1016/j.electacta.2015.03.085

Pozo, G., Pongy, S., Keller, J., Ledezma, P., Freguia, S. 2017. A novel bioelectrochemical system for chemical-free permanent treatment of acid mine drainage. *Water Res.* 126, 411–420. https://doi.org/10.1016/j.watres.2017.09.058

Ríos, C.A., Williams, C.D., Roberts, C.L. 2008. Removal of heavy metals from acid mine drainage (AMD) using coal fly ash, natural clinker and synthetic zeolites. *J. Hazard. Mater.* 156, 23–35.

Ryan, M.J., Kney, A.D., Carley, T.L. 2017. A study of selective precipitation techniques used to recover refned iron oxide pigments for the production of paint from a synthetic acid mine drainage solution. *Appl. Geochem.* 79, 27–35. https://doi.org/10.1016/j.apgeochem.2017.01.019

Saha, S., Sinha, A. 2018. A review on treatment of acid mine drainage with waste materials: a novel approach. *Global NEST J.* 20(3), 512–528.

Sampaio, R.M.M., Timmers, R.A., Kocks, N., Andr´e, V., Duarte, M.T., VanHullebusch, E.D., Farges, F., Lens, P.N.L. 2010. Zn–Ni sulfide selective precipitation: the role of supersaturation. *Sep. Purif. Technol.* 74, 108–118. https://doi.org/10.1016/J.SEPPUR.2010.05.013

Seo, E.Y., Cheong, Y.W., Yim, G.J., Min, K.W., Geroni, J.N. 2017. Recovery of Fe, Al and Mn in acid coal mine drainage by sequential selective precipitation with control of pH. *Catena* 148, 11–16.

Sheoran, A.S., Sheoran, V. 2006. Heavy metal removal mechanism of acid mine drainage in wetlands: a critical review. *Miner. Eng.* 19, 105–116. https://doi.org/10.1016/j.mineng.2005.08.006

Simate, G.S., Ndlovu, S. 2014. Acid mine drainage: challenges and opportunities. *J. Environ. Chem. Eng.* 2, 1785–1803. https://doi. org/10.1016/j.jece.2014.07.021

Svetlov, A.V., et al. 2015. Study of heap leaching of non-ferrous metals from sulfide materials from natural and technogenic objects of the Murmansk region. *Ecol. Ind. Prod.* 3, 65–70.

Trumm, D., Watts, M. 2010. Results of small-scale passive system trials to treat acid mine drainage, West Coast Region, South Island, New Zealand. *NZ J. Geol. Geophys.* 53(2–3), 227–237.

Vaziri Hassas, B., Rezaee, M., Pisupati, S.V. 2020. Precipitation of rare earth elements from acid mine drainage by CO_2 mineralization process. *Chem. Eng. J.* 399, 125716. https://doi.org/10.1016/j.cej.2020.125716

Zhang, W., Honaker, R. 2020. Process development for the recovery of rare earth elements and critical metals from an acid mine drainage. *Miner. Eng.* 153, 106382. https://doi.org/10.1016/j.mineng.2020.106382

9 Microorganism-Enhanced Recovery of Zinc from Mining Waste

Somanchi Venkata Ramalakshmi, Aradhana Basu, Sankha Chakrabortty, and Jayato Nayak

9.1 INTRODUCTION

In 2022, global zinc production from mines approached nearly 12.5 million metric tons, distinguishing it as one of the most widely employed nonferrous metals globally. Due to its ability to resist corrosion across diverse environments, zinc finds application in safeguarding steel through galvanization. The galvanizing industry accounts for approximately 57% of zinc consumption, with coatings utilizing 16%, die casting alloys 14%, oxides and chemicals 7%, and extruded products 6% (Indian Bureau of Mines, 2022). Therefore, with the increase in consumption, zinc production has also increased. It can be observed that the price of zinc rises when primary mining is unable to provide all of the metal, and secondary recovery from wastes provides the remaining amount. Zinc recovery from waste and secondary resources is crucial in the current circular economy framework. In addition, zinc is predominantly important in renewable energy technologies because it prevents wind turbines and solar panels from rusting; therefore, recycling is highly desired to fulfill the overall objective of reducing CO_2 emissions (International Zinc Association, 2022). Considering the aforementioned, there has been a rise in interest in establishing methods for recovering zinc from waste and secondary resources. Such secondary resources are often treated using hydrometallurgical and pyrometallurgical techniques (Jha et al., 2001; Mocellin et al., 2017; Wang et al., 2021). While pyrometallurgy and hydrometallurgy are considered effective procedures, they both have several disadvantages (Asghari et al., 2013). Two major drawbacks of pyrometallurgical processes are their high-temperature requirements (1,500°C–1,700°C), which result in considerable energy consumption, and the need for dust collection/gas cleaning equipment. Dust containing fluoride and chloride salts leads to serious corrosion issues and demands the use of expensive alloys for building components. On a modest scale, hydrometallurgical techniques are more cost-effective for treating materials with low zinc content. However, the usage of toxic acids and bases and complex downstream processing costs to manage the acidic waste that is resulting from hydrometallurgy are the main drawbacks of the process (Srichandan et al., 2019; Bharadwaj and Ting, 2013; Asghari et al., 2013).

DOI: 10.1201/9781003415541-9

The study on the microorganism-enhanced recovery or biorecovery of metals from mining waste reflects a paradigm shift toward sustainable practices in the mining industry. This innovative approach harnesses the power of microorganisms to extract metals from ores and waste materials. This evolution has been evidenced by two factors: increasing metal demand and a growing awareness of the environmental repercussions associated with conventional extraction methods (Barrie Johnson, 2014; Schippers et al., 2013). The commercial bioleaching of zinc minerals has the potential to considerably improve the handling of concentrated metal, which was difficult to manage using traditional methodologies. The advantage of zinc concentrate bioleaching over hydrometallurgy is that it eliminates the requirement for roasting, sulfuric acid (H_2SO_4) plants, and gaseous effluent washing (Viera et al., 2007). The characteristics of microorganisms, chemical reactions (precipitation formation), and the level of friction in the suspension are some of the factors that affect the bioleaching process (Olubambi et al., 2007). These factors also include pH, temperature, nutrient concentration, oxygen content, particle size, pulp density, stirring frequency (rpm), and total bioleaching duration (Rouchalova et al., 2020). This chapter overall covers the existing literature on various bioleaching and biomining studies which aim to illuminate a path toward a more sustainable and environmentally conscious approach to meeting our metal demands.

9.2 MINE WASTES AND CIRCULAR ECONOMY

In times of crisis, nothing may be regarded as waste. When discussing the circular economy, it is important to take waste minimization and its recovery from waste streams into account. A new economic paradigm called the "circular economy" has enormous promise and is centered on producing zero waste (Singh et al., 2020). Preserving the value of materials and extending the useful life of products and materials are the goals of the circular economy. Mining activities, which extract precious minerals and resources for various economic and societal requirements, inherently generate enormous wastes. From the beginning of the mining process, when exploration activities like drilling are carried out, to the end of the mine's life, waste is produced at regular intervals. Waste rock, tailings, and mine water are three of the many waste types generated during mining operations that are notable for their huge amount. Waste rock weathering has a role in the production of an acid drainage with high metal levels, often known as acid mine drainage (AMD), which is one of the most significant challenges facing the global mining industry (Yuan et al., 2022). Significant usage of energy (Boulamanti and Moya, 2016), generation of solid waste (Çoruh et al., 2013), emission of greenhouse gas, and the release of heavy metals (HMs; Lei et al., 2016; Karbassi et al., 2016) are all associated with the mining activities, including zinc mining. These factors pose a risk to the environment, human health, and resources. To address these challenges, researchers have conducted life cycle assessments (LCAs) to measure the environmental impact of zinc mining and beneficiation. LCAs identify key processes and substances contributing to environmental pollution, guiding the development of mitigation strategies (Tao et al., 2019). High concentrations of HMs in water sources exceed recommended standards, leading to potential health issues such as neurological disorders, cancer, and reproductive problems (Obiora et al., 2018, Obasi and Akudinobi, 2020). Additionally, there is a growing recognition of the need for effective contamination mitigation strategies and environmental remediation to safeguard human health and the environment (Du et al., 2020). The studies underscore the immediate need for sustainable mining practices and highlight the importance of integrating environmental considerations into mining operations. By adopting cleaner production (CP) strategies, implementing effective pollution control measures, and prioritizing environmental health, the mining industry can move toward a more sustainable and environmentally responsible future. Efforts to mitigate these challenges have led to the adoption of CP strategies aimed at maximizing efficiency and minimizing negative environmental impacts (Dong et al., 2019). CP involves a continuous application of integrated environmental strategies to processes, products, and services to reduce risks to humans and the environment. As one of the CP strategies for recovering metal, bioleaching, for instance, requires less capital investment, little energy usage, no production of hazardous waste, and minimal installation and operating costs (Rendón-Castrillón et al., 2023).

9.3 BIOLEACHING AND BIOTECHNOLOGY

Biohydrometallurgical routes present novel opportunities for recovering metals from low-grade mineral residues, thereby mitigating environmental concerns associated with landfill storage and inefficient land use (Williamson et al., 2020). By utilizing microbiologically produced organic acids, such as citric acid, bioleaching processes demonstrate high selectivity in extracting metals like zinc from iron oxide residues, with minimal co-extraction of unwanted elements like iron. Bioleaching, bioflocculation, bioprecipitation, biosorption, biooxidation, and bioreduction are among the various biohydrometallurgical processes (Minimol et al., 2021). Bioleaching, a cost-effective and eco-friendly alternative to conventional metallurgical methods, is gaining prominence in resource recovery from various metal-bearing wastes and by-products (Figure 9.1) (Kara et al., 2023). It leverages the metabolic activities of microorganisms to extract metals from materials like electronic wastes, tailings, and metallurgical residues. Through optimization of operational parameters such as pH, solid concentration, and microbial consortia, bioleaching offers promising avenues for efficient metal extraction. In the context of mineral extraction and metal recovery, bioleaching emerges as a key enabler for maximizing economic performance, particularly in treating low-grade sulfide ores (Kržanović et al., 2019). With different leaching mechanisms, bioleaching is facilitated by various microorganisms and their metabolites. Table 9.1 demonstrates a list of various microbial groups, leaching mechanisms, and biogenic leaching agents. Biogenic leaching agents are efficient reagents in terms of their metal leaching efficiency and low environmental impacts (Tezyapar Kara et al., 2023). As illustrated in Figure 9.2, biogenic leaching agents, such as hydrogen cyanide, organic acids, H_2SO_4, and ferric iron, are produced by microorganisms. Subsequently, metal-containing materials react with biogenic leaching agents, dissolving the metals. A metal in its elemental form is represented by M^0, and an ionic version of a metal is represented by M^{n+}. Laboratory-scale bioleaching

TABLE 9.1
Common Microorganisms Used for Bioleaching Applications

Microbial Groups	Species	Microbial Leaching Mechanism	pH Range and Temperature	Energy Source	Biogenic Leaching Agents	References
Acidophiles	*Acidianus infernus*, *Acidianus brierleyi*, *Acidithiobacillus ferrooxidans*, *Acidimicrobium* species, *Acidithiobacillus thiooxidans*, *Acidithiobacillus caldus*, *Brevibacillus* sp., *Ferroplasma acidiphilum*, *Ferrimicrobium* sp., *Ferroplasma acidarmanus*, *Leptospirillum ferrooxidans*, *Leptospirillum ferriphilum*, *Metallosphaera sedula*, *Sulfobacillus thermosulfidooxidans*, *Sulfobacillus thermotolerans*, *Sulfobacillus acidophilus*, *Sulfobacillus metallicus*, *Sulfobacillus acidocaldarius*, *Sulfolobus solfataricus*, *Sulfolobus brierleyi*	Thiosulfate pathway Polysulfide pathway	pH 0.8–2.5 Mesophiles: 28°C–37°C Moderate thermophiles: 40°C–60°C Thermophiles: 60°C–80°C	Fe^{+2} Reduced inorganic sulfur compounds, e.g., S_8, $S_2O_3^{2-}$, H_2S	F^{+3} H_2SO_4	Figueroa-Estrada et al. (2020), Ma et al. (2021), Rodrigues et al. (2019), Tian et al. (2022)

(Continued)

TABLE 9.1 (*Continued*)
Common Microorganisms Used for Bioleaching Applications

Microbial Groups	Species	Microbial Leaching Mechanism	pH Range and Temperature	Energy Source	Biogenic Leaching Agents	References
Fungi	*Aspergillus niger, Aspergillus flavus, Penicillium chrysogenum, Penicillium simplicissimum*	Acidolysis Complexolysis Redoxolysis Bioaccumulation	pH 3.0–7.0 25°C–35°C	Glucose Sucrose	Gluconic acid Citric acid Oxalic acid Malic acid	Amiri et al. (2012), Qayyum et al. (2019), Dusengemungu et al. (2021)
Cyanogens	Bacteria: *Pseudomonas fluorescens, Pseudomonas aeruginosa, Pseudomonas caldus, Pseudomonas putida, P. aeruginosa, Chromobacterium violaceum* Fungi: *Clytocybe* sp., *Polyporus* sp., *Marasmius oreades*	Glycine metabolism	pH 7.0–11.0 25°C–35°C	Glycine	Hydrogen cyanide	Arab et al. (2020), Liu et al. (2016), Srichandan et al. (2020)

Source: Adapted from Kara et al. (2023) licensed under CC BY 4.0. To view a copy of this license, visit https://creativecommons.org/licenses/by/4.0/.

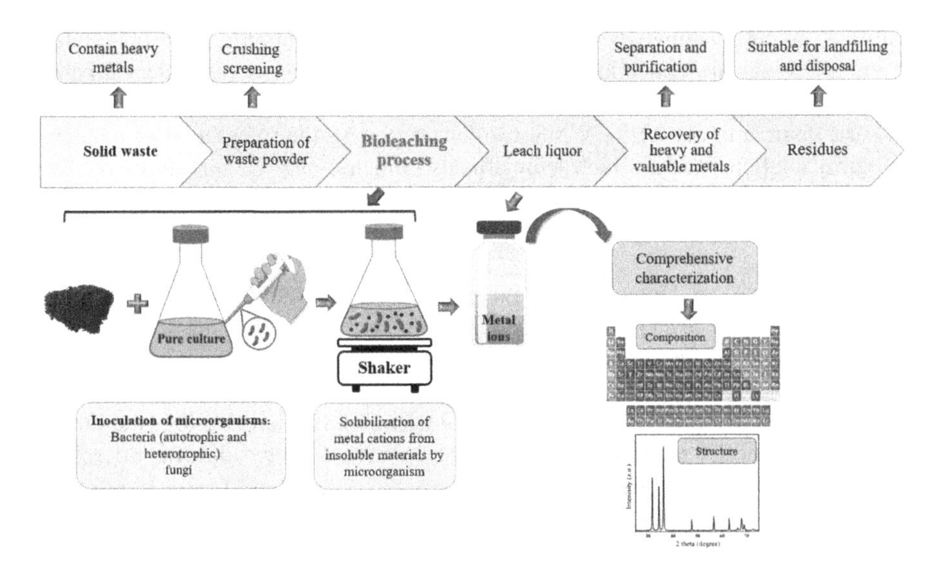

FIGURE 9.1 Flow sheet of the bioleaching process for metal recovery from solid waste. Reprinted with permission from Naseri et al. (2023) with license number 1473334-1.

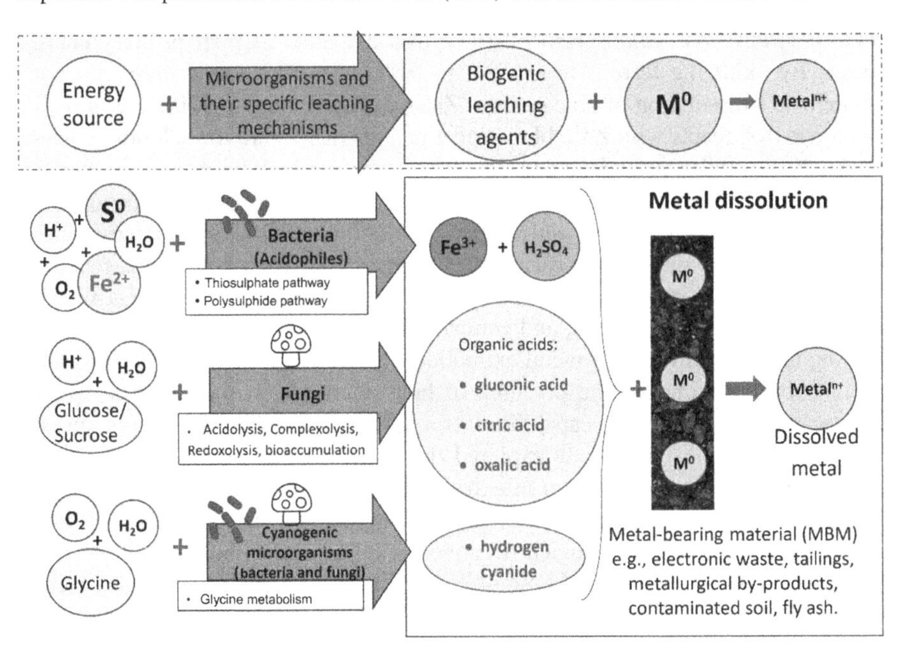

FIGURE 9.2 The bioleaching process. Adapted from Kara et al. (2023) licensed under CC BY 4.0. To view a copy of this license, visit https://creativecommons.org/licenses/by/4.0/.

experiments demonstrate significant metal recoveries, including copper, gold, silver, zinc, and lead, showcasing its potential as a viable processing technology. Economic feasibility studies further underscore the viability of bioleaching plants for metal extraction, affirming positive net present values and internal rates of return, even with low process volumes (Irrgang et al., 2021). The evolution and current status of biotechnological applications, particularly in the mining sector, reflect a paradigm shift toward sustainable resource management and environmental stewardship (Romero-García et al., 2019). While challenges persist, the integration of biotechnological solutions, such as bioleaching, holds immense promise for addressing the complexities of metal extraction from low-grade ores and waste materials, paving the way for a more environmentally conscious and economically viable future in the mining industry.

Contributing to the biotechnological landscape is the role of biotechnology in the management and recovery of valuable resources from metal-bearing solid wastes. Compared to more conventional techniques like roasting and smelting, which often require high temperatures and release pollutants, one of the main benefits of using biotechnology in bioleaching is its environmental friendliness. Biotechnology facilitates the enhancement and management of bioleaching procedures by adjusting microbial populations, substrate composition, and operational parameters. Bioleaching relies on the metabolic activities of various microorganisms, particularly acidophilic bacteria. These "acid-loving" microbes thrive in acidic environments generated during the bioleaching process (Sajjad et al., 2019). This occurs through microbial leaching, which is classified into direct and indirect leaching mechanisms. Direct bioleaching involves acidophilic bacteria like *Acidithiobacillus ferrooxidans* and *Leptospirillum ferrooxidans*, which utilize iron oxidation as their primary energy source. By oxidizing ferrous iron (Fe^{2+}) to ferric iron (Fe^{3+}), they indirectly contribute to the dissolution of zinc sulfide (ZnS), the most common zinc ore mineral. The ferric iron reacts with ZnS, liberating zinc ions (Zn^{2+}) into the solution (Ghassa et al., 2017). Indirect bioleaching involves some acidophilic bacteria, such as *Sulfobacillus thermosulfidooxydans*, which can directly oxidize sulfur compounds like elemental sulfur (S^0) or thiosulfate ($S_2O_3^{2-}$) to sulfate (SO_4^{2-}). This process generates H_2SO_4, which plays a key role in dissolving ZnS through a chemical reaction, releasing zinc ions into the solution (Tributsch et al., 1992). Three distinct mechanisms—redoxolysis, acidolysis, and complexolysis—emerge as key processes, each offering unique pathways for metal extraction (Sethurajan et al., 2018; Nancharaiah et al., 2016). Building on the potential of biotechnology, Vardanyan et al. (2018) investigated the bioleaching capabilities of a natural microbial consortium sourced from mine tailings. Their study explored the efficacy of *L. ferrooxidans* Teg and a naturally occurring consortium in extracting copper and zinc from concentrated tailings, providing promising results (Vardanyan et al., 2019). The study by Rathna and Nakkeeran explored biological treatment methods for mineral recovery from low-grade ores, emphasizing the significance of bioleaching and biohydrometallurgy. The focus on sustainable and environmentally friendly techniques echoes the industry's collective effort to address concerns related to conventional processing methods (Rathna and Nakkeeran, 2020). The critical analysis by Castro et al. (2019) compared thermophilic and mesophilic microorganisms in metal biorecovery and bioremediation, informing the selection of microorganisms for various applications. Considerations such as process stability, flexibility, and resistance to acidity and metal toxicity are meticulously examined, contributing to the optimization of biotechnological processes in metal recovery. This analysis is crucial for aligning microbial applications with specific mining environments (Castro et al., 2019). Moreover, in the domain of mining and metallurgical waste, Fomchenko et al. (2020) proposed an innovative biohydrometallurgical method. This approach involved the sequential bioleaching of pyritic tailings followed by ferric leaching of nonferrous slags, demonstrating effectiveness in recovering gold and nonferrous metals (Fomchenko et al., 2020). Levett et al. extended the exploration to innovation in mining from a microbial perspective, emphasizing microbe–mineral–metal interactions, molecular techniques for mineral exploration, and microbial strategies for mine remediation. The study acknowledged the increasing demand for metals in the transition to renewable energy resources and highlighted the potential of harnessing microorganisms for improving exploration strategies and expanding the range of economically viable biohydrometallurgical techniques (Levett et al., 2021). Expanding further on waste remediation, Begum et al. presented a comprehensive review of the applications of microbial communities in remediating toxic metal waste. The review emphasized microbial remediation mechanisms, including surface absorption, bioleaching, transformation, biomineralization, and bioaccumulation (Begum et al., 2021). Building on existing techniques and innovating in the field, Liao et al. (2021) introduced a novel mixed culture bioleaching process for simultaneous metal recovery and toxicity reduction in lead–ZnS mine tailings. Their study highlighted the advantages of mixed culture A, featuring *A. ferrooxidans* and *Sulfobacillus thermosulfidooxidans*, in shortening the bioleaching cycle and reducing ecological risks (Liao et al., 2021). In broader implications, Newsome and Falagán (2021) provided insights into the microbiology of metal mine waste, specifically focusing on its impact on planetary health. The article reviewed long-term field studies, underscoring the microbial influence on metal fate in mine wastes and proposing strategies like stimulating microbially reducing conditions for effective remediation (Newsome and Falagán, 2021).

9.4 CURRENT ADVANCEMENTS IN BIORECOVERY OF ZINC FROM MINING WASTE

In the quest for sustainable resource management and environmental conservation, innovative approaches are being explored to valorize metal-containing waste materials and extract valuable metals. The European project NEMO (Near-zero-waste recycling of low-grade sulfidic mining waste for critical-metal, mineral, and construction raw-material production in a circular economy) exemplifies this endeavor by developing methods to reprocess sulfidic tailings, a major source of metal-containing waste in Europe (Hubau et al., 2020). Addressing the positive impact of microorganism-enhanced recovery techniques on sustainable mining practices, challenges faced in the application of biomining and bioleaching for zinc recovery are discussed comprehensively. The latest developments and innovations are discussed, providing readers with a glimpse into the potential evolution of these technologies. In a recent comparative study, authors tried to investigate potential circular economy strategies for recovering metal from secondary resources using chemical as well as bioleaching process. According to the study by Swęd et al. (2023), 39% of zinc was extracted by chemical treatment (HCl and HNO_3), whereas 40% of zinc was extracted by utilizing a biorecovery (*Acidithiobacillus thiooxidans*) strategy from a Zn-bearing rock, which demonstrates that biotic processes can have approximately equal levels of efficiency as chemical processes. A ureolytic strain of *Lysinibacillus sphaericus* produces biogenic ammonia, which is tested for its capacity to recover Co and Zn from eight primary and secondary materials, including sludges, automotive shredder residues (ASRs), mining and metallurgical wastes, and sludges. For most materials, 1 mol/L total ammonia was required to provide moderate to high yields (30%–70%) and very high selectivity (>97% against iron) of copper and zinc (Williamson et al., 2021). In a pilot-scale study, Ahmadi et al. (2020) investigated the continuous biohydrometallurgical extraction of zinc from a bulk lead–zinc flotation concentrate, and their study employed a mixed culture of moderately thermophilic bacteria, showcasing its efficacy in bioleaching and subsequent solvent extraction for efficient zinc recovery. Deveci et al. incorporated an experimental analysis section focusing on the McArthur River complex Pb/Zn ore. The potential amenability of the ore to selective zinc extraction using strains of mesophilic (at 30°C), moderately thermophilic (at 50°C), and extremely thermophilic bacteria (at 70°C) was evaluated. Using selected bacterial strains, the influence of pH, precipitation of iron, and external ferrous iron addition on the bioleaching of the ore/concentrate and Zn extraction was investigated (Deveci et al., 2004). An additional dimension to waste remediation efforts can be added by looking at the biooxidation of pyritic flotation tailings of copper–zinc ores. The study explored continuous biooxidation processes at different temperature settings, showcasing the potential of biohydrometallurgical processes in utilizing mine waste for metal extraction (Muravyov, 2019). Spiess et al. (2023) contributed to the evolving landscape by investigating Zn recovery from bioleaching using a microbial electrolysis cell (MEC) and comparing it with selective precipitation methods. The study explored the efficiency of different recovery approaches, emphasizing energy consumption and recovery efficiency. Zn recovery was carried out at a synthetic wastewater-treating bioanode with a constant potential of −100 mV vs. Ag/AgCl. The recovery of zinc was first investigated from the bioleachate and then from the control zinc sulfate solution. After 4 days of operation, the Al content declined from 270 to 10 mg/L, while the Zn concentration reduced from 444 to 245 mg/L during MEC operation with bioleachate. The precipitation of aluminum hydroxide as a result of the catholyte pH increasing from 3.4 to 4.4 is most likely what caused the decrease in Al concentration. This technological dimension adds to the sustainable management and recovery of metals, showcasing the potential of innovative techniques such as MECs (Spiess et al., 2023). Without the aid of an outside energy source, a study showed that microbial fuel cells (MFCs) could effectively remove Zn from real industrial wastewaters. The elimination efficiency was determined to be 1.55 ± 0.35 mM, indicating a range of 96%–99%. Nonetheless, for the synthetic and industrial samples, the Zn recovery in terms of $Zn(OH)_2$ accumulation on the cathode surface was 83% and 42%, respectively (Lim et al., 2021). According to a study by Ye et al. (2017), 97.85% of Zn was recovered with a combined process of bioleaching and brine leaching from lead–zinc mine tailings. The findings established that initial pH and material concentration had a substantial impact on bioleaching efficiency (Ye et al., 2017). *Leptospirillum ferriphilum* isolate was used by Sundramurthy

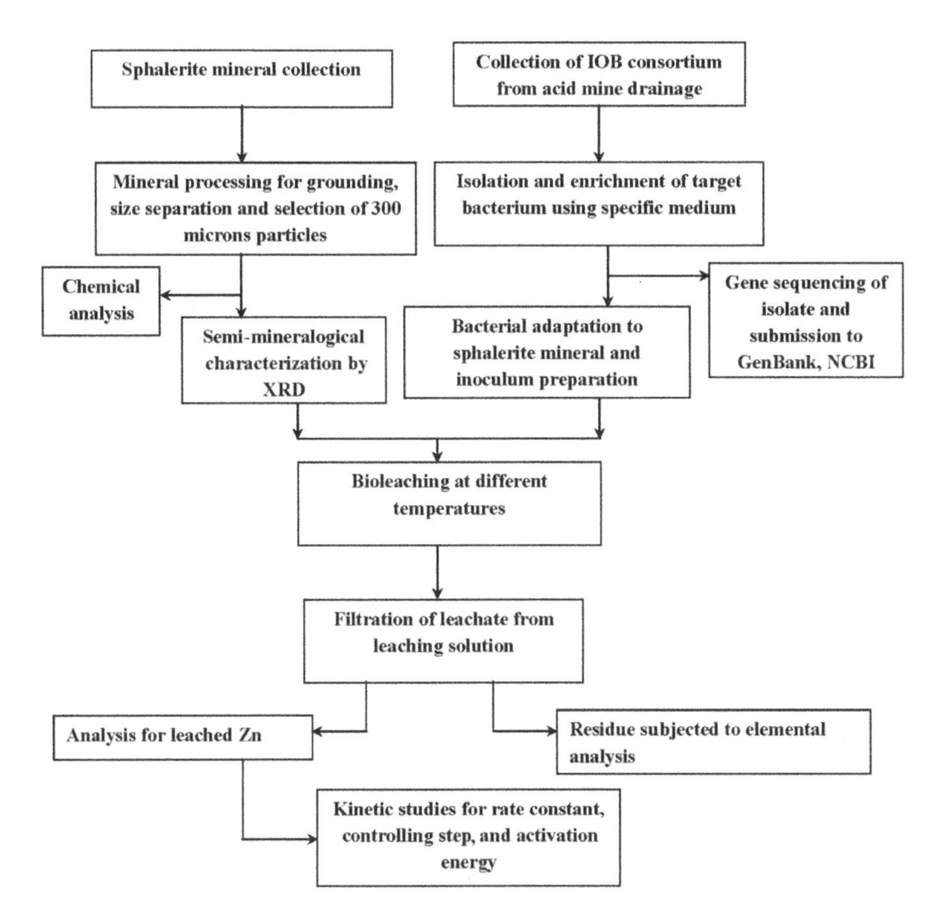

FIGURE 9.3 Schematic representation of the bioleaching process. Adapted from Sundramurthy et al. (2023) licensed under CC BY 4.0. To view a copy of this license, visit https://creativecommons.org/licenses/by/4.0/.

et al. to study the bioleaching of zinc from sphalerite. It has been observed that within 20 days, the leaching reaches its maximum leaching efficiency of 96.96% as the temperature rises to 313 K, the ideal temperature. The experimental steps that have been used in the above study are illustrated in Figure 9.3.

Sphalerite is the primary zinc mineral, and acidophilic autotrophic bacteria use the polysulfide process to break it down (Viera et al., 2007). The mechanism of the process is as follows:

$$\text{Zn S} + \text{H}^+ + \text{Fe}^{3+} \rightarrow \text{Zn}^{2+} + 0.5\ \text{H}_2\text{S}_n + \text{Fe}^{2+}\,(n \geq 2) \tag{1}$$

$$0.5\ \text{H}_2\text{S}_n + \text{Fe}^{3+} \rightarrow 0.125\ \text{S}_8 + \text{Fe}^{2+} + \text{H}^+ \tag{2}$$

$$0.125\ S_8 + 1.5\ O_2 + H_2O \rightarrow SO_4^{2-} + 2H^+ \quad (3)$$

$$2Fe^{2+} + 0.5\ O_2 + 2H^+ \rightarrow 2Fe^{3+} + H_2O \quad (4)$$

In reactions 1 and 2, *Leptospirillum*, *Sulfobacillus*, and other iron-oxidizing bacteria can reoxidize the ferrous iron to ferric iron. Reaction (2) indicates that when ferric ions oxidize ZnS, an elemental sulfur layer of product forms on the mineral's surface. The rate-limiting step is the diffusion of ferrous ions through this sulfur layer. Although sulfur-oxidizing bacteria can convert elemental sulfur to sulfate, elemental sulfur is generally stable. The surface reaction with ferric ions then turns into the step that limits the rate in this scenario. When sulfur-oxidizing bacteria were absent, the leaching process was significantly slowed down, confirming that a coating of elemental sulfur had passivated the sphalerite surface. Even in the presence of sulfur-oxidizing bacteria, diffusion across this layer may be the rate-limiting step if the rate of elemental sulfur oxidation is insufficient (reaction 3). In summary, microorganisms provide H_2SO_4 (reaction 3) to facilitate a proton attack, while iron remains in the oxidized ferric state to facilitate an oxidative attack on the mineral (reaction 4) (Viera et al., 2007).

9.5 CHALLENGES

The utilization of mineral coal, a significant natural resource, presents notable environmental challenges, particularly concerning mine tailings and coal fly ash. These residues contain valuable minerals such as zinc, molybdenum, vanadium, chromium, iron, and copper, yet their disposal in large tailing ponds or as ash poses environmental risks. Biomining and bioleaching emerge as promising technologies for extracting metals from these residues, aligning with the principles of green chemistry. However, limited data exist on the bioleaching of coal ash, necessitating further research (Valério et al., 2021). HM contamination remains a persistent environmental issue globally, primarily originating from industrial and mining activities. Despite efforts to address soil contamination, inadequate technologies persist, leading to continued HM persistence. Microbial bioremediation offers a cost-effective solution, leveraging microbial activity for HM degradation through processes such as biosorption, bioleaching, and biomineralization. Advances in genetically modified microorganisms and omics-based approaches enhance bioremediation effectiveness, yet challenges remain in implementing resilient ecotechnologies for widespread HM removal (Verma et al., 2021). As humanity ventures into space exploration, the demand for valuable metals intensifies, prompting interest in space mining. Biomining, which utilizes microorganisms to extract metals from minerals and wastes, presents a potential in situ resource utilization (ISRU) technology for space resource exploitation. However, adapting biomining processes for space environments poses unique challenges, including the effects of microgravity and cosmic radiation on biomining microbes and the design of space-based bioreactors. Integrating knowledge from various disciplines, such as synthetic biology and process engineering, is crucial for advancing biomining in space exploration (Gumulya et al., 2022). The mining industry faces increasing challenges, including rising metal demand, declining ore grades, and sustainability concerns. Biomining offers potential solutions by enabling the recovery of value from previously subeconomic mineral resources and mitigating environmental impacts. However, exploiting unconventional resources, such as deep subsurface minerals and extraterrestrial sources, presents technical and economic challenges. Emerging biotechnologies, including unconventional biolixiviants and biosorption, show promise but require critical evaluation within techno-economic limitations (Kaksonen and Petersen, 2022). Mining activities continue to pose significant environmental threats, necessitating alternative remediation techniques. Bioremediation, particularly employing microorganisms like sulfate-reducing bacteria (SRB), offers an economically viable and eco-friendly solution for treating mine wastes and recovering toxic HMs. However, successful implementation depends on various factors, including site conditions and the nature of the deposit. Therefore, integrating bioremediation with complementary techniques is advisable for effective mine waste management (Nayak, 2023). Rare earth elements (REEs) are crucial for advanced manufacturing but face supply challenges due to declining ore grades. Biomining emerges as a viable option for REE recovery from low-grade resources, offering reduced environmental impact and capital investment compared to traditional methods. Various biomining technologies, including bioleaching and biosorption, show promise for REE recovery, especially from coal ash, which demonstrates the highest profit potential among mining waste streams (Vo et al., 2024). The recovery of metals from secondary sources is essential for addressing metal shortages and environmental concerns. Microbial bioremediation offers a promising approach, but challenges remain in optimizing bioleaching processes, understanding microbial–mineral interactions, and improving leaching efficiency. Future research should focus on enhancing microbial performance through genetic manipulation and optimizing leaching system parameters for efficient metal recovery (Dong et al., 2023). Bioleaching, involving the solubilization of minerals from ores by microorganisms, presents a promising yet complex solution for metal extraction. Factors such as microbial activity, pH, temperature, and nutrient availability influence bioleaching efficiency, requiring careful optimization. Various bioleaching techniques, including dump, tank, heap, slope, and in situ bioleaching, offer versatility but necessitate proper environmental conditions and microbial populations for effective metal extraction. Despite its potential, the efficacy of bioleaching depends on addressing these factors systematically (Kumar and Yaashikaa, 2020).

9.6 CONCLUSION

The mining industry, a significant economic asset worldwide, generates substantial waste in the process. Long-term environmental sustainability requires effective and efficient waste management techniques. Bio-based technology has proven to be a sustainable way of recovering metals from mine wastes. There are various advantages to using microorganisms in metal leaching as compared to pyrometallurgical and hydrometallurgical processes. Microbe-based extraction requires less energy and is safe and environmentally beneficial. It lowers the cost of capital. Moreover, the type of microbial leaching makes it feasible to recover metals from low-quality wastes and secondary resources. The biotechnology provides an additional means of economically recovering metal values, which the mining sectors are recognized for. The bioleaching method has made significant progress in this century with the recovery of metals from various wastes, such as electronics waste, galvanic waste, and industrial sludge. Although the process is moderate, it has proven to be useful, and industries are currently using bioleaching to extract various metals from low-grade materials.

REFERENCES

Ahmadi, A., Hosseini, M., & Foroutan, A. (2020). Continuous bio-hydrometallurgical extraction of zinc from a bulk lead-zinc flotation concentrate on a pilot scale. *Minerals Engineering*, *156*, 106529. https://doi.org/10.1016/j.mineng.2020.106529

Amiri, F., Mousavi, S. M., Yaghmaei, S., & Barati, M. (2012). Bioleaching kinetics of a spent refinery catalyst using *Aspergillus niger* at optimal conditions. *Biochemical Engineering Journal*, *67*, 208–217. https://doi.org/10.1016/j.bej.2012.06.011

Arab, B., Hassanpour, F., Arshadi, M., Yaghmaei, S., & Hamedi, J. (2020). Optimized bioleaching of copper by indigenous cyanogenic bacteria isolated from the landfill of e-waste. *Journal of Environmental Management*, *261*, 110124. https://doi.org/10.1016/j.jenvman.2020.110124

Asghari, I., Mousavi, S. M., Amiri, F., & Tavassoli, S. (2013). Bioleaching of spent refinery catalysts: a review. *Journal of Industrial and Engineering Chemistry*, *19*(4), 1069–1081. https://doi.org/10.1016/j.jiec.2012.12.005

Barrie Johnson, D. (2014). Biomining—biotechnologies for extracting and recovering metals from ores and waste materials. *Current Opinion in Biotechnology*, *30*, 24–31. https://doi.org/10.1016/j.copbio.2014.04.008

Begum, S., Rath, S. K., & Rath, C. C. (2021). Applications of Microbial Communities for the Remediation of industrial and mining toxic metal waste: a review. *Geomicrobiology Journal*, *39*(3–5), 282–293. https://doi.org/10.1080/01490451.2021.1991054

Bharadwaj, A., & Ting, Y. P. (2013). Bioleaching of spent hydrotreating catalyst by acidophilic thermophile *Acidianus brierleyi*: leaching mechanism and effect of decoking. *Bioresource technology*, *130*, 673–680. https://doi.org/10.1016/j.biortech.2012.12.047

Boulamanti, A., & Moya, J. A. (2016). Production costs of the non-ferrous metals in the EU and other countries: copper and zinc. *Resources Policy*, *49*, 112–118. https://doi.org/10.1016/j.resourpol.2016.04.011

Castro, C., Urbieta, M. S., Cazón, J. P., & Donati, E. R. (2019). Metal biorecovery and bioremediation: whether or not thermophilic are better than mesophilic microorganisms. *Bioresource Technology*, *279*, 317–326. https://doi.org/10.1016/j.biortech.2019.02.028

ruh, S., Elevli, S., Ergun, O. N., & Demir, G. (2013). Assessment of leaching characteristics of heavy metals from industrial leach waste. *International Journal of Mineral Processing*, *123*, 165–171. https://doi.org/10.1016/j.minpro.2013.06.005

Deveci, H., Akçıl, A., & Alp, İ. (2004). Bioleaching of complex zinc sulphides using mesophilic and thermophilic bacteria: comparative importance of pH and iron. *Hydrometallurgy*, *73*(3–4), 293–303. https://doi.org/10.1016/j.hydromet.2003.12.001

Dong, L., Tong, X., Xi-Bing, L., Zhou, J., Wang, S., & Liu, B. (2019). Some developments and new insights of environmental problems and deep mining strategy for cleaner production in mines. *Journal of Cleaner Production*, *210*, 1562–1578. https://doi.org/10.1016/j.jclepro.2018.10.291

Dong, Y., Zan, J., & Lin, H. (2023). Bioleaching of heavy metals from metal tailings utilizing bacteria and fungi: mechanisms, strengthen measures, and development prospect. *Journal of Environmental Management*, *344*, 118511. https://doi.org/10.1016/j.jenvman.2023.118511

Du, B., Zhou, J., Lu, B., Zhang, C., Li, D., Zhou, J., Jiao, S., Zhao, K., & Zhang, H. (2020). Environmental and human health risks from cadmium exposure near an active lead-zinc mine and a copper smelter, China. *Science of the Total Environment*, *720*, 137585. https://doi.org/10.1016/j.scitotenv.2020.137585

Dusengemungu, L., Kasali, G., Gwanama, C., & Mubemba, B. (2021). Overview of fungal bioleaching of metals. *Environmental Advances*, *5*, 100083. https://doi.org/10.1016/j.envadv.2021.100083

Figueroa -Estrada, J. C., Aguilar-López, R., Rodríguez-Vázquez, R., & Neria-González, M. I. (2020). Bioleaching for the extraction of metals from sulfide ores using a new chemo-lithoautotrophic bacterium. *Hydrometallurgy*, *197*, 105445. https://doi.org/10.1016/j.hydromet.2020.105445

Fomchenko, N. V., & Muravyov, M. I. (2018). Two-step biohydrometallurgical technology of copper-zinc concentrate processing as an opportunity to reduce negative impacts on the environment. *Journal of Environmental Management*, *226*, 270–277. https://doi.org/10.1016/j.jenvman.2018.08.045

Fomchenko, N. V., & Muravyov, M. I. (2020). Sequential bioleaching of pyritic tailings and ferric leaching of nonferrous slags as a method for metal recovery from mining and metallurgical wastes. *Minerals*, *10*(12), 1097. https://doi.org/10.3390/min10121097

Ghassa, S., Noaparast, M., Shafaei, S. Z., Abdollahi, H., Gharabaghi, M., & Boruomand, Z. (2017). A study on the zinc sulfide dissolution kinetics with biological and chemical ferric reagents. *Hydrometallurgy*, *171*, 362–373. https://doi.org/10.1016/j.hydromet.2017.06.012

Gumulya, Y., Zea, L., & Kaksonen, A. H. (2022). In situ resource utilisation: the potential for space biomining. *Minerals Engineering*, *176*, 107288. https://doi.org/10.1016/j.mineng.2021.107288

Hubau, A., Guezennec, A., Joulian, C., Falagán, C., Dew, D., & Hudson-Edwards, K. A. (2020). Bioleaching to reprocess sulfidic polymetallic primary mining residues: determination of metal leaching mechanisms. *Hydrometallurgy*, *197*, 105484. https://doi.org/10.1016/j.hydromet.2020.105484

Indian Bureau of Mines. (2020). Indian minerals yearbook 2020 (Part-II: Metals and Alloys) Available at: https://ibm.gov.in/writereaddata/files/07142022173024Lead_Zinc_2020_AR.pdf

International Zinc Association. (2022). Zinc recycling 2050 demand + supply. Available at: https://www.zinc.org/wp-content/uploads/sites/30/2022/10/2050-DemandSupply_VF_11_22.pdf

Irrgang, N., Monneron-Enaud, B., Möckel, R., Schlömann, M., & Höck, M. (2021). Economic feasibility of the co-production of indium from zinc sulphide using bioleaching extraction in Germany. *Hydrometallurgy*, *200*, 105566. https://doi.org/10.1016/j.hydromet.2021.105566

Jha, M. K., Kumar, V., & Singh, R. J. (2001). Review of hydrometallurgical recovery of zinc from industrial wastes. *Resources, Conservation and Recycling*, *33*(1), 1–22. https://doi.org/10.1016/S0921-3449(00)00095-1

Kaksonen, A. H., & Petersen, J. (2022). *The Future of Biomining: Towards Sustainability in a Metal-Demanding World*. Springer, New York. (pp. 295–314). https://doi.org/10.1007/978-3-031-05382-5_17

Kara, I. T., Kremser, K., Wagland, S. T., & Coulon, F. (2023). Bioleaching metal-bearing wastes and by-products for resource recovery: a review. *Environmental Chemistry Letters*, *21*(6), 3329–3350. https://doi.org/10.1007/s10311-023-01611-4

Karbassi, S., Nasrabadi, T., & Shahriari, T. (2016). Metallic pollution of soil in the vicinity of national iranian lead and zinc (NILZ) company. *Environmental Earth Sciences*, *75*, 1–11. https://doi.org/10.1007/s12665-016-6244-7

Kržanović, D., Conić, V., Bugarin, D., Jovanović, I., & Božić, D. (2019). Maximizing economic performance in the mining industry by applying bioleaching technology for extraction of polymetallic mineral deposits. *Minerals*, *9*(7), 400. https://doi.org/10.3390/min9070400

Kumar, P., & Yaashikaa, P. (2020). *Recent Trends and Challenges in Bioleaching Technologies*. Elsevier, Amsterdam, The Netherlands (pp. 373–388). https://doi.org/10.1016/b978-0-12-817951-2.00020-1

Lei, K., Giubilato, E., Critto, A., Pan, H., & Lin, C. (2016). Contamination and human health risk of lead in soils around lead/zinc smelting areas in China. *Environmental Science and Pollution Research*, *23*, 13128–13136. https://doi.org/10.1016/j.chemosphere.2020.128909

Levett, A., Gleeson, S. A., & Kallmeyer, J. (2021). From exploration to remediation: a microbial perspective for innovation in mining. *Earth-Science Reviews*, *216*, 103563. https://doi.org/10.1016/j.earscirev.2021.103563

Liao, X., Ye, M., Li, S., Liang, J., Zhou, S., Fang, X., Gan, Q., & Sun, S. (2021). Simultaneous recovery of valuable metal ions and tailings toxicity reduction using a mixed culture bioleaching process. *Journal of Cleaner Production*, *316*, 128319. https://doi.org/10.1016/j.jclepro.2021.128319

Lim, S. S., Fontmorin, J. M., Pham, H. T., Milner, E., Abdul, P. M., Scott, K., & Yu, E. H. (2021). Zinc removal and recovery from industrial wastewater with a microbial fuel cell: experimental investigation and theoretical prediction. *Science of the Total Environment*, *776*, 145934. https://doi.org/10.1016/j.scitotenv.2021.145934

Liu, R., Li, J., & Ge, Z. (2016). Review on *Chromobacterium violaceum* for gold bioleaching from e-waste. *Procedia Environmental Sciences*, *31*, 947–953. https://doi.org/10.1016/j.proenv.2016.02.119

Ma, L., Huang, S., Wu, P., Xiong, J., Wang, H., Liao, H., & Liu, X. (2021). The interaction of acidophiles driving community functional responses to the re-inoculated chalcopyrite bioleaching process. *Science of the Total Environment*, *798*, 149186. https://doi.org/10.1016/j.scitotenv.2021.149186

Minimol, M., Shetty, V., & Saidutta, M. B. (2021). Biohydrometallurgical Methods and the Processes Involved in the Bioleaching of WEEE. In *Environmental Management of Waste Electrical and Electronic Equipment* (pp. 89–107). Elsevier, Amsterdam, The Netherlands. https://doi.org/10.1016/B978-0-12-822474-8.00005-2

Mocellin, J., Mercier, G., Morel, J. L., Charbonnier, P., Blais, J. F., & Simonnot, M. O. (2017). Recovery of zinc and manganese from pyrometallurgy sludge by hydrometallurgical processing. *Journal of Cleaner Production*, *168*, 311–321. https://doi.org/10.1016/j.jclepro.2017.09.003

Muravyov, M. (2019). Bioprocessing of mine waste: effects of process conditions. *Chemical Papers*, *73*(12), 3075–3083. https://doi.org/10.1007/s11696-019-00844-4

Nancharaiah, Y. V., Mohan, S. V., & Lens, P. N. L. (2016). Recent advances in nutrient removal and recovery in biological and bioelectrochemical systems. *Bioresource technology*, 215, 173–185.

Naseri, T., Beiki, V., Mousavi, S. M., & Farnaud, S. (2023). A comprehensive review of bioleaching optimization by statistical approaches: recycling mechanisms, factors affecting, challenges, and sustainability. *RSC Advances*, *13*(34), 23570–23589. https://doi.org/10.1039/D3RA03498D

Nayak, N. P. (2023). Microorganisms and their application in mining and allied industries. *Materials Today: Proceedings*, *72*, 2886–2891. https://doi.org/10.1016/j.matpr.2022.07.392

Newsome, L., & Falagán, C. (2021). The microbiology of metal mine waste: bioremediation applications and implications for planetary health. *Geohealth*, *5*(10), 380. https://doi.org/10.1029/2020gh000380

Obasi, P. N., & Akudinobi, B. B. (2020). Potential health risk and levels of heavy metals in water resources of lead–zinc mining communities of Abakaliki, southeast Nigeria. *Applied Water Science*, *10*(7), 1233. https://doi.org/10.1007/s13201-020-01233-z

Obiora, S. C., Chukwu, A., & Davies, T. (2018). Contamination of the potable water supply in the Lead–Zinc mining communities of Enyigba, southeastern Nigeria. *Mine Water and the Environment*, *38*(1), 148–157. https://doi.org/10.1007/s10230-018-0550-0

Olubambi, P. A., Ndlovu, S., Potgieter, J. H., & Borode, J. O. (2007). Effects of ore mineralogy on the microbial leaching of low grade complex sulphide ores. *Hydrometallurgy*, *86*(1–2), 96–104. https://doi.org/10.1016/j.hydromet.2006.10.008

Qayyum, S., Meng, K., Pervez, S., Nawaz, F., & Peng, C. (2019). Optimization of pH, temperature and carbon source for bioleaching of heavy metals by Aspergillus flavus isolated from contaminated soil. *Main Group Metal Chemistry*, 42(1), 1–7. https://doi.org/10.1515/mgmc-2018-0038

Rathna, R., & Nakkeeran, E. (2020). *Biological Treatment for the Recovery of Minerals from Low-Grade Ores*. Elsevier, Amsterdam, The Netherlands. (pp. 437–458). https://doi.org/10.1016/b978-0-444-64321-6.00022-7

Rendón-Castrillón, L., Ramírez-Carmona, M., Ocampo-López, C., & Gómez-Arroyave, L. (2023). Bioleaching techniques for sustainable recovery of metals from solid matrices. *Sustainability*, *15*(13), 10222. https://doi.org/10.3390/su151310222

Rodrigues, M. L., Santos, G. H., Leôncio, H. C., & Leão, V. A. (2019). Column bioleaching of fluoride-containing secondary copper sulfide ores: experiments with Sulfobacillus thermosulfidooxidans. *Frontiers in Bioengineering and Biotechnology*, *6*, 183. https://doi.org/10.3389/fbioe.2018.00183

Romero-García, A., Iglesias-González, N., Romero, R., Lorenzo-Tallafigo, J., Mazuelos, A., & Carranza, F. (2019). Valorisation of a flotation tailing by bioleaching and brine leaching, fostering environmental protection and sustainable development. *Journal of Cleaner Production*, *233*, 573–581. https://doi.org/10.1016/j.jclepro.2019.06.118

Rouchalova, D., Rouchalova, K., Janakova, I., Cablik, V., & Janstova, S. (2020). Bioleaching of iron, copper, lead, and zinc from the sludge mining sediment at different particle sizes, pH, and pulp density using *Acidithiobacillus ferrooxidans*. *Minerals*, *10*(11), 1013. https://doi.org/10.3390/min10111013

Sajjad, W., Zheng, G., Din, G., Ma, X., Rafiq, M., & Wang, X. (2019). Metals extraction from sulfide ores with microorganisms: the bioleaching technology and recent developments. *Transactions of the Indian Institute of Metals*, *72*(3), 559–579. https://doi.org/10.1007/s12666-018-1516-4

Schippers, A., Hedrich, S., Vasters, J., Drobe, M., Sand, W., & Willscher, S. (2013). Biomining: metal recovery from ores with microorganisms. In *Advances in Biochemical Engineering/Biotechnology* (pp. 1–47). Elsevier, Amsterdam, The Netherlands. https://doi.org/10.1007/10_2013_216

Sethurajan, M., Van Hullebusch, E. D., & Nancharaiah, Y. (2018). Biotechnology in the management and resource recovery from metal bearing solid wastes: recent advances. *Journal of Environmental Management*, *211*, 138–153. https://doi.org/10.1016/j.jenvman.2018.01.035

Singh, S., Sukla, L. B., & Goyal, S. K. (2020). Mine waste & circular economy. *Materials Today: Proceedings*, *30*, 332–339. https://doi.org/10.1016/j.matpr.2020.01.616

Spiess, S., Kučera, J., Vaculovič, T., Birklbauer, L., Habermaier, C., Conde, A. S., Mandl, M., & Haberbauer, M. (2023). Zinc recovery from bioleachate using a microbial electrolysis cell and comparison with selective precipitation. *Frontiers in Microbiology*, *14*, 1238853. https://doi.org/10.3389/fmicb.2023.1238853

Srichandan, H., Mohapatra, R. K., Parhi, P. K., & Mishra, S. (2019). Bioleaching approach for extraction of metal values from secondary solid wastes: a critical review. *Hydrometallurgy*, *189*, 105122. https://doi.org/10.1016/j.hydromet.2019.105122

Srichandan, H., Mohapatra, R. K., Singh, P. K., Mishra, S., Parhi, P. K., & Naik, K. (2020). Column bioleaching applications, process development, mechanism, parametric effect and modelling: a review. *Journal of Industrial and Engineering Chemistry*, *90*, 1–16. https://doi.org/10.1016/j.jiec.2020.07.012

Sundramurthy, V. P., Rajoo, B., Srinivasan, N. R., & Kavitha, R. (2023). Correction: bioleaching of Zn from sphalerite using Leptospirillum ferriphilum isolate: effect of temperature and kinetic aspects. *Applied Biological Chemistry*, *63*(1), 44. https://doi.org/10.1186/s13765-023-00812-3

Swęd, M., Potysz, A., Bartz, W., & Siepak, M. (2023). Element dissolution from Zn-bearing rocks treated with chemical and biotic agents: a prospective circular economy strategy for metal recovery from secondary resources. *Geochemistry*, *83*(4), 126008. https://doi.org/10.1016/j.chemer.2023.126008

Tao, M., Zhang, X., Wang, S., Cao, W., & Yi, J. (2019). Life cycle assessment on lead–zinc ore mining and beneficiation in China. *Journal of Cleaner Production*, *237*, 117833. https://doi.org/10.1016/j.jclepro.2019.117833

Tezyapar Kara, I., Kremser, K., Wagland, S. T., & Coulon, F. (2023). Bioleaching metal-bearing wastes and by-products for resource recovery: a review. *Environmental Chemistry Letters*, *21*(6), 3329–3350. https://doi.org/10.1007/s10311-023-01611-4

Tian, B., Cui, Y., Qin, Z., Wen, L., Li, Z., Chu, H., & Xin, B. (2022). Indirect bioleaching recovery of valuable metals from electroplating sludge and optimization of various parameters using response surface methodology (RSM). *Journal of Environmental Management*, *312*, 114927. https://doi.org/10.1016/j.jenvman.2022.114927

Tributsch, H. (1992). On the significance of the simultaneity of electron transfer and cooperation in electrochemistry. *Journal of Electroanalytical Chemistry*, 331(1-2), 783–800.

Valério, A., Maass, D., De Andrade, C. J., De Oliveira, D., & Hotza, D. (2021). Bioleaching from coal wastes and tailings: a sustainable biomining alternative. In *Environmental and microbial biotechnology* (pp. 203–224). Elsevier, Amsterdam, The Netherlands. https://doi.org/10.1007/978-981-15-9696-4_9

Vardanyan, A., Kafa, N., Konstantinidis, V., Shin, S. G., & Vyrides, I. (2018). Phosphorus dissolution from dewatered anaerobic sludge: Effect of pHs, microorganisms, and sequential extraction. *Bioresource technology*, 249, 464–472.

Vardanyan, N., Sevoyan, G., Navasardyan, T., & Vardanyan, A. (2019). Recovery of valuable metals from polymetallic mine tailings by natural microbial consortium. *Environmental Technology*, *40*(26), 3467–3472. https://doi.org/10.1080/09593330.2018.1478454

Verma, S., Bhatt, P., Verma, A., Mudila, H., Prasher, P., & Rene, E. R. (2021). Microbial technologies for heavy metal remediation: effect of process conditions and current practices. *Clean Technologies and Environmental Policy*, *25*(5), 1485–1507. https://doi.org/10.1007/s10098-021-02029-8

Viera, M., Pogliani, C., & Donati, E. (2007). Recovery of zinc, nickel, cobalt and other metals by bioleaching. In *Microbial Processing of Metal Sulfides* (pp. 103–119).Springer Netherlands, Dordrecht. https://doi.org/10.1007/1-4020-5589-7_5

Vo, H. N. P., Danaee, S., Hai, H. T. N., Huy, L. N., Nguyen, T. A., Nguyen, T. M. H., Kuzhiumparambil, U., Kim, M., Nghiem, L. D., & Ralph, P. J. (2024). Biomining for sustainable recovery of rare earth elements from mining waste: a comprehensive review. *Science of the Total Environment*, *908*, 168210. https://doi.org/10.1016/j.scitotenv.2023.168210

Wang, J., Zhang, Y., Cui, K., Fu, T., Gao, J., Hussain, S., & AlGarni, T. S. (2021). Pyrometallurgical recovery of zinc and valuable metals from electric arc furnace dust – a review. *Journal of Cleaner Production*, *298*, 126788. https://doi.org/10.1016/j.jclepro.2021.126788

Williamson, A., Folens, K., Van Damme, K., Olaoye, O., Atia, T. A., Mees, B., Nicomel, N. R., Verbruggen, F., Spooren, J., Boon, N., Hennebel, T., & Du Laing, G. (2020). Conjoint bioleaching and zinc recovery from an iron oxide mineral residue by a continuous electrodialysis system. *Hydrometallurgy*, *195*, 105409. https://doi.org/10.1016/j.hydromet.2020.105409

Williamson, A. J., Verbruggen, F., Rico, V. S. C., Bergmans, J., Spooren, J., Yurramendi, L., & Hennebel, T. (2021). Selective leaching of copper and zinc from primary ores and secondary mineral residues using biogenic ammonia. *Journal of Hazardous Materials*, *403*, 123842. https://doi.org/10.1016/j.jhazmat.2020.123842

Ye, M., Li, G., Yan, P., Ren, J., Zheng, L., Han, D., Sun, S., Huang, S., & Zhong, Y. (2017). Removal of metals from lead-zinc mine tailings using bioleaching and followed by sulfide precipitation. *Chemosphere*, *185*, 1189–1196. https://doi.org/10.1016/j.chemosphere.2017.07.124

Yuan, J., Ding, Z., Bi, Y., Li, J., Wen, S., & Bai, S. (2022). Resource utilization of acid mine drainage (AMD): a review. *Water*, *14*(15), 2385. https://doi.org/10.3390/w14152385

10 Bioleaching of Municipal Wastes
Recovery of Phosphorus

Chandrima Roy and Himabindu Vurimindi

10.1 CHAPTER HIGHLIGHTS

- Phosphorus is one of the most valuable and non-renewable resources on Earth
- Multiple recovery options for phosphorus from municipal waste
- Bioleaching of phosphorus from municipal solid waste/sewage sludge
- Hydrothermal, chemical, and hybrid technologies for efficient recovery
- Importance in the context of circular economy and techno-economic analysis

10.2 INTRODUCTION

10.2.1 Sources of Phosphorus in Nature and Its Usefulness

Recovery of phosphorus has become an important part of the world economy, and in 2014, phosphorus was declared as a critical raw material by the European Union. Phosphate rock reserve is not sufficient and not uniformly distributed all over the world (USGS, 2019). India has to import a huge quantity of urea and phosphate rock each year. In 2021, India imported over 11 million metric tons of urea and 8.1 million metric tons of phosphate rock. The urgent need for a countermeasure of phosphate mining can easily be understood from the fact that the existing phosphate reserve will be depleted within 100 years and the annual increase of phosphorus demand is 3%–4% as reported by Indian scientists in 2019.

Phosphorus is found in nature in the form of phosphate rock or phosphorites, which are mainly calcium salts with 4%–20% of available phosphorus pentoxides. Eighty percent of the phosphorus extracted from ores is used by fertilizer industries and the rest 20% as animal feeds. Ninety-five percent of global phosphorus available is used by agricultural fields and less than 5% is required for making phosphorus-based pesticides and animal feed supplements, in food industries, household appliances, and other industrial applications (Desmidt et al., 2015).

Phosphorus is one of the essential macronutrients for plant and animal growth and a major contributor to living cell functions being a primary component of DNA, RNA, and Adenosine triphosphate (ATP) involved in cell metabolism. Inorganic salts of

DOI: 10.1201/9781003415541-10

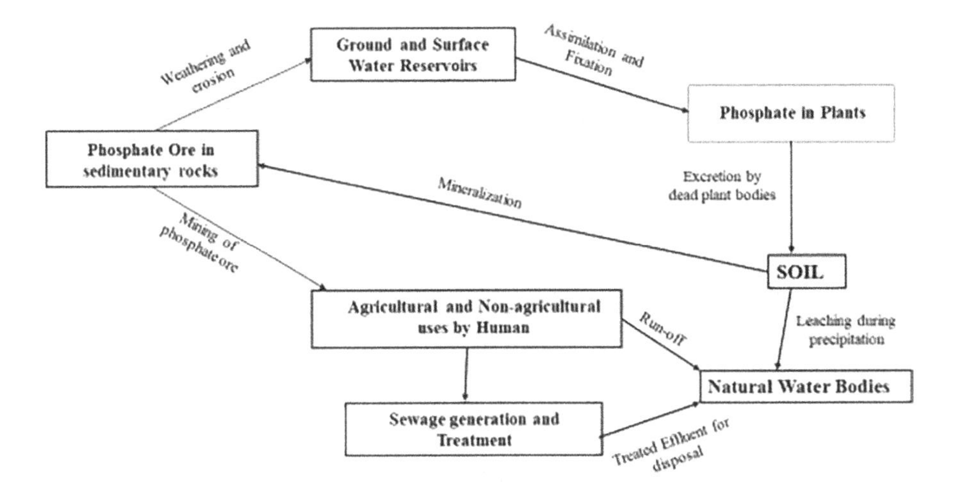

FIGURE 10.1 Sources and Applications: phosphorus cycle on earth.

phosphorus from natural deposits get solubilized in water and assimilated by plants, which have been converted to organic phosphates by plant cells and transferred to higher animals as the food chain grows. Apart from the phosphorus recycled by decomposing dead plant and animal cells, the addition of mineral fertilizers to soil also builds up the phosphorus level underground. Agricultural runoff and leaching out of phosphorus from soil to water bodies causes eutrophication, which is a major environmental concern. The Phosphorus open cycle in modern society was depicted by Desmidt et al in 2015. Figure 10.1 represents the sources and uses of phosphorus by animal and plants.

Therefore, to improve the bioavailability of immobilized P in agricultural soils, certain chemical, physical, and biological extraction technologies are being explored nowadays. This chapter will focus on biobased phosphorus recovery techniques from municipal waste in a comprehensive manner.

10.2.2 Recovery Techniques of Phosphorus from Municipal Waste/Sewage Sludge

Recovery of nutrients from domestic/industrial waste is an emerging topic of recent research (Mehta et al., 2015; Rao et al., 2017) from an environmental point of view, and a variety of techniques are being explored for economic and efficient recovery of them. Domestic and municipal/sewage wastewater and sewage sludge are promising sources of nutrients. Phosphorus recovery from municipal waste/sewage sludge is emerging as a promising focus area of research due to stringent environmental legislation on phosphate discharge to water bodies. The main challenge of recovering phosphorus is to bring it to a solubilized form from complex polyphosphate or other forms of metal phosphate present in waste biomass or sludge. Digested sewage sludge presents a promising source for phosphorus recovery, with magnesium ammonium phosphate (MAP)

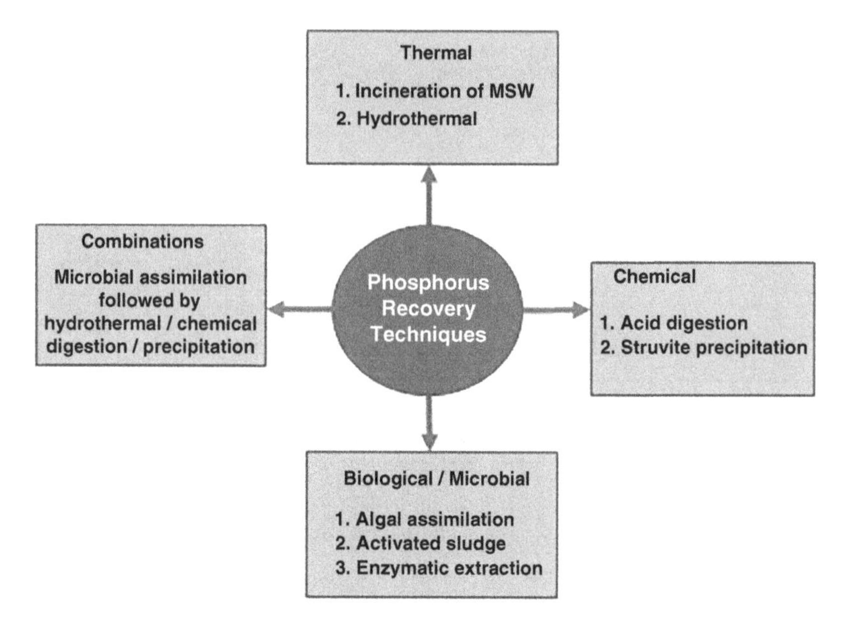

FIGURE 10.2 Phosphorus recovery options from solid waste/wastewater.

TABLE 10.1
Different Methods of Phosphorus Recovery from Sewage Sludge and Waste Biomass (Selected from Last 5 years)

Recovery Technology and Operating Conditions	Important Findings and Percentage of Phosphorus Recovery	Substrate and Extraction Media	Literature Reported
Wet chemical extraction, thermo-electrical, and thermochemical method	65%–98% depending on types of treatment technologies	Municipal solid waste, sewage sludge, and ash after incineration	Chrispim et al. (2019)
Adsorption of biochar from wood waste as a green adsorbent	Above 90%	Hydrolyzed urine and Mg-laden biochar from wood waste are low-cost and green adsorbents	Liu et al. (2020)
Both chemical leaching and bioleaching by bacteria (for the chemical method, 1.6M H_2SO_4 was added to 1 g of thermally treated ground sludge and placed in a shaker with 150 rpm speed and 30°C; for bioleaching, 1–20 g of ground sample + 0.5–5 g S powder + 1 mL of culture in a shaker under same conditions)	41%–51% of phosphorus recovery by chemical leaching method depends on the types of sample; 48%–62 % recovery by bioleaching method	Sewage sludge ash, municipal solid waste/biogenic sulfuric acid produced by a pure strain of *Acidithiobacillus thiooxidians*	Lee et al. (2020)

(*Continued*)

TABLE 10.1 (*Continued*)
Different Methods of Phosphorus Recovery from Sewage Sludge and Waste Biomass (Selected from Last 5 years)

Recovery Technology and Operating Conditions	Important Findings and Percentage of Phosphorus Recovery	Substrate and Extraction Media	Literature Reported
Thermochemical method Hydrothermal method (water at subcritical condition works as a catalyst and a solvent, temp 220°C, 200 bar pressure) Combined methods	Above 95% More than 98.37% (This review indicates ways to recover phosphorus from biowaste, with particular emphasis on wastewater, sewage sludge, manure, slaughter, and food waste. The importance of wastewater pretreatment is emphasized).	Sewage sludge, waste biomass from agrifood sector, manure, and wastewater	Timofeeva et al. (2022)
Wet oxidation (destructive oxidation under high pressure and temperature in the presence of steam and molecular oxygen; continuous mode, flow rate 50 L/h, 3,500°C, and 240 bar, 24 h)	More than 95% (optimization of two main reaction parameters shows maximum phosphorus recovery was achieved at pH 8.5 and Mg/P molar ratio of 1.8)	a. Sewage sludge after digestion, algal/coculture biomass. b. Sewage sludge from primary and secondary treatment from Spanish wastewater treatment plant	Cañas et al. (2023)
Hybrid ion exchange nanotechnologies (HIX-nano)	Adsorption and regeneration experiments were conducted and the regression model was developed and analyzed to test the best fitting of each experimental phase Above 90%	Synthetic wastewater effluent	Ownby et al. (2021)
Wet chemical treatment (acid/alkali leaching) Thermal treatment	Phosphorus purification and heavy metal separation are the challenges and the economics of phosphorus recovery has often been weak till now	Sewage Sludge Ash (SSA)	Xu et al. (2023)

being utilized. Additionally, metal recovery techniques from phosphorus-rich water have been applied to maximize yield and improve the quality of MAP. Phosphosphorus recovery from biowaste is a part of a conceptual model of "waste to BEST" toward sustainable biorefinery as reviewed in detail by Venkata Mohan et al. in 2016. The four possible routes of phosphorus recovery techniques from municipal/industrial waste are pictorially represented in Figure 10.2. A comparison table of recovery efficiency of different extraction technologies based on available literature is shown in Table 10.1.

10.3 BIOLEACHING OF MUNICIPAL SOLID WASTE FOR RECOVERY OF PHOSPHORUS

10.3.1 Responsible Microorganisms for Bioleaching

Leaching or extraction of nutrients (N, P) from municipal solid waste, industrial sludge, or wastewater with the aid of microbes (algae or bacteria) is called bioleaching, which is a cost-effective, cleaner technology compared to other conventional thermal treatment or chemical leaching. In bioleaching, nutrients and heavy metals get dissolved from solid substrates, either directly by the metabolism of leaching microorganisms or indirectly by the products of the metabolism (Mehta et al., 2015). One proteobacteria, *Acidithiobacillus* spp., can convert insoluble metal sulfides from solid materials to soluble metal sulfates. All acidophilic metal sulfide-oxidizing microorganisms are able to oxidize ferrous and/or sulfur compounds, and most of these organisms are chemolithoautotrophic (Schippers, 2013; Fang and Zhou, 2006). The most widely used bioleaching organisms are sulfur-oxidizing bacteria, *Acidithiobacillus thiooxidans;* iron-sulfur-oxidizing bacteria, *Acidithiobacillus ferrooxidans*; and iron-oxidizing bacteria, *Leptospirillium ferrooxidans* and *Leptospirillum ferriphilium* (Wen et al., 2013; Mahmoud et al., 2017).

10.3.2 Methods of Extraction

In one method, municipal solid waste can be segregated to separate the nonbiodegradables such as plastics/metal pieces, and a slurry is formed and microorganisms applied to this semisolid will take up nutrients for their metabolism. The nutrients taken up will depend on the type of microbial species, temperature, and light conditions (for algae). Highly efficient phosphate-solubilizing bacteria (PSB) (such as the most abundant *Pseudomonas sp.*) present in sewage/industry sludge can convert insoluble phosphorus to soluble phosphates, which can be taken up by plant/algal cells thereafter. Application of PSB to soil/sewage slurry prevents phosphorus leaching loss, thus making the nutrient more available for plants/algal cells' uptake. Thus, the phosphates accumulated in the living cells as polyphosphates are extracted in the aqueous phase by chemical digestion or by applying hydrothermal technique. The chemical method requires sulfuric acid/alkali solution for digestion/solubilization of phosphates in an aqueous media followed by chemical-induced precipitation. The hydrothermal technique employs only water as solvent (unlike the chemical extraction method) at high temperature (above 300°C) and pressure (220 bar), where the cell lysis happens and phosphate dissolves in water, which can be chemically precipitated or can be directly used as a nutrient supplement for algal/bacterial cultivation thereafter. Sewage sludge samples after anaerobic digestion collected from wastewater treatment plants can also be bioleached to recover phosphorus (Lee et al., 2020). *Acidithiobacillus thiooxidans* is generally used as the PSB, which requires sulfur supplement and definite bioleaching media during growth. Dehydrated ground sludge samples are mixed with the bacterial bioleaching media mixture and kept at 30°C and 150 RPM in an orbital shaker. Samples are taken at intervals of two days to analyze the amount of phosphorus leached out into the aqueous solution.

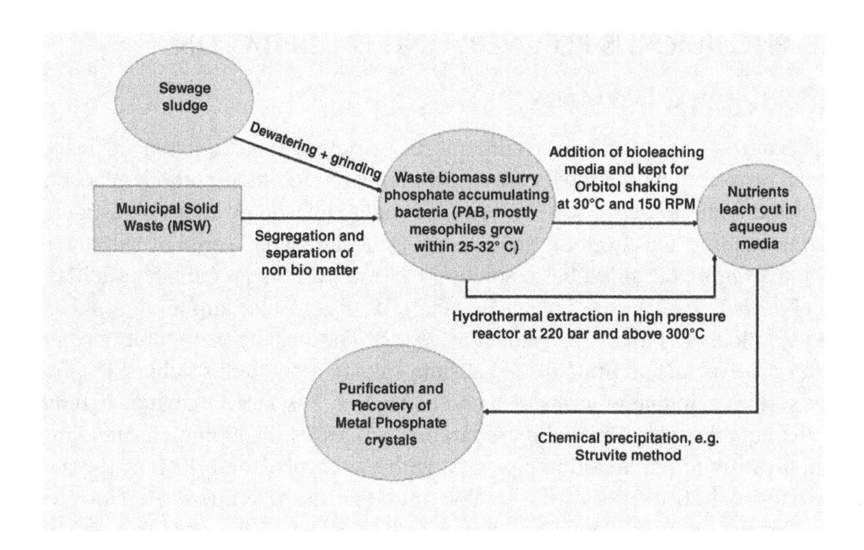

FIGURE 10.3 Scheme for phosphorus recovery methods from MSW and sewage sludge.

TABLE 10.2
Bioleaching of Phosphorus from Sewage Sludge Using Sulfur-Oxidizing Bacteria

Bioleaching Media	Bacterial Strain	Substrate	Process Conditions	Percentage Recovery of Phosphorus	Source
Growth media + Sulfur supplement + Mixed culture	Mixed culture of sulfur-oxidizing bacteria, e.g., *Acidithiobacillus ferrooxidans* and *A. thiooxidans*.	Sewage sludge ash + Sulfur powder	Temperature: 25°C–30°C Time:20–30 min pH 3	48%–61% depending upon the type of substrate	Lee et al. (2020)
Digested sewage sludge from wastewater treatment plant with added sulfur, later it was centrifuged and a supernatant was used for media	*Acidithiobacillus* sp. strains and polyphosphate (poly-P) accumulating bacteria	Sewage sludge incineration ash + sulfur added	Labscale bioleaching reactor Time: 15 days, Temperature: 22°C, pH 2.3–2.0	66% of mobilized phosphorus	Zimmermann et al. (2009)

The above discussion depicts that nutrient recovery from municipal solid waste (MSW) is a closed-loop circular economy approach to get wealth out of waste and a current trend being practiced worldwide. The bioleaching method of recovery has been proved to be more efficient compared to that of chemical extraction. Figure 10.3 shows the various phosphorus recovery methods from MSW and sewage sludge. Table 10.2 summarizes the process conditions for bioleaching of sewage sludge utilizing sulfur-oxidizing microorganisms.

10.4 PHOSPHORUS RECOVERY AND PRECIPITATION

10.4.1 Chemical Extraction

Phosphorus recovery from industrial/domestic wastewater can be employed onsite and has been proven to be the most convenient compared to other methods. Phosphorus is extracted from domestic waste or sewage sludge by acid or alkali digestion or by bacterial leaching, and from the aqueous phase, it can be recovered by salt precipitation by adding metal and a basic solution to form various precipitates such as calcium phosphate (CaP), iron phosphate (FeP), or magnesium ammonium phosphate (struvite) (Blöcher et al. 2012; Yuan et al., 2012). The struvite method of phosphorus removal involves the addition of magnesium salts to react with soluble phosphate to form a solid precipitate in terms of struvite ($MgNH_4\ PO_4\ 6(H_2O)$), which is removed from the aqueous phase by solids separation processes, including clarification and filtration. Struvite precipitation can remove between 80% and 90% of the soluble phosphates and 20%–30% of the soluble ammonia in the effluent (Le Corre et al., 2009). But the main challenge associated with phosphorus recovery from wastewater is that, if the phosphorus concentration is less than 50 mg/L (Mehta et al., 2015; Wong et al., 2013) and the concentration of suspended solids is above 2,000 mg/L (Drechsel et al., 2015), the recovery will not be effective. To avoid this, the struvite precipitation method can be applied to secondary streams generated in primary sludge thickening or after dewatering of the digested sludge, which has higher phosphorus concentrations. Other metal ions, mainly calcium interference, also affect the precipitation of struvite to some extent. Lavanya et al. (2020) studied the effect of pH on phosphate recovery from dairy wastewater and their findings showed that 95% of phosphate was recovered from solution at MgO:P molar ratio of 4:1 and pH 9.5. In 2023, Canas et al. (2023) recovered phosphate from both solid and liquid fractions. In the liquid phase, phosphorus was recovered (recovery efficiency greater than 95%) by chemical precipitate as struvite (($MgNH_4PO_4 \cdot 6H_2O$), and the solid fraction obtained after filtration and drying was acid leached to get up to 60% phosphorus recovery.

The struvite formation reaction can be shown in Eq. (10.1):

$$Mg^{2+} + NH_4^{+} + PO_4^{3-} + 6H_2O \rightarrow MgNH_4PO_4 6(H_2O) \tag{10.1}$$

10.4.2 Assimilation of Phosphorus by Microalgal/Coculture Biomass Cultivated in Wastewater

Algae/algal–bacterial cocultures are being cultivated in various types of industrial wastewater (dairy industry effluent, food industry wastewater, slaughter, and poultry industry wastewater) and municipal sewage/sewage-contaminated lake water in central and state university labs in India. Algae (e.g., *Chlorella vulgaris, Chlorella sorokiana, Botryococcus brauni, Scendesmus sp.,* etc.) are green cells and take up nitrogen and phosphorus just like plants for their metabolic activities. They can accumulate phosphorus in their cells as polyphosphates, and thus they remove phosphorus effectively from wastewater during cultivation within eight days of single batch treatment under the growth phase. However, the uptake of nutrients depends upon their photosynthetic activity, culture conditions, pH of the media, and type of culture and wastewater used. Thereafter, phosphorus can be significantly recovered from the cells by cell lysis, by the addition of enzymes/chemicals/bioleaching with bacteria/hydrothermal methods as already described in Section 10.5.

10.4.3 Hybrid Technologies

A hybrid process of low-pressure wet oxidation and nanofiltration to recover phosphorus from sewage sludge was investigated by Blöcher et al. in 2012, which recovered more than 50% of phosphorus, but the main concern was the presence of other metal ions which may affect the nanofiltration, thus decreasing the overall process efficiency. Some research works demonstrated the usefulness of hybrid ion-exchange nanotechnologies for phosphorus removal and recovery from wastewater (Ownby et al., 2021). They investigated four regeneration methods to desorb phosphorus from a commercially available HIX-Nano resin hybridized with iron oxide nanoparticles where they employed a less harmful and sustainable solution of KOH/H_2SO_4 blend or a recovered NH_4OH solution along with tap water. The latter showed better recovery efficiency of up to 90%. Another technology of selective sorption of phosphorus from treated municipal wastewater was studied by You et al. (2017) using hybrid anion exchangers containing hydrated ferric oxide nanoparticles and reported a recovery efficiency of up to 90%. Apart from the membrane systems, other hybrid technologies (Zeng et al., 2022) for recovering phosphate from sewage sludge are being attempted in our lab in JNTUH, Hyderabad. We are trying different pretreatments of sludge slurry like electrooxidation (applied voltage varied from 1 V to 12 V for 2 h) and ultrasonication or a combination of both followed by sulfuric acid digestion which improves the solubilization of phosphate salt in aqueous phase and increases % P recovery as well. The combined process of electrooxidation-ultrasonication followed by acid digestion can recover up to 95% of phosphorus from the source.

10.5 BIOLEACHING AND HYDROTHERMAL—GREEN METHODS OF PHOSPHORUS RECOVERY

Bioleaching of phosphorus from sewage sludge/MSW using sulfur-oxidizing bacterial strain is an example of the green method of phosphorus recovery as it doesn't require acid/alkali solution for digestion. Enhanced biological phosphorus removal (EPBR; Yuan et al., 2012) is related to wastewater treatment with an activated sludge system for the removal of phosphate. In place of chemical-induced leaching, it uses phosphate-accumulating organisms (PAOs) in the anaerobic tank prior to the aerobic treatment. This bioleaching technology has less adverse effects on human health and the environment as it reduces the risk of water or soil contamination with harmful chemicals. The hydrothermal method of phosphorus recovery is also considered as green as water under subcritical conditions is only used as solvent and media for leaching out phosphorus in aqueous solution. Therefore, the problem of solvent regeneration and spent solvent disposal can be avoided. Wet air oxidation (WAO) of sewage sludge slurry/waste biomass slurry is also considered a green method as water as a solvent with combined air and/or oxygen is used to break the bonds and solubilize the phosphate salts in the aqueous phase, which further can be precipitated as metal phosphate such as struvite.

10.6 A BRIEF NOTE ON TECHNO-ECONOMIC ANALYSIS OF PHOSPHATE RECOVERY

The chemical digestion method of phosphate extraction from biomass needs acids/alkali addition followed by the addition of calcium, magnesium, and ammonium salts for salt precipitation. The costs of chemicals required in this process will be compensated by the recovery of pure struvite precipitated out from the aqueous solution, which has a high demand as a nutrient in the fertilizer market. Hydrothermal extraction of phosphorus from biomass is a green technology which overcomes the usage of costly and noneco-friendly solvents and has been proven to be the best way of phosphorus recovery with more than 90% efficiency. However, the hydrothermal method employs high-pressure subcritical water as a solvent for extraction, where most of the variable cost is for electricity to run the high-pressure reactor at high temperatures. However, the fixed cost of land construction for wastewater treatment ponds can be minimized by utilizing onsite industrial effluent treatment facilities, thus resulting in phosphorus recovery from biowaste/wastewater, a more cost-effective and achievable goal approaching the circular economy. Aspen Plus modeling plays an important role in the techno-economic study as it provides the detailed mass and energy balance of each process involved (Bagheri et al., 2022). Economic feasibility is calculated keeping the minimum selling price of the products in consideration. The specific areas addressed in the article are how economic feasibility can be affected by the technical design, how much financial support is required, and existing market demand and supply chain of phosphorus fertilizer.

10.7 CHAPTER SUMMARY

Aspects of phosphate recovery from MSW and sewage sludge have been discussed. Chemical/thermochemical/hydrothermal/biological/hybrid technologies have been discussed giving a focus on bioleaching technology. Bioleaching technology as a green method of nutrient recovery has been discussed and a brief techno-economic analysis has been included at the end.

REFERENCES

Bagheri, M., Öhman, M., Wetterlund, E. (2022) Techno-economic analysis of scenarios on energy and phosphorus recovery from mono- and co-combustion of municipal sewage sludge. *Sustainability*, 14(5), 2603.

Blöcher, C., Niewersch, C., Melin, T. (2012) Phosphorus recovery from sewage sludge with a hybrid process of low pressure wet oxidation and nanofiltration. *Water Research*, 46, 6101–6108.

Cañas, J., Álvarez-Torrellas, S., Hermana, B., García, J. (2023) Phosphorus recovery from sewage sludge as struvite. *Water*, 15, 2382.

Chrispim, M.C., Scholz, M., Nolasco, M.A. (2019) Phosphorus recovery from municipal wastewater treatment: critical review of challenges and opportunities for developing countries. *Journal of Environmental Management*, 248, 109268.

Desmidt, E., Ghyselbrecht, K., Zhang, Y., Pinoy, Y., Van der Bruggen, B., Verstraete, W., Rabaey, K., Meesschaert, B. (2015) Global phosphorus scarcity and full-scale P-recovery techniques: a review. *Critical Reviews in Environmental Science and Technology*, 45(4), 336–384.

Drechsel, P., Qadir, M., Wichelns, D. (2015) *Wastewater: Economic Asset in an Urbanizing World*. Springer, Amsterdam.

Fang, D., Zhou, L.X. (2006) Effect of sludge dissolved organic matter on oxidation of ferrous iron and sulfur by Acidithiobacillus ferrooxidans and Acidithiobacillus thiooxidans. *Water, Air, & Soil Pollution*, 171, 81–94.

Lavanya, A., Thanga, R.S.K. (2020) Effective removal of phosphorous from dairy wastewater by struvite precipitation: process optimization using response surface methodology and chemical equilibrium modeling. *Separation Science and Technology*, 56(2), 395–410.

Le Corre, K.S., Valsami-Jones, E., Hobbs, P., Parsons, S.A. (2009) Phosphorus recovery from wastewater by struvite crystallization: a review. *Critical Reviews in Environmental Science and Technology*, 39(6), 433–477.

Lee, Y.J., Sethurajan, M., van de Vossenberg, J., Meers, E., van Hullebusch, E.D. (2020) Recovery of phosphorus from municipal wastewater treatment sludge through bioleaching using Acidithiobacillus thiooxidans. *Journal of Environmental Management*, 270, 336–384.

Liu, J., Zheng, M., Wang, C., Liang, C., Shen, Z., Xu, K. (2020) A green method for the simultaneous recovery of phosphate and potassium from hydrolyzed urine as value-added fertilizer using wood waste. *Resources, Conservation and Recycling*, 157, 104793.

Mahmoud, A., Cézac, P., Hoadley, A.F.A., Contamine, F., D'Hugues, P. (2017) A review of sulfide minerals microbially assisted leaching in stirred tank reactors. *International Biodeterioration & Biodegradation*, 119, 118–146.

Mehta, C.M., Khunjar, W.O., Nguyen, V., Tait, S., Batstone, D.J. (2015) Technologies to recover nutrients from waste streams: a critical review. *Critical Reviews in Environmental Science and Technology*, 45, 385–427.

Mohan, S.V., Nikhil, G.N., Chiranjeevi, P., Reddy, C.N., Rohit, M.V., Kumar, A.N., Sarkar, O. (2016) Waste biorefinery models towards sustainable circular bioeconomy: critical review and future perspectives. *Bioresource Technology*, 215, 2–12. https://doi.org/10.1016/j.biortech.2016.03.130.

Ownby, M., Desrosiers, D.A., Vaneeckhaute, C. (2021) Phosphorus removal and recovery from wastewater via hybrid ion exchange nanotechnology: a study on sustainable regeneration chemistries. *npj Clean Water*, 4(1), 6.

Rao, K.C., Otoo, M., Drechsel, P., Hanjra, M.A. (2017) Resource recovery and reuse as an incentive for a more viable sanitation service chain. *Water Alternatives*, 10, 493–512.

Schippers, A., Hedrich, S., Vasters, J., Drobe, M., Sand, W., Willscher, S. (2013) Biomining: metal recovery from ores with microorganisms. *Geobiotechnology*, 141, 1–47.

Timofeeva, A., Galyamova, M., Sedykh, S. (2022). Prospects for using phosphate-solubilizing microorganisms as natural fertilizers in agriculture. *Plants*, 11, 2119. https://doi.org/10.3390/plants11162119

USGS (2019) *Mineral Commodity Summaries 2019*. USGS, Reston.

Wen, Y.M., Cheng, Y., Tang, C., Chen, Z-L. (2013) Bioleaching of heavy metals from sewage sludge using indigenous iron-oxidizing microorganisms. *Journal of Soils and Sediments*, 13, 166–175.

Wong, P.Y., Cheng, K.Y., Kaksonen, A.H., Sutton, D.C., Ginige, M.P. (2013) A novel post denitrification configuration for phosphorus recovery using polyphosphate accumulating organisms. *Water Research*, 47, 6488–6495.

Xu, Y., Zhang, L., Chen, J., Liu, T., Li, N., Xu, J., Yin, W., Li, D., Zhang, Y., Zhou, X. (2023) Phosphorus recovery from sewage sludge ash (SSA): an integrated technical, environmental and economic assessment of wet-chemical and thermochemical methods. *Journal of Environmental Management*, 344, 118691.

You, X., Valderrama, C., Soldatov, V., Cortina, J.L. (2017) Phosphate recovery from treated municipal wastewater using hybrid anion exchangers containing hydrated ferric oxide nanoparticles. *Journal of Chemical Technology and Biotechnology*, 98(2), 358–364.

Yuan, Z., Pratt, S., Batstone, D.J. (2012) Phosphorus recovery from wastewater through microbial processes. *Current Opinion in Biotechnology*, 23(6), 878–883.

Zeng, Q., Huang, H., Tan, Y., Chen, G., Hao, T. (2022) Emerging electrochemistry-based process for sludge treatment and resources recovery: a review. *Water Research*, 209, 117939.

Zimmermann, J., Dott, W. (2009) Sequenced bioleaching and bioaccumulation of phosphorus from sludge combustion – a new way of resource reclaiming. *Advanced Materials Research*, 71–73, 625–628.

11 Bioleaching: A Feasible Option for Extraction of Essential Metals from Electroplating Sludge

Sujoy Chattaraj[1, 2*], *and Anoar Ali Khan*[3]
[1]Department of Chemical Engineering, Indian Institute of Technology Madras, Chennai, India
[2]Department of Chemical Engineering, Vignan's Foundation for Science, Technology & Research (Deemed to be University), Guntur, Andhra Pradesh, India
[3]Department of Chemical Engineering, Haldia Institute of Technology, Haldia, W.B, India

11.1 INTRODUCTION

The process of depositing a thin metal layer onto a conductive surface through an electrolytic method to provide protection to plated components is commonly referred to as electroplating (Mizushima et al., 2005). In this process, a metal ion present in a solution undergoes reduction by losing its electrons and is subsequently deposited onto a conductive surface, which acts as the cathode. Electroplating serves various purposes such as decorative coatings, corrosion protection, enhancement of electrical conductivity, and improvement of abrasion resistance. Industries such as jewelry, automotive, aerospace, and electronics extensively utilize electroplating. However, a drawback of the electroplating process is the presence of heavy metal content in the sludge and the generation of hazardous waste during effluent treatment (Celary and Sobik-Szołtysek, 2014; Wang et al., 2023).

Electroplating sludge (ES), a byproduct originating from the electroplating industry, falls under the hazardous waste category due to its substantial content of potentially harmful heavy metals (Peng and Tian, 2010). The effluent discharged by the electroplating industry often contains toxic heavy metal ions like chromium (Cr), zinc (Zn), copper (Cu), nickel (Ni), cadmium (Cd), and others (Ghorpade and Ahammed, 2017), rendering it universally recognized as hazardous waste. Presently, the primary methods for the safe disposal of electroplating sludge involve either landfilling or co-processing in cement kilns. Common procedures for heavy metal removal entail the precipitation of metal ions as hydroxides, sulfides, carbonates, or a combination of these, followed by the removal of precipitates through processes

DOI: 10.1201/9781003415541-11

such as clarification or microfiltration, ultimately resulting in electroplating sludge (Janson et al., 1982; Chmielewski et al., 1997).

Insufficiently addressing the disposal of electroplating sludge not only leads to significant secondary pollution of soil and groundwater, posing potential hazards to human health through the food chain but also results in substantial losses of vital metals and natural resources (Zhang et al., 2020). Health issues such as cancer, organ damage, and skin disorders may develop from the release of heavy metals into water and soil (Singh et al., 2023). Consequently, there has been a heightened focus on heavy metal removal from sludge in recent years. Traditional methods involve the use of pyrometallurgical and hydrometallurgical processes to extract metals from electroplating sludge (Zheng et al., 2020; Pinto et al., 2020). While pyrometallurgy is effective for large-scale sludge processing, it often incurs high energy costs and generates hazardous gases, demanding careful selection of smelting tools and enhanced safety precautions. On the contrary, hydrometallurgical treatment for the recovery of valuable metals from electroplating sludge is considered more viable due to its high metals leaching efficiency and easier separation, especially in small-scale processes.

Various methods, such as electrokinetic processes (Peng and Tian, 2010), supercritical fluid extraction, plant-based washing, chemical leaching, and advanced oxidation (Li et al., 2022), have been employed for metal recovery from electroplating sludge. Each method, however, comes with its merits and demerits. For instance, the electrokinetic method, utilizing a low-level direct electrical current and electrode array, exhibits high removal efficiency but is limited in universality. Supercritical fluid extraction faces limitations due to low universality and the generation of secondary pollutants. Chemical methods, while preferable for small-scale laboratory studies, involve complex extraction and purification techniques and the use of corrosive, strong acids and chemicals, posing potential hazards to health and the environment.

Hence, it becomes imperative to find a feasible and cost-effective method to enhance the extraction and recovery of heavy metals from hazardous electroplating sludge, ensuring sound environmental protection and efficient resource usage. The treatment of wastes and byproducts containing metals is an area where biological techniques are frequently recommended for sustainable growth. In this context, bioleaching by various microorganisms has gained importance for the recovery of metals from electroplating sludge without using corrosive reagents (Lee and Pandey, 2012; Zhang et al., 2019). Bio-extraction of metals from electroplating sludge using microbes, such as bacteria, fungi, and archaea, has gained popularity to meet the dual objectives of resource recycling and pollution reduction. While bioleaching technologies have been developed for a few decades, commercial use of these technologies has gained popularity more recently. This chapter provides a comprehensive review of the recovery of essential metals from electroplating sludge through the bioleaching process.

11.2 ELECTROPLATING PROCESS

Electroplating, a method for depositing a layer of metal through an electrodeposition process, is widely utilized in various industrial applications such as microfabrication,

TABLE 11.1
Different Types of Electroplating Processes with their Applications

Electroplating	Materials	Applications
Microfabrication	Cu	Application of dense conductive metallic layers
Microfabrication	Ni, Ni-Fe	Magnetic structural layer deposition
Engineering/functional coatings	Ag, Au, Ru, Pb, Pd, Rh, Os, Ir, and Pt	Enhancing specific properties of the surface (e.g., solderability)
Decorative/protective coatings	Au, Ag, Sn, Cu, Ni, Zn, and Cr	Appealing aesthetic
Sacrificial coatings	Zn and Cd	Preservation of the underlying metal
Alloy coating	Ni-Fe, Cu-Sn, Au-Cu-Cd, etc.	Diverse functions

alloy coating, and decorative or protective coatings (Schlesinger and Paunovic, 2011). This technique proves particularly challenging for generating metal films thicker than a few micrometers (μm) using methods like evaporation or sputtering due to the internal thermal stress induced during deposition. As a result, electroplating is frequently employed when thicker metal films, typically exceeding 5 μm, are required. Additionally, in applications where a robust magnetic force is needed, such as magnetic separation or magnetic force-driven actuators, electroplating is employed to deposit substantial coatings of magnetic material (Judy et al., 1995). The following are a few applications of electroplating (Table 11.1).

The electroplating process involves four distinct steps: surface preparation, immersion, plating, and finishing. During the surface preparation phase, the surface to be plated undergoes cleaning and polishing to eliminate any dirt, oil, or impurities that might interfere with the plating process. Following surface preparation, a solution containing ions of the metal intended for plating is applied to the surface. Additional chemicals may be introduced to the solution to enhance conductivity, pH, or other characteristics. The central plating process involves applying a direct current to the solution, attracting metal ions to the cathode's surface and causing them to deposit as a thin metal coating. Once the desired thickness of the plated layer is achieved, the surface undergoes cleaning, drying, and polishing to achieve the desired appearance and qualities.

The electroplating process utilizes an electric current to deposit a thin layer of metal onto a conductive object. In this process, a metal object is cleaned and designated as the cathode within an electrolytic cell. The anode, composed of the metal to be deposited, and both electrodes are immersed in a solution containing a suitable electrolyte (Figure 11.1). When an electric current passes through the solution, metal ions from the anode undergo oxidation and dissolve into the

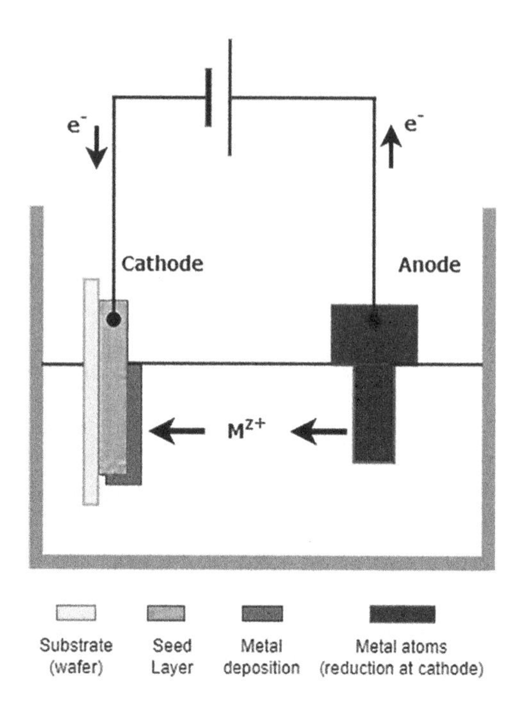

FIGURE 11.1 Generalized diagram for the electroplating process.

electrolyte. These positively charged metal ions then migrate toward the negatively charged cathode, where they are subsequently reduced back to their metallic form. As the metal ions accumulate on the cathode, they form a thin layer of metal, effectively electroplating the object. The overall reaction can be expressed as follows (Zhang and Hoshino, 2019):

$$M^{z+} \text{ [Metal ions]} + z\,e^- \text{ [electrons]} \rightarrow M\text{ (s) [Metal atoms (reduction at cathode)]} \quad (11.1)$$

$$M\text{ (s) [Metal atoms]} \rightarrow \text{Metal ions} + \text{electrons (oxidation at anode)} \quad (11.2)$$

11.3 ELECTROPLATING SLUDGE

The galvanization industry produces electroplating sludge as a byproduct during and after the electroplating process, particularly as a result of alkaline precipitation treatment applied to electroplating wastewater (Liu et al., 2021). It is a highly complex mixture with significant amounts of organic contaminants, hydroxides, pathogenic bacteria, and heavy metals. These hazardous components, particularly certain heavy metals, have highly biologically toxic effects and can be dangerous to human health. The disposal of heavy metal effluent, leading to the generation of significant amounts of electroplating sludge, is commonly addressed through various treatment methods including ion exchange, neutralization, precipitation, adsorption, electrochemical reduction, etc. (Fukuta et al., 2004). Typically, to recover metals from electroplating sludge, the waste is acidified, necessitating the addition of inorganic acids such as H_2SO_4, HCl, or HNO_3, followed by mixing. However, this process poses environmental concerns by generating toxic gases like SO_x, NO_x, etc. Therefore, it is imperative to optimize heavy metal recovery and recycling from electroplating sludge to simultaneously address environmental preservation and resource utilization goals.

11.3.1 Origins of Electroplating Sediment

The origins of electroplating sludge arise from various sources, encompassing the electroplating process itself and the treatment of electroplating wastewater. These sources include electroplating baths, rinsing operations, waste treatment processes, cleaning operations, maintenance activities, chemical spills, and similar events. The electroplating baths stand out as primary contributors to the generation of electroplating sludge. As metals are deposited onto the surface being plated, excess metal ions in the bath can form sludge. Further, rinsing operations are used to remove the excess plating solution from the surface of the part being plated. Contamination of rinsing water with metal ions can lead to the creation of electroplating sludge. In the treatment of wastewater from electroplating processes, electroplating sludge is commonly generated through the precipitation of metal ions. Cleaning operations, employed to eliminate contaminants from plated parts, may result in cleaning solutions becoming tainted with metal ions, further contributing to electroplating sludge formation. Additionally, accidental chemical spills containing metal ions can be a source of electroplating sludge formation.

As part of major industrial processes, electroplating sludge material mainly generates in the process of metal finishing operations, printed circuit board manufacturing, aerospace and defense industries, automotive industries, electronic industries, jewelry manufacturing, metalworking industries, etc. Industries that use metal finishing processes, such as plating, anodizing, and polishing, generate a large amount of electroplating sludge. These industries use electroplating processes to apply coatings to various components, parts, and equipment, which can generate significant amounts of electroplating sludge. By recovering precious metals and using materials, electroplating sludge can be reduced at the source. Additionally, by pre-treating the sludge, pollutants can be eliminated and recycled (Wang et al., 2024).

11.3.2 Physiochemical Characteristics of Electroplating Sludge

Electroplating sludge is classified as an alkaline material having varying pH in the range between 6.7 and 9.77 (Huang et al., 2022). As a byproduct of the electroplating process, electroplating sludge is often composed of a variety of metals, metal hydroxides, and other materials such as organic compounds and suspended particles. Electroplating sludge generally contains more than 76% ash content and roughly 10%–20% of volatile components on a dry basis (Peng et al., 2020). Apart from this, electroplating sludge typically contains approximately 75%–80% moisture, 8%–10% $CaCO_3$, 1.5%–3.0% heavy metals, and 5% contaminants. The heavy metal components commonly found in electroplating sludge include Cu, Cd, Cr, Al, Zn, Fe, Ni, and Cu. The specific composition of the sludge is influenced by the electroplating method used and the metals subjected to plating. Generally, high concentrations of major electroplating metals like Cr, Cu, Ni, Zn, and Pb are prevalent in electroplating sludge. Additionally, the sludge may contain metal hydroxides formed when the metal ions in the plating solution react with the alkali utilized in the process. Surfactants, diluents, and complexing agents, which have been employed to regulate the plating process, are additional elements that are frequently present in electroplating sludge. Sludge may also contain suspended particles like grease, oil, and dirt. The electroplating sludge contains irregular particles measuring approximately 50 nm in size. These nanoparticles lack distinct crystal shapes and are aggregated together. The EDS diagram shows that the amorphous nanoparticles include heavy metals that are uniformly distributed and doped with one another (Wang et al., 2024).

Nevertheless, the electroplating sludge contains a range of metal species that are not commonly encountered in organic ores and concentrates. Typically, hydrometallurgical methods, particularly those involving chemical leaching, are employed to extract metals from electroplating sludge (Yarzábal et al., 2004). Chemical extraction involves the use of both inorganic and organic acids. Solid metals dissolve when acids are introduced and convert to ionic when H^+ ions are introduced to electroplating sludge. The pH of the solution has a considerable impact on metal solubility in electroplating sludge. Multiple treatment approaches have been recorded for the elimination of heavy metals, encompassing thermal conversion, acid extraction, enhanced electrokinetic processes, plasma melting secondary recycling, and stabilization of metals using cement, clay, and/or lime, among others (Geng et al., 2020).

11.4 BIOLEACHING PROCESS

Bioleaching is a natural phenomenon involving the interaction of various ionic species, including iron, sulfate, and nutrient transfer to microorganisms (Bayat and Suri, 2010b). Specifically, it is the process of extracting metals from their parent materials, such as crystals suspended in the presence of a bacterial medium. Referred to as bio-solubilization, bioleaching utilizes microorganisms instead of chemicals for the solubilization of metallic oxides and sulfides (Sreedevi et al., 2022). The bioleaching process employs microbes to convert insoluble metal sulfides (or oxides) into water-soluble metal sulfates. A comprehensive understanding of the bioleaching process also relies heavily on the dynamics of the rate and identification of the rate-controlling phase. Often termed the bio-hydrometallurgical method, it aids in the recovery of precious metals like gold, uranium, copper, lead, vanadium, arsenic, antimony, zinc, cobalt, molybdenum, nickel, bismuth, and more. Apart from that, few valuable metals are present in complex combinations such as residual fractions and reducible fractions or as sulfides, making it difficult to fully extract by conventional leaching (Zhang et al., 2020). Valuable metal recovery from electronic wastes, including electroplating sludge, used batteries, and electronic printed circuit boards, has drawn attention to bio-based technology utilized for metal recovery from wastes (Mahmoud et al., 2017; Nikfar et al., 2020). In comparison with chemical leaching, bioleaching is considered a green method as there is no risk of causing environmental problems due to the handling and transportation of the hazardous strong acids. Additionally, bioleaching decreases the need for landfill areas and facilities (Liu et al., 2020).

11.4.1 Microorganisms Involved in Bioleaching for Metal Extraction

Most of the microorganisms employed in the bioleaching process are acidophilic, chemolithoautotrophic bacteria that specialize in the oxidation of iron and/or sulfur. These bacteria primarily fall within the *Leptospirillum, Halothiobacillus, Thiobacillus, Acidithiobacillus*, and *Sulfolobus genera* (Gu et al., 2017). *Chemolithotrophic* microorganisms are bacteria that actively engage in extracting metallic compounds from mineral sources through the bioleaching process. Previous research has shown that the synergy of acidophilic bacteria during bioleaching makes mixed prokaryotic cultures more efficient than pure cultures (Xin et al., 2009), which may successfully dissolve precious metals (Latorre et al., 2016). As a result, mixed microorganisms have been used extensively in the bioleaching of solid waste (Yang et al., 2016). The most well-known of these *chemolithotrophic* microbes is *Thiobacillus ferrooxidans*. Metal leaching is mostly carried out by *chemolithotrophic* mesophilic bacteria that need substances like ferrous sulfate, pyrite, and sulfur as metabolic energy sources. Generally, *Acidithiobacillus ferrooxidans* is used for iron-oxidizing and *Acidithiobacillus thiooxidans* bacteria is used for the sulfur-oxidizing process. *Acidithiobacillus* plays a pivotal role in extracting metals from sludge. These acidophilic microorganisms, which aid in dissolving metals from waste, are autotrophic in nature. They exhibit resilience to high concentrations of metal ions and thrive in inorganic media with low pH values (Lee et al., 2012). *Acidithiobacillus ferrooxidans* (*At. ferrooxidans*) and *Acidithiobacillus thiooxidans* (*At. thiooxidans*) are commonly utilized in studies related to sewage sludge because they can flourish at an acidic pH range of 1.0–3.0, which is conducive to the solubilization of heavy metals (Pathak et al., 2009). Acidophilic microorganisms are classified into three categories based on the extent to which species endure temperature, including mesophiles, moderate *Thermoacidophiles*, and extreme *Thermoacidophiles*. Microorganisms like *At. ferrooxidans* and *At. thiooxidans* come under mesophiles and can withstand the temperature up to 38°C, whereas moderate thermophile species such as *Sulfobacillus thermosulfidooxidans* can withstand the temperature up to 50°C, and extreme thermophiles such as *S. acidocaldarius* and *S. solfataricus* can be active at temperatures up to 70°C.

11.4.1.1 Mesophiles Microorganisms

Mesophiles are bacteria, such as sulfur-oxidizing *Acidithiobacillus thiooxidans* and iron-oxidizing *Acidithiobacillus ferrooxidans*, that derive their energy through the oxidation of ferrous iron or reduced sulfur compounds (Villar and Garcia, 2003). While *Acidithiobacillus ferrooxidans* can thrive at temperatures between 20°C and 40°C, its optimal growth temperature is approximately 33°C. Mesophiles exhibit growth at lower pH values within the range of 1.0–4.5. The ideal pH for mesophile growth is typically between 2.0 and 2.3 (Ruamsap et al., 2003). The metal extraction process begins with the inoculation of the sludge with the mesophilic microorganisms, which are then allowed to grow and metabolize in the sludge. As the microorganisms grow, they consume organic matter and produce acids, which lower the pH of the sludge. This acidic environment then facilitates the solubilization of metals, which can be further enhanced by the addition of nutrients such as sulfur or iron. After the dissolution of metal ions, they can be retrieved through different methods like precipitation, adsorption, or electroplating. The recovered metals can then be purified and processed further to obtain high-quality metal products. Mesophilic bacteria are often preferred for bioleaching over thermophilic bacteria (which grow optimally at high temperatures) because mesophilic bacteria are easier to maintain and operate, and they can achieve comparable metal extraction rates to thermophiles.

11.4.1.2 Moderate *Thermoacidophiles* Microorganisms

Moderate *Thermoacidophiles* are microorganisms that can grow optimally at moderate temperatures (around 40°C–70°C) and in acidic conditions (pH 1–5). These microorganisms are frequently employed in extracting metals from sludge because of their capacity to thrive in harsh environments and facilitate the solubilization of metals. The working principle of moderate *Thermoacidophiles* for metal extraction from sludge is similar to that of mesophiles. The microorganisms are inoculated into the sludge and allowed to grow, producing acids that lower the pH of the sludge and solubilize metals from the solid particles. However, moderate *Thermoacidophiles* are capable of tolerating higher temperatures and more extreme conditions than mesophiles, allowing them to catalyze reactions that mesophiles cannot.

Sulfolobus metallicus is an example of a moderate thermoacidophilic microorganism used for metal extraction from sludge. This microorganism can oxidize ferrous ions into ferric ions, which subsequently interact with sulfide minerals to liberate metal ions. Other examples include *Acidianus brierleyi* and *Acidianus sulfidivorans*, these microorganisms have the ability to oxidize sulfur and sulfur-containing compounds, resulting in the production of sulfuric acid and the solubilization of metals. The metal extraction process using moderate *Thermoacidophiles* is typically carried out at higher temperatures than mesophiles, which can result in faster reaction rates and higher metal extraction rates. However, the process may require more specialized equipment to maintain the higher temperatures and acidic conditions.

11.4.1.3 Extreme *Thermoacidophiles* Microorganisms

Extreme *Thermoacidophiles* are microorganisms that can grow optimally at very high temperatures (typically around 70°C) and extremely acidic conditions (pH 0–2). These microorganisms are used in bioleaching processes for the extraction of metals from ores and concentrates due to their ability to survive in extreme environments and catalyze the solubilization of metals. The microorganisms are inoculated into the ore or concentrate and allowed to grow, producing acids that lower the pH of the solution and solubilize metals from the solid particles.

Examples of extreme thermoacidophilic microorganisms used for metal extraction through bioleaching include *Sulfolobus solfataricus* and *Metallosphaera sedula*. These microorganisms are capable of oxidizing iron, sulfur, and sulfur-containing compounds, producing acids that solubilize metals and other elements from the ore or concentrate. The sole extreme *Thermoacidophiles* currently identified belong to the Crenarchaeotal class of Thermoprotei. This class encompasses several orders, namely

Desulfurococcales, Thermoproteales, Fervidococcales, Acidilobales, and *Sulfolobales.* Thus far, only a limited number of Sulfolobales species have been assessed for their potential in bioleaching (Schippers, 2007). Some extreme *Thermoacidophiles* can also use carbon dioxide as a carbon source, allowing them to grow in solutions containing low organic matter. The ability to operate at higher temperatures leads to faster reaction rates and higher metal extraction rates. The acidic conditions also help to reduce the formation of unwanted compounds such as sulfur dioxide. However, the use of extreme *Thermoacidophiles* also has some challenges. The high temperatures and acidic conditions required for their growth and metabolism can result in equipment corrosion and require specialized equipment for containment and handling.

11.4.1.4 Heterotrophic Microorganisms

Heterotrophic bacteria facilitate metal leaching through the production of particular organic acids, such as oxalic acid, malic acid, and citric acid. These acids serve the dual function of providing protons and acting as metal-complexing agents. The bioleaching process is further bolstered by linking the biological activity to the acidity generated by these organic metabolites (Varia et al., 2021). Further, heterotrophic microorganisms can extract the metal directly by breaking the metal-oxygen bond and consuming the organic components. These bacteria can be found in

TABLE 11.2
Bacterial Strains Engaged in Bioleaching to Extract Heavy Metals

Bacterial Species	Bioleaching of Metals	Reference
At. ferrooxidans	Cu, Pb, Ni, Zn, Cd	
At. ferrooxidans	Pb, Ni, Zn, Cd	Couillard and Mercier (1990)
At. ferrooxidans	Cu, NI, Zn, Cd, Mn	Couillard and Mercier (1991)
At. thiooxidans	Cu, Pb, Zn	Zhang et al. (2009)
Acidithiobacillus bacteria + Galactomyces sp. Z3	Cu, Zn	Zhou et al. (2013)
At. ferrooxidans + *At. thiooxidans*	Cu, Pb, Zn	Zhou et al. (2017)
Acidithiobacillus bacteria + *Meyerozyma guilliermondii*	Cu, Pb, Ni, Zn, Cd, Cr	Camargo et al. (2018)
S. thermosulfidooxidans, Sulfobacillus acidophilus, Acidithiobacillus caldus	Mn, Zn, Ni, Cu, Cr	Chen and Cheng (2019)
At. ferroxidans	Cu, Ni, Zn, Pb, Cr, Cd	Gu et al. (2019)
Streptomyces albidoflavus	Al, Cu, Cd, Fe, Ni, Zn, Ag, Pb	Kaliyaraj et al. (2019)
Leptospirillum sp, *Acidithiobacillus* sp, *Acidithiomicrobium sp, Sulfobacillus sp*	Ni, Co, Zn, Cu	Hubau et al. (2020)
At. ferrooxidans	Cu, Ni	Arshadi and Yaghmaei (2020)
Pseudomonas putida, P. fluorescens, P. azotoformans	Zn, Mn, Cu, Al	Williamson et al. (2021)

At., *Acidithiobacillus*; Cu, copper; Cd, cadmium; Al, aluminum; Zn, zinc; Co, cobalt; Ni, nickel; Cr, chromium; Fe, iron; Ag, silver; Mn, manganese.

a range of surroundings, such as sediments, water, and soil. The *Acidithiobacillus* genus bacteria is the most commonly employed heterotrophic microorganism in bioleaching. Apart from that genus, *Thiobacillus* and *Leptospirillum* are also used in bioleaching (Table 11.2).

11.4.2 Process and Mechanism of Bioleaching for Metal Extraction

Bioleaching is mainly accomplished by two methods: (1) direct and (2) indirect methods. This happens either as a direct result of microbial activity or indirectly through metabolic byproducts of bacteria. Metal leaching is mostly carried out by *chemolithotrophic*, mesophilic bacteria that need substances like ferrous sulfate, pyrite, and sulfur as metabolic energy sources. From the discussion of Section 11.3, it has been already observed that *At. thiooxidans* and *At. ferrooxidans* are the most commonly used microorganisms for the metal bioleaching process due to their ability to survive in an acidic environment. *Acidithiobacillus ferrooxidans*, a bacteria that oxidizes iron, and *Acidithiobacillus thiooxidans*, which oxides sulfur, are employed in bioleaching.

11.4.2.1 Sulfur-Based Bioleaching

Bacterial sulfur-based leaching encompasses both direct and indirect processes, resulting in the generation and dissolution of metal sulfides. In the direct approach, *chemolithotrophic* bacteria such as *Acidithiobacillus thiooxidans*, *Acidiphilium acidophilum*, and *A. caldus* transform insoluble metal sulfides into soluble metal sulfates (Eq. 11.3). Typically, this method is used to dissolve the metallic sulfides present in sludge, such as ZnS and CuS

$$MS + 2O_2 \xrightarrow{At.\ thiooxidans} MSO_4 \tag{11.3}$$

where M is the bivalent metal

On the other hand, in the indirect method, sulfur-oxidizing bacteria first convert elemental sulfur (S^0) to sulfuric acid (Eq. 11.4), after which metals are dissolved in sulfuric acid (Eq. 11.5). This technique decreases the pH level of the persistent medium, increasing the solubility of metals in solution. Reduced sulfur compounds or elemental sulfur are provided as an external energy source in indirect bacterial leaching

$$S^0 + 1.5O_2 + HO_2 \xrightarrow{At.\ thiooxidans} H_2SO_4 \tag{11.4}$$

$$H_2SO_4 + \text{Sludge} - M \rightarrow MSO_4 + \text{Sludge} - 2H \tag{11.5}$$

11.4.2.2 Iron-Based Bioleaching

Reduced sulfur and iron compounds undergo oxidation either directly or indirectly as components of an iron-based bioleaching process. Non-iron metal sulfides are readily oxidized into soluble metal sulfate by iron-oxidizing bacteria such as *Acidithiobacillus ferrooxidans*. Much like sulfur-based bioleaching, iron-based bioleaching processes are also divided into direct and indirect methods (Tyagi et al., 1988). In direct leaching, chemotactic behavior causes bacterial cells to attach only to the surface structure of minerals. In direct bacterial leaching, *At.*

ferrooxidans directly oxidizes non-ferrous metallic sulfide to produce soluble metal sulfate through the following reactions:

$$MS + 2O_2 \xrightarrow{At.\ ferroxidans} MSO_4 \tag{11.6}$$

The metal sulfates are subsequently extracted from the slurry by a number of techniques, including solvent extraction or precipitation.

Conversely, in the indirect approach of iron-based bioleaching, bacteria convert Fe^{2+} to Fe^{3+}, which in turn reacts chemically with metals and promotes leaching. Iron-based indirect bioleaching process occurs through four-step processes (Geng et al., 2020). In the first step, *At. ferrooxidans* convert Fe^{2+} to Fe^{3+} in the liquid phase (Eq. 11.7)

$$4FeSO_4 + O_2 + 2H_2SO_4 \xrightarrow{At.\ ferroxidans} 2Fe_2(SO_4)_3 + 2H_2O \tag{11.7}$$

In the 2nd step, metal sulfides are converted into metal sulfate by the action of Fe^{3+} as an electron acceptor by the following:

$$4Fe_2(SO_4)_3 + 2O_2 + 2MS + 4H_2O \rightarrow 8FeSO_4 + 2MSO_4 + 2H_2SO_4 \tag{11.8}$$

In the 3rd step, metal sulfides are oxidized into metal sulfates with the help of sulfuric acid generated in the 2nd step, and elemental sulfur is generated (Eq. 11.9). Finally, elemental sulfur is converted into sulfuric acid through oxidation with the help of *At. ferrooxidans* microorganisms (Eq. 11.10)

$$2MS + O_2 + 2H_2SO_4 \rightarrow 2MSO_4 + 2S + 2H_2O \tag{11.9}$$

$$2S + 3O_2 + 2H_2O \xrightarrow{At.\ ferroxidans} 2H_2SO_4 \tag{11.10}$$

In this indirect procedure, sulfuric acid is produced as a byproduct, which accelerates the solubilization process. In the indirect method, there is no requirement for direct contact between the bacterial cells and the minerals during this process. The efficiency of the procedure for removing heavy metals can be further increased by linking these reactions into a cyclic process.

11.5 EXTRACTION OF ESSENTIAL METALS FROM ELECTROPLATING SLUDGE VIA BIOLEACHING

Special measures, like a higher concentration of acid, high temperature, high pressure, are required to achieve the maximum leaching of heavy metals from electroplating sludge. Hence, in order to achieve higher metal extraction efficiency, there will be a major concern of higher energy consumption and negative environmental impact. To address these drawbacks, bioleaching emerges as a promising alternative method recently devised for extracting heavy metals from electroplating sludge and solid wastes in an environmentally friendly manner. This approach aims to recover different metals that can be more readily recycled from the treated residues and/or leachate (Prabhu and Bhaskar, 2015). The bioleaching process to extract metals from electroplating sludge is becoming increasingly common in order to meet the twin aims of resource regeneration and environmental protection. These bacteria might excrete complexing agents (which would result in the production of ligands), produce organic and inorganic acids (such as citric, oxalic, and sulfuric acids), and undergo oxidation and reduction reactions to solubilize the metals (Mishra and Rhee, 2014). Limited research has been documented regarding the extraction of valuable metals from electroplating sludge through bioleaching. Studies have explored that achieving optimal extraction yields by adjusting variables like initial pH, agitation duration, and pulp density. However, the operational condition is not only the major factor in achieving higher metal extraction efficiency; it is also highly dependent on the suitable system used for bioleaching. Most of the bioleaching studies for electroplating sludge find its limitation, even at low pulp density, due to improper selection of the bioleaching system (Rastegar et al., 2014).

Metal extraction from electroplating sludge through bioleaching can be conducted via two methods (Zhang et al., 2020). The first method is conventional, and the second one is by separately mixing sludge and microorganisms when they reach the logarithmic phase of growth.

11.5.1 Conventional Method

In conventional method, several mobilization methods, such as redoxolysis, acidolysis, and complexolysis, are employed by a diverse array of *chemolithotrophic* and *heterotrophic* bacteria, and fungi, to transform solid metallic compounds into soluble and retrievable ions (Gu et al., 2018). In the conventional bioleaching process, the culture medium is simultaneously infused with microorganisms and waste. Currently, bioleaching is widely employed for extracting gold, copper, cobalt, nickel, zinc, uranium, and other metals from electroplating sludge through conventional means.

11.5.1.1 Acidolysis

Microorganisms facilitate metal solubilization through the production of organic or inorganic acids. For instance, *Aspergillus niger* generates citric acid, *Penicillium simplicissimum* produces gluconic acid, and *At. ferrooxidans* and *At. thiooxidans* produce sulfuric acid. Acidolysis, a proton-driven process, utilizes protons from organic acids produced by fungi. These protons interact with the surface of the waste or sludge, destabilizing the bonds that hold metal ions on the surface. The following equation illustrates acidolysis reactions, where protons are acquired from the generated acids, and the quantity of enzyme available determines the amount of solubilized metal oxides (MO) (Eq. 11.11)

$$MO + 2H^+ \rightarrow M^{2+} + H_2O \tag{11.11}$$

11.5.1.2 Redoxolysis

In the redoxolysis process, oxidation and reduction of metals occur with the help of microorganisms. Metal mobility varies according to metal type and oxidation state. The mechanism of redoxolysis can be categorized as direct or indirect. Bacteria facilitate metal leaching through a redox process, where metals are made soluble through enzymatic reactions. This occurs through direct physical contact between the microbes and the leaching materials, leading to the breakdown of minerals in the redoxolysis process, as depicted in Eqs. (11.12) and (11.13)

$$4FeS_2 + 14O_2 + 4H_2O \xrightarrow{Thiobacillus} 4FeSO_4 + 4H_2SO_4 \tag{11.12}$$

$$4FeSO_4 + O_2 + 2H_2SO_4 \xrightarrow{At.\ ferroxidans} 2Fe_2(SO_4)_3 + 2H_2O \tag{11.13}$$

In the "indirect" process, the oxidation of reduced metals is facilitated by ferric (III) ions, which are produced through microbial oxidation of ferrous iron present in minerals. Ferric iron serves as an oxidizing agent, capable of oxidizing metal sulfides before being reduced back to ferrous iron. This ferrous iron, in turn, can undergo oxidation through microbial activity, with iron acting as an electron carrier in this context (Eq. 11.14)

$$4Fe_2(SO_4)_3 + 2MS \rightarrow 2FeSO_4 + MSO_4 + S^0 \tag{11.14}$$

11.5.1.3 Complexolysis

The metal ions brought into the solution through acidolysis are stabilized during complexolysis through chelation. Complexolysis is a process in which an organic acid leaches metals via complex formation. In comparison to acidolysis, complexolysis is a slower mechanism. The effectiveness of solubilizing metallic ions in complexolysis hinges on the substance's ability to interact with which a complex forms. Apart from organic acids, other metabolites like siderophores (chelating agents with low molecular weight) can create complexes through complexolysis, facilitating the solubilization of metals like ferric iron, magnesium, manganese, and chromium. Additionally, when there are substantial amounts of metals present, complexing heavy metals might lessen their toxicity to fungus.

Traditional approaches may encounter challenges when commercially applying detoxification to electroplating sludge due to relatively low extraction yields, even with a low pulp density of 2%. The metals under investigation were obtained for recycling after 7–20 days of bioleaching (Rastegar et al., 2014). The sludge produced from chrome electroplating settles in the bath following the plating process, typically comprising metals such as nickel, copper, cobalt, zinc, cadmium, vanadium, molybdenum, and substantial proportions of chromium (7%–11%) and iron (3%–5%). Bioleaching treatment is done on electroplating sludge by a conventional method with the help of *At. ferrooxidans* microorganisms, as reported by Bayat and Suri (2010a), found a fairly good recovery of zinc, copper, lead, and cadmium with a low recovery of chromium.

11.5.2 Logarithmic Phase of Microorganism Growth Method

Conversely, the second approach involves the separate stages of microbial cultivation and metal bioleaching. In this method, the sludge is introduced separately when the bacteria reach their maximum growth independently, during the logarithmic phase of growth. The approach of segregating microbial cultivation from sludge bioleaching has proven to be more efficient compared to the conventional method of bioleaching electroplating sludge, both in terms of heavy metal leaching rate and total treatment duration (Zhou et al., 2019). Study reported using second method is very limited. By using this technique, Zhang et al. (2020) conducted a bioleaching procedure on powdered electroplating sludge, achieving a substantial recovery of Cr (>95.6%), along with high percentages of copper, zinc, and nickel recovery (>95.6%). Moreover, they attained a 90.3% extraction rate for Cr under optimal experimental conditions, which included a pH of 2.0, pulp density of 15%, and Fe^{2+} concentration of 9.0 g/L. In order to resolve the problem of poor extraction yields and higher treatment time of conventional bioleaching of electroplating sludge, Yang et al. (2020) have introduced a hybrid and membrane bioreactor (MBR) and bioleaching technology for electroplating sludge. In a recent study, Wu et al. introduced a method to extract heavy metals from electroplating sludge by integrating acid leaching with bioelectrical reactors. Their findings revealed that the extraction of heavy metals was significantly influenced by the solution pH during acid leaching and the voltage applied in the bioleaching reactor.

11.6 COMPARATIVE ANALYSIS: BIOLEACHING OVER OTHER PROCESSES FOR METAL EXTRACTION FROM ELECTROPLATING SLUDGE

Electroplating sludge is currently being detoxified using a variety of chemical and biological processes, and its suitability depends on its technical and economic viability (Babel and Del Mundo Dacera, 2006). However, in most cases, acidification is being used to recover metal from the electroplating sludge. An inorganic acid such as H_2SO_4, HCl, or HNO_3 is introduced to the sludge and then mixed to lower the pH level to 1.5–2.0. In this low pH level, metal present in the sludge gets solubilized, known as chemical leaching. Chemical leaching necessitates the use of significant quantities of acid to decrease the pH value and also requires a substantial amount of alkali toward the end of the leaching process to neutralize the sludge (Bayat and Suri, 2010a). Acid treatment or chemical leaching is only favorable when significant metal content is present in the sludge and requires continuous monitoring.

Electrokinetic methods are being considered as a potential metal remediation technology because of its effective removal of heavy metals and lesser removal time. By placing electrode arrays inside the sludge and delivering a weak direct electrical current, the electrokinetic process is carried out. However, this method effectively extracts only low-permeability sludge contaminated with heavy metals or organic compounds (Gao et al., 2013). In addition, the energy requirements and costs associated with treating a large volume of sludge utilizing EK techniques are substantial (Xu et al., 2017). Metal extraction using supercritical fluid (CO_2) has been observed to offer advantages such as a low critical temperature and pressure, high diffusivity, low viscosity and surface tension, environmental friendliness, and low cost. However, sometimes it is considered unfeasible as heavy metal ions are strongly polar carries a positive charge and supercritical CO_2 is a non-polar solvent. To overcome this problem, the polarity of the supercritical fluid is increased by adding the methanol and ethanol as modifiers, or the polarity of the heavy metal ion is decreased by adding suitable chelating agents. Although metal extraction by organic and inorganic acids and chelating agents has higher removal efficiency and has been investigated widely, however, consumption of high amounts of chemical agents makes this process unfeasible in both ways, i.e., cost and environmental hazards. In this context, plant-based washing agents are regarded as a viable substitute for chemical agents due to their biodegradability, affordability, and versatility in sourcing from various origins. Plant-based washing studies are mainly implemented in metal-contaminated soil.

The utilization of the bioleaching process has notably risen compared to chemical leaching, as it offers advantages from both technical and economic standpoints in today's context. Bioleaching significantly enhances sludge dewaterability. According to Liu et al. (2012), the capillary suction time (CST) of sludge decreased from 48.9 s for fresh sludge to 10 s for bioleached sludge at a suitable pH. Wong et al. (2015) also provided evidence of an increase in sludge dewaterability brought about by bioleaching. The chemical leaching process is more expensive than the biological approach since it uses a lot of inorganic acids acid such as H_2SO_4, HCl, and HNO_3. Further, during the chemical leaching process, it also releases toxic gases like SO_x, NO_x etc., which creates environmental pollution. Apart from that, with its lower operating temperatures and improved environmental friendliness, bioleaching methods have been claimed to be superior to other technologies (Chen and Cheng, 2019). It is simple to use and uses 80% less chemicals than acid treatment (Pathak et al., 2009). More importantly, bioleaching technique is more viable in both technical and economic context for the extraction of metals from electroplating sludge because of low percentage of metal present in the sludge (less than 0.5 wt%), minimum generation of secondary pollutants and low cost. A comparative analysis between chemical and biological leaching for the removal of metal from electroplating sludge performed by Bayat and Suri (2010a) has found highest removal efficiency for bioleaching processes compared to ferric chloride and sulfuric acid leaching, and suggested it as an alternative of conventional physicochemical treatment of electroplating sludge. However, rate of leaching varies significantly with the metal to leach.

11.7 ECONOMIC CONSIDERATIONS OF THE BIOLEACHING PROCESS FOR METAL EXTRACTION FROM ELECTROPLATING SLUDGE

Since 1970, the idea of a circular economy has become increasingly popular. The researchers examined both the linear and open-ended dimensions of contemporary economic systems, elucidating how natural resources could impact the economy by providing inputs for production and consumption, while serving as repositories for waste outputs. The technologies for extracting valuable resources and energy can be treated as a part of circular economy. According to the European Commission (2011), "if waste is to become a resource to be fed back into the economy as a raw material, then reuse and recycling must be given much higher priority." Utilizing sludge as a raw material in various sectors presents a beneficial waste management strategy when embracing the concept of the circular economy (Eliche-Qusada et al., 2011).

The fundamental challenge in sludge treatment is to develop a financially viable operational strategy that can eliminate the need for extra water, produce less sludge, and transform the heavy chemicals into mild and/or advantageous compounds equally. Sludge disposal is considered as an important cost component of the electroplating industry, as it is generally disposed to agricultural land if it satisfies provincial heavy metal concentration regulations. If the concentrations exceed the standards, the sludge must be transported to a landfill, resulting in greater sludge disposal costs. Of the various processes available for disposing of heavy metal sludge, such as incineration, solidification, and stabilization, landfilling proves to be the most cost-effective option (Yang et al., 2015). Even though disposal at a landfill is the most cost-effective option, costs range from $298 to 745/ton (Zhang et al., 2020). However, disposal of electroplating sludge whether it to agricultural land or to landfill, prior to recovery of heavy metals, is a loss of precious resources.

The comprehensive cost assessment of metal leaching encompasses expenses associated with every aspect, including equipment, raw materials, and services required to execute the process effectively. The cost of equipment is calculated based on its capacity. The equipment capacity is dependent on the amount of time needed for each batch, the total amount of sludge that could be processed per day, and the sludge's solids concentration. The expense of raw materials is directly linked to factors such as plant capacity, sludge solids concentration, and their correlation with the acid needed to adjust pH or cultivate microorganisms in the case of bioleaching. The main component for utility expenses is the cost required for power consumption.

Among all factors, substrate type and concentration, initial sludge concentration, pH, and initial sludge concentration are considered the most crucial to optimize, as they directly affect both the process cost and the efficiency of metal removal. In general, the bioleaching process takes 16–20 days, with adequate aeration and mixing throughout. Therefore, the mainstream of the cost components for performing bioleaching includes cost of mixing and aeration, cost of chemicals, construction of holding tank and operational maintenance, etc. (Tyagi et al., 1988). Apart from that, the costs of dewatering and conditioning the sludge, and recovering the metal from the acid sludge filtrate, should also be considered in the overall cost of bioleaching process for metal extraction from electroplating sludge. To enhance the feasibility of bioleaching technology, minimizing the treatment duration and increasing the unit reaction rate and treatment volume are vital. This approach reduces the engineering scale and construction costs. The concentration of sludge is believed to heavily influence the leaching duration. In bioleaching, a high sludge concentration results in a strong buffering capacity and a gradual pH decline, which lengthens the bioleaching cycle. The substrate used in bioleaching is also an important factor for cost consideration. The dominating bacteria in the bioleaching process are constrained by the type and quantity, which also determines the effectiveness and cost of removal. Since bioleaching uses microbes rather than chemical agents to greatly lower process costs, it has attracted more interest than chemical leaching technology as an affordable and environmentally benign method of eliminating heavy metals (Marchenko et al., 2018). Instead of all these costs, the bioleaching process for metal extraction can be considered one of the economic processes compared to conventional chemical treatment for electroplating sludge, as it consumes only 20% of the chemicals as compare to chemical treatment (Pathak et al., 2009). Conversely, while the bioleaching process might entail lower chemical expenses, it could incur higher costs in terms of capital investment, energy consumption, and maintenance (Geng et al., 2020). Additionally, it is the most efficient in low-capacity plants with high solid content (Sreekrishnan and Tyagi, 1996). Regrettably, there is scant literature available on the economic dimensions of the bioleaching process, constraining the scope of thorough investigation. Thus, a more comprehensive cost analysis of the entire bioleaching process for electroplating sludge is necessary.

11.8 FUTURE PERSPECTIVE AND POSSIBLE SCALE-UP OF THE PROCESS

The electroplating solution formulation becomes more and more complex with the requirement of quality of plating layer increases. The primary knowledge gaps in this field include a lack of optimal techniques, a lack of integration of various approaches, and a lack of set standards for method evaluation that are systematic and objective. As a result, there are different types of pollutants are introduced in waste plating solution, making it challenging to treat waste solution while also balancing the demands of economy, environmental protection, and resource recovery. The treatment of electroplating sludge is a serious environmental problem since existing conventional procedures are mostly insufficient to reach the level of purity required to minimize detrimental environmental consequences. Not only that, but all the conventional treatment processes itself generate a certain amount of secondary pollutants, which create an environmental threat rather than remediation. The potential of bioleaching to recover important metals has led to its widespread use, although more research is needed to understand how it can remove contaminants and recover secondary resources. The need for less expensive, more sustainable techniques of metal solubilization will make bioleaching even more pertinent in the future.

However, the principal drawback of the mechanism continues to be the bioleaching-induced acid leakage into the nearby groundwater. Due to the permanent nature of the genetic damage caused by metal exposure, genome stability is negatively impacted. The consequences of these mutagenesis events are critical to understanding the toxicant's etiology and its impact on human health. Additionally, the microbiological characteristics of severe environments may offer a fresh perspective on how disease emerges, how adaptability and innovation play a part in the colonization of new environments, and how interactions between organisms occur within microbial communities. Furthermore, a significant concern or drawback of the bioleaching process is the loss of nutrient content, which can amount to as much as 75% of the sludge. This reduction diminishes the fertilizer value of the sludge. Additionally, during bioleaching, the pH of the electroplating sludge is lowered to a value of 2 or below, while the oxidation-reduction potential (ORP) is increased. Consequently, the combination of low pH and a highly oxidizing environment leads to the oxidation of organic matter, resulting in the dissolution of sludge-bound nutrients. Moreover, the breakdown of proteins by microorganisms in the sludge leads to a loss of nitrogen content. Addressing the loss of nutrient content, nitrogen content, and fertilizer value are critical issues that require immediate attention for scaling up the process.

Research in this field is crucial because it may progress the understanding of bioleaching toward the point where it may be utilized for mining and become economically viable for all applicable metals, since the microorganisms would assume all the risks and allow humans to work on safer surfaces. There is a need to put more efforts to find the microorganisms which can be stable beyond the temperature condition at which extreme *Thermoacidophiles* microbes can take part in bioleaching (>70°C), which would ultimately helpful for increasing the rate of leaching. Therefore, it is necessary to take use of the opportunities that bioleaching offers as the technology develops and incorporate a variety of disciplines ranging from material science, metallurgy, geology, microbiology that could be utilized going forward. Researchers are seeking for new microorganisms and optimized condition to enhance the bioleaching efficiency. In order to recover valuable metals effectively, researchers are also trying to develop an improvised hybrid method that combines physical, chemical, and biological leaching. To gain deeper insights into the mechanisms driving the bioleaching of metals from electroplating sludge, further research is particularly needed to comprehend the distribution of biological communities. It is anticipated that numerous new commercial-scale bioleaching plants will be established in the forthcoming years, becoming an integral component of the electroplating industry and capitalizing on this innovative technology.

REFERENCES

Arshadi, M., Yaghmaei, S. 2020. Bioleaching of basic metals from electronic waste PCBS. *Journal of Mining and Mechanical Engineering*, 1(2), 51–58 https://doi.org/10.32474/jomme.2019.01.000108

Babel, S., Del Mundo Dacera, D. 2006. Heavy metal removal from contaminated sludge for land application: a review. *Waste Management (New York, N.Y.)*, 26(9), 988–1004. https://doi.org/10.1016/j.wasman.2005.09.017

Bayat, B., Sari, B. 2010a. Comparative evaluation of microbial and chemical leaching processes for heavy metal removal from dewatered metal plating sludge. *Journal of Hazardous Materials*, 174(1–3), 763–769. https://doi.org/10.1016/j.jhazmat.2009.09.117

Bayat, S., Sari, B. 2010b. Bioleaching of dewatered metal plating sludge by *Acidithiobacillus ferrooxidans* using shake flask and completely mixed batch reactor. *African Journal of Biotechnology*, 9(44), 7504–7512. https://doi.org/10.5897/AJB10.1142

Camargo, F.P., Prado, P.F.D., Tonello, P.S., Dos Santos, A.C.A., Duarte, I.C.S. 2018. Bioleaching of toxic metals from sewage sludge by co-inoculation of *Acidithiobacillus* and the biosurfactant-producing yeast Meyerozyma guilliermondii. *Journal of Environmental Management*, 211, 28–35.

Celary, P.; Sobik-Szołtysek, J. 2014. Vitrification as an alternative to landfilling of tannery sewage sludge. *Waste Management*, 34, 2520–2527.

Chen, S.-Y. Cheng, Y.-K. 2019. Effects of sulfur dosage and inoculum size on pilot-scale thermophilic bioleaching of heavy metals from sewage sludge. *Chemosphere*, 234, 346–355, https://doi.org/10.1016/j.chemosphere.2019.06.084

Chmielewski, A., Urbański, T., Migdał, W. 1997. Separation technologies for metals recovery from industrial wastes. *Hydrometallurgy*, 45, 333–344.

Couillard, D., Mercier, G. 1990. Bacterial leaching of heavy metals from sewage sludge e bioreactors comparison. *Environmental Pollution*, 66, 237–252.

Couillard, D., Mercier, G. 1991. Optimum residence time (in CSTR and airlift reactor) for bacterial leaching of metals from anaerobic sewage sludge. *Water Research*, 25(2), 211–218.

Eliche-Quesada, D., Martínez-García, C., Martínez-Cartas, M.L., Cotes-Palomino, M.T., Pérez-Villarejo, L., Cruz-Pérez, N., Corpas-Iglesias, F.A. 2011. The use of different forms of waste in the manufacture of ceramic bricks. *Applied Clay Science*, 52(3), 270–276. https://doi.org/10.1016/j.clay.2011.03.003

European Commission. 2011. Communication from the Commission to the European Parliament, the Council, the European Economic and Social Committee and the Committee of the Regions. Roadmap to a Resource Efficient Europe. Brussels, 20.9.2011 COM, vol. 2011, 571 final. https://www.europarl.europa.eu/meetdocs/2009_2014/documents/com/com_com(2011)0571_/com_com(2011)0571_en.pdf.

Fukuta, T., Ito, T., Sawada, K., Kojima, Y., Bernardo, E.C., Matsuda, H. 2004. Separation of nickel from plating solution by sulfuration treatment. *Asia-Pacific Journal of Chemical Engineering*, 4, 24–31.

Gao, J., Luo, Q., Zhu, J., Zhang, C., Li, B. 2013. Effects of electrokinetic treatment of contaminated sludge on migration and transformation of Cd, Ni and Zn in various bonding states. *Chemosphere*, 93(11), 2869–2876.

Geng, H., Xu, Y., Zheng, L., Gong, H., Dai, L., Dai, X. 2020. An overview of removing heavy metals from sewage sludge: achievements and perspectives. *Environmental Pollution*, 266, 115375. https://doi.org/10.1016/j.envpol.2020.115375

Ghorpade, A., Ahammed, M.M. 2017. Water treatment sludge for removal of heavy metals from electroplating wastewater. *Environmental Engineering Research*, 23, 92–98.

Gu, T., Rastegar, S.O., Mousavi, S.M., Li, M., Zhou, M. 2018. Advances in bioleaching for recovery of metals and bioremediation of fuel ash and sewage sludge. *Bioresource Technology*, 261, 428–440.

Gu, W., Bai, J., Lu, L., Zhuang, X., Zhao, J., Yuan, W., Zhang, C., Wang, J. 2019. Improved bioleaching efficiency of metals from waste printed circuit boards by mechanical activation. *Waste Management*, 98, 21–28. https://doi.org/10.1016/j.wasman.2019.08.013.

Gu, X.Y., Wong, J.W.C., Tyagi, R.D. 2017. Bioleaching of heavy metals from sewage sludge for land application. *Current Developments in Biotechnology and Bioengineering: Solid Waste Management*, 90, 241–265.

Huang, Q.Y., Wang, Q.W., Liu, X.M., et al. 2022. Effective separation and recovery of Zn, Cu, and Cr from electroplating sludge based on differential phase transformation induced by chlorinating roasting. *Science of the Total Environment*, 820, 153260.

Hubau, A., Guezennec, A.-G., Joulian, C., Falagán, C., Dew, D., Hudson-Edwards, K. A. 2020. Bioleaching to reprocess sulfidic polymetallic primary mining residues: determination of metal leaching mechanisms. *Hydrometallurgy*, 197, 105484. https://doi.org/10.1016/j.hydromet.2020.105484

Janson, C.E., Kenson, R.E., Tucker, L.H. 1982. Treatment of heavy metals in wastewaters. What wastewater-treatment method is most cost-effective for electroplating and finishing operations? Here are the alternatives. *Environmental Progress*, 1, 212–216.

Judy, J.W., Muller, R.S., Zappe, H.H. 1995. Magnetic microactuation of polysilicon flexure structures. *Journal Micro Electro Mechanical System*, 4, 162–169.

Kaliyaraj, D., Rajendran, M., Angamuthu, V., Antony, A.R., Kaari, M., Thangavel, S., Venugopal, G., Joseph, J., Manikkam, R. 2019. Bioleaching of heavy metals from printed circuit board (PCB) by streptomyces albidoflavus tn10 isolated from insect nest. *Bioresources and Bioprocessing*, 6(1), 3. https://doi.org/10.1186/s40643-019-0283-3.

Latorre, M., Cortés, M.P., Travisany, D., Di Genova, A., Budinich, M., Reyes-Jara, A., Hödar, C., González, M., Parada, P., Bobadilla-Fazzini, R. A., Cambiazo, V., Maass, A. 2016. The bioleaching potential of a bacterial consortium. *Bioresource Technology*, 218, 659–666. https://doi.org/10.1016/j.biortech.2016.07.012

Lee, J.-C., Pandey, B. D. 2012. Bio-processing of solid wastes and secondary resources for metal extraction – a review. *Waste Management*, 32(1), 3–18. https://doi.org/10.1016/j.wasman.2011.08.010

Li, D., Shan, R., Jiang, L., Gu, J., Zhang, Y., Yuan, H., Chen, Y. 2022. A review on the migration and transformation of heavy metals in the process of sludge pyrolysis. In *Resources, Conservation and Recycling* (Vol. 185). Elsevier, Amsterdam, The Netherlands. https://doi.org/10.1016/j.resconrec.2022.106452

Liu, F., Zhou, L., Zhou, J., Song, X., Wang, D. 2012. Improvement of sludge dewnaterability and removal of sludge-borne metals by bioleaching at optimum pH. *Journal of Hazardous Materials*, 221–222, 170–177.

Liu, L., Huang, L., Huang, R., Lin, H., Wang, D. 2021. Immobilization of heavy metals in biochar derived from co-pyrolysis of sewage sludge and calcium sulfate. *Journal of Hazardous Materials*, 403, 123648. https://doi.org/10.1016/j.jhazmat.2020.123648

Liu, R., Mao, Z., Liu, W., Wang, Y., Cheng, H., Zhou, H., Zhao, K. 2020. Selective removal of cobalt and copper from Fe (III)-enriched high-pressure acid leach residue using the hybrid bioleaching technique. *Journal of Hazardous Materials*, 384, 121462. https://doi.org/10.1016/j.jhazmat.2019.121462

Liu, T., Zhou, H., Zhong, G., Yan, X., Su, X., Lin, Z. 2021. Synthesis of NiFeAl LDHs from electroplating sludge and their excellent supercapacitor performance. *Journal of Hazardous Materials*, 404, 124113.

Mahmoud, A., Cézac, P., Hoadley, A.F.A., Contamine, F., D'Hugues, P. 2017. A review of sulfide minerals microbially assisted leaching in stirred tank reactors. *International Biodeterioration & Biodegradation*, 119, 118–146. https://doi.org/10.1016/j.ibiod.2016.09.015.

Marchenko, O., Demchenko, V., Pshinko, G. 2018. Bioleaching of heavy metals from sewage sludge with recirculation of the liquid phase: a mass balance model. *Chemical Engineering Journal*, 350, 429–435.

Mishra, D., Rhee, Y.H. 2014. Microbial leaching of metals from solid industrial wastes. *Journal of Microbiology*, 52, 1–7.

Mizushima, I., Tang, P.T., Hansen, H.N., Somers, M.A. 2005. Development of a new electroplating process for Ni–W alloy deposits. *Electrochimica Acta*, 51, 888–896.

Nikfar, S., Parsa, A., Bahaloo-Horeh, N., Mousavi, S.M. 2020. Enhanced bioleaching of Cr and Ni from a chromium-rich electroplating sludge using the filtrated culture of Aspergillus niger. *Journal of Cleaner Production*, 264, 121622. https://doi.org/10.1016/j.jclepro.2020.121622.

Pathak, A., Dastidar, M.G., Sreekrishnan, T.R. 2009. Bioleaching of heavy metals from sewage sludge: a review. *Journal of Environmental Management*, 90(8), 2343–2353.

Peng, G., Tian G. 2010. Using electrode electrolytes to enhance electrokinetic removal of heavy metals from electroplating sludge. *Chemical Engineering Journal*, 165, 388–94.

Peng, G.L., Deng, S.B., Liu, F.L., et al. 2020. Calcined electroplating sludge as a novel bifunctional material for removing Ni(II)-citrate in electroplating wastewater. *Journal of Cleaner Production*, 262, 121416.

Pinto, F.M., Pereira, R.A., Souza, T.M., Saczk, A.A., Magriotis, Z.M. 2020. Treatment, reuse, leaching characteristics and genotoxicity evaluation of electroplating sludge. *Journal of Environmental Management*, 280, 111706.

Prabhu, S.V., Baskar, R. 2015. Kinetics of heavy metal biosolubilization from electroplating sludge: effects of sulfur concentration. *Journal of the Korean Society for Applied Biological Chemistry*, 58, 185–194.

Rastegar, S.O., Mousavi, S.M., Shojaosadati, S.A. 2014. Cr and Ni recovery during bioleaching of dewatered metal-plating sludge using *Acidithiobacillus ferrooxidans*. *Bioresource Technology*, 167, 61–68.

Ruamsap, N., Akaracharanya, A., Dahl, C. 2003. Pyritic sulfur removal from lignite by Thiobacilus *ferrooxidans*: strain improvement. *Journal of Scientific Research, Chulalongkorn University*, 28(1), 45–55.

Schippers, A. 2007. Microorganisms involved in bioleaching and nucleic acid-based molecular methods for their identification and quantification. In *Microbial Processing of Metal Sulfides*; Donati, E. R., Sand, W., Eds; Springer: Dordrecht, The Netherlands, pp. 3–33.

Schlesinger, M., Paunovic, M. (Eds.). 2011. *Modern Electroplating*. John Wiley & Sons, New York.

Singh, V., Singh, N., Rai, S.N., Kumar, A., Singh, A.K., Singh, M.P., Sahoo, A., Shekhar, S., Vamanu, E., Mishra, V. 2023. Heavy metal contamination in the aquatic ecosystem: toxicity and its remediation using eco-friendly approaches. *Toxics*, 11, 147. https://doi.org/10.3390/toxics11020147.

Sreedevi, P.R., Suresh, K., Jiang, G. 2022. Bacterial bioremediation of heavy metals in wastewater: a review of processes and applications. *Journal of Water Process Engineering*, 48, 102884. https://doi.org/10.1016/j.jwpe.2022.102884

Sreekrishnan, T.R., Tyagi, R.D. 1996. A comparative study of the cost of leaching out heavy metals from sewage sludges. *Process Biochemistry*, 31, 31–41.

Tyagi, R.D., Coullard, D., Fran, F.T. 1988. Heavy metal removal from anaerobically digested sludge by chemical and microbiological methods. *Environmental Pollution*, 50, 295–316.

Varia, J.C., Snellings, R., Hennebel, T. 2021. Sustainable metal recovery from secondary resources: screening and kinetic studies using analogue heterotrophic metabolites. *Waste Biomass Valorization*, 12, 2703–2721.

Villar, L.D., Garcia, O.J. 2003. Assessment of anaerobic sewage sludge quality for agricultural application after metal bioleaching. *Environmental Technology*, 24(12), 553–559.

Wang, H., Liu, X., Zhang, Z. 2024. Approaches for electroplating sludge treatment and disposal technology: reduction, pretreatment and reuse. *Journal of Environmental Management*, 349, 119535. https://doi.org/10.1016/j.jenvman.2023.119535

Wang, H-Y., Li, Y., Jiao, S-Q., Chou, K-C., Zhang, G-H. 2023. Recovery of Ni matte from Ni-bearing electroplating sludge. *Journal of Environmental Management*, 26, 116744.

Williamson, A. J., Folens, K., Matthijs, S., Paz Cortes, Y., Varia, J., Du Laing, G., Boon, N., Hennebel, T. 2021. Selective metal extraction by biologically produced siderophores during bioleaching from low-grade primary and secondary mineral resources. *Minerals Engineering*, 163, 106774. https://doi.org/10.1016/j.mineng.2021.106774.

Wong, J.W.C., Zhou, J., Kurade, M.B., Murugesan, K. 2015. Influence of ferrous ions on extracellular polymeric substances content and sludge dewaterability during bioleaching. *Bioresource Technology*, 179, 78–83.

Xin, B., Zhang, D., Zhang, X., Xia, Y., Wu, F., Chen, S., Li, L. 2009. Bioleaching mechanism of Co and Li from spent lithium-ion battery by the mixed culture of acidophilic sulfur-oxidizing and iron-oxidizing bacteria. *Bioresource Technology*, 100(24), 6163–6169. https://doi.org/10.1016/j.biortech.2009.06.086

Xu, Y., Lu, Y., Dai, X., Dong, B. 2017. The influence of organic-binding metals on the biogas conversion of sewage sludge. *Water Research*, 126, 329–341.

Yang, G., Zhang, G.M., Wang, H.C. 2015. Current state of sludge production, management, treatment and disposal in China. *Water Research*, 78, 60–73.

Yang, Y., Chu, H., Qian, C., Jia, C., Qi, S., Xin, B. 2020. Application of acidophilic microorganism in the metal enrichment of electroplating sludge use the membrane bioreactor. *Research Square*, Preprint version 1, https://doi.org/10.21203/rs.3.rs-54149/v1

Yang, Y., Liu, X., Wang, J., Huang, Q., Xin, Y., Xin, B. 2016. Screening bioleaching systems and operational conditions for optimal Ni recovery from dry electroplating sludge and exploration of the leaching mechanisms involved. *Geomicrobiology Journal*, 33(3–4), 179–184. https://doi.org/10.1080/01490451.2015.1068888

Yarzábal, A, Appia-Ayme, C., Ratouchniak, J. 2004. Regulation of the expression of the *Acidithiobacillus ferrooxidans* rus operon encoding two cytochromes c, a cytochrome oxidase and rusticyanin. *Microbiology*, 150(7), 2113–2123.

Zhang, J.X.J., Hoshino, K. 2019. Fundamentals of nano/microfabrication and scale effect. In *Molecular Sensors and Nanodevices* (pp. 43–111). Elsevier, Amsterdam, The Netherlands. https://doi.org/10.1016/b978-0-12-814862-4.00002-8

Zhang, L., Wu, B., Gan, Y., Chen, Z., Zhang, S. 2020. Sludge reduction and cost saving in removal of Cu(II)-EDTA from electroplating wastewater by introducing a low dose of acetylacetone into the Fe(III)/UV/NaOH process. *Journal of Hazardous Materials*, 382, 121107. https://doi. org/10.1016/j.jhazmat.2019.121107.

Zhang, L., Zhou, W., Liu, Y., Jia, H., Zhou, J., Wei, P., Zhou, H. 2020. Bioleaching of dewatered electroplating sludge for the extraction of base metals using an adapted microbial consortium: process optimization and kinetics. *Hydrometallurgy*, 191, 105227.

Zhang, P., Zhu, Y., Zhang, G., Zou, S., Zeng, G., Wu, Z. 2009. Sewage sludge bioleaching by indigenous sulfur-oxidizing bacteria: effects of ratio of substrate dosage to solid content. *Bioresource Technology*, 100(3), 1394–1398. https://doi.org/10.1016/j.biortech.2008.09.006.

Zhang, R., Neu, T.R., Blanchard, V., Vera, M., Sand, W. 2019. Biofilm dynamics and EPS production of a thermoacidophilic bioleaching archaeon. *New Biotechnology*, 51, 21–30.

Zhang, L., Zhou, W., Liu, Y., Jia H., Zhou, J., Wei P., Zhou, H., 2020. Bioleaching of dewatered electroplating sludge for the extraction of base metals using an adapted microbial consortium: process optimization and kinetics. *Hydrometallurgy*, 91, 105227

Zheng, J., Lv, J., Liu, W., Dai, Z., Liao, H., Deng, H., Lin, Z. 2020. Selective recovery of Cr from electroplating nanosludgevia crystal modification and dilute acid leaching. *Environmental Science: Nano*, 7, 1593–1601.

Zhou, J., Zheng, G., Wong, J.W.C., Zhou, L. 2013. Degradation of inhibitory substances in sludge by Galactomyces sp. Z3 and the role of its extracellular polymeric substances in improving bioleaching. *Bioresource Technology*, 132, 217–223.

Zhou, Q., Gao, J., Li, Y., Zhu, S., He, L., Nie, W., Zhang, R. 2017. Bioleaching in batch tests for improving sludge dewaterability and metal removal using *Acidithiobacillus ferrooxidans* and *Acidithiobacillus thiooxidans* after cold acclimation. *Water Science and Technology*, 76(6), 1347–1359.

Zhou, W., Zhang, L., Peng, J., Ge, Y., Tian, Z., Sun, J., Cheng, H., Zhou, H. 2019. Cleaner utilization of electroplating sludge by bioleaching with a moderately thermophilic consortium: a pilot study. *Chemosphere*, 232, 345–355.

12 Indium Recovery from Mining Discards Using Bioleaching

Sonai Dutta and Abhijit Bandyopadhyay

12.1 INTRODUCTION

The German scientists Ferdinand Reich and Hieronymus Theodor Richter made the unusual silvery-white metal indium in 1863 while examining samples of zinc ore [1]. It has a low melting point and is pliable and squishy. It is an element that is part of the periodic table's group 13, which has drawn attention from all over the world due to its special chemical properties. Estimates of indium's abundance in the earth's crust range from 0.05 to 0.24 ppm. Because of its dispersion throughout the crust, indium is difficult to generate as a primary product [2].

The extremely uncommon element indium is typically not included in geochemical datasets. It is regarded as a crucial metal and finds extensive use in the electronics sector, particularly in the manufacturing of liquid crystal displays (LCDs) and solar panels. Technology companies find it attractive due to its electrical and optical qualities. Indium also has a high boiling point of 2,353 K, a low melting point of 430 K, and a superconducting temperature of 3.37 K [3]. ITO is utilized in the creation of flat panel screens, which accounts for over 70% of indium production. Indium production accounts for 25% of the use of LED lights [4]. Technology companies find it attractive due to its electrical and optical qualities. Indium also has a high boiling point of 2,353 K, a low melting point of 430 K, and a superconducting temperature of 3.37 K. ITO is utilized in the creation of flat panel screens, which accounts for over 70% of indium production. Indium production accounts for 25% of the use of LED lights [5].

Indium is also becoming more and more necessary for the production of photovoltaic (PV) systems, which are used to create solar panels. Zinc, bauxite, tin, and silver ores are processed to produce indium [6]. China is the largest producer of primary indium, followed by South America, Canada, South Korea, and Japan. Japan consumes 60% of the world's indium supply, making it by far the greatest consumer of indium [7,8]. Given the shortage of indium, it seems imperative to concentrate on recycling as a supplementary supply of indium [9]. Spent ITO (90% In_2O_3, 10% SnO_2, with less than 20 parts per million) appears to be the most appealing material. It is applied as thin conductive coatings (approximately 150 nm thick) over LCD glass [10].

Currently, indium is the most commonly used in the production of indium tin oxide (ITO), which is 90% indium oxide and 10% tin oxide. 70% of the world's indium usage is accounted for by ITO [11]. Given the shortage of indium, it seems

DOI: 10.1201/9781003415541-12

imperative to concentrate on recycling as a secondary supply of indium. ITO is applied as thin conductive sheets, varying in thickness from 150 to 200 nm (manufacturer-dependent), onto LCD glass in order to use it in LCDs. Silicon, aluminum, and calcium—which are present in the panel's glass as SiO_2, Al_2O_3, and CaO, respectively—are the elements that are most frequently found in LCD panels [7,12]. LCDs can have a variety of other metals in addition to indium, including Zn, Fe, and Cu. However, indium is the most sought-after metal in terms of recovery because it can be found in greater quantities in LCD waste than other minerals [13]. While the amount of indium in LCD panels ranges from 100 to 300 mg In/kg glass, the concentration of indium in sphalerite and chalcopyrite is between 10 and 20 mg/kg.

According to the United States Geological Survey (USGS) bureau, due to indium's geochemical characteristics, base metals including copper, tin, lead, and zinc are the most common combinations with it [14]. It happens with silver, cadmium, and bismuth. Indium is mostly generated as a byproduct of processing zinc concentrates, primarily for financial reasons. Indium is mostly employed in the electronics sector as indium tin oxide (ITO), which is used in the production of solar panels, LCDs, and flat panel displays (FPDs) due to its physicochemical characteristics.

Countries such as the United States, Japan, Australia, and Canada support initiatives to secure material supply through diversification, development, and recycling. The agricultural, industrial, and domestic sectors are economically interested in recovering valuable minerals, processing them, and converting them into marketable products [15]. Metal recovery through recycling is the process of converting trash back into new metal goods in order to control energy use, preserve natural resources, and lower greenhouse gas emissions. Recycling metal from household solid trash is essential to sustainable waste management in Europe [16]. Sorting, separation procedures, and garbage collection systems are some of the ways that nations like the United States, Canada, and the United Kingdom support the recycling sector. Achieving a circular economy and sustainable growth requires effective waste management. Governments are placing additional pressure on businesses to develop more sustainably and ecologically friendly products by tightening regulations on trash generation and creating new landfills.

The recycling market, which was valued at 217.0 billion dollars in 2020 and is expected to reach 368.7 billion dollars by 2030, is expected to contribute to the global generation of waste, including electronic waste, which was valued at 42 million tons in 2014, 53.6 million tons in 2019, and an estimated 74.7 million tons by 2030 [17]. As a result, as the market for FPDs grew in the 1990s, so did the demand for indium metal; however, supply did not keep up with demand. In 2019, recoverable material profit from e-waste was projected to be USD 57 billion. In the same year, 3.6% more steel scrap was used by the nation and region, while 70.9% of steel cans and other steel packaging, such as strapping and drums, were recycled [18]. In the process of burning municipal solid garbage, mixed waste is recovered to extract certain metals. As a result of the steel scrap's lower quality and significant oxidation and melting of the aluminum, the recoverability is constrained. Furthermore, organic elements, carbon, and other materials, such as graphite anodes, are lost during the smelting furnace's burning process and cannot be recovered when pyrometallurgical techniques are employed to recover metal.

The use of hazardous chemicals in the presence of extremely high temperatures is a fundamental component of conventional mineral extraction techniques, which frequently result in environmental damage, disease, and fatalities [19]. Therefore, efforts are being made to develop safer and greener techniques for mineral extraction that are more in line with the world's growing trend toward environmentally friendly solutions. While "bioleaching," a branch of biotechnology, relies entirely on natural biological processes for its effectiveness, it has emerged as a potentially useful solution [20]. The microbes that are naturally affixed to the mineral ores can be activated with minimal intervention. It does not require an external fuel source and does not emit any harmful byproducts. Bacterial leaching, or bioleaching, is the collective term for two related microbial processes called bio-oxidation and bioleaching. In general, leaching refers to the solubilization of one or more complex solid constituents by interaction with a liquid. The solubilization in bacterial leaching is mediated by bacteria.

Leaching is the most used technique for extracting metals from e-waste at various concentrations, per research by Inman et al. [21]. Organic acids such as oxalic acid, formic acid, citric acid, tartaric acid, maleic acid, ascorbic acid, and malic acid are used in leaching, along with inorganic acids such as hydrochloric acid, nitric acid, sulfuric acid, and phosphoric acid. Although inorganic acids are less expensive and have a high leaching efficiency, they can also cause secondary contamination and equipment degradation. Organic acids are more expensive than inorganic acids, which limits their widespread industrial use, although they can leach with the same effectiveness in a gentler environment [22].

Hydrometallurgy offers benefits for material recycling, including increased productivity, superior metal selectivity, lower energy consumption, and a decrease in harmful petrol emissions [23]. However, because of its multiple stages and associated costs, it also presents challenges, including wastewater production and increased costs. These challenges serve as a strong incentive to research more sustainable formulations.

Bioleaching, for instance, is regarded as one of the green technologies for metal recovery, offering low cost in terms of installation and operation, low energy consumption, no toxic waste generation, and low capital investment [24]. In contrast, biohydrometallurgy adds value to these processes by recovering quantities of metals using aqueous solutions and biological metabolites produced by certain microorganisms.

During the bioleaching process, metals are obtained from insoluble solid substrates by means of direct microbial metabolism or indirectly through the products of microorganism metabolism, which include the synthesis of organic acids, chelating agents, amino acids, and complexing agents from heterotrophic bacteria or fungi [25]. In many industrial sectors, bioleaching has drawn interest, particularly in mineral and solid industrial waste materials (such as fly ash, sewage sludge, galvanic sludge, fly ash, electronic waste, spent petrochemical catalysts, medical waste, spent batteries, and residual slag). These materials may have low metal concentrations, contain elements that could harm smelters, or be treated biologically due to environmental concerns. Along with producing concentrated solutions of metal salts that might be recycled, this procedure also makes it possible to recover metals from low-grade sulfide ores and concentrates that cannot be profitably processed using traditional methods [26]. According to existing research, some metals ions, including copper, iron, lead, aluminum, manganese, zinc, nickel, and cadmium from minerals and electronic waste, could be bioleached using microorganisms such as *Acidithiobacillus ferrooxidans*, *Acidithiobacillus thiooxidans*, silicate bacteria, and *Aspergillus niger* [27].

12.2 RECOVERY OF INDIUM FROM SOLID MATRIX

Waste materials originating from various sources, including industrial, municipal, and agricultural operations, fall under the wide category of solid matrices. These solid matrices usually contain high quantities of valuable metals, but it might be challenging to recover the metals from them due to their complex composition and low metal concentrations. One of the increasingly popular approaches for recovering metals is bioprocessing waste materials. The biological recovery of metals of interest involves three steps: the solute is extracted from the solvent using bioleaching in the first stage; concentration is achieved through the use of biosorption/desorption in the second stage; and the metal is deposited as a species in concentrated solution as a precipitated solid phase in the third stage. The first stage is to remove the precious metal indium from primary ores and other solid matrices. We call this procedure "bioleaching." Solid matrices with metal content (SMMC) are the basic materials used in this bioprocess, which is headed toward a circular economy. These comprise agricultural trash, municipal solid waste, low-grade ore from mining tailings, and consumer garbage like electronic junk and used batteries. SMMC is a more environmentally friendly and energy-efficient alternative to conventional metallurgical processes. Employing the biosorption/desorption process, the metals in the solution are concentrated in the second step. The work of Ramírez et al. shows how packed bed columns containing microorganisms immobilized in porous matrices are used to accomplish biosorption in continuous systems. It is recommended to use leftover yeast or fungus from the industry, especially from the brewing sector, due to their large generation volume, ease of procurement, and inexpensive cost. The goal of the desorption phase that follows the adsorption phase is to recover and separate the metals that were contained in the packed bed column, creating a concentrated solution of those metals. The desorption process makes use of a variety of eluting agents or extracting solutions, such as chelating agents (EDTA), acidic solutions (HCl, H_2SO_4, HNO_3), inorganic and organic salts ($NaNO_3$, $Ca(NO_3)_2$, sodium citrate), and others. When the active sites of the immobilized microorganisms in the column are protonated, the bacteria get reactivated, hence increasing the efficacy of the subsequent adsorption process and enabling the reuse of the packed column. Using common chemical processes including electrodeposition, precipitation, and reduction, the metal is deposited as a stable solid phase and as a concentrated solution species in the third phase. These methods lower the redox potential by using reducing chemicals such as $SnCl_2$, $FeSO_4$, and SO_2. Recently, there has been research on organic reducing agents as these are "greener" than inorganic reductants because they release fewer hazardous gases and are less toxic. These agents can be present in a variety of biomasses and include thiourea, glucose, sucrose, lactose, and even cellulose, hemicellulose, and lignin [28].

12.3 SOURCE OF THE METAL

Since metal availability varies depending on specific geological, physical, and industrial conditions, no single supply statistic can be applied to all metals. Waste generation creates an increasing store of renewable resources that can replace decreasing metals from enriched reserves in the geosphere. Waste is defined as something that is thrown away after being used only once. There are three types of solid waste: municipal, industrial, and agricultural waste. Solid trash is the most common type of garbage. Over 12 billion tons of solid waste are produced annually worldwide [29], with over 2 billion tons coming from cities and 1.55 billion tons from agriculture. Industrial garbage is created at a rate more than 18 times higher than that of municipal solid trash. The aforementioned wastes are good candidates to be used as matrices in the bioleaching process, which recovers metals by dissolution, because they contain metals in a variety of concentrations [30–33].

Industrial waste is any material rendered useless during manufacturing, including those from mines, factories, industries, and mills. Industrial waste includes, among other things, electronic waste, dirt, and gravel, as well as masonry and concrete. American industrial enterprises produce and dispose of about 7.6 billion tons of industrial solid waste annually. Moreover, it is projected that the annual production of e-waste, which includes electronic devices like phones, TVs, and laptops, will be 54 million tons by 2030 and will increase to 75 million tons. It was stated that in 2019, just 17% of e-waste was appropriately gathered and recycled [34]. Handling e-waste and its components improperly can have detrimental effects on development and health, especially for young children. Metallurgical slag is among the most common byproducts generated by these industries. The purpose of producing this waste is to recover iron. On the other hand, although titanium is concentrated in the blast furnace titanium slag as vanadium titanomagnetite, which comprises SiO_2, Al_2O_3, CaO, and MgO, metals like Ba and Cd are found in the solid residue left behind after the purification of silica sand. Vanadium products are manufactured using fly ash, coal, petroleum coke, vanadium slag, and leftover catalysts as raw materials. One chemical method for recovering important metals is leaching. Using a potentially soluble component, a solid solute together with extra undesirable solids are leaching processes typically consist of three parts:

A: solid solute that goes into the solution;
B: inert solid (insoluble in S);
S: extracting solvent. Extracted in all three processes—leaching, peroration, and solid/liquid extraction.

12.4 BIOLEACHING

Microorganisms, be it fungi or bacteria, participate in the biogeochemical cycle of minerals and hence play the role of the solvent in bioleaching, either directly through their metabolism or indirectly through the byproducts of their metabolism [35]. Stated differently, the technique of extracting metals from insoluble solid substrates is known as bioleaching. By using the mineral as a substrate, the bacteria are able to obtain electrons for their metabolic activities and emit heat and metals on their own, all without the need for outside energy.

Organic acids produced by fungi can be used to remove metals from solid matrices. Microorganisms can also release ligands that stabilize the metal by building complexes rich in metal. Metals can be made more soluble by the use of cyanide, thiosulfate, and amino acids produced by biological processes. Additionally, microorganisms can participate in the redox cycling of iodine, which is a potential alternative leachate for obtaining the metal and reducing metal solubility, by consuming ligands bound to the metal or by biosorption, enzymatic reduction, precipitation, and using the metal as a micronutrient. Depending on the objectives of the bioprocess, either fungus or bacteria can be used in SMMC metal bioleaching processes in solid matrices containing metal. Bacterial bioleaching is used when recovering a desired metal is the objective and preserving the properties of the solid matrix is not. On the other hand, if the solid matrix is a mineral (SMMC), fungal bioleaching is used when it is essential to preserve the properties of the solid matrix, especially its crystalline features. Furthermore, while bacterial bioleaching often involves only one phase and minimizes exposure to external contaminants, it is not necessary to sterilize and sanitize, but fungal bioleaching does require these steps due to the former's resistance to contamination. Fungal bioleaching, on the other hand, is carried out in two steps utilizing either a direct or indirect method. Bacterial bioleaching requires more time than the other method to finish because the bacteria are actively participating in the process. Conversely, the length of time that fungal bioleaching varies according to the SMMC/bacteria system and the method used. Fungal bioleaching is an indirect technology that requires more time to manufacture the leaching solvents (fermented broth), but the bioleaching process itself proceeds more quickly.

12.4.1 Bacterial Bioleaching

Bacterial leaching, also referred to as bioleaching, biohydro-metallurgy, or bio-oxidation of sulfides, is the term used to describe the natural dissolution process that occurs when a group of bacteria, primarily from the genus *Thiobacillus*, oxidize sulfide SMMC and release the metals that are contained in it. By means of direct oxidation, indirect chemical oxidation brought on by corrosive metabolic byproducts produced by electrochemistry, or a combination of the two, microorganisms can convert solid compounds into soluble and extractable elements or expose metals contained in ores and concentrates [36]. Different processes, which rely on the sulfur matrix of the matrix, carry out the assault and solubilization of an SMMC by microbes. Similar to this, microorganisms play a catalytic role in the bioleaching process when certain SMMC dissolves. For instance, bacteria are known to biocatalyze the leaching processes of metal compounds by converting them into forms that are soluble in water [37]. Without requiring energy from outside sources, the bacterium consumes SMMC as fuel, transfers electrons for living, and releases metals. The fact that the reactions in this sort of process occur at low pressure and, in some cases, at low temperature, depending on the type of microbe, is evidence that large activation energies are not required, whereas other methods require harsh conditions for their growth and operation. Bacterial bioleaching, often known as leaching, is defined by Rodríguez et al. [38] as the assault and solubilization of an SMMC by the direct or indirect action of several microbes. Two types of microorganisms have evolved to withstand bioleaching processes [39].

Autotrophic: The inorganic substance around them provides these microbes with the nutrients and energy they need for their life cycles.

Heterotrophic: In order to finish their life cycles, they need the presence of organic substances.

Certain types of bacteria function best in the presence of oxygen within both groupings or categories. These aerobic bacteria have the ability to convert metal sulfides into soluble sulfates and are responsible for initiating the oxidation processes [39]. Anaerobic bacteria are similar in that they can survive and complete their life cycles without oxygen; these batteries first go through reduction processes. The ability of the microorganisms utilized in metal recovery to adapt to extremely harsh or high temperatures, pH levels, and living circumstances is their primary trait, regardless of the type of bacteria involved. A portion of the bacteria in question are acidophilic microbes, which means they can endure high temperatures, low pH values, and high concentrations of metallic elements. They obtain their energy from the reduction of sulfur compounds and the oxidation of Fe^{2+} to Fe^{3+}. The microbial oxidation of sulfide (SMMC) is the most important field within the biotechnological process of metals. Aqueous conditions associated with SMMC discharges are characterized by low pH, high amounts of metals, and occasionally high temperatures. Nonetheless, in these settings and with these circumstances, certain bacteria may survive, grow, and procreate. In these conditions, the main energy source for microbes is reduced sulfur species and certain metals in solution, which causes precious metals to become soluble. Rodriguez et al. [38] state that the primary function of microorganisms in recovery via bioleaching is either indirect through regeneration of the leaching agent Fe^{3+} of SMMC or direct oxidation through an enzymatic assault.

In certain instances, the bacteria aid in the weathering of the gangue by freeing the precious mineral and making it easier for it to be attacked later, because the bacteria's sulfuric acid causes the Si-O and Al-O bonds in the gangue's alumino-silicates to break.

Through the catalysis of redox processes, microorganisms may change a wide variety of metals that exist in various valence states. The majority of bacterial strains utilized in industrial and laboratory bioleaching applications are from the genus *Acidithiobacillus*. Both the ferrous ion and reduced sulfur compounds can be obtained as the oxidizing energy. A growing body of research is being done on the use of acidophilic microorganisms in the bioleaching of different kinds of metal-containing wastes (printed circuit boards, Ni-Cd batteries, Li-ion batteries, spent refinery catalysts), with an impressive bioleaching efficiency of metals (>90%), including the valuable extraction of indium from LCDs. In the past, indium was recovered using both pure and mixed cultures of *Acidithiobacillus thiooxidans* and *Acidithiobacillus ferrooxidans*. These organisms create sulfuric acid by the chemical oxidation of elemental sulfur or iron (II). Compared to chemical leaching, indium bioleaching has demonstrated a strong potential for recovering metal from discarded LCD. With adapted *A. thiooxidans* (LCD density 1.6% w/v) or, even faster, with adapted sulfur *Acidithiobacillus* (LCD density 1.5% w/v), indium was recovered from waste LCD at 100% in 15 days, or even less quickly in 6 days, with a 74% and

8% chemical leaching, respectively. Because of the toxicity of the LCD powder, the development of non-adapted *A. ferrooxidans* bacteria was inhibited in these settings, and the percentage of indium recovered did not surpass 10%. Although findings from different authors suggest that great efficiency was achieved in the presence of pure bacteria that acquire energy from the bio-oxidation of inorganic compounds containing decreased S and Fe^{2+}, adaptation is a crucial component in guaranteeing the efficacy of indium leaching. Which oxidizing agents and bacterial strains are most important in indium bioleaching is still up for debate.

Willner et al. looked into the effects of pure and mixed cultures of *A. ferrooxidans* and *A. thiooxidans*, as well as varying LCD panel pulp densities (1% and 2%), on bioleaching efficiency [3]. The pure strain and a combined culture of *A. ferrooxidans* and *A. thiooxidans* were used for biological leaching. The microbial oxidation of elemental sulfur to sulfuric acid and ferrous ions to ferric ions happens as follows throughout the bioleaching process:

$$S^0 + H_2O + 1.5O_2 \rightarrow H_2SO_4$$

$$2Fe^{2+} + 1/2O_2 + 2H^+ \xrightarrow{A.\ ferrooxidans} 2Fe^{3+} + H_2O$$

The pH drops more quickly in the presence of bacteria that promote the oxidation of S^0 to H_2SO_4 (reaction 1) than it does in the presence of *A. ferrooxidans*, particularly when the pulp density is lower (1%), as seen by the ultimate pH value of 1.7. The precipitation of iron (III) compounds in the form of jarosite (particularly in the pH range of 1.8–2.7), where the precipitation of jarosite is an acid-forming reaction, is the primary source of the acidification of the environment with *A. ferrooxidans*

$$3Fe^{3+} + X^+ + 2HSO_4^- + 6H_2O \rightarrow XFe_3(SO_4)_2(OH)_6 + 8H^+$$

where X is a $K+, Na+, NH_4+, or\ H3O+$.

At 1% pulp density, however, the Fe^{2+} bio-oxidation process proceeds a little more quickly. When the bacteria catalyze the bio-oxidation of Fe^{3+}, the ferrous ions present in the bacterial leaching solutions regenerate. In addition to NH^{4+} (ions in the 9 K medium), monovalent cations such as Na^+ and K^+ that are present in the thin-film transistor liquid crystal display (TFT-LCD) panels as Na_2O and K_2O help to form ferric hydroxyl salts, which were visible in the leaching system as a yellow-brown phase. The control samples' pH ranged from 2.1 to 2.4 at the same time. In the presence of *A. thiooxidans*, Xie et al. and Jowkar et al. obtained 100% leaching efficiency of indium after 8 and 15 days, respectively. Although it has been established that H^+ ions are crucial to the dissolving of indium, successful indium leaching requires bacterial engagement and cannot be achieved only by acid leaching, which is unable to completely remove the indium from LCD powder [40]. Therefore, 1% w/v pulp density is a suitable number for effective bioleaching of In from LCD material. Higher pulp densities impeded the kinetics and rates of metal extraction when employing the pure strain. Compared to a single pure culture, mixed cultures including bacteria that oxidize sulfur and iron were more productive. *A. ferrooxidans* and *A. thiooxidans* removed 84.7% of In, 97.3% of

Sn, 71% of In, and 66.9% of Sn, respectively, whereas mixed bacteria (1% w/v) removed 98.2% of Sn and 94.7% of In. Compared to what would be predicted with *A. thiooxidans*, Sn is extracted more quickly and completely when *A. ferrooxidans* and Fe^{3+} ions are present. As a result, the quick solubility of tin raises the prospect that, at the start of bioleaching, it might be successfully separated and recovered using stronger oxidizing agents, such iron oxidizing bacteria.

Cui et al. conducted experiments to recover indium from used LCD panels, the bioleaching process of *A. niger* fermentation and its optimization technique were studied. In the optimized fermentation system, the indium bioleaching efficiency increased from 12.3% to 100% when the starting pH, shaking speed, and sucrose addition were optimized from 7.0, 200 rpm, 100 g/L to 4.0, 125 rpm, and 50 g/L, respectively [41]. Iron oxide and indium oxide mostly battled for H^+ ions throughout the leaching process, and fermentation broth can further enhance the indium bioleaching effects. The study showed that fermentation by *A. niger* has great promise for bioleaching precious metals from electronic trash.

REFERENCES

1. Alfantazi, A., & Moskalyk, R. (2003). Processing of indium: a review. *Minerals Engineering*, 16(8), 687–694. https://doi.org/10.1016/s0892-6875(03)00168-7
2. Werner, T. T., Mudd, G. M., & Jowitt, S. M. (2017). The world's by-product and critical metal resources part III: a global assessment of indium. *Ore Geology Reviews*, 86, 939–956. https://doi.org/10.1016/j.oregeorev.2017.01.015
3. Willner, J., Fornalczyk, A., Gajda, B., & Saternus, M. (2018). Bioleaching of indium and tin from used LCD panels. *Physicochemical Problems of Mineral Processing*, 54(3), 639–645.
4. Da Silveira, A. V. M., Fuchs, M., Pinheiro, D. K., Tanabe, E. H., & Bertuol, D. A. (2015). Recovery of indium from LCD screens of discarded cell phones. *Waste Management*, 45, 334–342. https://doi.org/10.1016/j.wasman.2015.04.007
5. Ciacci, L., Werner, T. T., Vassura, I., & Passarini, F. (2018). Backlighting the European Indium recycling potentials. *Journal of Industrial Ecology*, 23(2), 426–437. https://doi.org/10.1111/jiec.12744
6. Fthenakis, V., Wang, W., & Kim, H. C. (2009). Life cycle inventory analysis of the production of metals used in photovoltaics. *Renewable & Sustainable Energy Reviews*, 13(3), 493–517. https://doi.org/10.1016/j.rser.2007.11.012
7. Lin, S., Mao, J., Chen, W., & Shi, L. (2019). Indium in mainland China: insights into use, trade, and efficiency from the substance flow analysis. *Resources, Conservation and Recycling*, 149, 312–321. https://doi.org/10.1016/j.resconrec.2019.05.028
8. O'Neill, B. (2010). "Indium Market Forces, A Commercial Perspective," *2010 35th IEEE Photovoltaic Specialists Conference*, Honolulu, HI, USA, pp. 000556–000559. https://doi.org/10.1109/PVSC.2010.5616842.
9. Rhodes, C. J. (2019). Endangered elements, critical raw materials and conflict minerals. *Science Progress*, 102(4), 304–350. https://doi.org/10.1177/0036850419884873
10. Wang, H. (2009). A study of the effects of LCD glass sand on the properties of concrete. *Waste Management*, 29(1), 335–341. https://doi.org/10.1016/j.wasman.2008.03.005
11. Senthilkumar, V., Vickraman, P., Jayachandran, M., & Sanjeeviraja, C. (2010). Structural and optical properties of indium tin oxide (ITO) thin films with different compositions prepared by electron beam evaporation. *Vacuum*, 84(6), 864–869. https://doi.org/10.1016/j.vacuum.2009.11.017

12. Sayehi, M., Tounsi, H., Garbarino, G., Riani, P., & Busca, G. (2020). Reutilization of silicon and aluminum-containing wastes in the perspective of the preparation of SiO_2-Al_2O_3 based porous materials for adsorbents and catalysts. *Waste Management*, 103, 146–158. https://doi.org/10.1016/j.wasman.2019.12.013
13. Zhang, K., Wu, Y., Wang, W., Li, B., Zhang, Y., & Zuo, T. (2015). Recycling indium from waste LCDs: a review. *Resources, Conservation and Recycling*, 104, 276–290. https://doi.org/10.1016/j.resconrec.2015.07.015
14. Jowitt, S. M., Mudd, G. M., Werner, T. T., Weng, Z., Barkoff, D., & McCaffrey, D. (2018). *The Critical Metals: An Overview and Opportunities and Concerns for the Future*. Society of Economic Geologists (SEG), McLean, VA. https://doi.org/10.5382/sp.21.02
15. Yadav, V. K., Yadav, K. K., Tirth, V., Gnanamoorthy, G., Gupta, N., Algahtani, A., Islam, S., Choudhary, N., Modi, S., & Jeon, B. (2021). Extraction of value-added minerals from various agricultural, industrial and domestic wastes. *Materials*, 14(21), 6333. https://doi.org/10.3390/ma14216333
16. Kuusiola, T., Wierink, M., & Heiskanen, K. (2012). Comparison of collection schemes of municipal solid waste metallic fraction: the impacts on global warming potential for the case of the Helsinki Metropolitan Area, Finland. *Sustainability*, 4(10), 2586–2610. https://doi.org/10.3390/su4102586
17. Van Yken, J., Boxall, N. J., Cheng, K. Y., Nikoloski, A. N., Moheimani, N. R., & Kaksonen, A. H. (2021). E-waste recycling and resource recovery: a review on technologies, barriers and enablers with a focus on Oceania. *Metals*, 11(8), 1313. https://doi.org/10.3390/met11081313
18. Nithya, R., Sivasankari, C., & Thirunavukkarasu, A. (2021). Electronic waste generation, regulation and metal recovery: a review. *Environmental Chemistry Letters*, 19, 1347–1368. https://doi.org/10.1007/s10311-020-01111-9
19. Kaya, M. (2016). Recovery of metals and nonmetals from electronic waste by physical and chemical recycling processes. *Waste Management*, 57, 64–90. https://doi.org/10.1016/j.wasman.2016.08.004
20. Vera, M., Schippers, A., & Sand, W. (2013). Progress in bioleaching: fundamentals and mechanisms of bacterial metal sulfide oxidation – part A. *Applied Microbiology and Biotechnology*, 97(17), 7529–7541. https://doi.org/10.1007/s00253-013-4954-2
21. Inman, G., Nlebedim, I. C., & Prodius, D. (2022). Application of ionic liquids for the recycling and recovery of technologically critical and valuable metals. *Energies*, 15(2), 628. https://doi.org/10.3390/en15020628
22. Liu, M., Ma, W., Zhang, X., Zhao, Q., & Zhao, Q. (2022). Recycling lithium and cobalt from LIBs using microwave-assisted deep eutectic solvent leaching technology at low-temperature. *Materials Chemistry and Physics*, 289, 126466. https://doi.org/10.1016/j.matchemphys.2022.126466
23. Guo, H., Min, Z., Hao, Y., Wang, X., Fan, J., Shi, P., Min, Y., & Xu, Q. (2021). Sustainable recycling of LiCoO2 cathode scrap on the basis of successive peroxymonosulfate activation and recovery of valuable metals. *Science of the Total Environment*, 759, 143478. https://doi.org/10.1016/j.scitotenv.2020.143478
24. Mahmoud, A., Cézac, P., Hoadley, A., Contamine, F., & D'Hugues, P. (2017). A review of sulfide minerals microbially assisted leaching in stirred tank reactors. *International Biodeterioration & Biodegradation*, 119, 118–146. https://doi.org/10.1016/j.ibiod.2016.09.015
25. Dusengemungu, L., Kasali, G., Gwanama, C., & Mubemba, B. (2021). Overview of fungal bioleaching of metals. *Environmental Advances*, 5, 100083. https://doi.org/10.1016/j.envadv.2021.100083
26. Rene, E.R., Sahinkaya, E., Lewis, A., & Lens, P.N.L. (2017). Sustainable heavy metal remediation. In *Environmental Chemistry for a Sustainable World*. Springer International Publishing, Cham. https://doi.org/10.1007/978-3-319-61146-4.

27. Gopikrishnan, V., Vignesh, A., Radhakrishnan, M., Joseph, J., Shanmugasundaram, T., Doble, M., & Balagurunathan, R. (2020). Microbial Leaching of Heavy Metals from E-Waste. Elsevier, Amsterdam, The Netherlands (pp. 189–216). https://doi.org/10.1016/b978-0-12-817951-2.00010-9
28. Benavente, Ó., Hernández, M. C., Melo, E., Němec, D., Quezada, V., & Zepeda, Y. (2019). Copper dissolution from black copper ore under oxidizing and reducing conditions. *Metals*, 9(7), 799. https://doi.org/10.3390/met9070799
29. Kuthiala, T., Thakur, K., Sharma, D., Singh, G., Khatri, M., & Arya, S. K. (2022). The eco-friendly approach of cocktail enzyme in agricultural waste treatment: a comprehensive review. *International Journal of Biological Macromolecules*, 209, 1956–1974. https://doi.org/10.1016/j.ijbiomac.2022.04.173
30. Medici, F. (2022). Recovery of waste materials: technological research and industrial scale-up. *Materials*, 15(2), 685. https://doi.org/10.3390/ma15020685
31. Ippolito, N. M., Medici, F., Pietrelli, L., & Piga, L. (2021). Effect of acid leaching pre-treatment on gold extraction from printed circuit boards of spent mobile phones. *Materials*, 14(2), 362. https://doi.org/10.3390/ma14020362
32. Sydow, M., Chrzanowski, Ł., Leclerc, A. S. C., Laurent, A., & Owsianiak, M. (2018). Terrestrial ecotoxic impacts stemming from emissions of Cd, Cu, Ni, Pb and Zn from manure: a spatially differentiated assessment in Europe. *Sustainability*, 10(11), 4094. https://doi.org/10.3390/su10114094
33. Mavakala, B. K., Sivalingam, P., Laffite, A., Mulaji, C. K., Giuliani, G., Mpiana, P. T., & Poté, J. (2022). Evaluation of heavy metal content and potential ecological risks in soil samples from wild solid waste dumpsites in developing country under tropical conditions. *Environmental Challenges*, 7, 100461.
34. Sajid, M., Syed, J. H., Iqbal, M., Abbas, Z., Hussain, I., & Baig, M. A. (2019). Assessing the generation, recycling and disposal practices of electronic/electrical-waste (E-waste) from major cities in Pakistan. *Waste Management*, 84, 394–401.
35. Sousa, R., Silva, A. F., Fiúza, A., Vila, M. C., & De Lurdes Dinis, M. (2018). Bromine leaching as an alternative method for gold dissolution. *Minerals Engineering*, 118, 16–23. https://doi.org/10.1016/j.mineng.2017.12.019
36. Potysz, A., & Kierczak, J. (2019). Prospective (bio)leaching of historical copper slags as an alternative to their disposal. *Minerals*, 9(9), 542. https://doi.org/10.3390/min9090542
37. Mikoda, B., Potysz, A., & Kmiecik, E. (2019). Bacterial leaching of critical metal values from polish copper metallurgical slags using Acidithiobacillus thiooxidans. *Journal of Environmental Management*, 236, 436–445. https://doi.org/10.1016/j.jenvman.2019.02.032
38. Rodríguez, Y., Ballester, A., Blázquez, M.L., González, F., & Muñoz, J.A. (2001). Mecanismo de biolixiviación de sulfuros metálicos. *Revista de Metalurgia*, 37, 665–672.
39. Rendón-Castrillón, L., Ramírez-Carmona, M., Ocampo-López, C., & Gómez-Arroyave, L. (2023). Bioleaching techniques for sustainable recovery of metals from solid matrices. *Sustainability*, 15(13), 10222. https://doi.org/10.3390/su151310222
40. Xie, Y., Wang, S., Tian, X., Che, L., Wu, X., & Zhao, F. (2019). Leaching of indium from end-of-life LCD panels via catalysis by synergistic microbial communities. *Science of the Total Environment*, 655, 781–786. https://doi.org/10.1016/j.scitotenv.2018.11.141
41. Cui, J., Zhu, N., Mao, F., Wu, P., & Dang, Z. (2021). Bioleaching of indium from waste LCD panels by Aspergillus niger: method optimization and mechanism analysis. *Science of the Total Environment*, 790, 148151. https://doi.org/10.1016/j.scitotenv.2021.148151

13 Chemical and Biological Extraction of Chromium from Industrial Wastes

Moumita Sharma, Souptik Bhattacharya, Sanjukta Banik, and Shaoli Das

13.1 INTRODUCTION

13.1.1 Overview of Industrial Heavy Metals

The sustainability of the food chain depends on the availability of water. In addition to lowering the quality of water, pollutants alter its physicochemical characteristics (Briggs, 2003). Domestic sewage, industrial waste effluent, air deposition, radioactive waste, marine dumping, pesticides, fertilizer, and other pollutants are some of the foremost grounds of water contamination (Singh et al., 2022). The wastewater generated from these resources contains a diversity of contaminants, such as suspended particles, metallic compounds, toxic chemicals, paints and pigments, dyes, oils and emulsifiers, and more (Owa, 2013). When these materials are released without proper treatment, it seriously compromises both human and animal health. Major immunological suppression, acute poisoning, reproductive issues, gastroenteritis, vomiting, diarrhea, dermatological and renal issues, as well as several infectious diseases including cholera and typhoid, are all consequences of it (Ho et al., 2012). As stated in Table 13.1, different heavy metals need careful consideration that can induce toxicity in the environment (Table 13.1) (Mazumder et al., 2020). Copper, manganese, zinc, molybdenum, cobalt, and chromium are a few essential metals for natural life forms. Natural processes including soil and rock erosion, volcanic eruptions, uncontrollable forest fires, and man-made processes like excavating, metal processing, tanning, industrial process and landfills discharge toxic metals in the biosphere. They are generally added to particles that are recognizable everywhere or similarly bonded in organic or inorganic substances. The focus has been on significant levels of toxic metals and how they harm numerous biological systems for decades. Because of human involvement, several biogeochemical cycles have changed substantially (GracePavithra et al., 2019).

13.1.2 Properties, Speciation, and Sources of Chromium

The utilization of chromium is wide, mainly in industrial processes such as metallurgy, electroplating, dyeing, tanning, and printing. Chromium is a glossy, hard,

 DOI: 10.1201/9781003415541-13

TABLE 13.1
Sources of Distinct Heavy Metals, Their Effects on Human Life, and Their Acceptable Limit in Drinking Water According to the WHO

Heavy Metals	Sources	Ill Effects	Permissible Limit (mg/L)
As	Pesticides, fungicides, metal smelters	Skin and vascular diseases, visceral cancer	0.01
Cd	Welding, Cd, and Ni batteries, electroplating, pesticides, fertilizer, nuclear fission plant	Renal disorders and damage, carcinogenic	0.003–0.005
Cr	Mines, mineral sources, electroplating, leather tanning, paints, and pigments	Headache, diarrhea, nausea, carcinogenic	0.05
Cu	Mining, pesticide, chemical industry, metal piping	Liver damage, Wilson's disease, insomnia	2
Hg	Pesticides, batteries, paper industries	Rheumatoid arthritis, nervous disorders	0.006
Ni	Metal plating industries, mining, combustion of fossil fuel, electroplating	Dermatitis, chronic asthma, carcinogenic	0.02–0.07
Pb	Paint, pesticide, automobile emission, mining, smoking, burning of coal	Cerebral disorders, renal, nervous disorders	0.01
Zn	Refineries, brass manufacture, metal plating, plumbing	Depression, lethargy, neurological signs	3

steel-diminished metal that is widely dispersed in our surroundings and comes in a variety of oxidation states (GracePavithra et al., 2019; Malaviya & Singh, 2016). Depending on how the oxidation takes place, chromium can evolve into acidic, ant-acid, or amphoteric oxides. Environmental damage is caused by these enterprises' inappropriate wastewater disposal, though. Because it damages tissue and has teratogenic, mutagenic, and carcinogenic effects on plants, animals, and the environment, it is life-threatening. In the year 1798, chromium was isolated from mineral crocoite ($PbCrO_4$) by reducing CrO_3 with charcoal at a high temperature. Its various colored compounds led to the suggestion that the term be originated from the Greek word "Chroma," which means "color." The oxidation levels of +3 and +6 are where chromium is most stable. When exposed to air, the process causes the production of green colored chromic oxide. The melting point and breaking limits of the afore mentioned metal are 1,907°C and 2,671°C (Haynes, 2014), and it possesses metallic brightness, steely-dark shading, resists staining, is hard, and is delicate. Oxidation reactions that passivate chromium create a thin coating on top of the basic metal to prevent oxygen from spreading (Wallwork, 1976). Large amounts of metals are produced by some businesses, including tannery processing, chromate preparation, production of the

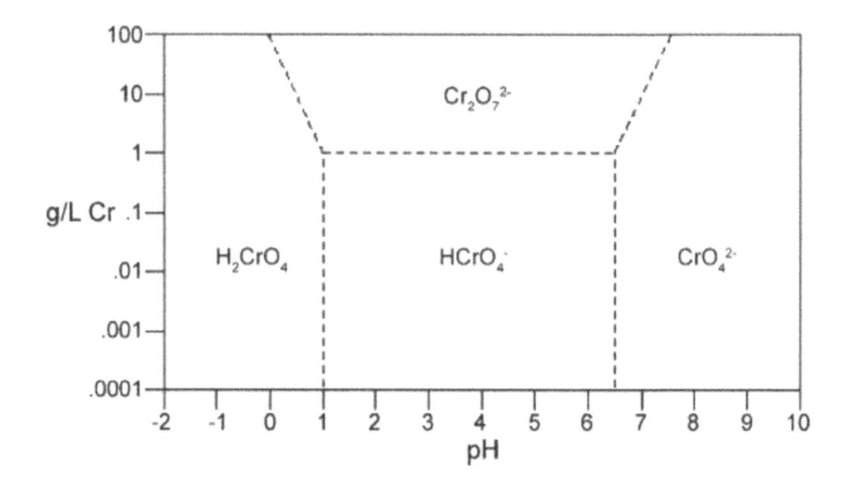

FIGURE 13.1 Water's pH and the relative distribution of Cr(VI) species. Adapted from Chigondo et al. (2022) under the terms of the Creative Commons Attribution License (CC BY 4.0).

color ferrochrome and chrome, and temper steel welding. Numerous authoritative and non-managerial organizations have said that hexavalent chromium is considered a human malignant growth-causing agent. Cr(VI) is one of the 17 chemicals identified by the United States Environmental Protection Agency as being harmful to individual.

When compared to Cr(VI) in water, Cr(III) mixtures dissolve less readily. Cr(III) frames hydrated mixtures that are less soluble at pH levels below 5. The pH and overall concentration of the solution affect the speciation of chromium ions. Figure 13.1 shows the distribution of hexavalent species at various pH levels (Tandon et al., 1984). When the pH is below 10.0 and above 6.0, the main species are dihydrogen chromate (H_2CrO_4)and chromate (CrO_4^{2-}), respectively. Dichromate ($Cr_2O_7^{2-}$) and hydrogen chromate ($HCrO_4^-$) are both present in substantial amounts in the pH range of 2.0–6.0. The most stable and significant oxidation state out of the wide variety of oxidation states is Cr(III). Cr(III) hydrolysis is challenging. It generates polynuclear, neutral, and mononuclear species. It is safe to use because Cr(III) is practically immobile in the environment and stable and insoluble at neutral pH levels (Das et al., 2008). Enterprises must maintain the level of contaminants in the effluent below a set threshold before releasing it to aquatic bodies (Chigondo et al., 2022). In this chapter various extraction and remediation methods of chromium are discussed, which utilizes the chemical and biological pathways.

13.1.3 Toxicity Caused by Chromium

For the digestion of glucose, lipids, and amino acids, Cr(III) is a necessary component that is also used as a dietary supplement. According to a study, chromium was found in the tissues of human newborns and hatchlings. Chromium is most abundant in hair, where it can harm cell structures even at low levels. Immunological and psychological changes emerge with extreme Cr(VI) intake. Inhaled materials containing Cr(VI) can irritate the liver and larynx, as well as cause bronchitis, asthma, nasal septal fissures, and other respiratory issues (Shanker & Venkateswarlu, 2011). Skin contact with Cr(VI) mixtures can cause skin hypersensitivity, dermal putrefaction, dermatitis, and dermal ingestion. The double-stranded deoxyribonucleic acid can be tied by Cr(VI), which alters gene replication, duplication, and repair. Workers in the chromium sector were not the only ones affected. Significant amounts of chromium are released into the soil, water, and air by the manufacturing industries. The material shape affects chromium's adaptability and bioavailability. In contrast to Cr(III), which is only marginally soluble, Cr(VI) is extremely bioavailable and transportable. Intracellular Cr(III) and Cr(VI) both can be mutagenic and can reframe the amino acid coding nucleotide structures (Costello et al., 2019). Zhitkovich (2011) asserts that Cr(VI) is formed from phenotypic and anthropogenic causes and is very adaptive to environmental conditions. It is believed that Cr(VI), which is discharged from industrial sources or depleted into the soil and groundwater, acts as an inducer of malignant development. At low doses, it has been demonstrated to be poisonous and soluble. Cr(III) is a component that supports the digestion of lipids, carbohydrates, and proteins and is somewhat harmless in high concentrations. It is known that Cr(VI) has both beneficial and harmful impacts on health because of its mutagens and toxic properties. The amount of chromium in our bodies is approximately 0.03 mg/L, and a daily intake of 15–200 g is both healthy and sufficient (GracePavithra et al., 2019).

Cr(VI) is the primary contaminant found in the wastewater of diverse industries, including electroplating, tanneries, paint manufacturing, petroleum production, and textile dyeing. The textile industry, for instance, employs synthetic dyes that contain various toxic heavy metals, with chromium being a prominent one (Ahsan et al., 2019). When these industrial effluents, laden with Cr(VI), are released, they pose a significant risk to aquatic ecosystems, where it can accumulate and subsequently magnify in fish at the top of the aquatic food chain. Fish affected by Cr(VI) contamination display noticeable physical symptoms such as increased mortality, scale deterioration, changes in coloration, excessive mucus production, erratic swimming patterns, and disruptions in their osmoregulatory functions. As a result, Cr(VI) enters the terrestrial food chain, undergoing substantial biomagnification before reaching human consumption (Bhattacharya et al., 2022). This poses a range of health risks, including liver and kidney damage, compromised immune function, genetic damage, skin ulcers, neurotoxic effects, asthma, and an elevated risk of developing cancer, most commonly lung cancer (Putatunda et al., 2019). In laboratory studies, rats exposed to Cr(VI) intraperitoneally exhibited anatomical and functional abnormalities in their pituitary and thyroid glands. Rats given subcutaneous injections of potassium dichromate displayed hyaline casts in the renal tubule lumen, cellular degeneration in the proximal convoluted tubule, and significant chromium accumulation, primarily in the renal cortex (Mishra & Mohanty, 2008). Therefore, the development of an effective method for treating chromium-contaminated wastewater is an urgent necessity.

13.1.4 Chromium Removal Techniques

To remove heavy metals from wastewater, a number of treatment systems have been developed. Chemical precipitation, coagulation-flocculation, adsorption, extraction of liquids, ion exchange, biological techniques, membrane separation, electrochemical treatment, improved oxidation process, etc., are examples of common processes. Different cleanup strategies have different benefits and drawbacks depending on how the pollutants are applied and concentrated. These traditional methods, however, have a number of shortcomings, such as membrane fouling, excessive reagent and energy requirements, sludge generation, insufficient metal removal, and metal precipitate aggregation. In order to eliminate dangerous substances from wastewater and aquatic habitats, scientists are working to create new, inventive, useful, economical, efficient, environmentally friendly, and sustainable methods (Suresh Kumar et al., 2015).

As a result, attention has switched to biological remediation techniques, which are less expensive, environmentally safe, and have a relatively high removal efficiency. Utilizing microorganisms' or plants' capacity for heavy metal adsorption or transformation, biotransformation and biosorption are two often employed technologies (Pradhan et al., 2017). Technologies for the biological removal of heavy metals are affordable, safe for the environment, and simple to apply; they use biomass to extract chromium from industrial waste (Pradhan et al., 2017). The biological method of chromium biosorption is an alternate and ecologically friendly method. Biosorption, bioaccumulation, oxidation/reduction, leaching, precipitation, volatilization, decay, and phytoremediation are examples of potential biological techniques that have been used to eliminate and detoxify harmful heavy metals from contaminated water and sediments. Furthermore, approaches for recovering metals sequestered by microbial biomass and biosorbents (with appropriate desorbing agents) are identified (Pradhan et al., 2017).

Due to its appealing qualities, bioleaching has garnered more interest than other approaches of a similar nature (Muddanna & Baral, 2019). Bioleaching is a desired and green process due to its low energy and cost consumption, environmental friendliness, ease of operation, and proper recovery, among other reasons (Naseri et al., 2019). Based on the capacity of microorganisms to change insoluble solid materials into soluble and extractable components, bioleaching is a bio-hydrometallurgical process. The first microorganisms to be employed and investigated for bioleaching from diverse sources are fungi and bacteria through various cellular and chemical mechanisms (Bahaloo-Horeh & Mousavi, 2017). Even though a variety of chemical and physical techniques have been used for removal, the biological cure utilizing microorganisms has garnered interest because it is simple, effective, and produces little to no secondary waste.

In this chapter, various extraction and remediation methods of chromium are discussed, which utilize the chemical and biological pathways.

13.2 DIFFERENT TECHNIQUES FOR TREATMENT

13.2.1 Extraction

Liquid–liquid extraction finds extensive use in various industrial applications, including the recovery and elimination of heavy metals from wastewater, as well as in the hydrometallurgical, pharmaceutical, and biochemical industries, among others. This method is favored over traditional purification processes due to its simplicity, high efficiency, and loading capacity, as highlighted by Wei et al. (2016). In the case of removing Cr(VI) metal, a range of commercial extractants are available in the market, such as isobutyl monoethyl ether, ethyl acetate, ethyl glycol, hexane, tri-*n*-butyl phosphate, ketones, amine compounds, chloroform, diaminopimelic acid, methyl violet, and trioctyl methylammonium compounds, to name a few. Notably, the high molecular weight amines Aliquat 336 and Alamine 336, as researched by Saw et al. in 2018, are among the most commonly used (Saw et al., 2018). Numerous studies have been conducted to continuously remove Cr(VI) from wastewater using various types of extractors, including packed columns, spray columns, mechanically agitated columns, centrifugal extractors, mixer settlers, hollow fiber contactors, and impinging stream extractors.

Packed columns: One of the hollow towers filled with various types of packing materials is the packed columns. This kind of column was created to lessen the issue of back mixing within the spray column. Common packing materials include glass beads, wire mesh, pall rings, Raschig rings, Berl saddles, and Intralox saddles. This equipment is built with a grid at the bottom of the column to support packing material and an appropriate liquid distributor at the top and bottom. The dispersed phase can consolidate and re-disperse with the aid of packing materials. Ajmani et al. (2020) examined the biosorption capability of zinc chloride-activated feedstock of *Phaneravahlii* fruit in a column with a packed bed to provide constant elimination of hexavalent chromium from synthesized water solution (Ajmani et al., 2020).

Scheibel extractor: This is one of the earliest mechanically agitated extractors, as shown in Figure 13.2. It operates on the mixer-settler principle. On a vertical shaft that is fixed in the middle, the agitators are positioned at specific intervals. To enhance phase separation, wire mesh packing is offered along the vertical column. Fouling in the separation area, poor disassembly characteristics, and poor efficiency for high column diameter are some of the main drawbacks of this column.

Kühni extractor: One stator disk comprised of perforated plates and a single agitator makes up the Kühni extractor. The agitator, which is mounted on a central axis, is made up of a few blades that create an interior eddy current pattern like that of a rotating disk column. Each level inside the column is separated by fixed perforated plates. The Kühni column's adaptable hole configuration promotes design flexibility for various column applications. Wang et al. (2020) investigated the elimination of chromium (III) from liquid wastewater using solvent extraction employing "2-ethylhexyl phosphonic acid mono-2-ethylhexyl (P507)" as an extractant in an innovative rotor-stator spinning disk reactor having similarity with the Kühni extractor. According to the research findings, the highest yield of extraction and volumetric mass transfer coefficient are 97.2% and 0.40 L/s, respectively (Wang et al., 2020).

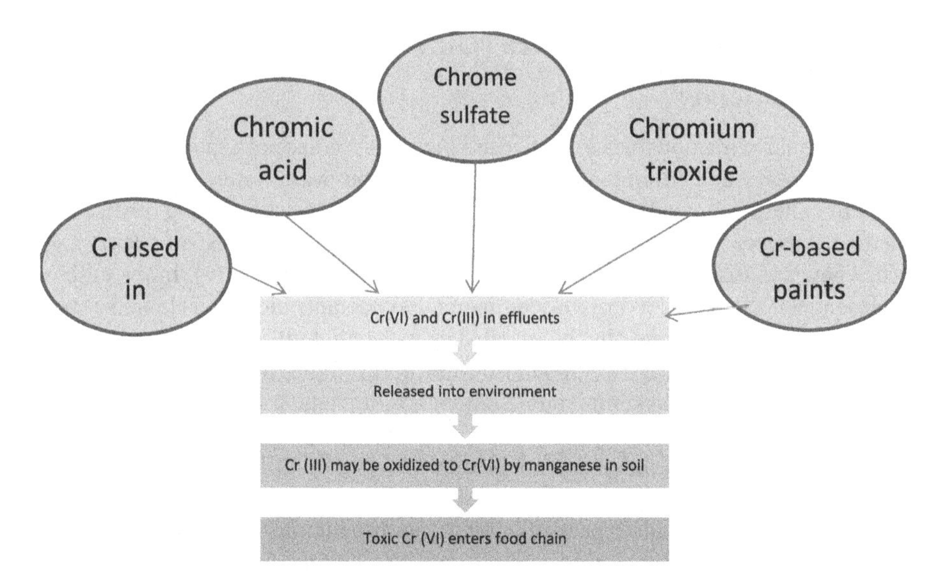

FIGURE 13.2 Effluents released from various industries rich in chromium.

Spinning disk extractor: The spinning disk contactor is made up of a shaft that is supported in the middle and a few disks. These disks serve as the agitation component and are positioned horizontally on the shaft. On the column wall are fixed stationary disks. The stator rings offset the agitator disks because they are greater in diameter than they are. The phases in the apparatus are inter-dispersed by the shearing action caused by the revolving disks (Wang et al., 2020).

13.2.2 Adsorption

Adsorption is the term used to describe the surface deposition of subatomic particles. Adsorption has many advantages, including low formation of chemical and biological sludge, simple maintenance, adaptability in design, and superior efficiency. However, selectivity, volume contemplations, and adsorbent replacement are some of the limits of adsorption (Kumar & Yashwanthraj, 2017; Suganya & Senthil Kumar, 2018).

Due to its great efficacy and low operating cost, adsorption is one of the most effective technologies for the removal of pollutants. The competence of this method depends on some factors, including the pH of the solution, the temperature, the amount of reactant present, the kind and size of the adsorbent, the duration of the contact, the adsorbent's surface properties, any surface modifications, the ionic strength, etc. Hence, (1) clay materials, including bentonite, kaolinite, siliceous material, zeolites, (2) agricultural wastes (peels, coconut shell, bagasse pith, rice husk, tamarind fruit shell, etc.) (Singh et al., 2018), (3) activated carbon, (4) biosorbents, (5) industrial waste products (carbon slurries from industrial waste deposition, metal hydroxide sludge, etc.), and many other adsorbents have all been considered for the control of heavy metal pollution. Most often, "activated carbon" is used to remove metal particles from effluent. The expense of adsorbents as a company is a barrier to adsorption. For instance, fly fiery remains, rice husk carbon, used activated clay, sawdust, wollastonite, raw rice bran, and *Strychnous potatorum* seed. Dolochar, unburned coal from wiping press plants at no cost, was used as an adsorbent for the expulsion of chromium (VI) in a study conducted by Panda et al. (2011). For the evacuation of Cr(VI), dolochar particles that have been warmed to a temperature of 100°C and treated with tap water were used. Furthermore, Tiadi et al. (2022) used an inexpensive absorbing material (dolochar) to remove chromium from artificial solution and mine wastewater (Tiadi et al., 2022).

13.2.3 Membrane-Based Techniques

One of the current effective separation methods used in the treatment of wastewater is membrane separation technology. As a narrow barrier separating two phases or a medium, a semipermeable membrane serves this purpose. The membrane allows one or more elements to move from one medium to another when the right driving force is present. This process can be categorized into microfiltration, ultrafiltration, nanofiltration, electrodialysis, and reverse osmosis based on the separation technique, type of driving force, membrane materials, and membrane pore size (Shrestha et al., 2021). In tabular form, a brief explanation of these procedures is provided. The membrane separation method has several key benefits, including excellent selectivity, minimal energy requirements (because phase transitions are not necessary), straightforward operating procedures, etc. Operational problems including low fluxes, fouling, impurity clogging, and expensive capital costs are among the process's biggest drawbacks (Shrestha et al., 2021). "Cellulose triacetate and poly(butylene adipate-co-terephthalate)-based polymer inclusion membranes containing ionic liquid (tricapryl methylammonium chloride)" as the carrier extractant was used by Sellami et al. (2019) for the efficient and selectively transport of Cr(VI) ions for its remediation. It was found that the membrane transported >99% of Cr(VI) within 6 h (Sellami et al., 2019) (Table 13.2).

13.2.4 Electrocoagulation

An extensive water treatment invention called electrocoagulation eliminates suspended particles, harmful metals, emulsified oils, bacteria, and other impurities. Effluent treatment sectors nowadays merge the functions of coagulation and electrochemistry. Coagulant development is now being emerged as an important step for the water treatment sectors and it is increasing at a faster rate. The idea behind electrochemistry and coagulation/flotation is to get rid of the electrostatic forces that keep the particles suspended in the solution by diminishing or eliminating the water or wastewater's electrostatic forces. The suspended particulate matters will encircle the larger particles when the negative charges are vanquished. The metal anodes and potential are frequently linked (Grace Pavithra et al., 2017). A metal particle is pushed through the water when water or wastewater enters an electrocoagulation cell, and the water is hydrolyzed into hydrogen gas and hydroxyl groups at the cathode. The benefits of electrocoagulation include the exclusion of complex

TABLE 13.2
Common Membrane Separation Techniques

Characteristics	Microfiltration	Ultrafiltration	Nanofiltration	Electrodialysis	Reverse Osmosis
Pore size	50 nm–5 μm	5–100 nm	1–5 nm	<5 nm	<5 nm
Driving force	Pressure gradient (<2 bar)	Pressure gradient (1–10 bar)	Pressure gradient (10–40 bar)	Pressure gradient (10–100 bar)	Applied pressure
Mechanism	Convection	Convection	Diffusion/convection	Ion exchange	Solubilization/diffusion
Membrane material	Cellulose nitrate/acetate, polyamides, polysulfone, metal oxides, etc.	Poly sulfone, polypropylene, nylon 6, acrylic copolymer	Polyethylene terephthalate (PTA)	Sulfonated crosslinked polystyrene	Thin-film composite, cellulose acetate, aromatic polyamide
Membrane type	Symmetric, microporous	Asymmetric, microporous	Thin film	Cationic and anionic exchange membrane	Asymmetric, skin type
Application	Clarification, bacteria, and cyst filtration, RO pretreatment, potable water treatment	RO pretreatment, treatment of potable water, the macromolecular solution	Removal of hardness and desalting	Desalting of ionic solution	Separation of salts and microsolutes from solution

chemicals, a basic hardware layout requiring less maintenance, economic effectiveness, and energy efficiency (Bhatti et al., 2011). However, high amount of conductivity, turbidity, and the presence of suspended solids in the wastewater are a few of the electrocoagulation's constraints. Moreover, the process also needs specialized labor (Putatunda et al., 2019). The presence of abundant oxygen and hydrogen increases thc productivity of the electrofloatation detachment method. Electrolytic cells with a cathode of metallic electrodes and an anode made of anodes are electrically connected to a DC power source to form the electrocoagulation unit. Due to their abundance, extended lifespans, and slower rates of deterioration, iron and aluminum are utilized as electrodes. When compared to coagulation/flotation, electrocoagulation is seen to have a comparable mechanism, although it differs from the intermediate aspect. When a direct current is given to an electrocoagulation cell, the anode serves as a coagulant and breaks down to form metallic cations that abide by Faraday's rule, in accordance with the equipment (Moussa et al., 2017). Das et al. (2020) successfully reduced the Cr(VI) concentration of effluent from 40 to <0.05 mg/L post 60 min of conducting electrocoagulation (Das & Nandi, 2020).

13.2.5 Electrochemical Reduction

Recently, electrochemical reduction has been emerged as an efficient technology that is being used for the treatment of Cr(VI) contaminated wastewater. This process has been marked as a clean technology with a high capacity for chromium removal. Previously, steel electrodes were used under high pH condition, where both electrochemical reduction and precipitation occurred simultaneously which facilitated the removal of Cr(VI). The process was simple and needed no other costly chemicals (Golder et al., 2011). Similarly, another research investigated the unipolar removal of Cr(VI) by using iron rods. This process converted Cr(VI) to Cr(III) with high efficiency while using various types of electrolytes. Electrochemical reduction has been used to treat effluents including suspended particles, oil, grease, colloids in natural waterways, algae, and microbes, as well as organic colors and heavy metals. Because it serves as a pretreatment approach before other methods, electrochemical reduction therapy is suitable at greater pollutant concentrations (Cao et al., 2020). High energy costs are a significant drawback of this procedure. However, much research was done using this process to remediate chromium (Stern et al., 2021; Yao et al., 2020).

13.2.6 Advanced Oxidation Process

The advanced oxidation process is primarily used to treat wastewater (which typically contains organic compounds and bacteria) and is a very competitive non-waste-generating method. According to its definition, it is an oxidation process that produces extremely reactive oxygen species, particularly hydroxyl radicals. The hydroxyl radical is regarded as the strongest among the several oxidizing species due to its high redox potential (2.8 V). Hydroxyl radicals have the capability to attack almost all organic molecules by a variety of different methods, including radical addition or combination, hydrogen abstraction, and electron transfer. By destroying organic molecules, this approach reduces the concentration of pollutants from about 100 ppm to less than 5 ppb. Typically, the removal of a hydrogen atom by hydroxyl radicals causes organic molecules to change into organic radicals (Khan et al., 2023). These organic radical experiences a series of reactions before being mineralized into CO_2 and water. Because of the high rate of oxidation involved with this method, pollutant elimination takes less time. The success of the techniques depends on how quickly and effectively hydroxyl (OH^-) is produced, which is the catalyst for advanced oxidation processes. Advanced oxidation methods that are predominantly used and their reactive species include the Fenton method, photocatalytic reaction, photo-Fenton method, ultrasonic treatment, and radiolysis, among others (Yin et al., 2020).

13.3 CHROMIUM EXTRACTION BY MICROORGANISMS

Parasites, bacteria, yeast, green growth, and a few plants are examples of living things that have demonstrated the ability to clean up after themselves. In addition, fungi and bacteria showed greater efficacy in the cleanup. Due to the advantages, such as minimal energy requirements and operating costs, negligible or no environmental and health risks, high proficiency, and metal recovery (Bhattacharya et al.,

2022). Microorganisms use metal as a means of progress and metabolize harmful substances through an enzyme-catalyzed process, where they are ultimately converted to methane, carbon dioxide, water, and biomass. The following factors need to be taken into consideration or viewed as obstacles to the successful completion of the bioremediation process: nutritional condition, moisture content, soil structure, and hydrogeology, temperature, the microbial community at a contaminated location, and pollutants' chemical composition (Bhattacharya et al., 2020; Tang et al., 2007).

Research groups nowadays focused on the biosorption, biotransformation, and bioaccumulation of the atomic component of Cr. Biosorption is the process by which different kinds of latent, inactive microbial biomass adhere to and concentrate on heavy metals. The vitality of a living thing is what propels the metabolically dynamic process of bioaccumulation. An further mechanism is called biotransformation, which is the process by which Cr(VI) becomes Cr(III) with the help of a biological system (Gunasundari & Senthil Kumar, 2017). The advantages of biological treatment include the eradication of heavy metals using a wide range of temperatures and pH, low operational costs, use of less expensive growth media, among others. The disadvantages include the lengthy period required to remove pollutants and the need for trained labor.

13.3.1 Bacteria

In 1977, Romanenko and Koren'Kov made an initial observation of the capability of *Pseudomonas* species to reduce Cr(VI) in anaerobic conditions, marking the inception of microbial Cr(VI) reduction (Romanenko & Koren'kov, 1977). Subsequently, numerous researchers have identified various microorganisms capable of catalyzing the conversion of Cr(VI) to Cr(III) in diverse environments. These microorganisms, possessing the capacity to reduce Cr(VI) concentrations, are commonly referred to as chromium-reducing bacteria (Somasundaram et al., 2009). In the dominion of metal removal, consortium cultures have demonstrated enhanced metabolic abilities and proved to be more suitable for practical field applications. Recognizing these advantages, researchers have determined that consortium cultures, when isolated from their environmental context, exhibit heightened effectiveness in Cr(VI) reduction. However, it is important to note that the toxicity of Cr(VI) can be detrimental to these bacteria, making the biological treatment of Cr(VI)-contaminated wastewater a challenging task. Consequently, some researchers have adopted cell immobilization techniques to safeguard bacterial cells. Unlike planktonic cells, these methods can tolerate larger concentrations of Cr(VI) and make it simple to separate the treated liquid from the biomass. Because Cr(VI) toxicity is so harmful to bacterial cells, this approach has been used to overcome the difficulty of biologically treating wastewater contaminated with Cr(VI). Furthermore, lowering Cr(VI) toxicity has been the focus of current study in an effort to investigate alternatives and lessen the negative consequences connected to specific physicochemical techniques. Growing interest has been shown in "plant growth promoting rhizobacteria (PGPR)," which are soil bacteria that actively colonize plant roots and stimulate plant development. The concept of rhizoremediation, which combines the abilities of plants and specific microbes to enhance the efficiency of pollutant extraction, is being explored as an alternative (Guo et al., 2021).

13.3.2 Fungi

Through a process known as "adsorption mechanism," positively charged functional groups on the dead fungal biomass, such as amines, are bound by anionic chromate ions to remove Cr(VI) from aqueous solutions. In the case of *Trichoderma* species, it has been observed that carboxyl and amine groups serve as the primary binding sites on the surfaces of fungal cells (Vankar & Bajpai, 2008).

As per Park et al. (2005), their research suggests that *Aspergillus niger* can facilitate the conversion of Cr(VI) to Cr(III) through a redox process that does not depend on enzymatic activity. Additionally, it was revealed that hazardous Cr(VI) can be transformed into its less toxic, benign Cr(III) form by utilizing the deceased fungal biomass of various fungal strains, including *Aspergillus niger, Saccharomyces cerevisiae*, *Rhizopus oryzae,* and *Penicillium chrysogenum.* The reduction of Cr(VI) to Cr(III) is also a viable process in other fungal species, such as *Paecilomyces lilacinus* and *Hypocrea tawa* (Medfu Tarekegn et al., 2020). Notably, the adsorption of heavy metals is a phenomenon observed in both living and non-living fungal cells. In fact, there are several advantages to utilizing dormant deceased cells for this purpose (Morales-Barrera et al., 2008; Sharma & Adholeya, 2011).

13.3.3 Algae

Growing photosynthetic algal cells possess the capacity to mitigate Cr(VI) contamination. The attachment of Cr(VI) ions to the cell surfaces of algae species is the first stage in this process. Importantly, this binding process occurs rapidly and is not contingent on cellular metabolic activity. The subsequent phase of intracellular metal accumulation is instigated by a combination of surface biosorption and growth-related effects, which is notably a much slower process reliant on cellular metabolic energy (Ociński et al., 2021).

The process of biosorption of Cr(VI) onto algal biomass in green algae, such as *Chlorella miniata* and *Cladophora albida,* is characterized by the bioreduction of Cr(VI) and the subsequent biosorption of Cr(III). Algae offer several advantages over alternative biological materials. These advantages include cost-effective recyclability, the potential for metal recovery, the generation of reduced biological sludge, effectiveness in treating diluted effluents, and a high surface area-to-volume ratio (Joutey et al., 2015).

13.3.4 Reduction and Uptake Mechanism of Chromate

The U.S. Environmental Protection Agency has designated chromium as a priority contaminant, with historical perceptions favoring hexavalent chromium's substantially greater hazard compared to trivalent chromium. Current methods for treating chromium-containing waste often involve the reduction of Cr(VI) to Cr(III) in an aqueous solution using a reducing agent at an acidic pH to facilitate the precipitation of the less soluble Cr species (Gheju & Balcu, 2011). More recently, the biological conversion of Cr(VI) to Cr(III) has come to light, with Cr(VI) potentially acting as an electrophile in the oxidation of organic molecules during this process. Under specific conditions, numerous facultative anaerobic bacteria have demonstrated the capability to convert Cr(VI) into Cr(III). These chromium-reducing bacteria are found in several genera, including *Agrobacterium, Bacillus, Desulfovibrio, Enterobacter, Escherichia*, and Pseudomonas (Chardin et al., 2003). Bacteria can employ both direct and indirect enzymatic methods to reduce Cr(VI), utilizing compounds such as glutathione, cysteine, sulfite, and thiosulfates in the chemical reduction of Cr(VI). Numerous aerobic, facultative, and anaerobic bacteria have an enzyme called reductases, which can be found in both soluble and membrane-bound forms. These reductases appear to be involved in the enzymatic reduction of Cr(VI). Anaerobic biological reduction is generally sluggish; however, abiotic reduction, which is aided by Fe(II) or hydrogen sulfide, frequently occurs first. In anaerobic bacteria, chromate reduction is often attributed to the presence of membrane-bound enzymes, whereas in most aerobic bacteria, chromate-reducing enzymes are typically found as soluble cytosolic proteins (McLean & Beveridge, 2001).

The term "bioaccumulation" refers to all procedures through which living cells take up accessible metal ions. It consists of bioprecipitation, intracellular accumulation, and biosorption processes. Hexavalent chromium ions may get caught in cellular architecture and then biosorb to the binding sites. It is known as biosorption or passive absorption, since it does not require any energy. Furthermore, Cr(VI) can enter cells in ways that need metabolic energy. The phrase "active uptake" refers to this membrane transfer. The ingested metal may bioaccumulate through both active and passive ways of metal absorption. Pollutants that are difficult to biodegrade, such as metals, can be removed from water through biosorption (Tang et al., 2021).

13.3.4.1 Biosorption

Pollutants that are difficult to biodegrade, such as metals, can be removed from water through biosorption. Activated carbon, synthetic resins, inorganic sorbent substances, or so-called biosorbents made from non-living biomaterials have all been used in the sorption-based methods that several researchers have created. Biosorbents are often the most affordable, plentiful, and ecologically beneficial choice among them. It is recognized that many different biomaterials, such as dead bacteria, fungi, algae, seaweed, industrial leftovers, and agricultural wastes, may bind contaminants (Tang et al., 2021).

Anionic adsorption to cationic functional groups involves the electrostatic attraction of negatively charged chromium species, such as chromate (CrO_4^{2-}) and dichromate ($Cr_2O_7^{2-}$), in the medium to positively charged functional groups present on the surface of biosorbents. The mechanism behind this process hinges on the observation that the adsorption of Cr(VI) is more pronounced under acidic conditions, while it diminishes under alkaline conditions. At lower pH levels, the biosorbent's functional groups become protonated, facilitating the attraction of negatively charged chromium ions. Conversely, at higher pH levels, deprotonation occurs, leading to the functional groups acquiring a negative charge, which repels negatively charged chromium species. Meanwhile, functional groups like carbonyl and amide facilitate absorption of reduced Cr(III) on the surface of microbes like *Pseudomonas putida* (Guo et al., 2021). In the adsorption-coupled reduction process, biomass is responsible for reducing Cr(VI) to Cr(III) in an acidic environment, following which the

Cr(III) is adsorbed onto the biomass. The extent of adsorption depends on the specific biosorbent used. A part of Cr(VI) is changed into Cr(III) by the anionic and cationic adsorption processes. After that, chromium (Cr(VI) and Cr(III)) in both its anionic and cationic forms are adsorbed onto biosorbents. A biosorbent can reduce some of the Cr(VI) to Cr(III) during the reduction and anionic adsorption process, which permits Cr(VI) to adhere mostly to the biomass and leave Cr(III) in the solution.

13.3.4.2 Bioaccumulation

Biomembranes typically act as barriers against Cr(III). However, in aqueous solutions, Cr(III) promptly develops complexes with a variety of physiologically relevant ligand molecules, which can then be taken up by cells. The prevalent form of Cr(VI) is the tetrahedral CrO_4^{2-}, which closely resembles physiological anions like SO_4^{2-} and PO_4^{2-}. Cr(VI) can be taken up by "sulfate transporters," where Cr(VI) and sulfate compete for entry, or it can be taken up by non-selective anion channels, which facilitate transport. The addition of sulfate can mitigate chromate toxicity. Furthermore, studies have demonstrated that chromate significantly inhibits the assimilation of sulfate and impacts sulfur metabolite pools, indicating a shortage of sulfur within cells (Sharma et al., 2022). Due to the fact that Cr(VI) is swiftly reduced to Cr(III) inside cells, providing the cells are equipped with sufficient reducing capacity and the proportion of Cr in the Cr(VI) oxidation state on the inside and outside of the cell's plasma membrane is not comparable. The primary mechanism by which Cr(VI) accumulates in cells is through its reduction. In a number of investigations, microbial cells—both living and dead—have been employed in biosorption and bioaccumulation methods to extract Cr(VI) from aqueous solutions. For the purpose of maintaining cellular metabolic activity and overcoming obstacles associated with metal toxicity, the use of dead biomass is beneficial (Sharma et al., 2022). Moreover, the adsorbed metal is readily recovered, and the biomass is recyclable. The absence of responses in dried cells, however, limits the use of this strategy. It is possible to recover metals without causing bacterial growth, drying out, or biomass storage by using living biomass. Nonetheless, if the environmental metal content is excessively high when using live biomass, it may be detrimental to the biomass's growth. As a result, wherever feasible, employ microorganisms that possess a great deal of tolerance to increased Cr(VI) concentrations or are already pre-adapted to chromium.

13.3.4.3 Bioleaching

One of the fungus most frequently employed in bioleaching is *Aspergillus niger*. It is used in the manufacture of oxalic, citric, and gluconic acids, among other organic acids. These naturally occurring organic acids are among the most significant lixiviants for removing heavy metals from hazardous wastes and minerals (Muddanna & Baral, 2019). A compelling benefit of using organic acids as leaching agents is their ability to facilitate recovery under mildly acidic conditions (pH 3–5). Furthermore, in natural ecosystems, microbes release low molecular weight organic acids into the soil, which makes these acids biodegradable. Since organic acids may complex other compounds, thereby lowering the amounts of harmful metals, they are less poisonous to the majority of natural species than inorganic acids. Furthermore, compared to synthetic lixiviants, these biogenic organic acids generated by microbes pose less of a threat to the environment (Muddanna & Baral, 2019). Because the chemical synthesis of organic acid requires expensive raw ingredients and intricate procedures that provide low yields, it is ineffective. The microbiological synthesis is a more sophisticated technique with lower overheads than chemical procedures (Bahaloo-Horeh & Mousavi, 2017).

There are two approaches of bioleaching heavy metals: (1) adding the waste to the fungus-containing medium and allowing them to come into direct contact, or (2) allowing the fungi to produce as much organic acid as possible before filtering. The solid waste is then added to the fungal filtrated culture, which is considered to be perfect for industrial use. Microbial biomass does not contaminate the solid waste when using the fungal filtrated culture method, which also allows for the recycling of fungi. Additionally, if there is no waste, the fermentation process can be improved individually. Additionally, in the event that waste toxicity does not cause bacteria or cell wall damage, there is no restriction on the use of high pulp densities. When fungi are present, higher pulp densities can thus be reached relative to bioleaching (Bahaloo-Horeh & Mousavi, 2017).

13.3.5 Application of Mixed Culture Bioreactors

Conducting large-scale remediation of Cr(VI) using a sole microbial species proves impractical due to the difficulties associated with maintaining consistent laboratory conditions for pure cultures. In an effort to address this challenge, researchers often employ bioreactors with indigenous microbial consortia. These bioreactors were run in three distinct modes: recirculating sequencing batch reactor (SBR), batch, and continuous, mode. Their findings indicated that the SBR-recirculation mode proved to be more proficient in Cr(VI) eradication compared to the batch mode. In contrast to free cell cultures, cell immobilization represents a considerably more advanced approach for the reduction and biosorption of Cr(VI). Immobilized cells offer advantages such as ease of regeneration, greater reusability, and enhanced stability when compared to free-floating cells. Moreover, they do not necessitate a continuous supply of nutrients for growth and are easily separable from water, preventing clogging in continuous flow systems (Sharma et al., 2022). Bed cultured and mixed bed cultured bioreactors have demonstrated increased effectiveness and have gained strong support from the scientific community for the reasons outlined above. Consequently, the utilization of bioreactors containing Cr(VI)-reducing bacteria for removing Cr(VI) from water is deemed an economical, straightforward, safe, and technologically viable approach. Various materials, including gravel, elemental sulfur, glass beads, wood husk, rubber wood sawdust, calcium alginate, PVC bed, granular activated carbon, and alginate-carboxymethyl cellulose with activated carbon, have been employed to promote the biofilm formation of chromium-reducing bacteria. The possibility of reuse depends on the microbial species, as some are susceptible to cell lysis. One of the ongoing scientific mysteries that continues to captivate researchers worldwide is the quest for more useful and adaptable microbial species (Naz et al., 2021).

13.3.6 Case Studies

Numerous investigations have explored the utility of microbial species derived from diverse sources such as serpentine soil, tannery waste, effluent treatment plants, and chrome liquor for the purpose of Cr(VI) removal. Here are some notable examples.

Verma et al. (2009) conducted research involving the isolation of various *Bacillus* species, including *B. subtilis* and *B. brevis*, with the aim of assessing their effectiveness in reducing Cr(VI) within the concentration range of 160–180 mg/L. Their findings indicated that, among the strains tested, *B. coagulans* was less efficient in Cr(VI) reduction in that specific range (Verma et al., 2009).

Ahmad et al. (2010) designed a pilot-scale bioreactor known as "Chrom-BacTM," which was inoculated with the microbial strain *Acinetobacter haemolyticus*. Through their study, they discovered that the Cr(VI) content in electroplating water was completely reduced within the range of 17–81 mg/L. Notably, they observed that factors like pH variation (ranging from 6.2 to 8.4) and fluctuations in nutrient temperature (30°C–38°C) had no discernible impact on the efficacy of the "Chrom-BacTM" system. Their findings suggested that the "Chrom-BacTM" system was technically viable (Ahmad et al., 2010).

Wani et al. (2019) isolated the *Pseudomonas entomophila* MAI4 bacterial strain from industrial effluent. According to their research, this strain exhibited significant Cr(VI) removal capabilities under controlled laboratory conditions at a concentration dose of 100 ppm. Moreover, it displayed an impressive 80% Cr(VI) removal efficiency when applied to original industrial wastewater (Wani et al., 2019).

More recent works are listed in Table 13.3.

13.4 FUTURE SCOPE

In several industrial processes, chromium is still widely used, and waste from these operations is regularly dumped into the environment. Due to its environmental persistence and lack of biodegradability, this heavy metal may be harmful to both human health and the environment (Naz et al., 2021). Hexavalent chromium biosorption is still mostly limited to laboratory research today. Future research should consider several aspects.

- Non-ionic surfactants could be helpful in this regard. Non-functional surfactants provide "preferential partitioning" of the reactants in hydrophobic and hydrophilic layers, which will improve comprehension of the reaction process. The right choice of adsorbent can accelerate biosorption; in this regard, utilization of food waste like ascorbic acid-pertaining peels can be a wise option (Guo et al., 2021).
- When translating technology from a lab setting to a larger application, models are crucial. More focus needs to be placed on the creation of dynamic models that replicate the biosorption process and provide valuable data for its use in real-world scenarios. To comprehend the molecular basis of the biosorption mechanism, molecular biotechnology is required. Designing modified organisms with improved sorption capacity and targeting certain metal ions.

TABLE 13.3
Various Case Studies on the Removal/Remediation of Chromium by Different Methods

Heavy Metals	Adsorbent	Initial Concentration (mg/L)	pH	Adsorption Capacity (mg/g)	Adsorbent Characteristics	Temp (°C)	Remarks	References
Adsorption								
Cr^{6+}	CS-GO/CMC composite	–	–	127.4	–	25	• Synthetic wastewater was employed. • The adsorption isotherm was well suited to the Langmuir model. • One fabrication step was used to create the adsorbent. • Adsorption took 75 min.	Huang et al. (2019)
Cr^{6+}	KOH-activated porous biochar	100.0 each	6.5	116.97	d_p = 1–2 nm Vp = 0.4991 cc/g As = 2,183.80 m^2/g	20	• Ion exchange, complexation, and electrostatic attraction were the key causes of the adsorption process. • The adsorption kinetics were well-fitted by the Avrami fractional-order model. • The data were well-fit by the Freundlich model since the adsorption was multi-layer.	Qu et al. (2021)

(Continued)

TABLE 13.3 (*Continued*)
Various Case Studies on the Removal/Remediation of Chromium by Different Methods

Heavy Metals	Adsorbent	Initial Concentration (mg/L)	pH	Adsorption Capacity (mg/g)	Adsorbent Characteristics	Temp (°C)	Remarks	References
Cr^{6+} and Pb^{3+}	KOH-activated carbon	30–180 each	2.0	113.63 and 232.56, respectively	$d_p = 1.764$ nm Vp = 0.4991 cc/g As = 1,085 m^2/g	25	• The wastewater was synthetic. • The data were well-fit by the Langmuir model. • One hour was spent in contact.	Kharrazi et al. (2020)
Cd^{2+}, Pb^{2+}, and Cr^{6+}	Graphitic carbon nitride nanosheets	500 each	8.0, 6.0, and 2.0	123.21, 136.57, and 684.45, respectively	$d_p = 1.6$ nm $A_s = 111.2$ m^2/g	45	• The adsorbent was made using a one-step calcination procedure. • The wastewater was artificial. • Adsorption took 1 h. • Adsorbent dosage of 1 g/L was employed. • For ten consecutive cycles, the adsorbent demonstrated good reusability.	Xiao et al. (2019)

(*Continued*)

TABLE 13.3 (*Continued*)
Various Case Studies on the Removal/Remediation of Chromium by Different Methods

Heavy Metals	Adsorbent	Initial Concentration (mg/L)	pH	Adsorption Capacity (mg/g)	Adsorbent Characteristics	Temp (°C)	Remarks	References
Cr^{6+}	KOH-activated porous biochar (PBCKOH)	100.0 each	6.5	116.97	d_p = 1–2 nm Vp = 0.4991 cc/g As = 2,183.80 m^2/g	20	• The wastewater was synthetic. • The data were well-fit by the Freundlich model since the adsorption was multi-layer. • The adsorption kinetics were well-fitted by the Avrami fractional-order model. • Ion exchange, complexation, and electrostatic attraction were the key causes of the adsorption process.	Qu et al. (2021)
Cr^{6+} and Pb^{3+}	KOH-activated carbon	30–180 each	2.0	113.63 and 232.56, respectively	d_p = 1.764 nm Vp = 0.4991 cc/g As = 1,085 m^2/g	25	• The wastewater was artificial. • The data were well-fit by the Langmuir model. • One hour passed during the contact.	Kharrazi et al. (2020)

(*Continued*)

TABLE 13.3 (*Continued*)
Various Case Studies on the Removal/Remediation of Chromium by Different Methods

Heavy Metals	Adsorbent	Initial Concentration (mg/L)	pH	Adsorption Capacity (mg/g)	Adsorbent Characteristics	Temp (°C)	Remarks	References
Cr^{6+}	Honeycomb carbon material	40	2.0	332.5	$d_p = 1.68$ nm. $V_p = 0.57$ cc/g As = 1,436.21 m^2/g	25	• The Langmuir model was employed. • The wastewater was synthetic. • At a pH of 2.0, the greatest removal was noted. • Time at equilibrium was 12 h. • Glucose used the synthetic hydrothermal process to create the adsorbent.	Liang et al. (2019)
Cr^{6+}	*p*TSA-Pani nanocomposite	200	2.0	166.66	–	50	• Over the course of three cycles, the adsorbent's adsorption capacity remained nearly constant. • Synthetic wastewater was employed. • The results were well-fit by pseudo-second-order and Langmuir models.	Kumar et al. (2019)

(*Continued*)

TABLE 13.3 (*Continued*)
Various Case Studies on the Removal/Remediation of Chromium by Different Methods

Heavy Metals	Adsorbent	Initial Concentration (mg/L)	pH	Adsorption Capacity (mg/g)	Adsorbent Characteristics	Temp (°C)	Remarks	References
Cu^{2+} and Cr^{6+}	Carbonaceous nanofiber/Ni-Al layered double hydroxide	40 and 25, respectively	5.0	219.6 and 341.2, respectively	d_p = 2–50 nm. A_s = 43.7 m^2/g	25	• The Freundlich model and the isotherm were in agreement. • Simulated wastewater was employed. • The 2-h contact time was followed. • After 2 h, the adsorption efficiency was almost reached.	Yu et al. (2018)
Cr^{6+}	Crosslinked chitosan beads	500 each	2.0	325.2	A_s = 5.31 m^2/g	50	• The contact time was 5 h. • The wastewater was synthetic. • Langmuir isotherm and pseudo-second-order models fit the data. • After five cycles, the adsorption capacity decreased.	Vakili et al. (2018)

(Continued)

TABLE 13.3 (*Continued*)
Various Case Studies on the Removal/Remediation of Chromium by Different Methods

Heavy Metals	Adsorbent	Initial Concentration (mg/L)	pH	Adsorption Capacity (mg/g)	Adsorbent Characteristics	Temp (°C)	Remarks	References
				Biosorption				
Cu^{2+} and Cr^{6+}	Straw-based adsorbent	500 and 90, respectively	3.0 and 5.0	17.82 and 53.88, respectively	$A_{s<}1.0\,m^2/g$	25	• The data were well-fit by the Langmuir model. • A straw-based adsorbent is called WS-CA-AM. • Methylene blue (MB) has an adsorbent capacity of 120.84 and methyl orange (MO) of 3053.48 mg/g.	Liu et al. (2020)
Cr^{3+}	Waste fish scale	150	5.0	18.35	–	21	• The time of adsorption was 1.5 h. • The dosage of the adsorbent was 0.8 g. • There was a 99.75% removal efficiency found. • The Cr solution was synthetic. • The experimental data were fitted using the Langmuir and pseudo-second-order kinetic models.	Teshale et al. (2020)

(*Continued*)

TABLE 13.3 (*Continued*)
Various Case Studies on the Removal/Remediation of Chromium by Different Methods

Heavy Metals	Adsorbent	Initial Concentration (mg/L)	pH	Adsorption Capacity (mg/g)	Adsorbent Characteristics	Temp (°C)	Remarks	References
Cd^{2+}, Pb^{2+}, and Cr^{3+}	Arrowhead plant stalk	100 each	5.0	38.2, 97.1, and 23.5, respectively	–	35	• The wastewater was synthetic. • The 2-h contact time. • The pseudo-second-order model and the Langmuir model both adequately fit the data.	Zhang et al. (2018)
Cr^{6+} and Pb^{2+}	Activated avocado seed	30 and 20, respectively	5.0	26.6 and 5.1, respectively	$V_p = 0.0001$ cc/g $A_s = 0.1276\ m^2/g$	25	• The wastewater was synthetic. • As the starting concentration and adsorption duration increased, the removal efficiency also did so. • The Langmuir model and the data were compatible.	Boeykens et al. (2019)

(*Continued*)

TABLE 13.3 (*Continued*)
Various Case Studies on the Removal/Remediation of Chromium by Different Methods

Heavy metal	Membrane	Initial concentration (mg/L)	Surfactant/ complexing agent (concentration)	pH	Removal efficiency (%)	Remarks	Reference
Membrane Processes (Ultrafiltration)							
Cu^{2+}, Cr^{3+}	Polyethersulfone	100 mg/L	SDS (7 mM)	10	66.64 and 26.02, respectively	• Artificial wastewater was employed. • CMC was initially assessed at several pH levels (pH 3, pH 10, and without pH modification). • Compared to neutral and alkaline circumstances, CMC is less effective in an acidic environment (pH 3). • At neutral pH, the maximum selectivity of Cu^{2+} over Cr^{3+} was achieved.	Sum et al. (2021)

(*Continued*)

TABLE 13.3 (*Continued*)
Various Case Studies on the Removal/Remediation of Chromium by Different Methods

Heavy metal	Membrane	Initial concentration (mg/L)	Surfactant/ complexing agent (concentration)	pH	Removal efficiency (%)	Remarks	Reference
Cr^{3+}, Cr^{6+}	Regenerated cellulose	0.60 mmol/L	Sodium alginate (SA) (10 mmol/L)	2–5	100	• Fabricated wastewater was utilized. • The electrolytic medium used to reduce Cr^{6+} to Cr^{3+} was acidic SA solution. • During the PEUF procedure, acidic SA was also utilized as an extracting agent. • Nitric acid SA solution demonstrated to be an effective extracting agent for the extraction of Cr^{3+} by the PEUF process and a good electrolytic medium for the complete reduction of Cr^{6+} by electrolysis.	Butter et al. (2021)

(*Continued*)

TABLE 13.3 (*Continued*)
Various Case Studies on the Removal/Remediation of Chromium by Different Methods

Heavy metal	Membrane	Initial concentration (mg/L)	Surfactant/ complexing agent (concentration)	pH	Removal efficiency (%)	Remarks	Reference
Cr^{3+} and Cr^{6+}	Polysulfone	10–400	Rhamnolipid (JBR 425) (1%–2% vol/vol.)	6.0	96.2 and 98.7, respectively	• Artificial wastewater was employed. • The initial Cr^{6+} concentration was 10 mg/L. • The permeate flux was significantly influenced by transmembrane pressure and temperature. • The maximum permeate flux achieved was 63.5 L/m^2h. • Various concentrations were tried, and 10 mg/L produced the best results.	Abbasi-Garravand and Mulligan (2014)

(*Continued*)

TABLE 13.3 (*Continued*)
Various Case Studies on the Removal/Remediation of Chromium by Different Methods

Heavy metal	Membrane	Initial concentration (mg/L)	Surfactant/ complexing agent (concentration)	pH	Removal efficiency (%)	Remarks	Reference
			Membrane Processes (Nanofiltration)				
Cr^{3+} and Pb^{2+}	Thin-film composite polyamide with polyamide active layer embedded with black TiO_2 nanoparticles	$Na_2Cr_2O_7$ and $Pb(NO_3)_2$	150 ppm	8 bar (pressure), pH 2–10	Maximum > 90 and 80, respectively	• Synthetic wastewater was employed. • The greatest performance for the rejection of Cr^{3+} and Pb^{2+} was shown by TFN0.05 with 0.05 wt% black TiO_2 nanoparticles (BNP) loading. • For both Cr^{3+} and Pb^{2+} FO using a $MgCl_2$ draw solution achieved a rejection rate of about 100%. • The addition of BNPs improved the membrane surface's hydrophilicity and roughness. • The maximum permeate flux was recorded by the TFN0.1 membrane at 66 L/m²h.	Mohammad Gheimasi et al. (2021)

(*Continued*)

TABLE 13.3 (*Continued*)
Various Case Studies on the Removal/Remediation of Chromium by Different Methods

Heavy metal	Membrane	Initial concentration (mg/L)	Surfactant/ complexing agent (concentration)	pH	Removal efficiency (%)	Remarks	Reference
Cr^{6+}	Hybrid chitosan/ polyvinyl alcohol/ montmorillonite clay	$HCrO_4^-$	50.0 ppm	1.0 bar (pressure), pH 7	88.34	• Artificial wastewater was employed. • Peak permeate flow and porosity for the used membrane were 25.72 L/m^2h and 81.22%, respectively. • A pH of 7 and a membrane thickness of 0.2 mm were found to be optimal for removing the most metal from chromium.	Sangeetha et al. (2019)
Cr^{3+}, Ni^{2+}, Ba^{2+} and NaCl (metal is Na^+)	Polyamide	Effluent of sludge dewatering process in oil and gas well drilling industries	5.3, 6.2, 209.0, and 14,180.0	11.72 bar (pressure), pH 4	58.5, 77.4, 85.3, and 79.6, respectively	• Wastewater from industry was utilized. • Taguchi approach was used to design the experiment. • The ideal pH value is 4, and the ideal temperature is 25, respectively. • With a feed concentration of 61,500 mg/L, the removal efficiency of total dissolved solids was found to be 56.3%. • The recovery rate for the NF membrane process is 47.17%.	Hedayatipour et al. (2017)

(Continued)

TABLE 13.3 (*Continued*)
Various Case Studies on the Removal/Remediation of Chromium by Different Methods

Heavy metal	Membrane	Initial concentration (mg/L)	Surfactant/ complexing agent (concentration)	pH	Removal efficiency (%)	Remarks	Reference
Cr^{6+}	Thin-film composite polyamide on polysulfone	Synthetic samples containing Cr^{6+}	1.0–120.0 ppm	40.0 bar (pressure), pH 2–11	> 95	• Synthetic wastewater was treated by chelating and using a surfactant (SDS). • The extraction rate of Cr^{6+} increases with an increase in operating pressure. • The penetrated water's Cr^{6+} concentration ranged between 0.04 and 1.9 ppm. • The extraction rate of Cr^{6+} increased as feed concentration rose.	Otero et al. (2012)

(*Continued*)

TABLE 13.3 (*Continued*)
Various Case Studies on the Removal/Remediation of Chromium by Different Methods

Heavy metal	Membrane Material	Feed Solution type	Feed Solution concentration (mg/L)	Draw Solution type	Draw Solution concentration (mg/L)	pH	Removal efficiency (%)	Remarks	Reference
Membrane Processes (Forward Osmosis)									
Cr^{6+}, Cd^{2+}, and Pb^{2+}	Thin film nanocomposite polyamide	chromate ($Na_2Cr_2O_7$), cadmium ($CdCl_2$) and Lead ($Pb(NO_3)_2$)	100.0	NaCl	0.5 – 2.0	7	98.3, 99.7, and 99.9, respectively	• Artificial wastewater was employed. • Using a phase inversion method and an interfacial polymerization reaction, a novel TFN-FO membrane was created. • From aqueous solutions, TFN membranes produced greater chromium, cadmium, and lead rejection efficiencies than TFC. • Compared to commercial FO membrane and TFC membrane, whose water permeate flux reach 15.0 and 12.5 L/m2.h, respectively, the recorded water permeate flux of TFN membrane was 34.3 L/m2.h. • A 96% increase in flux recovery ratio. • Raising the concentration of the draw solution improved the heavy metals' ability to be rejected.	Saeedi-Jurkuyeh et al., 2020

(*Continued*)

TABLE 13.3 (*Continued*)
Various Case Studies on the Removal/Remediation of Chromium by Different Methods

Heavy metal	Membrane Material	Feed Solution type	Feed Solution concentration (mg/L)	Draw Solution type	Draw Solution concentration (mg/L)	pH	Removal efficiency (%)	Remarks	Reference
Cr3+ and Pb2+	Thin-film composite polyamide with polyamide (PA) active layer embedded with black TiO2	$Na_2Cr_2O_7$ and $Pb(NO_3)_2$	150 ppm	NaCl, $CaCl_2$, $MgCl_2$ and Na_2SO_4	1000 ppm	2 –10	Nearly 100% each	• The wastewater was synthetic. • The hydrophilicity and roughness of the membrane surface are improved by the addition of BNPs. • The permeate flux measured at the highest membrane was 66 L/m2h. • The highest performance for the rejection of Cr^{3+} and Pb^{2+} • TFN0 was shown by TFN0.05 with 0.05 wt% black TiO_2 nanoparticles	[Mohammad Gheimasi et al., 2021]

(*Continued*)

TABLE 13.3 (*Continued*)
Various Case Studies on the Removal/Remediation of Chromium by Different Methods

Heavy metal	Initial concentration (mg/L)	Anode/ Cathode material	Operating conditions	Initial pH	Removal efficiency (%)	Remarks	Reference
				Electrocoagulation			
Cr^{6+} and Pb^{2+}	55.3 and 3.5	SS/SS	Current density (j) = 73.5 A/m^2, t = 1.5 h, Salinity (S) = 3 mS/cm, and Distance between electrodes (d) = 1 cm	3.5	91.7 and 91.3, respectively	• Indian electroplating wastewater was utilized. • MP-P electrodes, each having a surface area of 68 cm^2. • At pH 6.5, the lowest anode consumption was recorded.	D. Sharma et al., 2020
Cr^{6+}	50	Fe/Fe	j = 20 A/m^2, S = 10 mS/cm, d = 15 cm, and t = 1 h	6.0	99.9	• At pH 9.5, settling was discovered to be optimal. • Wastewater made artificially was employed. • When positive single pulse current electrocoagulation was used, better results were seen. • A sludge of 1.3024 g/L was produced.	Zhou et al., 2020
Cu^{2+}, Ni^{2+}, Zn^{2+}, and Cr^{3+}	20 each	Fe/Fe	NaCl (100 mg/L), j = 10 A/m2, and t = 40 min	9.0	100	• Wastewater made artificially was employed. • Compared to Al, Fe electrodes were more effective in removing Cr ions. • A sludge mass of 0.68–2.50 kg/m^3 was generated. • In the presence of CN– ions, there was a decrease in the removal of Ni and Cu. • The energy usage ranged from 0.37 to 2.78 kW h/m^3.	Kim et al., 2020

(*Continued*)

TABLE 13.3 (*Continued*)
Various Case Studies on the Removal/Remediation of Chromium by Different Methods

Heavy metal	Initial concentration (mg/L)	Electrodes material	Operating conditions	pH	Removal efficiency (%)	Remarks	Reference
Electrochemical reduction							
Cr^{6+}	550 mg/mg	Magnetite	Electric potential (V) = −0.2 V	2	93.7	• Wastewater synthesized was employed. • Over the cathode surface, Cr^{6+} was converted to Cr^{3+}. • Magnetite produced Fe^{2+}, which improved the removal of Cr^{3+}. • The addition of K_2SO_4 sped up the reduction of Cr^{6+}.	Yang et al., 2019

Metal	Initial concentration (mg/L)	Precipitant	pH level	Removal efficiency (%)	Remarks	Reference
Advanced oxidation process						
Zn^{2+}, Fe^{3+}/Fe^{2+}, Ni^{2+}, Pb^{2+}/Pb^{4+}, Cd^{2+}, and Cr^{3+}/Cr^{6+}	91.13, 205.0, 1.58, 27.52, 0.45, and 0.69, respectively	H_2O_2, NaOH, and HCl	12 – 14	99.96, 99.97, 95.62, 99.62, 94.11, and 96.79, respectively	• Low-grade vehicle shredder residue was treated using it. • Temperature, liquid/solid ratio, and nitric acid (HNO_3) concentration all increased leaching efficiency. • The process was run at 35 degrees Celsius.	Qasem et al., 2021

13.5 CONCLUDING REMARKS

The current chapter delves into the complex and pressing issue of chromium contamination in the environment from various sources. Because of its non-biodegradability and ecological persistence, chromium, a versatile metal widely employed in several industrial processes, poses a considerable environmental and human health danger. This chapter has investigated numerous ways of tackling this issue, with a particular emphasis on chemical and biological extraction methods. The importance of this problem cannot be emphasized, since poor chromium contamination treatment can have far-reaching implications. Environmental contamination and related health risks highlight the importance of developing effective remediation solutions. We have reviewed the fundamental characteristics of chromium throughout this chapter, emphasizing the critical contrast between Cr(III) and Cr(VI) with reference to their environmental effect. One of the key takeaways from this chapter is the growing preference for bioremediation as an environmentally friendly and cost-effective approach to dealing with chromium-contaminated sites. This "natural" method uses microorganisms in eradicating Cr(VI) from water. The procedure consists of several phases, beginning with chromium binding to microbial cell surfaces and progressing to the reduction of Cr(VI) to the less dangerous Cr(III). The adaptability of microorganisms in lowering Cr(VI), whether directly through particular enzymes or indirectly through metabolites, demonstrates the possibility of biological remedies for chromium pollution. Furthermore, we investigated the dual-phase method of Cr ion absorption by microorganisms, which included both biosorption and bioaccumulation. These procedures provide information on the efficiency and kinetics of microbial chromium removal. Understanding how these systems interact is critical for improving bioremediation techniques.

As the urgency of tackling chromium pollution grows, so does the need for more research and development in this sector. Understanding the fundamental principles of microbial resistance and Cr(VI) elimination is crucial for designing and executing efficient bioremediation techniques. We have the opportunity to develop more efficient, sustainable, and targeted ways to mitigating chromium pollution in a variety of industrial scenarios with continued innovation. *Ergo*, this chapter is an essential resource for chromium contamination researchers, environmentalists, and industry experts. It emphasizes the need of resolving the issues created by chromium pollution in our industrial environment with sustainable and eco-friendly methods, notably bioremediation. We may advance toward a cleaner, safer, and more ecologically responsible industrial future by promoting a broader understanding of the chemical and biological extraction processes presented in this chapter.

ACKNOWLEDGMENT

The authors would like to thank GNIT, JIS group, and the MODROB-REG scheme of AICTE, Ministry of Education, Government of India.

REFERENCES

Abbasi-Garravand, E., & Mulligan, C. N. (2014). Using micellar enhanced ultrafiltration and reduction techniques for removal of Cr(VI) and Cr(III) from water. *Sep. Purif. Technol.*, *132*, 505–512. https://doi.org/10.1016/j.seppur.2014.06.010

Ahmad, W. A., Zakaria, Z. A., Khasim, A. R., Alias, M. A., & Ismail, S. M. H. S. (2010). Pilot-scale removal of chromium from industrial wastewater using the ChromeBacTM system. *Bioresour. Technol.*, *101*(12), 4371–4378. https://doi.org/10.1016/j.biortech.2010.01.106

Ahsan, M. A., Satter, F., Siddique, M. A. B., Akbor, M. A., Ahmed, S., Shajahan, M., & Khan, R. (2019). Chemical and physicochemical characterization of effluents from the tanning and textile industries in Bangladesh with multivariate statistical approach. *Environ. Monit. Assess.*, *191*(9), 575. https://doi.org/10.1007/s10661-019-7654-2

Ajmani, A., Patra, C., Subbiah, S., & Narayanasamy, S. (2020). Packed bed column studies of hexavalent chromium adsorption by zinc chloride activated carbon synthesized from Phanera vahlii fruit biomass. *J. Environ. Chem. Eng.*, *8*(4), 103825. https://doi.org/10.1016/j.jece.2020.103825

Bahaloo-Horeh, N., & Mousavi, S. M. (2017). Enhanced recovery of valuable metals from spent lithium-ion batteries through optimization of organic acids produced by Aspergillus niger. *Waste Manag.*, *60*, 666–679. https://doi.org/10.1016/j.wasman.2016.10.034

Bhattacharya, S., Mazumder, A., Sen, D., & Bhattacharjee, C. (2020). Bioaugmentation technology to remove oil from oily wastewater using isolated actinobacteria. In Daniels, J. A. (ed.) *Advances in Environmental Research* (Vol. 79). Nova Science Publishers, Inc, New York.

Bhattacharya, S., Mazumder, A., Sen, D., & Bhattacharjee, C. (2022). Bioremediation of dye using mesophilic bacteria: mechanism and parametric influence. In: Muthu, S. S. & Khadir, A. (eds.) *Dye Biodegradation, Mechanisms and Techniques. Sustainable Textiles: Production, Processing, Manufacturing & Chemistry* (pp. 67–86). Springer, Singapore. https://doi.org/10.1007/978-981-16-5932-4_3

Bhatti, M. S., Reddy, A. S., Kalia, R. K., & Thukral, A. K. (2011). Modeling and optimization of voltage and treatment time for electrocoagulation removal of hexavalent chromium. *Desalination*, *269*(1–3), 157–162. https://doi.org/10.1016/j.desal.2010.10.055

Boeykens, S. P., Redondo, N., Obeso, R. A., Caracciolo, N., & Vázquez, C. (2019). Chromium and lead adsorption by avocado seed biomass study through the use of total reflection x-ray fluorescence analysis. *Appl. Radiat. Isot.*, *153*, 108809. https://doi.org/10.1016/j.apradiso.2019.108809

Briggs, D. (2003). Environmental pollution and the global burden of disease. *Br. Med. Bull.*, *68*(1), 1–24. https://doi.org/10.1093/bmb/ldg019

Butter, B., Santander, P., Pizarro, G. del C., Oyarzún, D. P., Tasca, F., & Sánchez, J. (2021). Electrochemical reduction of Cr(VI) in the presence of sodium alginate and its application in water purification. *J. Environ. Sci.*, *101*, 304–312. https://doi.org/10.1016/j.jes.2020.08.033

Cao, P., Quan, X., Zhao, K., Chen, S., Yu, H., & Niu, J. (2020). Selective electrochemical H_2O_2 generation and activation on a bifunctional catalyst for heterogeneous electro-Fenton catalysis. *J. Hazard. Mater.*, *382*, 121102. https://doi.org/10.1016/j.jhazmat.2019.121102

Chardin, B., Giudici-Orticoni, M.-T., De Luca, G., Guigliarelli, B., & Bruschi, M. (2003). Hydrogenases in sulfate-reducing bacteria function as chromium reductase. *Appl. Microbiol.* Biotechnol., *63*(3), 315–321. https://doi.org/10.1007/s00253-003-1390-8

Chigondo, M., Nyamunda, B., Maposa, M., & Chigondo, F. (2022). Polypyrrole-based adsorbents for Cr(VI) ions remediation from aqueous solution: a review. *Water Sci. Technol.*, *85*(5), 1600–1619. https://doi.org/10.2166/wst.2022.050

Costello, R. B., Dwyer, J. T., & Merkel, J. M. (2019). Chromium supplements in health and disease. In *The Nutritional Biochemistry of Chromium (III)* (pp. 219–249). Elsevier, Amsterdam, The Netherlands. https://doi.org/10.1016/B978-0-444-64121-2.00007-6

Das, D., & Nandi, B. K. (2020). Removal of hexavalent chromium from wastewater by electrocoagulation (EC): parametric evaluation, kinetic study and operating cost. *Trans. Indian Inst. Met.*, *73*(8), 2053–2060. https://doi.org/10.1007/s12666-020-01962-4

Das, S. K., Mukherjee, M., & Guha, A. K. (2008). Interaction of chromium with resistant strain aspergillus versicolor : investigation with atomic force microscopy and other physical studies. *Langmuir*, *24*(16), 8643–8650. https://doi.org/10.1021/la800958u

Gheju, M., & Balcu, I. (2011). Removal of chromium from Cr(VI) polluted wastewaters by reduction with scrap iron and subsequent precipitation of resulted cations. *J. Hazard. Mater.*, *196*, 131–138. https://doi.org/10.1016/j.jhazmat.2011.09.002

Golder, A. K., Chanda, A. K., Samanta, A. N., & Ray, S. (2011). Removal of hexavalent chromium by electrochemical reduction–precipitation: investigation of process performance and reaction stoichiometry. *Sep. Purif. Technol.*, *76*(3), 345–350. https://doi.org/10.1016/j.seppur.2010.11.002

Grace Pavithra, K., Senthil Kumar, P., Carolin Christopher, F., & Saravanan, A. (2017). Removal of toxic Cr(VI) ions from tannery industrial wastewater using a newly designed three-phase three-dimensional electrode reactor. *J. Phys. Chem. Solids*, *110*, 379–385. https://doi.org/10.1016/j.jpcs.2017.07.002

GracePavithra, K., Jaikumar, V., Kumar, P. S., & SundarRajan, P. (2019). A review on cleaner strategies for chromium industrial wastewater: present research and future perspective. *J. Clean. Prod.*, *228*, 580–593. https://doi.org/10.1016/j.jclepro.2019.04.117

Gunasundari, E., & Senthil Kumar, P. (2017). Higher adsorption capacity of Spirulina platensis alga for Cr(VI) ions removal: parameter optimisation, equilibrium, kinetic and thermodynamic predictions. *IET Nanobiotechnology*, *11*(3), 317–328. https://doi.org/10.1049/iet-nbt.2016.0121

Guo, S., Xiao, C., Zhou, N., & Chi, R. (2021). Speciation, toxicity, microbial remediation and phytoremediation of soil chromium contamination. *Environ. Chem. Lett.*, *19*(2), 1413–1431. https://doi.org/10.1007/s10311-020-01114-6

Haynes, W. M. (ed.). (2014). *CRC Handbook of Chemistry and Physics*. CRC Press, Boca Raton, FL. https://doi.org/10.1201/b17118

Hedayatipour, M., Jaafarzadeh, N., & Ahmadmoazzam, M. (2017). Removal optimization of heavy metals from effluent of sludge dewatering process in oil and gas well drilling by nanofiltration. *J. Environ. Manage.*, *203*, 151–156. https://doi.org/10.1016/j.jenvman.2017.07.070

Ho, Y. C., Show, K. Y., Guo, X. X., Norli, I., Alkarkhi, F. M., & Mor, N. (2012). Industrial Discharge and Their Effect to the Environment. In: Show, K. Y. & Guo, X. (eds.) *Industrial Waste*. InTech, New York. https://doi.org/10.5772/38830

Huang, T., Shao, Y., Zhang, Q., Deng, Y., Liang, Z., Guo, F., Li, P., & Wang, Y. (2019). Chitosan-cross-linked unction oxide/carboxymethyl cellulose aerogel globules with high structure stability in liquid and extremely high adsorption ability. *ACS Sustain. Chem. Eng.*, *7*(9), 8775–8788. https://doi.org/10.1021/acssuschemeng.9b00691

Joutey, N. T., Sayel, H., Bahafid, W., & El Ghachtouli, N. (2015). Mechanisms of hexavalent chromium resistance and removal by microorganisms. In: Whitacre, D. (ed.) *Reviews of Environmental Contamination and Toxicology* (Vol. 233, pp. 45–69). Springer, Cham. https://doi.org/10.1007/978-3-319-10479-9_2

Khan, Z. U. H., Gul, N. S., Sabahat, S., Sun, J., Tahir, K., Shah, N. S., Muhammad, N., Rahim, A., Imran, M., Iqbal, J., Khan, T. M., Khasim, S., Farooq, U., & Wu, J. (2023). Removal of organic pollutants through hydroxyl radical-based advanced oxidation processes. *Ecotoxicol. Environ. Saf.*, *267*, 115564. https://doi.org/10.1016/j.ecoenv.2023.115564

Kharrazi, S. M., Mirghaffari, N., Dastgerdi, M. M., & Soleimani, M. (2020). A novel post-modification of powdered activated carbon prepared from lignocellulosic waste through thermal tension treatment to enhance the porosity and heavy metals adsorption. *Powder Technol.*, *366*, 358–368. https://doi.org/10.1016/j.powtec.2020.01.065

Kim, T., Kim, T.-K., & Zoh, K.-D. (2020). Removal mechanism of heavy metal (Cu, Ni, Zn, and Cr) in the presence of cyanide during electrocoagulation using Fe and Al electrodes. *J. Water Process Eng.*, *33*, 101109. https://doi.org/10.1016/j.jwpe.2019.101109

Kumar, P. S., & Yashwanthraj, M. (2017). Sequestration of toxic Cr(VI) ions from industrial wastewater using waste biomass: a review. *Desalin. Water Treat.*, *68*, 245–266. https://doi.org/10.5004/dwt.2017.20322

Kumar, R., Ansari, M. O., Alshahrie, A., Darwesh, R., Parveen, N., Yadav, S. K., Barakat, M. A., & Cho, M. H. (2019). Adsorption modeling and mechanistic insight of hazardous chromium on para toluene sulfonic acid immobilized-polyaniline@CNTs nanocomposites. *J. Saudi Chem. Soc.*, *23*(2), 188–197. https://doi.org/10.1016/j.jscs.2018.06.005

Liang, H., Song, B., Peng, P., Jiao, G., Yan, X., & She, D. (2019). Preparation of three-dimensional honeycomb carbon materials and their adsorption of Cr(VI). *Chem. Eng. J.*, *367*, 9–16. https://doi.org/10.1016/j.cej.2019.02.121

Liu, Q., Li, Y., Chen, H., Lu, J., Yu, G., Möslang, M., & Zhou, Y. (2020). Superior adsorption capacity of unctionalized straw adsorbent for dyes and heavy-metal ions. *J. Hazard. Mater.*, *382*, 121040. https://doi.org/10.1016/j.jhazmat.2019.121040

Malaviya, P., & Singh, A. (2016). Bioremediation of chromium solutions and chromium containing wastewaters. *Crit. Rev. Microbiol.*, *42*(4), 607–633. https://doi.org/10.3109/1040841X.2014.974501

Mazumder, A., Bhattacharya, S., & Bhattacharjee, C. (2020). Role of nano-photocatalysis in heavy metal detoxification. *Front. Environ. Sci.*, 12, pp. 1–33. https://doi.org/10.1007/978-3-030-12619-3_1

McLean, J., & Beveridge, T. J. (2001). Chromate reduction by a pseudomonad isolated from a site contaminated with chromated copper arsenate. *Appl. Environ. Microbiol.*, *67*(3), 1076–1084. https://doi.org/10.1128/AEM.67.3.1076-1084.2001

Medfu Tarekegn, M., Zewdu Salilih, F., & Ishetu, A. I. (2020). Microbes used as a tool for bioremediation of heavy metal from the environment. *Cogent Food Agric.*, *6*(1), 1783174. https://doi.org/10.1080/23311932.2020.1783174

Mishra, A. K., & Mohanty, B. (2008). Acute toxicity impacts of hexavalent chromium on behavior and histopathology of gill, kidney and liver of the freshwater fish, Channa punctatus (Bloch). *Environ. Toxicol. Pharmacol.*, *26*(2), 136–141. https://doi.org/10.1016/j.etap.2008.02.010

Mohammad Gheimasi, M. H., Lorestani, B., Kiani Sadr, M., Cheraghi, M., & Emadzadeh, D. (2021). Synthesis of novel hybrid NF/FO nanocomposite membrane by incorporating black TiO_2 nanoparticles for highly efficient heavy metals removal. *Int. J. Environ. Res.*, *15*(3), 475–485. https://doi.org/10.1007/s41742-021-00317-1

Morales-Barrera, L., Guillén-Jiménez, F. de M., Ortiz-Moreno, A., Villegas-Garrido, T. L., Sandoval-Cabrera, A., Hernández-Rodríguez, C. H., & Cristiani-Urbina, E. (2008). Isolation, identification and characterization of a Hypocrea tawa strain with high Cr(VI) reduction potential. *Biochem. Eng. J.*, *40*(2), 284–292. https://doi.org/10.1016/j.bej.2007.12.014

Moussa, D. T., El-Naas, M. H., Nasser, M., & Al-Marri, M. J. (2017). A comprehensive review of electrocoagulation for water treatment: potentials and challenges. *J. Environ. Manage.*, *186*, 24–41. https://doi.org/10.1016/j.jenvman.2016.10.032

Muddanna, M. H., & Baral, S. S. (2019). A comparative study of the extraction of metals from the spent fluid catalytic cracking catalyst using chemical leaching and bioleaching by Aspergillus niger. *J. Environ. Chem. Eng.*, *7*(5), 103335. https://doi.org/10.1016/j.jece.2019.103335

Naseri, T., Bahaloo-Horeh, N., & Mousavi, S. M. (2019). Bacterial leaching as a green approach for typical metals recovery from end-of-life coin cells batteries. *J. Clean. Prod.*, *220*, 483–492. https://doi.org/10.1016/j.jclepro.2019.02.177

Naz, A., Chowdhury, A., & Mishra, B. K. (2021). An insight into microbial remediation of hexavalent chromium from contaminated water. In: Kumar, M., Snow, D., Honda, R., & Mukherjee, S. (eds.) *Contaminants in Drinking and Wastewater Sources. Springer Transactions in Civil and Environmental Engineering* (pp. 209–224). Springer, Singapore. https://doi.org/10.1007/978-981-15-4599-3_9

Ociński, D., Augustynowicz, J., Wołowski, K., Mazur, P., Sitek, E., & Raczyk, J. (2021). Natural community of macroalgae from chromium-contaminated site for effective remediation of Cr(VI)-containing leachates. *Sci. Total Environ.*, *786*, 147501. https://doi.org/10.1016/j.scitotenv.2021.147501

Otero, J. A., Mazarrasa, O., Otero-Fernández, A., Fernández, M. D., Hernández, A., & Maroto-Valiente, A. (2012). Treatment of wastewater. Removal of heavy metals by nanofiltration. Case study: use of TFC membranes to separate Cr(VI) in industrial pilot plant. *Procedia Eng.*, *44*, 2020–2022. https://doi.org/10.1016/j.proeng.2012.09.029

Owa, F. D. (2013). Water pollution: sources, effects, control and management. *Mediterr. J. Soc. Sci.*, *8*, 8. https://doi.org/10.5901/mjss.2013.v4n8p65

Panda, L., Das, B., Rao, D. S., & Mishra, B. K. (2011). Application of dolochar in the removal of cadmium and hexavalent chromium ions from aqueous solutions. *J. Hazard. Mater.*, *192*(2), 822–831. https://doi.org/10.1016/j.jhazmat.2011.05.098

Park, D., Yun, Y.-S., Hye Jo, J., & Park, J. M. (2005). Mechanism of hexavalent chromium removal by dead fungal biomass of Aspergillus niger. *Water Res.*, *39*(4), 533–540. https://doi.org/10.1016/j.watres.2004.11.002

Pradhan, D., Sukla, L. B., Sawyer, M., & Rahman, P. K. S. M. (2017). Recent bioreduction of hexavalent chromium in wastewater treatment: a review. *J. Ind. Eng. Chem.*, *55*, 1–20. https://doi.org/10.1016/j.jiec.2017.06.040

Putatunda, S., Bhattacharya, S., Sen, D., & Bhattacharjee, C. (2019). A review on the application of different treatment processes for emulsified oily wastewater. *Inter. J. Environ. Sci. Technol.*, *16*(5), 2525–2536. https://doi.org/10.1007/s13762-018-2055-6

Qasem, N. A. A., Mohammed, R. H., & Lawal, D. U. (2021). Removal of heavy metal ions from wastewater: a comprehensive and critical review. *NPJ Clean Water*, *4*(1), 36. https://doi.org/10.1038/s41545-021-00127-0

Qu, J., Wang, Y., Tian, X., Jiang, Z., Deng, F., Tao, Y., Jiang, Q., Wang, L., & Zhang, Y. (2021). KOH-activated porous biochar with high specific surface area for adsorptive removal of chromium (VI) and naphthalene from water: affecting factors, mechanisms and reusability exploration. *J. Hazard. Mater.*, *401*, 123292. https://doi.org/10.1016/j.jhazmat.2020.123292

Romanenko, V. I., & Koren'kov, V. N. (1977). Pure culture of bacteria using chromates and bichromates as hydrogen acceptors during development under anaerobic conditions. *Mikrobiologiia*, *46*(3), 414–417. https://www.ncbi.nlm.nih.gov/pubmed/895551

Saeedi-Jurkuyeh, A., Jafari, A. J., Kalantary, R. R., & Esrafili, A. (2020). A novel synthetic thin-film nanocomposite forward osmosis membrane modified by graphene oxide and polyethylene glycol for heavy metals removal from aqueous solutions. *React. Funct. Polym.*, *146*, 104397. https://doi.org/10.1016/j.reactfunctpolym.2019.104397

Sangeetha, K., Sudha, P. N., & Sukumaran, A. (2019). Novel chitosan based thin sheet nanofiltration membrane for rejection of heavy metal chromium. *Int. J. Biol. Macromol.*, *132*, 939–953. https://doi.org/10.1016/j.ijbiomac.2019.03.244

Saw, P. K., Prajapati, A. K., & Mondal, M. K. (2018). The extraction of Cr(VI) from aqueous solution with a mixture of TEA and TOA as synergic extractant by using different diluents. *J. Mol. Liq.*, *269*, 101–109. https://doi.org/10.1016/j.molliq.2018.07.115

Sellami, F., Kebiche-Senhadji, O., Marais, S., Couvrat, N., & Fatyeyeva, K. (2019). Polymer inclusion membranes based on CTA/PBAT blend containing Aliquat 336 as extractant for removal of Cr(VI): efficiency, stability and selectivity. *React. Funct. Polym.*, *139*, 120–132. https://doi.org/10.1016/j.reactfunctpolym.2019.03.014

Shanker, A. K., & Venkateswarlu, B. (2011). Chromium: environmental pollution, health effects and mode of action. In *Encyclopedia of Environmental Health* (pp. 650–659). Elsevier, Amsterdam, The Netherlands. https://doi.org/10.1016/B978-0-444-52272-6.00390-1

Sharma, S., & Adholeya, A. (2011). Detoxification and accumulation of chromium from tannery effluent and spent chrome effluent by Paecilomyces lilacinus fungi. *Int. Biodeterior. Biodegradation*, *65*(2), 309–317. https://doi.org/10.1016/j.ibiod.2010.12.003

Sharma, D., Chaudhari, P. K., & Prajapati, A. K. (2020). Removal of chromium (VI) and lead from electroplating effluent using electrocoagulation. *Sep. Sci. Technol.*, *55*(2), 321–331. https://doi.org/10.1080/01496395.2018.1563157

Sharma, P., Singh, S. P., Parakh, S. K., & Tong, Y. W. (2022). Health hazards of hexavalent chromium (Cr(VI)) and its microbial reduction. *Bioengineered*, *13*(3), 4923–4938. https://doi.org/10.1080/21655979.2022.2037273

Shrestha, R., Ban, S., Devkota, S., Sharma, S., Joshi, R., Tiwari, A. P., Kim, H. Y., & Joshi, M. K. (2021). Technological trends in heavy metals removal from industrial wastewater: a review. *J. Environ. Chem. Eng.*, *9*(4), 105688. https://doi.org/10.1016/j.jece.2021.105688

Singh, N. B., Nagpal, G., Agrawal, S., & Rachna, A. (2018). Water purification by using adsorbents: a review. *Environ. Technol. Innov.*, *11*, 187–240. https://doi.org/10.1016/j.eti.2018.05.006

Singh, R., Andaluri, G., & Pandey, V. C. (2022). Cities' water pollution – challenges and controls. In *Algae and Aquatic Macrophytes in Cities* (pp. 3–22). Elsevier, Amsterdam, The Netherlands. https://doi.org/10.1016/B978-0-12-824270-4.00015-8

Somasundaram, V., Philip, L., & Bhallamudi, S. M. (2009). Experimental and mathematical modeling studies on Cr(VI) reduction by CRB, SRB and IRB, individually and in combination. *J. Hazard. Mater.*, *172*(2–3), 606–617. https://doi.org/10.1016/j.jhazmat.2009.07.043

Stern, C. M., Jegede, T. O., Hulse, V. A., & Elgrishi, N. (2021). Electrochemical reduction of Cr(VI) in water: lessons learned from fundamental studies and applications. *Chem. Soc. Rev.*, *50*(3), 1642–1667. https://doi.org/10.1039/D0CS01165G

Suganya, S., & Senthil Kumar, P. (2018). Influence of ultrasonic waves on preparation of active carbon from coffee waste for the reclamation of effluents containing Cr(VI) ions. *J. Ind. Eng. Chem.*, *60*, 418–430. https://doi.org/10.1016/j.jiec.2017.11.029

Sum, J. Y., Kok, W. X., & Shalini, T. S. (2021). The removal selectivity of heavy metal cations in micellar-enhanced ultrafiltration: a study based on critical micelle concentration. *Mater. Today Proc.*, *46*, 2012–2016. https://doi.org/10.1016/j.matpr.2021.02.683

Suresh Kumar, K., Dahms, H.-U., Won, E.-J., Lee, J.-S., & Shin, K.-H. (2015). Microalgae – a promising tool for heavy metal remediation. *Ecotoxicol. Environ. Saf.*, *113*, 329–352. https://doi.org/10.1016/j.ecoenv.2014.12.019

Tandon, R. K., Crisp, P. T., Ellis, J., & Baker, R. S. (1984). Effect of pH on chromium(VI) species in solution. *Talanta*, *31*(3), 227–228. https://doi.org/10.1016/0039-9140(84)80059-4

Tang, C. Y., Fu, Q. S., Criddle, C. S., & Leckie, J. O. (2007). Effect of flux (transmembrane pressure) and membrane properties on fouling and rejection of reverse osmosis and nanofiltration membranes treating perfluorooctane sulfonate containing wastewater. *Environ. Sci. Technol.*, *41*(6), 2008–2014. https://doi.org/10.1021/es062052f

Tang, X., Huang, Y., Li, Y., Wang, L., Pei, X., Zhou, D., He, P., & Hughes, S. S. (2021). Study on detoxification and removal mechanisms of hexavalent chromium by microorganisms. *Ecotoxicol. Environ. Saf.*, *208*, 111699. https://doi.org/10.1016/j.ecoenv.2020.111699

Teshale, F., Karthikeyan, R., & Sahu, O. (2020). Synthesized bioadsorbent from fish scale for chromium (III) removal. *Micron*, *130*, 102817. https://doi.org/10.1016/j.micron.2019.102817

Tiadi, N., Dash, R. R., & Mohanty, C. R. (2022). Utilization of industrial waste for Cr(VI) adsorption from aqueous solution: a statistical modeling approach. *Arab. J. Geosci.*, *15*(3), 260. https://doi.org/10.1007/s12517-022-09433-4

Vakili, M., Deng, S., Li, T., Wang, W., Wang, W., & Yu, G. (2018). Novel crosslinked chitosan for enhanced adsorption of hexavalent chromium in acidic solution. *Chem. Eng. J.*, *347*, 782–790. https://doi.org/10.1016/j.cej.2018.04.181

Vankar, P. S., & Bajpai, D. (2008). Phyto-remediation of chrome-VI of tannery effluent by Trichoderma species. *Desalination*, *222*(1–3), 255–262. https://doi.org/10.1016/j.desal.2007.01.168

Verma, T., Garg, S. K., & Ramteke, P. W. (2009). Genetic correlation between chromium resistance and reduction in Bacillus brevis isolated from tannery effluent. *J. Appl.* Microbiol., *107*(5), 1425–1432. https://doi.org/10.1111/j.1365-2672.2009.04326.x

Wallwork, G. R. (1976). The oxidation of alloys. *Reports Prog. Phys.*, *39*(5), 401–485. https://doi.org/10.1088/0034-4885/39/5/001

Wang, Y., Li, J., Jin, Y., Chen, M., & Ma, R. (2020). Extraction of chromium (III) from aqueous waste solution in a novel rotor-stator spinning disc reactor. *Chem. Eng. Process. - Process Intensif.*, *149*, 107834. https://doi.org/10.1016/j.cep.2020.107834

Wani, P. A., Wahid, S., Khan, M. S. A., Rafi, N., & Wahid, N. (2019). Investigation of the role of chromium reductase for Cr(VI) reduction by Pseudomonas species isolated from Cr(VI) contaminated effluent. *Biotechnol. Res. Innov.*, *3*(1), 38–46. https://doi.org/10.1016/j.biori.2019.04.001

Wei, W., Cho, C.-W., Kim, S., Song, M.-H., Bediako, J. K., & Yun, Y.-S. (2016). Selective recovery of Au(III), Pt(IV), and Pd(II) from aqueous solutions by liquid–liquid extraction using ionic liquid Aliquat-336. *J. Mol. Liq.*, *216*, 18–24. https://doi.org/10.1016/j.molliq.2016.01.016

Xiao, G., Wang, Y., Xu, S., Li, P., Yang, C., Jin, Y., Sun, Q., & Su, H. (2019). Superior adsorption performance of graphitic carbon nitride nanosheets for both cationic and anionic heavy metals from wastewater. *Chinese J. Chem. Eng.*, *27*(2), 305–313. https://doi.org/10.1016/j.cjche.2018.09.028

Yang, X., Liu, L., Zhang, M., Tan, W., Qiu, G., & Zheng, L. (2019). Improved removal capacity of magnetite for Cr(VI) by electrochemical reduction. *J. Hazard. Mater.*, *374*, 26–34. https://doi.org/10.1016/j.jhazmat.2019.04.008

Yao, F., Jia, M., Yang, Q., Luo, K., Chen, F., Zhong, Y., He, L., Pi, Z., Hou, K., Wang, D., & Li, X. (2020). Electrochemical Cr(VI) removal from aqueous media using titanium as anode: simultaneous indirect electrochemical reduction of Cr(VI) and in-situ precipitation of Cr(III). *Chemosphere*, *260*, 127537. https://doi.org/10.1016/j.chemosphere.2020.127537

Yin, H., Guo, Q., Lei, C., Chen, W., & Huang, B. (2020). Electrochemical-driven carbocatalysis as highly efficient advanced oxidation processes for simultaneous removal of humic acid and Cr(VI). *Chem. Eng. J.*, *396*, 125156. https://doi.org/10.1016/j.cej.2020.125156

Yu, S., Liu, Y., Ai, Y., Wang, X., Zhang, R., Chen, Z., Chen, Z., Zhao, G., & Wang, X. (2018). Rational design of carbonaceous nanofiber/Ni-Al layered double hydroxide nanocomposites for high-efficiency removal of heavy metals from aqueous solutions. *Environ. Pollut.*, *242*, 1–11. https://doi.org/10.1016/j.envpol.2018.06.031

Zhang, D., Wang, C., Bao, Q., Zheng, J., Deng, D., Duan, Y., & Shen, L. (2018). The physicochemical characterization, equilibrium, and kinetics of heavy metal ions adsorption from aqueous solution by arrowhead plant (Sagittaria trifolia L.) stalk. *J. Food Biochem.*, *42*(1), e12448. https://doi.org/10.1111/jfbc.12448

Zhitkovich, A. (2011). Chromium in drinking water: sources, metabolism, and cancer risks. *Chem. Res. Toxicol.*, *24*(10), 1617–1629. https://doi.org/10.1021/tx200251t

Zhou, R., Liu, F., Wei, N., Yang, C., Yang, J., Wu, Y., Li, Y., Xu, K., Chen, X., & Zhang, C. (2020). Comparison of Cr(VI) removal by direct and pulse current electrocoagulation: implications for energy consumption optimization, sludge reduction and floc magnetism. *J. Water Process Eng.*, *37*, 101387. https://doi.org/10.1016/j.jwpe.2020.101387

14 Recovery of Uranium from Mine Discards Through Bioleaching

Zunipa Roy

14.1 INTRODUCTION

The rising global human population has confiscated all the possible energy resources to meet the ubiquitous demand. Combustion of coal, oil, and other fossil fuels in course of time has encountered an alarming phase down schedule owing to its greenhouse gas emission. This has aggravated the exploitation of nuclear energy which has duly consumed the high-grade uranium reserve (Cahill and Burkhart, 1990; Mudd, 2009). Excavation of the existing reserve of uranium all across the globe has not produced rich content ores. Therefore, to avoid the extractive metallurgical industries managing low ore grades with mounting operation cost of mining and grinding, the awareness of the researchers has drifted toward the relatively economic yet energy and environment friendly up-surging bioleaching process. Present practices of bio-hydrometallurgy have been implemented in uranium ore leaching commercially (McCready and Gould, 1990). Uranium (U) contamination poses a significant health risk, necessitating effective and sustainable remediation methods. Microbial reduction of soluble U(VI) in bioremediation produces high fractions (>50%) of insoluble non-crystalline U(IV). However, this U(IV) can be remobilized by sulfur-oxidizing bacteria. *Acidithiobacillus ferrooxidans* and *Thiobacillus denitrificans* have the capability to mobilize non-crystalline U(IV) and examine the associated U isotope fractionation. *A. ferrooxidans* mobilized between 74% and 91% of U within a week, with mobilization occurring in both living and inactive cells. Contrary to previous findings, *T. denitrificans* did not exhibit any mobilization. Additionally, uranium mobilization by *A. ferrooxidans* did not result in U isotope fractionation, indicating that U isotope ratio determination is unsuitable as a direct proxy for bacterial U remobilization. Environmental uranium (U) contamination arises from various sources, such as mining and milling operations, military applications, and illegal or improper disposal (Dienemann and Utermann, 2012; Zammit et al., 2014). In oxidizing environments, highly soluble uranyl compounds and U complexes are formed (Abdelouas, 2006), which can be transported over long distances by water. Further environmental U enrichment can occur due to naturally high U content in bedrock and soils, the application of U-containing mineral phosphate fertilizers, and groundwater intrusion into permanent nuclear waste repository sites

 DOI: 10.1201/9781003415541-14

(Dienemann and Utermann, 2012). Addressing such contamination requires effective in-situ treatment. Physical and chemical remediation methods are often cost-prohibitive and lack sustainability (Lovley and Phillips, 1992). In contrast, bioremediation through the (micro)biological reduction of soluble U(VI) to sparingly soluble U(IV), such as by *Shewanella oneidensis*, shows great promise (Finneran et al., 2002; Wall and Krumholz, 2006; Newsome et al., 2014, 2015). A significant challenge in bacterial bioremediation is the formation of biomass-associated non-crystalline U(IV), a mixture of compounds potentially coordinated to carboxylic, phosphate, or silicate moieties (Wang et al., 2013; Alessi et al., 2014). The proportion of non-crystalline U(IV) formed by *S. oneidensis* MR-1 can range from 50% to nearly 100%, depending on the presence of dissolved solutes like phosphate, silicate, and sulfate, which naturally occur in sediments, soils, or groundwater, and on the U concentration (Stylo et al., 2013). Non-crystalline U(IV) is more labile than uraninite, and its rapid oxidation during oxygen exposure, by nitrate, or through ligand complexation (Cerrato et al., 2013; Newsome et al., 2015; Roebbert et al., 2021) necessitates detailed investigations into its environmental stability to assess the efficiency and vulnerability of bioremediation efforts.

The declining grades of uranium has come up with sequential leaching of uranium and phosphate (Makinen et al., 2019), thorium (Desouky et al., 2016), rare earth elements (Nancucheo et al., 2017), and base metal (Lecomte et al., 2014) and has drawn widespread attention. The process of extraction of specific metals from their ores by means of bacteria defines the process of bioleaching. Since the excavation of uranium started on large scale, major developments have been observed in the arenas of heap bioleaching of run-off-mine (ROM) and crushed ore as well as agitated tank bioleaching of concentrates. The pyritic heap leaching operation has effectively commercialized bioleaching process, thereby suspending in-situ blasting. Bioleaching has been used commercially for pyritic heap leaching and in-place leaching of low-grade underground mine stopes, including those broken by blasting and old mine stopes. Since the last uranium boom, significant advancements have been made in both heap bioleaching of ROM or crushed ore and agitated tank bioleaching of concentrates. These advancements include the use of aeration pipes, the addition of nutrients, more efficient agitators, and new ultrafine grinding equipment. These innovations are likely to increase the use of bioleaching in future uranium projects involving sulfidic ores. Heap or in-place systems will be more suitable for lower-grade ores, while tank bioleaching will be better for uranium-bearing sulfidic concentrates. Figure 14.1 shows the roles of microbes in uranium recovery and Figure 14.2 represents the activities of different types of bacteria that take active part in the bioleaching of uranium.

Bioleaching of uranium from low-grade ores has been commercially practiced since the 1960s. Notable in-situ leaching operations occurred in the underground uranium mines in the Elliot Lake district of Canada, including the Stanrock, Milliken, and Denison Mines. During that time, the Stanrock Mine produced about 50,000 kg of U_3O_8 annually, while the Milliken Mine produced 60,000 kg of U_3O_8 after improving leaching conditions. However, uranium production significantly declined in the early 1980s. Denison Mines resumed activities in 1984, and by 1988–1990, flood leaching stopes were in various stages of operation or preparation. These operations produced 347 tons of uranium, valued at over US$25 million.

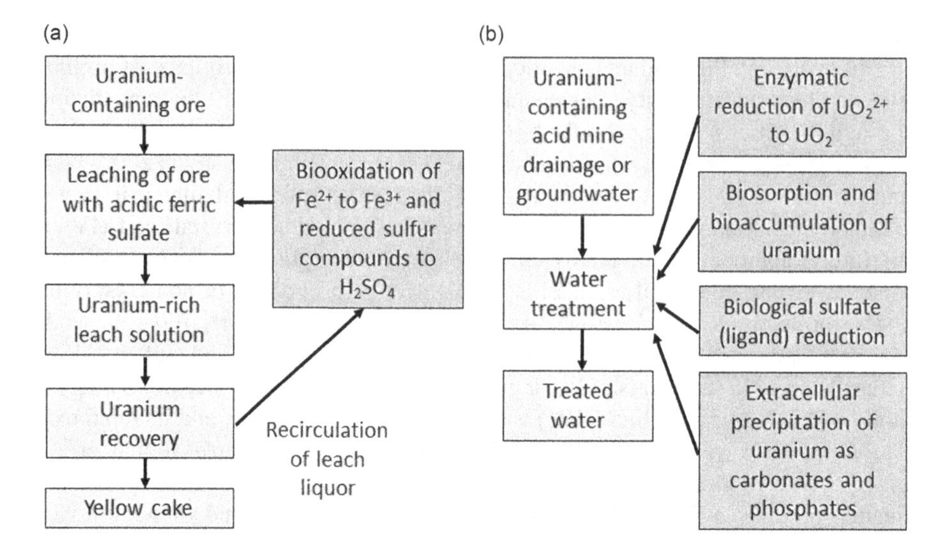

FIGURE 14.1 Potential roles of microorganisms in (a) acid and ferric sulfate leaching of uranium ores and (b) treatment of uranium-containing acid mine drainage and groundwater. Reproduced with copyright from Kaksonen et al. (2020).

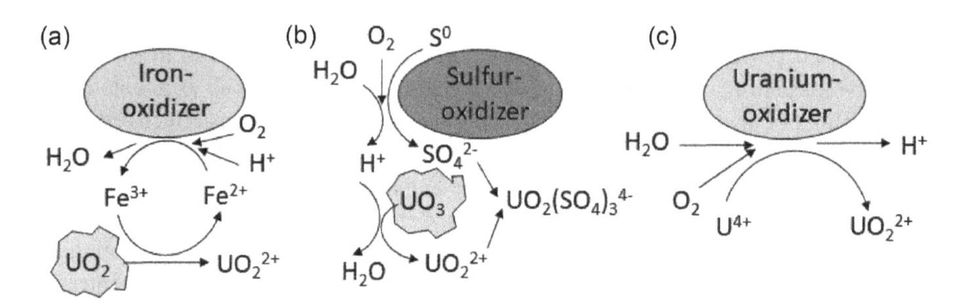

FIGURE 14.2 A schematic diagram of the roles of (a) iron-oxidizing, (b) sulfur-oxidizing, and (c) uranium-oxidizing microorganisms in the bioleaching or uranium ores. Reproduced with copyright from Kaksonen et al. (2020).

14.2 OCCURRENCE OF URANIUM ORE

The occurrence of uranium in Earth's crust, which is about two parts per million, is predominantly in the form of oxides and silicate minerals and in the phosphate rock in alliance with various other metals, while a rough estimation states that the oceans contain around 4.5×10^9 tons. Among ~200 minerals containing uranium, a few namely, autunite, betafite, brannerite, carnotite, coffinite, pitchblende, uranite, uranophane, uranothorianite, uranothorite, etc. are of economic significance and tested for bioleaching (Edwards and Oliver, 2000; Pohl, 2005; Pownceby and Johnson, 2014; Bhargava et al., 2015). As per the data obtained from the World Nuclear Association, around 63% of the global uranium production comes from the mines of Kakastan, Australia, and Canada.

TABLE 14.1
Classification of Uranium Ores (Patra et al., 2011)

Source	ppm U
Very high-grade ore (Canada)	200,000
High-grade ore	20,000
Low-grade ore	1,000
Very low-grade ore (Namibia)	100
Granite	4–5
Sedimentary rock	2
Earth's continental crust	2.8
Seawater	0.003

Primary uranium minerals, found in magmatic hydrothermal veins and in pegmatites, include uraninite and pitchblende. The uranium in these two ores occurs in the form of uranium dioxide, which—owing to oxidation—can vary in exact chemical composition from UO_2 to $UO_{2.67}$. Generally, the oxide minerals contain uranium in the form of triuranium octoxide (U_3O_8) which comprises 70%–90% (w/w) of the main component and has the appearance of yellow cake which is both physically and chemically stable that is handled in the refineries and mills. Basically, U_3O_8 is a composite of uranyl sulfate $(UO_2)_x(SO_4)_y(OH)_{2(x-y)}$, uranium trioxide ($UO_3$), sodium p-uranate ($Na_6U_7O_{24}$), uranyl hydroxide ($UO_2(OH)_2.nH_2O$), uranyl peroxide ($UO_4$), and inconsequential quantity of other uranium oxides (Hausen, 1998). Table 14.1 depicts the classification of uranium ores.

14.3 RECOVERY OF URANIUM USING MICROORGANISMS FOR LEACHING

The leaching process overshadows the ore dressing process for extraction of high purity uranium from low-grade ores. During execution of the bioleaching extraction process, significant amounts of pyrite and other ferric sulfide surfaces along with the uranium ores get oxidized to acidic ferric sulfate. Depending on the grade of deposit and economical factors, the newer wings of leaching like heap, in situ, or bioleaching may be implemented. For the past four decades, ground ore has been leached mostly with agitated acid and with alkali for ores where acid consumption is very high. Later on, further modification of technology has led to pressure leaching which essentially treats the refractory ores in the presence of pyrites that further get oxidized to form the acid and ferric ion required for further progression. Due to the current reduction in global demand for uranium, prices have dropped, and Denison Mines has ceased production. Harrison et al. (1966) reported the role of the iron-oxidizing *A. ferrooxidans* in uranium leaching. At the Elliot Lake Mine in Ontario, Canada, uranium ore was stacked in heaps similar to the dump leaching of low-grade copper ore and leached using an acidic ferric sulfate solution. The bacteria in the heaps played a crucial role in maintaining oxidizing conditions by converting ferrous to ferric iron for uranium extraction.

A unique commercial application for extracting uranium from underground low-grade ore was demonstrated at the Denison Mine in Ontario, Canada. This system involved intermittent flooding of blasted ore in an underground stope sealed with a concrete wall, showcasing the potential of a biohydrometallurgical process for uranium extraction. This process also considered the microorganisms' needs for optimal activity by providing nutrients and aeration to promote bacterial growth. However, the economics of uranium have hindered further use and development of microbial processes for uranium extraction. Iron-oxidizing acidophilic bacteria from coal mine samples, initially named *Thiobacillus ferrooxidans*, were later reclassified as *A. ferrooxidans* (Temple and Colmer, 1951). Sulfur-oxidizing bacteria, initially identified as *Thiobacillus thiooxidans*, were also described (Leathen et al., 1953a,b, 1956), with the original identification of *A. thiooxidans* dating back to the 1920s (Harrison, 1988). *Ferrobacillus ferrooxidans*, a novel isolate capable of oxidizing Fe^{2+} but not inorganic sulfur compounds, was later renamed *T. ferrooxidans* (and subsequently *A. ferrooxidans*) after recognizing that its apparent inability to grow on sulfur compounds was due to flawed experimental conditions (Kelly and Wood, 2000). Similarly, *Ferrobacillus sulfooxidans*, identified from coal mine drainage, was eventually reclassified under *T. ferrooxidans*.

However, post 1960s saw the discovery and characterization of numerous new genera and species, including *Acidithiobacillus ferrivorans*, *Acidithiobacillus ferridurans*, *Acidithiobacillus ferriphilus*, *Acidithiobacillus ferrianus*, *Acidithiobacillus sulfuriphilus*, *Leptospirillum* spp., the moderately thermophilic *Acidithiobacillus caldus*, *Acidimicrobium*, *Sulfobacillus* spp., as well as mesophilic and thermophilic archaea such as *Ferroplasma*, *Sulfolobus*, *Acidianus*, and *Metallosphaera* spp., among others, including iron-oxidizing heterotrophs (Johnson and Hallberg, 2008; Schippers et al., 2010, 2014; Johnson, 2012; Mahmoud et al., 2017; Falagán et al., 2019; Norris et al., 2020). The initial impetus for bioleaching approaches stemmed from laboratory studies that demonstrated higher yields of metal leaching from sulfide minerals in the presence of iron- and sulfur-oxidizing bacteria compared to abiotic chemical controls. As the potential for bioleaching of sulfide ores became better recognized, these findings extended to uranium leaching from ores. Acidophilic bacteria were employed to produce lixiviants for uranium leaching, utilizing ferric iron and sulfuric acid from Fe-sulfides and elemental sulfur (Hamidian et al., 2009) and Hamidian (2012). Microbial diversity in active and abandoned uranium mine sites has been examined using cultivation-dependent and molecular ecological methods. Dhal (2018) reviewed several studies on cultivable and uncultivable bacteria in uranium mine tailings, concluding that microbial diversity encompasses many physiological and ecological groups of aerobes and anaerobes. These analyses revealed not only the ubiquitous presence of acidithiobacilli but also considerable microbial diversity (Coral et al., 2018). Proteobacteria typically dominate with Acidobacteria and Firmicutes also invariably present. At the genus level, dominant microorganisms usually include *Acidithiobacillus*, *Leptospirillum*, *Sulfobacillus*, *Alicyclobacillus*, and *Ferroplasma*. *Alicyclobacillus* and *Sulfobacillus* spp. have dominated some microbial communities in laboratory column leaching of weathered low-grade uranium ore from the Ranger Uranium Mine (Vázquez-Campos et al., 2017). Fungi have also been found in uranium mine water and raffinate samples (Vázquez-Campos et al., 2014;

Coelho et al., 2020). Microbial composition varies with specific mine locations, such as heap interior layers, leach solutions, and raffinates, which can host very different microbial populations. *Acidithiobacillus ferrooxidans* (*A. ferrooxidans*) is a facultatively anaerobic, Gram-negative, obligate chemolithoautotrophic, extremely acidophilic bacterium with optimal growth at 30°C–35°C and pH 2.5 (Johnson and Quatrini, 2020). It is frequently isolated from mining environments and significantly impacts the biogeochemical cycles of Fe, S, and H in low-pH conditions (Quatrini and Johnson, 2020). This bacterium also plays a role in the solubilization of uranium from ores. *A. ferrooxidans* can directly oxidize U(IV), using the energy conserved from this process for carbon dioxide fixation (Schippers et al., 2014).

T. denitrificans is another facultatively anaerobic, Gram-negative, obligate chemolithoautotrophic bacterium, with optimal growth around pH 7 and 30°C. This bacterium inhabits diverse environments such as soil, mud, freshwater and marine sediments, domestic sewage, and industrial waste-treatment systems (Kelly and Wood, 2000). *T. denitrificans* uses both oxygen and nitrate as electron acceptors and couples thiosulfate oxidation to nitrate reduction during anaerobic growth. Beller et al. (2005, 2009) demonstrated its ability for anaerobic, nitrate-dependent oxidative dissolution of synthetic and biogenic U(IV) oxides, with nitrate reduction coupled to the presence of another electron donor like hydrogen. To evaluate the oxidation efficacy of non-crystalline U(IV) by sulfur-oxidizing bacteria, mobilization experiments were conducted using both *A. ferrooxidans* and *T. denitrificans*. Additionally, U isotope fractionation associated with microbial U mobilization was investigated as a potential monitoring tool. Previous studies on U reduction using sulfate or metal-reducing bacteria, such as *Geobacter sulfurreducens*, *Anaeromyxobacter dehalogenans*, *Shewanella* sp., *Desulfitobacterium* sp., and *Desulfovibrio brasiliensis*, showed significant U isotope fractionation ($\varepsilon = 1{,}000\% * (\alpha - 1)$) ranging from 0.65‰ to 0.99‰, indicating that bacterial U(VI) reduction generally induces isotopic fractionation with U^{238} enrichment in the product U(IV) (Basu et al., 2014; Stirling et al., 2015; Stylo et al., 2015, 2013). Wang et al. (2015a, b) observed that oxidation of dissolved U(IV) with oxygen at acidic pH results in isotopically lighter U(VI), whereas oxidation of solid U(IV) showed only a limited isotope effect. Recent research by Roebbert et al. (2021) indicated isotope fractionation during the complexation of non-crystalline U(IV) by organic ligands, leading to U^{238} enrichment in the mobilized fraction with $\delta^{238}U$ ranging from 0.2% to 0.7%.

14.4 MECHANISM OF BIOLEACHING

The selected microorganisms for bioleaching are iron- and sulfur-oxidizing acidophiles, including both bacteria and archaea with mesophilic and thermophilic temperature ranges. Uranium is bioleached from ores using an acidic ferric sulfate lixiviant. In this process, ferric iron oxidizes tetravalent uranium to the hexavalent form, reducing itself to ferrous iron. Microorganisms involved in bioleaching oxidize ferrous iron back to ferric iron, regenerating ferric sulfate. Oxygen acts as the electron acceptor in this iron oxidation within the leach solution. The acidity of the solution keeps ferric iron soluble and enhances the solubilization of the hexavalent uranium. Ancillary sulfide minerals like pyrite aid the bioleaching process by releasing ferrous iron and reduced sulfur compounds, which are essential for the biological generation of ferric iron and sulfuric acid.

The main mining engineering methods for uranium leaching include heap, dump, stope, in-situ, and in-place leaching. The efficiency of uranium bioleaching is influenced by various mineralogical, physicochemical, microbial, and process factors. Although bioinformatics and synthetic biology are advancing the research on bioleaching microorganisms, these developments have yet to be widely adopted in the industrial practice of uranium mining. Future applications of uranium bioleaching may increasingly target deposits where additional products, such as rare earth elements or base metals, can be recovered alongside uranium.

The bacterial leaching mechanism occurs both by the direct physical contact of organism with insoluble sulfide or by indirect involvement in the ferric–ferrous cycle.

$$UO_2 + Fe_2(SO_4)_3 \rightarrow UO_2SO_4 + 2Fe_2SO_4$$

In ferric sulfate leaching, these oxidants regenerate ferric iron if it has been reduced to ferrous iron through contact with U^{4+} and sulfide minerals in the ore. Bioleaching processes using acidic ferric sulfate and iron- and sulfur-oxidizing bacteria eliminate the need for additional chemical oxidants like chlorate ($NaClO_3$), pyrolusite (MnO_2), or hydrogen peroxide (H_2O_2) to achieve effective ferric iron regeneration rates (Venter and Boylett, 2009). Uranium ores often contain both tetravalent and hexavalent uranium as admixtures. U^{6+} as in paraschoepite UO_3, xH$_2$O, is solubilized to uranyl ion (UO_2^{2+}) in dilute sulfuric acid without the need of an oxidant:

$$UO_3 + 2H^+ \rightarrow UO_2^{2+} + H_2O$$

Tetravalent uranium (UO_2) is insoluble in acidic, sulfate-rich solutions. As a semiconductor (Habashi, 2020), its dissolution necessitates oxidation to hexavalent uranium. This process is significantly accelerated by a chemical oxidant like Fe^{3+}, resulting in the formation of uranyl ions:

$$2UO_2 + O_2 + 4H^+ \rightarrow 2UO_2^{2+} + 2H_2O$$

$$UO_2 + 2Fe^{3+} \rightarrow UO_2^{2+} + 2Fe^{2+}$$

Acidophilic Fe- and S-oxidizing microorganisms drive uranium leaching by producing soluble ferric iron and sulfuric acid from Fe^{2+} and reduced sulfur compounds like elemental sulfur:

$$4Fe^{2+} + O_2 + 4H^+ \rightarrow 4Fe^{3+} + 2H_2O$$

$$2S^0 + 3O_2 + 2H_2O \rightarrow 2SO_4^{2-} + 4H^+$$

The key steps in uranium bioleaching involve producing the lixiviant and contacting it with the ore to solubilize the uranium. Both of these steps can be enhanced through rigorous testing and optimization. In the two-stage bioleaching process (also known as indirect bioleaching), the creation of bioleach solutions (acidic ferric iron-based lixiviants) is separate from the ore contact phase. For the biological oxidation of Fe^{2+} and the regeneration of Fe^{3+} in sulfuric acid solutions, microbes require a supply of dissolved O_2 and CO_2. Beyond the ferric iron-mediated oxidation, tetravalent uranium can also be directly oxidized to the hexavalent form by *A. ferrooxidans.* DiSpirito and Tuovinen (1982a,b) demonstrated that washed cell suspensions of *A. ferrooxidans* took up O_2 and fixed CO_2 with uranous sulfate $[U(SO_4)_2]$ as the substrate, even in the absence of iron.

$$2U^{4+} + O_2 + 2H_2O \rightarrow 2UO_2{}^{2+} + 4H^+$$

Metallosphaera prunae, a thermophilic acidophilic archaeon initially discovered in a smoldering refuse pile at a uranium mine, has been documented to solubilize U_3O_8 into its soluble hexavalent form (Mukherjee et al., 2012). U_3O_8 comprises both tetravalent and hexavalent uranium. The mechanism—whether direct oxidation of U^{2+} to U^{6+} by biomolecules or indirect chemical action—remains unclear. The facultative anaerobe *T. denitrificans* has demonstrated the ability to oxidize uraninite using it as an electron donor, paired with nitrate consumption, with c-type cytochromes seemingly involved (Beller, 2005; Beller et al., 2009). These findings, derived from pure culture experiments with limited uranium quantities, make it difficult to assess their significance for the bioleaching of uranium ores.

14.5 CONCLUSIONS

Since ancient times, when ore bodies were excavated and exposed to air, humidity, and rain, iron- and sulfur-oxidizing bacteria have been prevalent in mine water and on mineral surfaces. Approximately a decade after the discovery of acidophilic iron- and sulfur-oxidizers (*A. ferrooxidans*), commercial uranium bioleaching using heap, dump, and stope technology began around the 1960s. These acidophiles produce sulfuric acid through the oxidation of pyrite and sulfur, maintaining a high redox potential in the ferric iron-based lixiviant by oxidizing ferrous iron. Historical uranium leaching practices show that microbes aided in the oxidation and dissolution of uranium minerals even without intentional augmentation of their role.

In the 1960s, bacteria involved in uranium and sulfide leaching were generally recognized as iron- and sulfur-oxidizers because modern genetic and phylogenetic analysis methods were not yet available, and archaea were unknown. Information on prokaryotic diversity in the environment expanded rapidly in subsequent decades, revealing that microbial life in uranium and sulfide mine environments involves complex interactions among biological, chemical, and physical factors, including cells, solutes, and mineral surfaces. Molecular and biochemical studies of acidophiles have since become prominent, focusing on genome sequences, gene regulation and expression, and bioinformatics. These resources can be used to select traits that improve bioprocess conditions. Key research areas include the study of transmissible metabolic traits and gene regulation in free-swimming and biofilm-associated acidophiles in environmental contexts. Genetic modification for strain improvement is also possible but requires containment in a controlled environment.

Research in uranium bioleaching shares many challenges with the bioleaching of copper, nickel, and zinc from sulfide ores. Optimizing bioleaching processes is specific to each ore type, necessitating interdisciplinary approaches and expertise at every research stage. Studies on sulfide mineral bioleaching have progressively led to multiple commercial-scale bioprocesses. Ongoing investigations are exploring potential bioleaching applications for extracting metals from electronic and other metal-containing waste streams. New applications of uranium bioleaching are expected to emerge alongside the extraction of other commodities, such as rare earths, base metals, and phosphate. Long-term environmental mitigation, monitoring, and public opposition to uranium mining are significant challenges. However, global uranium demand and the need for national self-sufficiency in uranium supply could justify opening new mine sites. The role of microorganisms is now so well understood that mining operations can be designed to incorporate bioleaching steps tailored to their metabolic, physiological, and environmental requirements.

REFERENCES

Abdelouas, A., 2006. Uranium mill tailings: geochemistry, mineralogy, and environmental impact. *Elements*, 2, 335–341. https://doi.org/10.2113/gselements.2.6.335.

Alessi, D.S., Lezama-Pacheco, J.S., Stubbs, J.E., Janousch, M., Bargar, J.R., Persson, P., et al., 2014. The product of microbial uranium reduction includes multiple species with U(IV)–phosphate coordination. *Geochim. Cosmochim. Acta*, 131, 115–127. https://doi.org/10.1016/j.gca.2014.01.005.

Basu, A., Sanford, R. A., Johnson, T. M., Lundstrom, C. C., and Löffler, F. E., 2014. Uranium isotopic fractionation factors during U(VI) reduction by bacterial isolates. *Geochim. Cosmochim. Acta*, 136, 100–113 https://doi.org/10.1016/j.gca.2014.02.041.

Beller, H.R., 2005. Anaerobic, nitrate-dependent oxidation of U(IV) oxide minerals by the chemolithoautotrophic bacterium Thiobacillus denitrificans. *Appl. Environ. Microbiol.*, 71, 2170–2174.

Beller, H.R., Legler, T.C., Bourguet, F., Letain, T.E., Kane, S.R., Coleman, M.A., 2009. Identification of c-type cytochromes involved in anaerobic, bacterial U(IV) oxidation. *Biodegradation*, 20, 45–53.

Bhargava, S.K., Ram, R., Pownceby, M., Grocott, S., Ring, B., Tardio, J., Jones, L., 2015. A review of acid leaching of uraninite. *Hydrometallurgy*, 51, 10–24.

Cahill, A.E., Burkhart, L.E., 1990. Continuous precipitation of uranium with hydrogen peroxide. *Metall. Trans. B*, 21(8), 91–116.

Cerrato, J.M., Ashner, M.N., Alessi, D.S., Lezama-Pacheco, J.S., Bernier-Latmani, R., Bargar, J.R., et al., 2013. Relative reactivity of biogenic and chemogenic uraninite and biogenic noncrystalline U(IV). *Environ. Sci. Technol.*, 47, 9756–9763. https://doi.org/10.1021/es401663t.

Coelho, E., Reis, T.A., Cotrim, M., Mullan, T.K., Correa, B., 2020. Resistant fungi isolated from contaminated uranium mine in Brazil shows a high capacity to uptake uranium from water. *Chemosphere*, 248, 126068.

Coral, T., Descostes, M., De Boissezon, H., Bernier-Latmani, R., de Alencastro, L.F., Rossi, P., 2018. Microbial communities associated with uranium in-situ recovery mining process are related to acid mine drainage assemblages. *Sci. Total Environ.*, 628–629, 26–35.

Desouky, O.A., El-Mougith, A.A., Wesam, A.H., Awadalla, G.S., Hussien, S.S., 2016. Extraction of some strategic elements from thorium-uranium concentrate using bioproducts of *Aspergillus ficuum* and *Pseudomonas aeruginosa*. *Arab. J. Chem.*, 9, S795–eS805.

Dhal, P.K., 2018. Bacterial communities of uranium-contaminated tailing ponds and their interactions with different heavy metals. In: Adhya, T. K., Lal, B., Mohapatra, B., Paul, D., Das, S. (Eds.), *Advances in Soil Microbiology: Recent Trends and Future Prospects*. Springer Nature, Singapore, pp. 109–127.

Dienemann, C., Utermann, J., 2012. Uran in Boden and Wasser. Umweltbundesamt Available at: https://www.umweltbundesamt.de/publikationen/uran-in-boden-wasser (Accessed June 2, 2024).

DiSpirito, A.A., Tuovinen, O.H., 1982a. Uranous ion oxidation and carbon dioxide fixation by Thiobacillus ferrooxidans. *Arch. Microbiol.*, 133, 28–32.

DiSpirito, A.A., Tuovinen, O.H., 1982b. Kinetics of uranous ion and ferrous ion oxidation by Thiobacillus ferrooxidans. *Arch. Microbiol.*, 133, 33–37

Edwards, A.J., Oliver, C.R., 2000. Uranium processing: a review of current methods and technology. *J. Met.*, 52(9), 12–20.

Falagan, C., Moya-Beltran, A., Castro, M., Quatrini, R., Johnson, D.B., 2019. Acidithiobacillus sulfuriphilus sp. nov.: an extremely acidophilic sulfur-oxidizing chemolithotroph isolated from a neutral pH environment. *Int. J. System. Evol. Microbiol.*, 69, 2907–2913.

Finneran, K.T., Housewright, M.E., Lovley, D.R., 2002. Multiple influences of nitrate on uranium solubility during bioremediation of uranium-contaminated subsurface sediments. *Environ. Microbiol.*, 4, 510–516. https://doi.org/10.1046/j.1462–2920.2002.00317.x.

Habashi, F., 2020. Dissolution of uraninite. *Hydrometallurgy*, 194, 105329.

Hamidian, H., 2012. Microbial leaching of uranium ore. In: Tsvetkov, P. (Ed.), *Nuclear Power e Deployment, Operation and Sustainability*. InTech, Rijeka, Croatia, pp. 291–304.

Hamidian, H., Rezai, B., Milani, S.A., Vahabzade, F., Shafaie, S.Z., 2009. Microbial leaching of uranium ore. *Asian J. Chem.*, 21, 5808–5820.

Harrison Jr., A.P., 1988. The acidophilic thiobacilli and other acidophilic bacteria that share their habitat. *Annu. Rev. Microbiol.*, 8, 265–292.

Harrison, V.F., Gow, W.A., Hughson, M.R., 1966. I. Factors influencing the application of bacterial leaching to a Canadian uranium ore. *JOM*, 18, 1189–1194. https://doi.org/10.1007/BF03378508.

Hausen, D. 1998. Characterizing and classifying uranium yellow cakes: a back- ground. *J. Met.* 50(12), 45–47.

Johnson, D.B., 2012. Geomicrobiology of extremely acidic subsurface environments. *FEMS (Fed. Eur. Microbiol. Soc.) Microbiol. Ecol.*, 81, 2–12.

Johnson, D.B., Hallberg, K.B., 2008. Carbon, iron and sulfur metabolism in acidophilic micro-organisms. *Adv. Microb. Physiol.*, 54, 202–256.

Johnson, D.B., Quatrini, R., 2020. Acidophile microbiology in space and time. *Curr. Issues Mol. Biol.*, 39, 63–76.

Kaksonen, A.H., Lakaniemi, A.-M., Tuovinen, O.H. 2020. Acid and ferric sulfate bioleaching of uranium ores: a review. *J. Cleaner Prod.*, 264, 121586. https://doi.org/10.1016/j.jclepro.2020.121586.

Kelly, D.P., Wood, A.P., 2000. Reclassification of some species of Thiobacillus to the newly designated genera Acidithiobacillus gen. nov., Halothiobacillus gen. nov. and Thermithiobacillus gen. nov. *Int. J. System. Evol. Microbiol.*, 50, 511–516.

Leathen, W.W., Braley Sr., S.A., McIntyre, L.D., 1953a. The role of bacteria in the formation of acid from certain sulfuritic constituents associated with bituminous coal. I. Thiobacillus thiooxidans. *Appl. Microbiol.*, 1, 61–64.

Leathen, W.W., Braley Sr., S.A., McIntyre, L.D., 1953b. The role of bacteria in the formation of acid from certain sulfuritic constituents associated with bituminous coal. II. Ferrous iron oxidizing bacteria. *Appl. Microbiol.* 1, 65–68.

Leathen, W.W., Kinsel, N.A., Braley, S.A., 1956. Ferrobacillus ferrooxidans: a chemosynthetic autotrophic bacterium. *J. Bacteriol.*, 72, 700–704.

Lecomte, A., Cathelineau, M., Deloule, E., Brouand, M., Peiffert, C., Loukola- Ruskeeniemi, K., Pohjolainen, E., Lahtinen, H., 2014. Uraniferous bitumen nodules in the Talvivaara NieZneCueCo deposit (Finland): influence of meta morphism on uranium mineralization in black shales. *Miner. Deposita*, 49, 513–533.

Lovley, D.R., Phillips, E.J.P., 1992. Bioremediation of uranium contamination with enzymatic uranium reduction. *Environ. Sci. Technol.*, 26, 2228–2234. https://doi.org/10.1021/es00035a023.

Mahmoud, A., Cezac, P., Hoadley, A.F.A., Contamine, F., D'Hugues, P., 2017. A review of sulfide minerals microbially assisted leaching in stirred tank reactors. *Int. Biodeterior. Biodegrad.*, 119, 118e146.

Makinen, J., Wendling, L., Lavonen, T., Kinnunen, P., 2019. Sequential bioleaching of phosphorus and uranium. *Minerals*, 9, 331. https://doi.org/10.3390/min9060331.

McCready, R.G.L., Gould, W.D., 1990. Bioleaching of uranium. In: Ehrlich, H. L. and Brierley, C., (Eds.), *Microbial Mineral Recovery*. McGraw-Hill, New York, NY, pp. 107–125.

Mudd, G.M., 2009. The Sustainability of Mining in Australia: Key Production Trends and Their Environmental Implications for the Future. Research Report No. RR5. Department of Civil Engineering, Monash University and Mineral Policy Institute. Revised - April 2009.

Mukherjee, A., Wheaton, G.H., Blum, P.H., Kelly, R.M., 2012. Uranium extremophily is an adaptive, rather than intrinsic, feature for extremely thermoacidophilic Metallosphaera species. *Proc. Natl. Acad. Sci. U.S.A.* 109, 16702–16707

Nancucheo, I., Johnson, D.B., Lopes, M., Oliveira, G., 2017. Reductive dissolution of a lateritic ore containing rare earth elements (REE) using *Acidithiobacillus species. Solid State Phenom.*, 262, 299–302. https://doi.org/10.4028/www.scientific.net/ssp.262.299

Newsome, L., Morris, K., Lloyd, J.R., 2014. The biogeochemistry and bioremediation of uranium and other priority radionuclides. *Chem. Geol.*, 363, 164–184. https://doi.org/10.1016/j.chemgeo.2013.10.034.

Newsome, L., Morris, K., Shaw, S., Trivedi, D., Lloyd, J.R., 2015. The stability of microbially reduced U(IV); impact of residual electron donor and sediment ageing. *Chem. Geol.*, 409, 125–135. https://doi.org/10.1016/j.chemgeo.2015.05.016.

Norris, P.R., Falagan, C., Moya-Beltran, A., Castro, M., Quatrini, R., Johnson, D.B., 2020. Acidithiobacillus ferrianus sp. nov.: an ancestral extremely acidophilic and facultatively anaerobic chemolithoautotroph. *Extremophiles*, 24, 329–337.

Patra, A.K., Pradhan, D., Kim, D.J., Ahn, J.G., Yoon, H.S., 2011. Review on bioleaching of uranium from low-grade ore. *J. Korean Inst. Resour. Recycling*, 20(2), 30–44.

Pohl, W.L., 2005. *Economic Geology: Metals, Minerals, Coal and Hydrocarbons e Introduction to Formation and Sustainable Exploitation of Mineral Deposits*. Wiley Blackwell, Chichester, U.K.

Pownseby, M.I., Johnson, C., 2014. Geometallurgy of Australian uranium deposits. *Ore Geol. Rev.*, 56, 25–44.

Roebbert, Y., Rosendahl, C.D., Brown, A., Schippers, A., Bernier-Latmani, R., Weyer, S., 2021. Uranium isotope fractionation during the anoxic mobilization of noncrystalline U(IV) by ligand complexation. *Environ. Sci. Technol.* 55, 7959–7969. https://doi.org/10.1016/10.1021/acs.est.0c08623

Schippers, A., Breuker, A., Blazejak, A., Bosecker, K., Kock, D., Wright, T.L., 2010. The biogeochemistry and microbiology of sulfidic mine waste and bioleaching dumps and heaps, and novel Fe(II)-oxidizing bacteria. *Hydrometallurgy*, 104, 342–350.

Schippers, A., Hedrich, S., Vasters, J., Drobe, M., Sand, W., Willscher, S., 2014. Biomining: metal recovery from ores with microorganisms. In: Schippers, A., Glombitza, F., Sand, W. (Eds.), Geobiotechnologogy I: Metal-Related Issues. Springer, Berlin; *Adv. Biochem. Eng. Biotechnol.*, 141, 1–47.

Stirling, C.H., Andersen, M.B., Warthmann, R., Halliday, A.N., 2015. Isotope fractionation of 238U and 235U during biologically-mediated uranium reduction. *Geochim. Cosmochim. Acta*, 163, 200–218. https://doi.org/10.1016/j.gca.2015.03.017

Stylo, M., Alessi, D.S., Shao, P.P., Lezama-Pacheco, J.S., Bargar, J.R., Bernier-Latmani, R., 2013. Biogeochemical controls on the product of microbial U(VI) reduction. *Environ. Sci. Technol.*, 47, 12351–12358. https://doi.org/10.1021/es402631w.

Stylo, M., Neubert, N., Wang, Y., Monga, N., Romaniello, S.J., Weyer, S., et al., 2015. Uranium isotopes fingerprint biotic reduction. *Proc. Natl. Acad. Sci. U.S.A.*, 112, 5619–5624. https://doi.org/10.1073/pnas.1421841112.

Temple, K.L., Colmer, A.R., 1951. The autotrophic oxidation of iron by a new bacterium: Thiobacillus ferrooxidans. *J. Bacteriol.*, 62, 605–611.

Vazquez-Campos, X., Kinsela, A.S., Waite, T.D., Collins, R.N., Neilan, B.A., 2014. Fodinomyces uranophilus gen. nov. sp. nov. and Coniochaeta fodinicola sp. nov., two uranium mine-inhabiting Ascomycota fungi from northern Australia. *Mycologia*, 106, 1073–1089.

Vazquez-Campos, X., Kinsela, A.S., Collins, R.N., Neilan, B.A., Waite, T.D., 2017. Uranium extraction from a low-grade, stockpiled, non-sulfidic ore: impact of added iron and the native microbial consortia. *Hydrometallurgy*, 167, 81–91.

Venter, R., Boylett, M., 2009. The evaluation of various oxidants used in acid leaching of uranium. In: Hydrometallurgy Conference 2009. Southern African Institute of Mining and Metallurgy, Johannesburg, pp. 445–455.

Wall, J.D., Krumholz, L.R., 2006. Uranium reduction. *Annu. Rev. Microbiol.*, 60, 149–166. https://doi.org/10.1146/annurev.micro.59.030804.121357.

Wang, X., Johnson, T.M., Lundstrom, C.C., 2015a. Isotope fractionation during oxidation of tetravalent uranium by dissolved oxygen. *Geochim. Cosmochim. Acta*, 150, 160–170. https://doi.org/10.1146/10.1016/j.gca.2014.12.007.

Wang, X., Johnson, T.M., Lundstrom, C.C., 2015b. Low temperature equilibrium isotope fractionation and isotope exchange kinetics between U(IV) and U(VI). *Geochim. Cosmochim. Acta*, 158, 262–275. https://doi.org/10.1016/j.gca.2015.03.006.

Wang, Y., Frutschi, M., Suvorova, E., Phrommavanh, V., Descostes, M., Osman, A.A.A., et al., 2013. Mobile uranium(IV)-bearing colloids in a mining-impacted wetland. *Nat. Commun.*, 4, 2942. https://doi.org/10.1038/ncomms3942.

Zammit, C.M., Brugger, J., Southam, G., Reith, F., 2014. In situ recovery of uranium – the microbial influence. *Hydrometallurgy*, 150, 236–244. https://doi.org/10.1016/j.hydromet.2014.06.003.

15 Bioleaching of Spent Automobile Catalyst

Recovery of Palladium

Tripti De

15.1 INTRODUCTION

Metal recovery from wastes and converting it into new metal products helps in reducing green house gases and also managing the energy consumption. Many manufacturing companies use recycled materials to produce new metal products. Moreau et al. (2019) studied the usage of some recycled energy technologies emphasizing the preservation of their original properties of the recycle materials and effectiveness of the process. According to Inman et al.'s (2022) studies, the process for recovering and recycling the waste materials into some new metals products is the basic leaching method. In the leaching process, use of some inorganic acids like HNO_3, HCl, and H_2SO_4 are very low cost and high efficiency in metal recovery in spite of high corrosive and high contamination properties, whereas with organic acids, it yields same efficiency but they costlier than inorganic acids. There are many methods like chemical precipitation, ion-exchange, membrane filtration, and photocatalytic methods for the removal of platinum group of metals (PGMs) from electronic wastes and disposals. The ion-exchange method proves to be the most technologically simple for selective removal of heavy metal ions like Pb(II), Hg(II), Cd(II), Ni(II), V(IV,V), Cr(III,VI), Cu(II), and Zn(II) from contaminated waste water by using ion exchangers (Dabrowski et al., 2004). The adsorption process is extensively used for removal of heavy metal ions from wastewater by using low-cost adsorbents from wastes of agricultural and industrial fields (Renge et al., 2012). On the other hand, due to excellent properties like high efficiency, easy operation, and low space requirements, the membrane filtration method, including reverse osmosis, ultrafiltration, and nanofiltration, proves to be potentially viable for heavy metal ions removal (Xiang et al., 2022). In some studies, it is demonstrated that advanced oxidation processes (AOPs) (Fenton's reagent/CaO) methods are more efficient than conventional methods for sludge dewatering and removal of heavy metal ions from contaminated water (Liang Zhang et al., 2020).

The removal of palladium, a precious heavy metal extensively used in various industrial applications such as catalytic converters, electronics manufacturing, and as a catalyst in chemical processes, has garnered significant attention due to its economic and environmental implications. Palladium recovery and recycling

DOI: 10.1201/9781003415541-15

from industrial waste streams are critical, not only for resource sustainability but also for minimizing the environmental impact associated with raw palladium mining and processing. Various techniques have been developed for palladium recovery, including leaching, ion exchange, adsorption, and membrane filtration, each with its own set of efficiencies, challenges, and suitable applications. Leaching, particularly with strong acids such as hydrochloric and nitric acids or their mixture in aqua regia, is commonly used due to its effectiveness in dissolving palladium from solid waste matrices. The chemistry of palladium leaching involves the dissolution of palladium into a liquid medium forming soluble complexes, which can then be recovered through precipitation (Al-Brboot et al., 2011; Park et al., 2014) or other chemical reactions. However, the process conditions such as acid concentration, temperature, and agitation play crucial roles in determining the leaching kinetics and overall recovery efficiency. Innovations in leaching include ultrasonic-assisted leaching and electrochemical leaching, which aim to enhance the leaching efficiency while reducing environmental impacts by utilizing less aggressive conditions or achieving better selectivity for palladium over other metals. On the other hand, adsorption involves the capture of palladium ions on the surface of materials like activated carbon or specialized resins, which can be particularly effective when dealing with lower concentrations of palladium (Renge et al., 2012). Ion-exchange and membrane filtration techniques (Zaheri et al., 2015; Tenório and Espinosa, 2001) offer high selectivity and the ability to handle large volumes of effluents, making them suitable for both industrial-scale and fine-tuning applications where purity is crucial. Each of these methods comes with environmental considerations, primarily revolving around the handling and disposal of palladium-containing wastes and chemicals used in the extraction process. Proper waste management and adherence to environmental regulations are imperative to mitigate the potential impacts of palladium recovery processes. As research continues to advance in this field, there is a growing emphasis on developing more sustainable and economically viable techniques for palladium recovery, reflecting a broader trend toward greener and more sustainable practices in the metals recovery industry. This focus not only addresses the environmental and safety concerns associated with traditional recovery methods but also aligns with global sustainability goals aimed at reducing the footprint of critical material extraction and processing. This sustainable process also known as bioleaching or biohydrometallurgy process by various fungi (*Thiobacillus* sp.) or bacterial species (like *Aspergillus* sp. or *Penicillium* sp.). However, bioleaching by fungi is more advantageous than bacteria because of numerous factors like leaching rate is faster than bacteria, they have a shorter lag phase than bacteria, they have a higher toxicity tolerance level than bacteria, and also they can grow at various pH levels (Hosseinzadeh et al., 2021). The recovery of palladium can be executed through different methods depending on the process parameters and specific requirements like chemical precipitation, ion exchange, membrane filtration, and photocatalytic degradation. Figure 15.1 gives a schematic diagram of different types of treatment methods for the recovery of heavy metal ions.

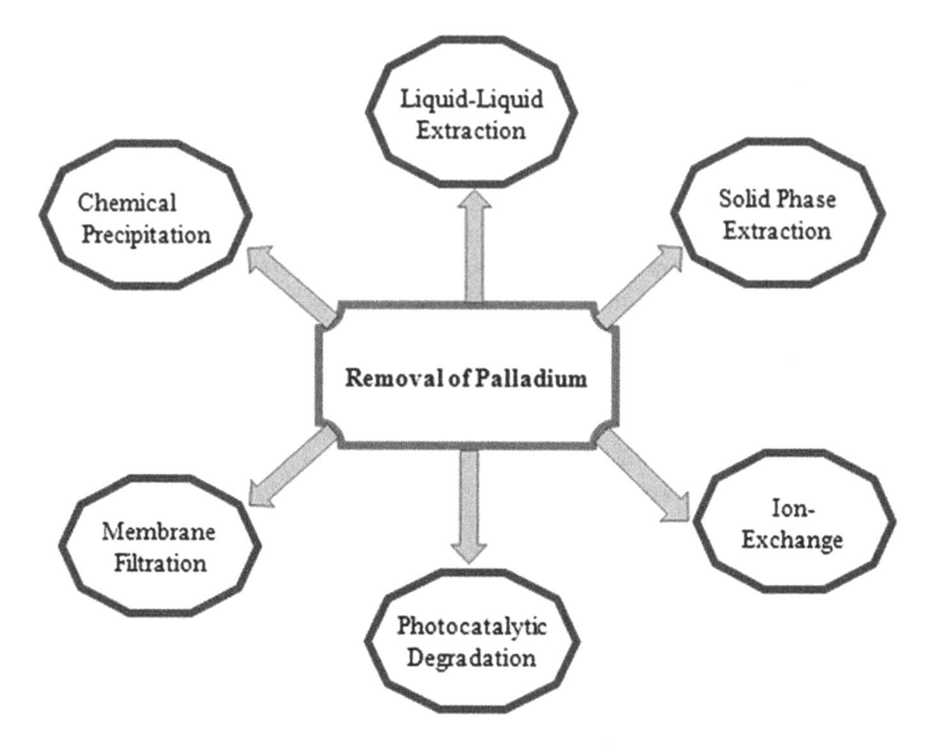

FIGURE 15.1 Removal of palladium by various methods. Reproduced with copyright from Sharma et al. (2017).

15.2 HISTORY OF PALLADIUM – A PLATINUM GROUP OF METAL (PGM)

Palladium, a precious metal, is also one of the six PGMs, which also includes platinum, rhodium, ruthenium, iridium, and osmium. This lustrous silvery-white metal is distinguished by its unique physical and chemical properties, making it highly sought after in various industrial applications, from automotive catalysts to electronics and jewelry (Adelekun, 2023). Here, we delve into the characteristics, uses, sources, market dynamics, and environmental aspects surrounding palladium, providing a comprehensive overview of this valuable metal. Palladium boasts several distinctive physical and chemical properties that enhance its utility across various sectors. It is extremely ductile and malleable, easily formable in an annealed state, and retains significant mechanical strength at higher temperatures. With a melting point of about 1,554°C and a relatively low density compared to other platinum group metals, palladium is both practical and efficient for industrial use. Chemically, palladium is notable for its excellent corrosion resistance, particularly to oxygen, and thus does not tarnish in the air. It can absorb hydrogen up to 900 times its own volume, making it useful in hydrogen storage and purification technologies. This ability also plays a critical role in catalytic processes, where hydrogen is a key reactant. Palladium is primarily mined as a by-product of nickel and copper mining and is also extracted alongside other PGMs in platinum ores. The major palladium producers include Russia and South Africa, with significant contributions from Canada and the United States. Russia's Norilsk Nickel is one of the largest producers of palladium, while South Africa is known for its PGM mining operations in the Bushveld Complex (Murray, 2012).

Palladium has many applications in many fields such as catalysis, electronics, jewelry making, and medicines. The largest use of palladium today is in catalytic converters, which are used in car exhaust systems to reduce the harmful gases from engine emissions into less harmful substances. Palladium helps in converting hydrocarbons, carbon monoxide, and nitrogen oxides into nitrogen, carbon dioxide, and water vapor. It is favored over platinum for this application due to its superior performance in oxidizing environments and its (historically) lower cost (Joudeh et al., 2022). In the electronics industry also, palladium is used in the manufacture of multi-layer ceramic capacitors and in connector platings, where its excellent conductivity and resistance to oxidation are crucial. As a precious metal, palladium is also used in jewelry, serving as a hypoallergenic, tarnish-resistant alternative to white gold or platinum (Kielhorn et al., 2002). Palladium and its alloys are components in dental equipment and materials, where they are valued for their biocompatibility, strength, and resistance to oxidation. Palladium has lots of medicinal-based applications based on toxicity of nanoparticles. Biogenic palladium nanoparticles offer some antimicrobial and anticancer activities from different types of leaf extract and stem extract (Vijilvani et al., 2020; Anjana et al., 2019). Due to its unique ability to absorb hydrogen, palladium is explored for use in energy applications, particularly in hydrogen purification and storage systems. Palladium has strong affinity toward hydrogen, and thus, it has good hydrogen-absorbing properties (Adams and Chen, 2011). PGMs are used in different reactions as tabulated in Table 15.1.

The price of palladium has experienced significant volatility due to its limited supply and growing demand, particularly from the automotive sector. This demand is fueled by stricter global emissions regulations, which increase the need for more catalytic converters and thus more palladium. The market is also influenced by geopolitical factors, trade policies, and technological advancements that affect mining output and recycling rates.

15.2.1 Environmental Impact of Palladium

Mining operations for palladium, like other metals, pose environmental challenges, including habitat destruction, water pollution, and carbon emissions (Agboola et al., 2020). However, palladium's role in pollution control through catalytic converters highlights its environmental benefits. Moreover, recycling palladium from electronic waste and spent catalytic converters is becoming increasingly important in reducing the environmental impact associated with primary production. While palladium is less toxic than some other heavy metals (such as mercury or lead), its compounds,

TABLE 15.1
Applications of PGM Catalysts

Catalyst	Reaction
$Pt/Al_2O_3/SiO_2$	Disproportionation of toluene to benzene
Pd/H–Y-Zeolite	Obtaining fuel by cracking of vacuum distillates
Pt/Zeolite	Xylene isomerization
Pt/Pd/Rh	Industrial exhaust
Pt, Pd/oxide supports	Volatile organic compounds removal
Pt	Hydrodesulphurization
Pt/Zeolite	Naphtha reforming
Pd	Caprolactam synthesis
Pd/supported oxides	Telomerization of 1,3-dienes
Pd	Production of toluene diisocyanate
Pd suspension	Production of H_2O_2
Pd	Bio-oils hydrogenation
Pt/Pd/Rh	Oxidation of ammonia
Pt, Pd	Ketones/aldehydes to alcohol
Pt/Pd/Rh	Production of nitric acid
Rh, Pd/SiO_2	Acetic acid synthesis
Rh	Production of citronella
$PdCl_2$	Acetaldehyde synthesis
Pd	2,5-dichloropyridine amidocarbonylation
Pt	Hydrogenation
$PdCl_2$	Substituted alcohol carbonylation
Pd	Aldehydes and ketones amination
Pd	Production of 1-octene

Source: Reproduced with copyright from Chidunchi et al. (2024).

particularly those that are soluble, can pose health risks to aquatic and terrestrial life forms. Prolonged exposure can cause various adverse effects, including allergic reactions in humans and other animals. Ensuring its removal from effluents before they are discharged into the environment is crucial for protecting ecosystem health (Worlanyo & Jiangfeng, 2021). Palladium has the potential to bioaccumulate in aquatic organisms and plants. Although not as extensively studied as mercury or lead, the bioaccumulation of palladium could pose long-term ecological risks, affecting food chains and biodiversity, which is shown cyclically in a process chain in Figure 15.2.

Looking ahead, the demand for palladium is expected to remain robust, driven by its applications in emission control technologies. However, the push toward electric vehicles (EVs), which do not require catalytic converters, could alter demand dynamics significantly in the long term (Alshahrani et al., 2019). Furthermore, ongoing research into alternative materials and catalysts that could potentially replace palladium poses a continuous source of market uncertainty.

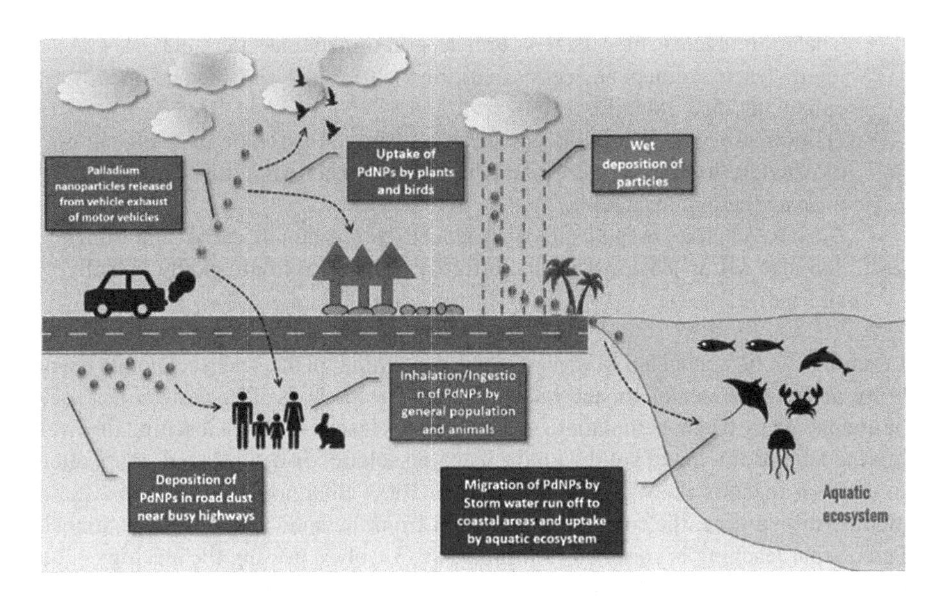

FIGURE 15.2 Toxicity assessment of palladium on environment. Reproduced with copyright from Aarzoo and Samim (2022).

15.3 CHEMICAL LEACHING PROCESS FOR REMOVAL OF HEAVY METALS

Leaching is a widely used process for the removal of heavy metals from various matrices, such as contaminated soils, industrial sludges, and mineral ores (Gunarathne et al. 2022; Coruh et al., 2013; Derakhshan Nejad et al., 2018). The method involves the use of solvents, typically aqueous solutions, which dissolve the metals, allowing them to be separated from their physical substrate and subsequently recovered or stabilized. Leaching is particularly important in the fields of environmental remediation, mining, and recycling because it offers an effective way to handle pollutants and recover valuable materials. This detailed overview will examine the principles, methodologies, challenges, and advancements in the field of heavy metal leaching.

The basic principle behind leaching is the transfer of solute constituents (heavy metals) from a solid phase (waste material) into a liquid phase (leachant). This process is driven by the concentration gradient that exists between the two phases. The efficiency of leaching depends on several factors, including the chemical properties of the heavy metal, the characteristics of the solid matrix, and the type of leaching agent used. The choice of leaching agent is critical and can vary widely depending on the specific heavy metals and the matrix from which they are being removed (Tran et al., 2017), Common leaching agents include:

- Acids like sulfuric acid, hydrochloric acid, and nitric acid are frequently used due to their ability to effectively solubilize many metals.
- Bases: Sodium hydroxide or potassium hydroxide can be used to leach acidic or amphoteric metals.

- Chelating agents like EDTA (ethylenediaminetetraacetic acid), DTPA (diethylenetriaminepentaacetic acid), and citric acid can bind to metals, enhancing their solubility.
- Oxidizing agents like oxidizers such as hydrogen peroxide or manganese dioxide are used to convert metals into their higher oxidation states, often making them more soluble.
- Biosolvents like organic acids produced by certain bacteria and fungi can also act as leaching agents, offering a more environmentally friendly alternative.

Leaching can be conducted in several ways, depending on the scale, economic feasibility, and environmental impact considerations. The methods of leaching process can be conducted by batch, percolation, and continuous leaching. Batch leaching involves mixing a finite amount of solid with the leaching solution in a controlled setup, often in tanks or reactors (DeWindt & Badreddine, 2007; Bhargava et al., 2015). After a predetermined time, the solution is separated from the solids for further treatment. Percolation leaching is common in mining; this involves passing the leaching solution through a bed of solids. The leachant percolates through the material, dissolving the metals as it goes. This can be conducted in situ for contaminated soils or in large heaps for ore processing, whereas continuous leaching is an industrial-scale operation where the solid and liquid phases are continuously fed and discharged, allowing for ongoing metal recovery.

The leaching process generally follows one or a combination of these mechanisms like dissolution (direct solubilization of metal from the solid phase into the liquid phase), complexation (formation of soluble complexes with the leaching agent, which facilitates metal dissolution), oxidation/reduction (chemical reactions that change the oxidation state of the metals, often enhancing their solubility), and physical displacement (physical movement of particles can sometimes contribute to the leaching by exposing new surface areas) (Liao et al., 2022). The kinetics of leaching are influenced by several factors which include temperature, particle size, agitation, and pH of the solution. Higher temperatures generally increase reaction rates of leaching. Smaller particles increase the surface area exposed to the leaching agent and thus enhancing the leaching rate. Stirring or agitating the mixture can increase contact between the solid and liquid phases, facilitating faster leaching, whereas the pH of the solution can significantly affect the solubility of certain metals and can determine the speciation of the metals in solution. Wang et al. (2022) have conducted some experiments on leaching behavior of lead from cement solidified contaminated soil. The results concluded that leaching amount of lead at 55°C was found to be 5.81 times greater than that at 25°C after 11days of leaching process. The leaching index is larger than 9 which showed that temperature affects the leaching process rate. The temperature variation also has effects on effectiveness behavior of the solidification process. Zhang et al. (2018) has experimented on the removal of Cd, Ni, and Cu metal ions from river sediment at different pHs (in the range of 0–4 for Cd ions, 0–5 for Ni ions, and 0–9 for Cu ions) using river water or deionized water as leachants and the leaching results showed the results of 10.2–27.3, 80.5–140.1, and 6.1–30.8 mg/kg, respectively, removal efficiency. While leaching is effective for extracting heavy metals, it poses environmental challenges, primarily related to the disposal of the used leachants and the potential release of toxic metals into the environment if not adequately managed. Treatment of the leachate to remove or recover the dissolved metals is necessary before discharge or reuse of the leachant.

15.3.1 Advancements in Leaching

Recent advances aim to make the leaching process more sustainable and efficient. Research is ongoing into development of leachants that selectively dissolve specific metals, reducing unwanted dissolution. Techniques such as ion exchange, precipitation, and electrochemical deposition are used post-leaching for efficient metal recovery. Recently, leaching processes are increasingly designed to minimize waste and maximize reagent recycling (Trinh et al., 2020). So, as a whole, leaching is a critical process for the removal and recovery of heavy metals from various matrices. Its effectiveness hinges on the careful selection of leaching agents and conditions tailored to the specific metal and matrix involved. Ongoing research and technological development continue to enhance the efficiency, selectivity, and environmental sustainability of leaching processes (Li et al., 2013).

15.4 REMOVAL OF PALLADIUM BY USING THE LEACHING PROCESS

The removal of palladium from various matrices through the leaching process is a specialized and technically sophisticated area of metallurgy and waste management. Palladium, a precious metal found in electronics, automotive catalysts, and jewelry, often needs recovery and recycling due to its high economic value and the environmental risk associated with its waste. The leaching process for palladium involves several steps, each designed to optimize the extraction of palladium while minimizing the environmental impact. This detailed exploration will cover the fundamentals, methodologies, conditions, and technological advances in the leaching process specifically aimed at palladium recovery. Leaching is a process by which soluble constituents are dissolved from a solid material into a liquid medium, typically using chemical solutions known as leachants. In the case of palladium, leaching is often part of a broader recovery strategy that includes initial separation, leaching, solution concentration, and metal recovery stages. The goal is to effectively transfer palladium from used materials such as electronic scrap or automotive catalysts into a solution from which it can be further processed and purified (Paiva et al., 2017).

The choice of leaching agent is crucial to the effectiveness of the palladium leaching process (Behnamfard et al., 2013). Common leaching agents used include:

1. *Aqua regia*: A mixture of nitric acid and hydrochloric acid, aqua regia is one of the most effective solvents for palladium and other precious metals. It dissolves palladium to form chloropalladic acid, a soluble form of palladium.

2. *Thiourea*: An alternative to cyanide, thiourea is used in acidic solutions to dissolve palladium. It forms a stable complex with palladium, which is soluble in water.
3. *Cyanide*: Cyanide leaching, despite its toxicity and environmental risks, is very effective for palladium. It forms soluble complexes with palladium, facilitating its extraction.
4. *Halide leaching*: Chlorides, bromides, and iodides can also be used to dissolve palladium, especially in the presence of oxidizing agents that can form soluble palladium halides.

The method of applying these leaching agents can vary based on the type of material being processed and the specific setup of the recovery plant. Some common leaching techniques include batch leaching, percolation leaching, and heap leaching (Dhawan et al., 2013). Batch leaching is often used in smaller-scale or laboratory settings; batch leaching involves adding a fixed amount of solid waste to a vessel where it is treated with the leaching solution. After sufficient reaction time, the solution is separated and treated to recover the dissolved palladium. Percolation leaching is suitable for larger volumes, such as in industrial applications; this method involves passing the leaching solution through a bed of palladium-containing material. The solution gradually percolates through the material, dissolving the palladium as it moves along. Whereas heap leaching is similar to percolation leaching but conducted on a large scale where the waste material is piled in large heaps. The leaching solution is sprayed over the top and percolates through the pile to a collection point.

The efficiency of the palladium leaching process depends on various factors such as temperature, pH, redox conditions, and agitation, which must be optimized for effective recovery. Higher temperatures generally increase the rate of chemical reactions, including those in leaching processes. However, excessive temperatures might lead to the decomposition of the leaching agents or unwanted side reactions. The pH of the solution significantly affects the solubility of palladium complexes. For example, leaching recovery of 98% palladium, 96% platinum and 86% rhodium can be achieved by applying optimal process condition of a 0.8 vol% hydrogen peroxide and 9.0 M HCl solution mixture at 60°C for 2.5 hours (Yousif, 2019). The effect of suitable process parameters on palladium extraction from PGM oxide ore that also contains traces of Ni and Cu which shows that maximum palladium ions extraction efficiency is 76.6% at 10 g/L of ferricyanide, pH value of 11, dissolved oxygen at 15 ppm at 80°C at 15% solid concentration (Li et al., 2023). Oxidation-reduction potential is crucial, especially when using oxidizing agents like halides or when managing the stability of certain palladium complexes. Stirring or agitating the leach solution can enhance contact between the solid waste material and the leaching agents, increasing the efficiency of the palladium extraction.

Once the palladium is dissolved in the leachate, it must be recovered from the solution by various methods – chemical precipitation, solvent extraction, and adsorption. Chemicals can be added to the palladium-containing solution to precipitate palladium as a solid compound, which can then be filtered and refined, and this process is known as precipitation. Solvent extraction involves transferring palladium from the aqueous phase into an organic phase using selective solvents, followed by re-extraction into an aqueous phase. Using activated carbon or other materials, palladium can be adsorbed from the solution and later desorbed for recovery. Barakat et al. (2006) has investigated and obtained that palladium can be recovered from spent catalyst in hot HCl acid with traces of Al ions. The results configured that complete dissolution of palladium is possible (>99% removal) in leaching process with optimum process conditions like a solution contains 7% HCl and 5% H_2O_2 at 60°C for 2h with a liquid/solid feed ratio of 10/1.

15.4.1 Challenges and Environmental Considerations of the Palladium Leaching Process

The major challenges in palladium leaching include the handling and disposal of hazardous chemicals, especially when using highly toxic agents like cyanide or aqua regia. The environmental implications are significant, necessitating strict controls and recycling of chemicals where possible. Advances in green chemistry aim to develop less harmful leaching agents and processes that minimize waste and environmental impact.

15.4.2 Technological Advances and Future Prospects of Palladium Leaching

Recent technological advances focus on improving the efficiency and environmental sustainability of palladium leaching. These include the development of biodegradable leachates, closed-loop systems that minimize waste and allow for the recycling of leaching agents, and more selective extraction techniques that target palladium specifically, reducing unwanted byproducts and enhancing overall recovery rates.

Wang et al. (2023) investigated palladium recovery from palladium or aluminum oxide catalysts using ultrasound-assisted leaching at 200W of leaching power, an L-S ratio of 5:1, a time of 1h, at 600°C, and 60% sulfuric concentration and 0.1 mole of sodium chloride. The recovery rate reached to 99%, and it was found that with activation reaction energy of 28.7 KJ/mol, the reaction is a diffusion reaction. Later, the results are analyzed by scanning electron microscopy (SEM) which directly increased the leaching rate by increasing the leaching surface area.

Lee (2018) proposed PGMs like platinum (Pt) and palladium (Pd) can be recovered from a mixture of concentrated HCl, TBP (tributyl phosphate) in kerosene, and H_2O_2 by the continuous leaching and extraction process with Pt and Pd ion concentration in the stripped sections to be 99%. So, this selective dissolution process proved to be efficient than aqueous system of HCl and H_2O_2.

15.4.3 Advantages and Disadvantages of Palladium Removal by Leaching Methodology

The leaching process, particularly in the context of extracting valuable metals such as palladium from various matrices (e.g., spent catalysts, electronic waste), plays a pivotal role in resource recovery and environmental management (Jadhao et al.,

2021). This process involves solubilizing the target metals from solid material into a liquid phase, facilitating their subsequent recovery or stabilization. While leaching offers several benefits, it also presents some challenges, especially when compared to other extraction and recovery techniques. This comprehensive review explores the advantages, disadvantages, and overall importance of the leaching process in the context of palladium recovery.

Leaching is highly effective for processing complex materials, such as electronic components and catalytic converters, which contain a mix of different metals and other materials (Llamas, 2020; Makuza et al., 2021). Chemical leaching can selectively dissolve palladium, allowing for its separation from base metals and other impurities. Compared to pyrometallurgical processes (e.g., smelting, roasting), leaching generally requires less energy. Many leaching operations can be conducted at ambient or slightly elevated temperatures, significantly reducing the energy consumption associated with the recovery of metals. Leaching processes can be adapted to a range of scales, from small batch operations to large-scale continuous systems. This flexibility allows for the optimization of operations based on the specific requirements of the palladium-bearing material and the economic considerations of the operation (Moskalyk & Alfantazi, 2003). For many applications, especially where ore grades are low or the material is already in a finely divided state, leaching can be more cost-effective than high-temperature metallurgical processes. The ability to use cheaper chemicals or recycled acids can also reduce costs. Although leaching processes involve chemicals, they can be designed to minimize environmental impact through controlled operations and the recycling of solvents. Especially when non-toxic leaching agents are used, the process can be relatively more environmentally friendly than those involving high temperatures and significant emissions (Danouche et al., 2024). Besides the advantageous prospects, the most effective leaching agents for palladium, such as aqua regia and cyanide, are highly toxic and corrosive. The handling, storage, and disposal of these chemicals require stringent safety and environmental controls, adding to the complexity and cost of operations. Improper management of leachates, especially those containing toxic metals and chemicals, can lead to severe environmental contamination. This includes risks to water sources, soil contamination, and impacts on biodiversity. Depending on the material and the specific leaching agent used, the leaching process can be slower than pyrometallurgical processes. Extended contact times between the material and the leaching agent are often necessary to achieve acceptable recovery rates. After leaching, additional processing steps are required to recover the dissolved palladium and convert it into a usable form. These steps can be complex and costly, especially if the purity requirements for the recovered palladium are high. While leaching is flexible, scaling up the process can present challenges, particularly in maintaining consistent process conditions and managing large volumes of hazardous chemicals and leachates (Brewer et al., 2022; Krishnan et al., 2021).

15.4.4 Importance of Leaching in Palladium Recovery

Despite the disadvantages, the leaching process remains crucial for several reasons. Leaching facilitates the recovery of palladium from secondary sources like electronic waste and automotive catalysts. This not only reduces the demand for primary mining but also contributes to the sustainability of palladium as a resource (Pollmann et al., 2016). Palladium is a highly valuable metal, used in numerous high-tech applications, including electronics and catalysis. Efficient recovery processes like leaching contribute significantly to the economic viability of recycling operations. With increasing global attention on environmental sustainability, leaching provides a method that can be adapted to meet stringent environmental standards, especially when using more benign leaching agents and implementing thorough waste treatment systems. Ongoing research and development in leaching technologies focus on improving the selectivity and efficiency of the process, reducing environmental impacts, and lowering costs through innovations such as ionic liquids, microbial leaching, and membrane technologies.

15.5 BIOLEACHING PROCESS – OVERVIEW

Bioleaching is a process to remove heavy metals from their ores or waste materials by using microorganisms. It has traditionally been related with the mining of base metals like copper, zinc, and gold; bioleaching has also been explored for its potential to recover precious metals such as palladium from various industrial wastes, including wastewater and spent catalysts used in gas treatment processes (Benalia et al., 2021; Pohl, 2020; Anotai et al., 2007). It is also known as microbial leaching or biomining or biohyrometallurgy process.

Bioleaching involves the use of naturally occurring or specially tailored microorganisms that can oxidize the metal components of a solid material, thereby solubilizing the metal and making it available for recovery. The process is environmentally friendly compared to traditional methods such as chemical leaching, as it uses less energy and generates fewer pollutants (Nguyen et al., 2021). Recently, removal by bioleaching process has gathered more attention for decontaminating the toxic metals from the solid wastes present in the contaminated soil (Yang et al., 2018). Recently, this process is used in different fields of Engineering for metal recovery and also some non-hazardous disposal generation which is shown in Figure 15.3. Bioleaching just like depending on process parameters like stirring rate, pH value, solid concentration, aeration rate, and inoculum concentration also depends on the diversity of the microorganisms and the methods followed for microbial growth activity. This approach needs a future detailed research for conducting an efficient leaching process (Kara et al., 2023).

15.5.1 Role of Microorganisms in Bioleaching

The bioleaching process relies on the metabolic activities of bacteria, fungi, or archaea. The most common bacteria used in bioleaching belong to the genera *Acidithiobacillus* and *Leptospirillum*. These organisms oxidize sulfides or iron-containing minerals, releasing protons and creating an acidic environment which solubilizes metals, including palladium. These microorganisms thrive in the bioleaching environment affected by several factors like temperature, carbon source, oxygen availability, and pH value of the nutrient broth. Thiooxidans species bacteria

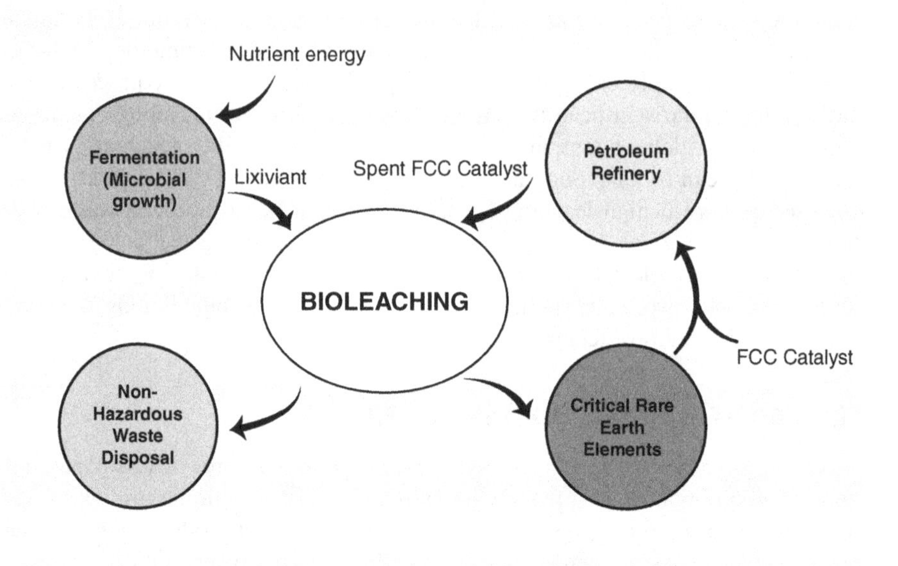

FIGURE 15.3 Overview of bioleaching. Reproduced with copyright from Mukherjee et al. (2023).

can be very dominant at low pH value and thus be advantageous for leaching iron metal ions from iron minerals (Marrero et al., 2020).

15.5.2 Mechanism of Bioleaching

Microorganisms produce acids (such as sulfuric acid) and other compounds (like ferric sulfate) that directly or indirectly attack the metal-bearing minerals. In the case of palladium, which might be present as tiny particles within electronic waste or as a component in spent catalysts, the microbial activity helps to break down the matrix enclosing the metal, allowing the palladium to go into solution (Jones and Santini, 2023).

15.5.3 Process Conditions/Factors Dependable for Bioleaching

The conditions or factors for optimal bioleaching include control of pH, temperature, oxygen supply, and nutrient availability (Bosecker, 1997). Typically, bioleaching is carried out at low pH levels (around 1.5–3.0), which are favorable for acidophilic bacteria. Temperature (of about 28°C–30°C are required for oxidation process by T. ferrooxidans), and oxygen supply must be maintained within specific ranges to support the active metabolism of the microorganisms involved. Nutrients like iron and sulfur compounds together with ammonium, phosphate, and magnesium salts are used for optimum microbial growth. The mineral composition of the substrate is very important to maintain lower pH value of the substrate by addition of acids.

The leaching of metal sulfides is accompanied by an increase in metal concentration in the leachate. Different microbial strains show different sensitivities to heavy metals. Mineral bioleaching is generally effective for the removal of minerals like copper and uranium which largely depends on microbial activity. Nowadays, bacterial leaching is coupled with solvent extraction for removal of heavy metals and recirculation of aqueous solution back to the leaching process. Many factors like temperature, pressure, pH, and chemical density of the solution also affect the efficiency of the bioleaching process. Many studies were investigated and showed that temperature and diurnal temperature range has inhibitory effect on the bioleaching process (Das et al., 1999).

15.6 BIOLEACHING OF PALLADIUM FROM WASTEWATER AND SPENT GASES

15.6.1 From Wastewater

In industrial fields, palladium can end up in wastewater streams as a result of manufacturing processes like those found in the automotive and electronics industries. Bioleaching can be applied to treat such wastewater in several mechanisms like collection, operation, and palladium recovery (Kaksonen et al., 2011). Wastewater containing palladium is collected and pre-treated to adjust pH and remove inhibitors that might adversely affect microbial activity. Microorganisms are introduced into bioreactors containing the wastewater. Over time, they produce acidic conditions that lead to the solubilization of palladium. The solution containing dissolved palladium is processed to recover the metal, often using methods like precipitation, adsorption, or ion exchange.

15.6.2 From Spent Gas Treatment Catalysts

Spent catalysts from gas treatment processes often contain valuable metals like palladium. The process of recovery from spent catalysts includes pre-treatment, bioleaching process, and metal recovery. Spent catalysts are usually crushed and ground to increase the surface area accessible to microorganisms. The processed catalyst is mixed with a culture of suitable microorganisms in a bioreactor. The microbial activity leads to the leaching of palladium into the solution. Similar to wastewater treatment, the leachate is treated to recover palladium in a usable form (Aung, 2005; Aung, 2005; Gómez Bolívar, 2023).

15.6.3 Advantages and Limitations of Bioleaching Process

As discussed above, bioleaching is less polluting than chemical leaching methods and reduces the carbon footprint associated with metal recovery. It can be more economical (low – costly), especially for low-grade ores or complex wastes that are not economically viable with traditional methods. The bioleaching process operates at ambient or near-ambient temperatures, saving significant energy costs (Bosecker, 1997). Besides being advantageous, the bioleaching process has some of the following limitations. Bioleaching generally takes longer than chemical processes, which can be a limitation for time-sensitive projects. Maintaining optimal conditions for microbial growth and activity can be challenging, especially on a large scale, and the process may need to be specifically tailored to effectively target palladium, especially when present in low concentrations or as part of a complex waste stream.

Advancements in biotechnology, including genetic engineering of microorganisms and better understanding of microbial metabolisms, hold promise for improving the efficiency and selectivity of bioleaching processes for palladium recovery (Islam & Awual et al., 2022). Due the growing demand for palladium, coupled with stricter environmental regulations, bioleaching could become an increasingly attractive option for industries looking to recover palladium from waste streams efficiently and sustainably (Schippers, 2014).

15.6.4 Advantages of Palladium Removal by Bioleaching

Six metals, which are platinum, palladium, rhodium, ruthenium, iridium, and osmium, comprise PGMs. These metals are precious and expensive because of high values in melting points, heat resistivity, corrosion resistivity, and also due to its

TABLE 15.2
Table for PGMs Production per Year

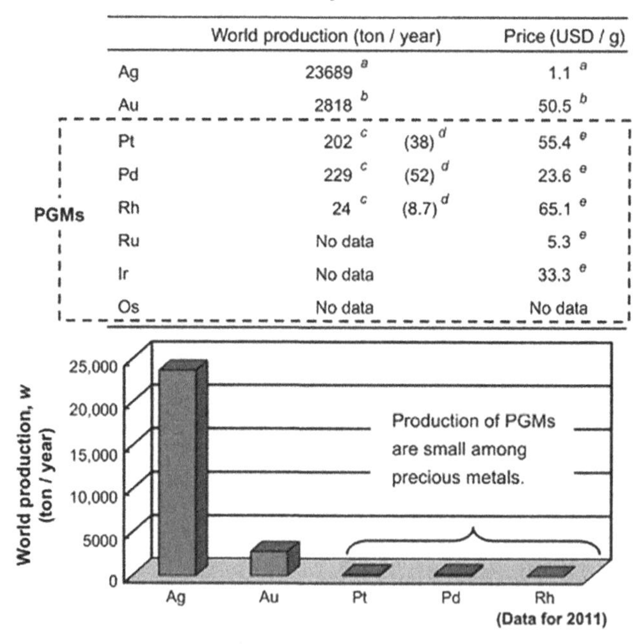

		World production (ton / year)		Price (USD / g)
	Ag	23689 [a]		1.1 [a]
	Au	2818 [b]		50.5 [b]
PGMs	Pt	202 [c]	(38) [d]	55.4 [e]
	Pd	229 [c]	(52) [d]	23.6 [e]
	Rh	24 [c]	(8.7) [d]	65.1 [e]
	Ru	No data		5.3 [e]
	Ir	No data		33.3 [e]
	Os	No data		No data

a Thomson Reuters GFMS Ltd.: *World Silver Survey 2012* (2012)
b Thomson Reuters GFMS Ltd.: *Gold Survey 2012* (2012)
c Johnson Matthey Plc.: *Platinum 2012* (2012)
d Johnson Matthey Plc.: *Platinum 2012* (2012), Recovery from auto catalyst
e Annual average. http://www.platinum.matthey.com/prices/price_charts.html

Source: Reproduced with copyright from Nose and Okabe (2014).

superior catalytic properties. Due to their above unique properties, they are precious and expensive and most have applications in industrial fields like as catalysts in automobile exhaust. Table 15.2 depicts the world production of different heavy metal ions, including platinum group metal ions.

Karim and Ting (2022) experimentally studied the removal of platinum group of metals from the automotive catalyst with the help of two types of microorganisms, i.e. *Pseudomonas fluorescens* and *Bacillus megaterium*. Both these microorganisms form a water-soluble complex (cyanide which is generated as a secondary metabolite) with the PGMs. *Pseudomonas* sp. showed greater efficiency than *Bacillus* sp. in the extraction of platinum, palladium, and rhodium of 58%, 65%, and 97%, respectively, on only a day at 0.5% (w/v) pulp density and at 10 pH value. Chipise et al. (2023) proposed the bioleaching process for extraction of PGMs using three methods, i.e., base metal removal method, biodecomposition of silicates method, and biogenic cyanide extraction method. Under oxidative reaction, the base metals form stable complexes with cyanide; the kinetics for removal of base metals is increased by using mixed thermophilic microorganisms which results in a residue free from base metal for PGM bioleaching. A double approach was involved including biogenic cyanide production, followed by PGM bioleaching. Mukherjee et al. (2023) reviewed extraction of PGMs and rare earth elements (REEs) by the bioleaching process of industrial waste effluent and mining sludge by microalgae. PGMs have recently become highly important for their economic cost. They also suggest to the future scope of renewal usage of those heavy metals. Table 15.3 illustrates microorganisms responsible for the bioleaching process of different metals from different metal sources.

Mohapatra et al. (2018) emphasized the use of salt-tolerant bacteria from different types of water for bioremediation of various toxic metals from various marine environment. Hosseinzadeh e,t al. (2021) demonstrated in his investigation that the removal efficiency of cerium, aluminum, and lanthanum was 25.9%, 43%, and

TABLE 15.3
Critical Metals, Sources, and the Microorganisms Helping in Bioleaching

Critical Metals	Metal Source	Organism
Gold (Au)	$HAuCl_4$	*Bacillus megaterium, Bacillus subtilis, Escherichia coli*
Silver (Ag)	$AgNO_3$	*E. coli, Klebsiella pneumoniae, Lactobacillus fermentum*
Copper (Cu^{+2})	Chalcocite, chalcopyrite, and covellite	*Chlorella* spp., Phormidium spp., *Microcystis* spp.
Cobalt (Co)	CoAs	*Chlamydomonas reinhardtii, Spirogyra hyaline*
Uranium (U)	$UO_2(NO_3)$	*Paenibacillus* sp., *Pseudomonas putida*
Zinc (Zn)	$ZnCO_3$	*Aspergillus niger, Pseudomonas* sp.
Nickel (Ni)	Iron ore limonite	*Spirulina* spp., *Aulosira fertilissima*
Iron (Fe^{+3})	FeS_2	*Chlorella vulgaris, Microcystis* sp.
Platinum (Pt)	$PtCl_6^{2-}$	*Desulfovibrio desulfuricans, Shewanella algae, Pseudomonas* sp.
Lithium	Petalite ($LiAl(Si_2O_5)_2$	*A. niger, Penicillium purpurogenum*

Source: Reproduced with copyright from Mukherjee et al. (2023).

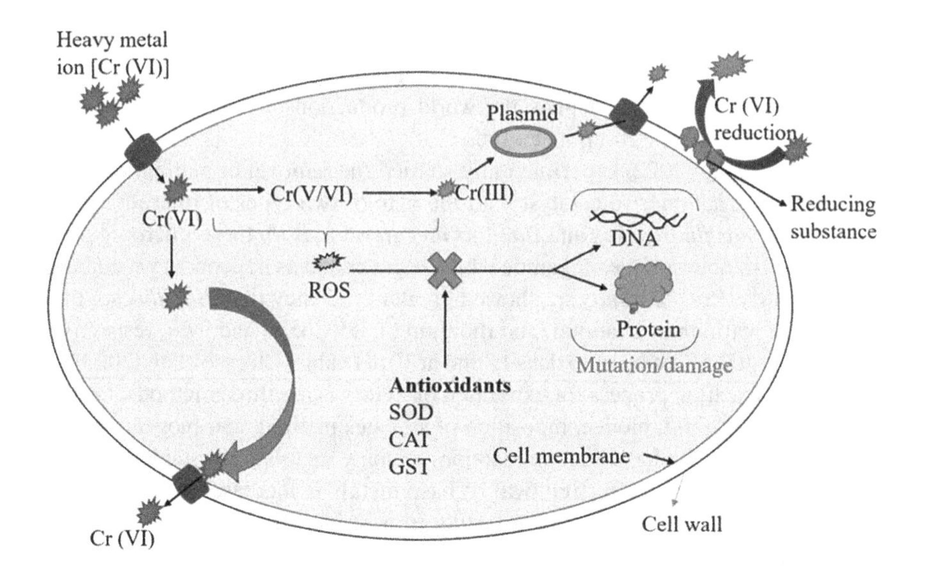

FIGURE 15.4 Mechanisms of bioremediation of heavy metals (Cr) using bacteria. Reproduced from Zhou et al. (2023).

23.9%, respectively for 2 days. The reaction conditions maintained during the process are 5 pH value, volume of inoculums 2.5%, and volume density of solid to liquid at 1%. The bioleaching process also proved to be efficient for 100% removal of platinum and palladium metals by testing with SEM, Fourier-Transform Infrared Spectroscopy (FTIR), High-performance liquid chromatography (HPLC), and X-ray diffraction (XRD) analysis. Therefore, bioleaching technique is processed in many bacterial bioremediation steps for chromium metal which is schematically shown in Figure 15.4.

REFERENCES

Aarzoo, N. and Samim, M., 2022. Palladium nanoparticles as emerging pollutants from motor vehicles: an in-depth review on distribution, uptake and toxicological effects in occupational and living environment. *Science of the Total Environment*, *823*, p.153787.

Adams, B.D. and Chen, A., 2011. The role of palladium in a hydrogen economy. *Materials Today*, *14*(6), pp.282–289.

Adelekun, A.B., 2023. Extraction of Platinum Group Metals from Spent Catalytic Converter Using Sulphur & Selenium Ligands (Master's thesis, Lamar University-Beaumont).

Agboola, O., Babatunde, D.E., Fayomi, O.S.I., Sadiku, E.R., Popoola, P., Moropeng, L., Yahaya, A. and Mamudu, O.A., 2020. A review on the impact of mining operation: monitoring, assessment and management. *Results in Engineering*, *8*, p.100181.

Alshahrani, S., Khalid, M. and Almuhaini, M., 2019. Electric vehicles beyond energy storage and modern power networks: challenges and applications. *IEEE Access*, *7*, pp.99031–99064.

Anjana, P.M., Bindhu, M.R., Umadevi, M., and Rakhi, R.B. (2019). Antibacterial and electrochemical activities of silver, gold, and palladium nanoparticles dispersed amorphous carbon composites. *Applied Surface Science*, *479*, pp.96–104. DOI: 10.1016/j. apsusc. 2019.02.057

Anotai, J., Tontisirin, P. and Churod, P., 2007. Integrated treatment scheme for rubber thread wastewater: sulfide precipitation and biological processes. *Journal of Hazardous Materials*, *141*, 1–7.

Aung, K.M.M., 2005. Bioleaching of metals from spent catalysts for metal removal/recovery. Thesis submitted in National University of Singapore, https://core.ac.uk/download/pdf/48632705.pdf

Barakat, M., Mahmoud, M.H.H. and Mahrous, Y.S., 2006. Recovery and separation of palladium from spent catalyst. *Applied Catalysis A General*, *310*(2), pp.182–186.

Behnamfard, A., Salarirad, M.M. and Veglio, F., 2013. Process development for recovery of copper and precious metals from waste printed circuit boards with emphasize on palladium and gold leaching and precipitation. *Waste Management*, *33*(11), pp.2354–2363.

Benalia, M.C., Leila, Y., Bouaziz, M.G., Samia, A. and Hayet M., 2021. Removal of heavy metals from industrial wastewater by chemical precipitation: mechanisms and sludge characterization. *Arabian Journal for Science and Engineering*, *47*, pp.5587–5599. DOI: 10.1007/s13369-021-05525-7.

Bhargava, S.K., Ram, R., Pownceby, M., Grocott, S., Ring, B., Tardio, J. and Jones, L., 2015. A review of acid leaching of uraninite. *Hydrometallurgy*, *151*, pp.10–24.

Bosecker, K., 1997. Bioleaching: metal solubilization by microorganisms. *FEMS Microbiology Reviews*, 20(3–4), pp.591–604.

Brboot, M.M., Abid, B.A. and Al-Shuwaik, N.M., 2011. Removal of heavy metals using chemical precipitation. *Journal of Engineering Technology*, *29*(3), pp.595–612.

Brewer, A., Florek, J. and Kleitz, F., 2022. A perspective on developing solid-phase extraction technologies for industrial-scale critical materials recovery. *Green Chemistry*, *24*(7), pp.2752–2765.

Chidunchi, I., Kulikov, M., Safarov, R. and Kopishev, E., 2024. Extraction of platinum group metals from catalytic converters. *Heliyon*, *10*(3), p.e25283

Chipise, L., Ndlovu, S., Shemi, A., Moodley, S.S., Kumar, A., Simate, G.S. and Yah, C.S., 2023. Towards bioleaching of PGMS. *Minerals Engineering*, *202*(6), p.108291

Dabrowski, A., Hubicki, Z., Podkoscielny, P. and Robens, E., 2004. Selective removal of the heavy metal ions from waters and industrial wastewaters by ion-exchange method. *Chemosphere*, *56*(2), pp.91–106.

Danouche, M., Bounaga, A., Oulkhir, A., Boulif, R., Zeroual, Y., Benhida, R. and Lyamlouli, K., 2024. Advances in bio/chemical approaches for sustainable recycling and recovery of rare earth elements from secondary resources. *Science of the Total Environment*, *912*, p.168811.

Das, T., Ayyappan, S. and Chaudhury, G.R., 1999. Factors affecting bioleaching kinetics of sulfide ores using acidophilic micro-organisms. *BioMetals*, *12*, pp.1–10.

DeWindt, L. and Badreddine, R., 2007. Modelling of long-term dynamic leaching tests applied to solidified/stabilised waste. *Waste Management*, *27*(11), pp.1638–1647.

Derakhshan Nejad, Z., Jung, M.C. and Kim, K.H., 2018. Remediation of soils contaminated with heavy metals with an emphasis on immobilization technology. *Environmental Geochemistry and Health*, *40*, pp.927–953.

Dhawan, N., Safarzadeh, M.S., Miller, J.D., Moats, M.S. and Rajamani, R.K., 2013. Crushed ore agglomeration and its control for heap leach operations. *Minerals Engineering*, *41*, pp.53–70.

Gómez Bolívar, J., 2023. Synthesis of palladium and ruthenium nanoparticles from metal solutions using bacteria with applications as nanocatalyst. University of Granada, https://hdl.handle.net/10481/80970

Gunarathne, V., Rajapaksha, A.U., Vithanage, M., Alessi, D.S., Selvasembian, R., Naushad, M., You, S., Oleszczuk, P. and Ok, Y.S., 2022. Hydrometallurgical processes for heavy metals recovery from industrial sludges. *Critical Reviews in Environmental Science and Technology*, *52*(6), pp.1022–1062.

Hosseinzadeh, F., Rastegar, S.O. and Ashengroph, M., 2021. Bioleaching of rare earth elements from spent automobile catalyst as pretreatment method to improve Pt and Pd recovery: Process optimization and kinetic study. *Process Biochemistry*, *105*, pp.1–7.

Inman, G., Nlebedim, I.C. and Prodius, D., 2022. Application of ionic liquids for the recycling and recovery of technologically critical and valuable metals. *Energies*, *15*(2), p.628.

Islam, A., Swaraz, A.M., Teo, S.H., Taufiq-Yap, Y.H., Vo, D.V.N., Ibrahim, M.L., Abdulkreem-Alsultan, G., Rashid, U. and Awual, M.R., 2021. Advances in physiochemical and biotechnological approaches for sustainable metal recovery from e-waste: a critical review. *Journal of Cleaner Production*, *323*, p.129015.

Jadhao, P.R., Pandey, A., Pant, K.K. and Nigam, K.D.P., 2021. Efficient recovery of Cu and Ni from WPCB via alkali leaching approach. *Journal of Environmental Management*, *296*, p.113154

Jones, S. and Santini, 2023. Mechanisms of bioleaching: iron and sulfur oxidation by acidophilic microorganisms. *Essays Biochemistry*, *67*(4), pp.685–699.

Joudeh, N., Saragliadis, A., Koster, G., Mikheenko, P. and Linke D., 2022. Synthesis methods and applications of palladium nanoparticles: a review. *Frontiers in Nanotechnology*, *4*, p.1062608. DOI: 10.3389/fnano.2022.1062608.

Kaksonen, A.H., Lavonen, L., Kuusenaho, M., Kolli, A., Närhi, H., Vestola, E., Puhakka, J.A. and Tuovinen, O.H., 2011. Bioleaching and recovery of metals from final slag waste of the copper smelting industry. *Minerals Engineering*, *24*(11), pp.1113–1121.

Kara, I.T., Kremser, K., Wagland, S.T. and Coulon, F., 2023. Bioleaching metal – bearing wastes and by-products for resource recovery: a review. *Environmental Chemistry Letters*, *21*, pp.3329–3350.

Karim, S. and Ting, Y.-P., 2022. Bioleaching of platinum, palladium, and rhodium from spent automotive catalyst using bacterial cyanogenesis. *Bioresource Technology Reports*, *18*, pp.101069. DOI: 10.1016/j.biteb.2022.101069

Kielhorn, J., Melber, C., Keller, D. and Mangelsdorf, I., 2002. Palladium – a review of exposure and effects to human health. *International Journal of Hygiene and Environmental Health*, *205*(6), pp.417–432.

Krishnan, S., Zulkapli, N.S., Kamyab, H., Taib, S.M., Din, M.F.B.M., Abd Majid, Z., Chaiprapat, S., Kenzo, I., Ichikawa, Y., Nasrullah, M. and Chelliapan, S., 2021. Current technologies for recovery of metals from industrial wastes: an overview. *Environmental Technology & Innovation*, *22*, p.101525.

Lee, Y.N., 2018. *The Biosorption of Platinum and Palladium in Chloride Real Leach Solution by Modified Biomass* (Doctoral dissertation, Univcrsity of Saskatchewan).

Li, H., Oraby, E., Bezuidenhout, G.A. and Eksteen, J., 2023. The leaching of palladium from polymetallic oxide ores using alkaline ferricyanide solutions. *Mineral Processing and Extractive Metallurgy Review*, *2023*, pp. 1–10. DOI: 10.1080/08827508.2023.2243013.

Li, J., Zhang, B., Yang, M. and Lin, H., 2021. Bioleaching of vanadium by Acidithiobacillusferrooxidans from vanadium-bearing resources: performance and mechanisms. *Journal of Hazardous Materials*, *416*, p.125843.

Li, L., Dunn, J.B., Zhang, X.X., Gaines, L., Chen, R.J., Wu, F. and Amine, K., 2013. Recovery of metals from spent lithium-ion batteries with organic acids as leaching reagents and environmental assessment. *Journal of Power Sources*, *233*, pp.180–189.

LiangZhang, H.L., Ye, M., Guan, Z., Huang, J., Liu, J., Li, L., Huang, S. and Sun, S., 2020. Evaluation of the dewaterability, heavy metal toxicity and phytotoxicity of sewage sludge in different advanced oxidation processes. *Journal of Cleaner Production*, *265*, p.121839.

Llamas, A., Bartie, N.J., Heibeck, M., Stelter, M. and Reuter, M.A., 2020. Resource efficiency evaluation of pyrometallurgical solutions to minimize iron-rich residues in the roast-leach-electrowinning process. In PbZn 2020: 9th International Symposium on Lead and Zinc Processing (pp. 351–364). Springer International Publishing, Chem.

Makuza, B., Tian, Q., Guo, X., Chattopadhyay, K. and Yu, D., 2021. Pyrometallurgical options for recycling spent lithium-ion batteries: a comprehensive review. *Journal of Power Sources*, *491*, p.229622.

Malik, L. and Hedrich, S., 2022. Ferric iron reduction in extreme acidophiles. *Frontiers in Microbiology*, *12*, p.818414.

Marrero, J., Coto, O. and Schippers, A., 2020. Metal bioleaching: fundamentals and geobiotechnical application of aerobic and anaerobic acidophiles. *Biotechnological Applications of Extremophilic Microorganisms by Jeannette Marrero,Orquidea Coto and Axel Schippers published by De Gruyter 2020* pp. 261–287. (https://doi.org/10.1515/9783110424331-011).

Moreau, V., Dos Reis, P. and Vuille, F., 2019. Enough metals? resource constraints to supply a fully renewable energy system. *Resources*, *8*(1), p.29.

Moskalyk, R.R. and Alfantazi, A.M., 2003. Review of copper pyrometallurgical practice: today and tomorrow. *Minerals Engineering*, *16*(10), pp.893–919.

Mukherjee, S., Paul, S., Bhattacharjee, S., Nath, S., Sharma, U. and Paul, S., 2023. Bioleaching of critical metals using microalgae. *AIMS Environmental Science*, *10*(2), pp.226–244.

Mukherjee et.al (2023) not Sonali Paul et. al (2023),. from refernce list "Mukherjee, S., Paul, S., Bhattacharjee, S., Nath, S., Sharma, U. and Paul, S., 2023. Bioleaching of critical metals using microalgae. AIMS Environmental Science, 10(2), pp.226–244".

Murray, A.J., 2012. Recovery of platinum group metals from spent furnace linings and used automotive catalysts (Doctoral dissertation, University of Birmingham).

Nguyen, T.H., Won, S., Ha, M.-G., Nguyen, D.D., Kang, H.Y., 2021. Bioleaching for environmental remediation of toxic metals and metalloids: a review on soils, sediments, and mine tailings. *Chemosphere*, *282*, p.131108.

Nose, K., and Okabe, T.H., 2014. *Treatise on Process Metallurgy: Industrial Processes*. Elsevier, Oxford.

Paiva, A.P., 2017. Recycling of palladium from spent catalysts using solvent extraction – some critical points. *Metals*, *7*(11), p.505.

Park, J.-H., Choi, G.-J. and Kim, S.-H., 2014. Effects of pH and slow mixing conditions on heavy metal hydroxide precipitation. *Journal of the Korea Organic Resources Recycling Association*, *22*, pp.50–56.

Pohl, A., 2020. Removal of heavy metal ions from water and wastewater by sulphur coating precipitation agents. *Water, Air and Soil Pollution*, *231*, 503.

Pollmann, K., Kutschke, S., Matys, S., Kostudis, S., Hopfe, S. and Raff, J., 2016. Novel biotechnological approaches for the recovery of metals from primary and secondary resources. *Minerals*, *6*(2), p.54.

Renge, V.C., Khedkar, S.V. and Pande, S.V., 2012. Removal of heavy metals from wastewater using low cost adsorbents: a review. *Scientific Reviews and Chemical Communications*, *2*(4), pp.580–584.

Schippers, A., Hedrich, S., Vasters, J., Drobe, M., Sand, W. and Willscher, S., 2014. Biomining: metal recovery from ores with microorganisms. *Geobiotechnology I: Metal-Related Issues*, *2014*, pp.1–47.

Sharma, S., Santhana Krishna Kumar, A. and Rajesh, N., 2017. A perspective on diverse adsorbent materials to recover precious palladium and the way forward. *Royal Scociety of Chemistry Advances*, *7*(82), pp.52133–52142.

Tenório, J.A.S. and Espinosa, D.C.R., 2001. Treatment of chromium plating process effluents with ion exchange resins. *Waste Management*, 21(7), pp.637–642.

Tran, T.-K., Leu, H.-J., Chiu, K.-F. and Lin, C.-Y., 2017. Electrochemical treatment of heavy metal-containing wastewater with the removal of COD and heavy metal ions. *Journal of the Chinese Chemical Society*, *64*(5), pp.493–502. DOI: 10.1002/jccs.201600266.

Trinh, H.B., Lee, J.C., Suh, Y.J. and Lee, J., 2020. A review on the recycling processes of spent auto-catalysts: towards the development of sustainable metallurgy. *Waste Management*, *114*, pp.148–165.

Vijilvani, C., Bindhu, M.R., Frincy, F.C., AlSalhi, M.S., Sabitha, S., Saravanakumar, K., et al., 2020. Antimicrobial and catalytic activities of biosynthesized gold, silver and palladium nanoparticles from Solanum nigurum leaves. *Journal of Photochemistry and Photobiology B: Biology*, *202*, p.111713. DOI: 10.1016/j.jphotobiol.2019.111713.

Wang, J., Zhu, X., Fan, J., Xue, K., Ma, S., Zhao, R., Wu, H. and Gao, Q., 2023. Improved palladium extraction from spent catalyst using ultrasound-assisted leaching and sulfuric acid–sodium chloride system. *Seperations*, *10*(6), pp.355–367.

Wang, P., Liu, X., Zeng, G., Ma, J. and Xia, F., 2022. Effects of temperature on the leaching behavior of Pb from cement stabilization/solidification-treated contaminated soil. *Seperations*, *9*(12), pp.402.

Worlanyo, A.S. and Jiangfeng, L., 2021. Evaluating the environmental and economic impact of mining for post-mined land restoration and land-use: a review. *Journal of Environmental Management*, *279*, p.111623.

Xiang, H., Min, X., Tang, C.-J., Sillanpää, M. and Zhao, F., 2022. Recent advances in membrane filtration for heavy metal removal from wastewater: a mini review. *Journal of Water Process Engineering*, *49*, p.103023.

Yang, Z., Shi, W., Yang, W., Liang, L., Yao, W., Chai, L., et al., 2018. Combination of bioleaching by gross bacterial biosurfactants and flocculation: a potential remediation for the heavy metal contaminated soils. *Chemosphere*, *206*, pp.83–91.

Yousif, A.M., 2019. Recovery and then individual separation of platinum, palladium, and rhodium from spent car catalytic converters using hydrometallurgical technique followed by successive precipitation methods. *Journal of Chemistry*, *7*, pp.1–7, DOI: 10.1155/2019/2318157

Zaheri, P., Mohammadi, T., Abolghasemi, H. and Ghannadi Maraghe, M., 2015. Supported liquid membrane incorporated with carbon nanotubes for the extraction of Europium using Cyanex272 as carrier. *Chemical Engineering Research and Design*, *100*, pp.81–88.

Zhang Y., Zhang, H., Zhang, Z., Liu, C., Sun, C., Zhang, W. and Marhaba, T., 2018. pH effect on heavy metal release from a polluted sediment. *Journal of Chemistry*, *2018*, p.7597640. DOI: doi.org/10.1155/2018/7597640.

Zhou, B., Zhang, T. and Wang, F., 2023. Microbial-based heavy metal bioremediation: toxicity and eco-friendly approaches to heavy metal decontamination. *Applied Sciences*, *13*(14), pp.8439.

16 Microorganism-Based Leaching of Industrial Waste for the Recovery of Cobalt

Satyabrata Si and Sankha Chakrabortty

16.1 INTRODUCTION

The electronic market of the entire world depends on various metals for the configuration of devices like computers, laptops, mobile phones, and storage devices and is still moving toward more advanced technologies [1,2]. Furthermore, metals play a significant role in making various household goods and materials of industrial importance. Therefore, the increased demand for metals by the whole society puts stress on the mining industry for non-stop delivery of these precious metals, which are basically non-renewable resources, and accordingly, there is a high chance of their extinction [3,4]. Furthermore, excessive mining results in the dumping of industrial waste contaminated with toxic heavy metal, which is a major concern for the environment, thereby affecting living organisms in terms of toxicity, reactivity, and corrosivity [5,6]. Although waste management is generally practiced in most of the developing countries like India, a lack of certain parameters like energy economy and minimization of toxic by-products still remains a major challenge [7–9]. Various processes of metal recovery such as pyrometallurgical and hydrometallurgical have been practiced for centuries to extract metal values from metal ores or industrial wastes [10,11]. However, the major drawbacks of these strategies are their high cost of energy consumption in pyrometallurgical processes and expensive chemical waste management in hydrometallurgical processes [12,13]. The economy of a particular process becomes significant only when the recovery cost is much lower than the value of the recovered metal. Furthermore, due to stringent environmental regulations toward waste disposal, it is utmost necessary to develop eco-friendly technologies for efficient metal recovery.

The Earth's crust, with typical cobalt deposits in the range of 25–30 ppm, ranks as the 33rd most abundant element, making it rarer than most other transition metals, with the exception of scandium [14,15]. The majority of cobalt extraction typically occurs as a secondary outcome of mining other metals, principally Ni and Cu, from a diverse array of deposits, including sediment-hosted Cu-Co deposits, Co-Cu-Ni sulfides, Co-Ni laterites, and volcanogenic deposits [16]. There is a noticeable deficiency in basic understanding of cobalt minerals and their amenability to processing.

DOI: 10.1201/9781003415541-16

Moreover, cobalt recovery rates are usually low, especially in processes that involve flotation and smelting, resulting in substantial losses of cobalt to mine tailings or smelter slags [16]. Thus, the biorecovery of cobalt ions can be particularly important for several reasons [17–23]. Firstly, cobalt is a crucial component in various industrial processes, including the production of rechargeable batteries, catalysts, and superalloys. Its recovery can significantly contribute to reducing the dependency on mining, which can have environmental and social impacts. Secondly, cobalt resources are not as abundant as some other metals and much of the world's cobalt supply comes from politically unstable regions or conflict zones, leading to concerns about supply security. Thirdly, industrial wastes contain trace amounts of cobalt and their recovery can be economically feasible only through bioleaching processes compared to traditional methods of extraction, where biotechnology offers the potential for lower operating costs, energy savings, and reduced waste generation. Moreover, biorecovery processes are often more environmentally friendly and can help reduce the environmental footprint associated with cobalt production. These factors make cobalt biorecovery a compelling area of research and application in the field of biotechnology and resource management.

Therefore, the development of efficient, cost-effective, and environmentally benign processes is a major thrust area of research nowadays, and in this regard, the focus toward bioleaching to recover valuable metals from both ores and industrial wastes is increasing [24–27]. In fact, bioleaching is a novel approach to the leaching process that utilizes microorganisms for the extraction of metals from low-quality minerals and secondary waste and is popular due to its minimum energy consumption and operational cost [20,24,28–32]. Most of the bioleaching processes utilize microorganisms, which generate sulfuric acid by the oxidation of S and reduced S compounds or Fe^{3+}/Fe^{2+} ions [33,34]. In this chapter, various aspects of bioleaching to recover cobalt from various industrial wastes, like mine tailing and electronic industry, are discussed.

16.2 IMPORTANCE OF COBALT

The elemental cobalt (Co) is classified under group VIII of the periodic table, which is mostly used for heat-resistant and magnetic alloys [35,36]. It can be found in various forms in plants, animals, air, water, soil, and rocks. Despite its wide dispersion, cobalt makes up just 0.001% of Earth's crust, which is available in a variety of minerals, such as cobaltite, smaltite, linnaeite, erythrite, heterogenite, and skutterudite. Additionally, it occurs as a trace element in animals, where it plays a crucial role in the nutrition of ruminants (such as cattle and sheep) and acts as a metabolic cofactor in vitamin B12, also known as cobalamin, the only vitamin known to contain such a significant element. The world's leading producers of cobalt are from Congo, Russia, Australia, Canada, etc., whereas Congo ranks as the top producer of cobalt in the Earth as per the U.S. Geological Survey, 2017 [37]. However, China leads as the largest producer of refined cobalt all over the world and the largest exporter as well. Exceptionally, the mining of cobalt ore is not because of its cobalt content; rather, cobalt is extracted as a by-product from the mining of various ores, like copper, zinc, nickel, iron, and manganese, which contain trace amounts of cobalt as impurities.

Thus, there always remain challenges to extract the trace amount of cobalt present along with other mineral ores [16].

The importance of cobalt recovery relies on its applicability in many sectors [36,38]. The primary application of cobalt lies in the manufacturing of alloys and superalloys utilized for crafting components, which require high strength and durability. A significant portion of the global production is dedicated to magnetic alloys like alnicos, employed for permanent magnets. The stable isotope cobalt-59 is the natural form of cobalt, serving as the starting material for the creation of cobalt-60, which is the longest-lived synthetic radioactive isotope. Gamma radiation emitted by cobalt-60 has found utility in industrial material inspection, uncovering internal structures, flaws, or foreign objects. Furthermore, cobalt-60 has been harnessed in cancer therapy, sterilization studies, and both biological and industrial applications as a radioactive tracer. Additionally, it serves as a catalyst in the chemical and petrochemical sectors.

16.3 VARIOUS METHODS OF METAL EXTRACTION

Although cobalt is mostly extracted from its primary resources, there are huge opportunities to exploit it from various secondary wastes, like electronic waste, spent catalyst, and battery waste, since they are composed of various heavy metals, like gold, silver, copper, nickel, cobalt, zinc, and rare Earth elements [10,11,39,40]. The advantages of using secondary waste that acts like an artificial ore to recover valuable metals, e.g., cobalt, include not only recycling the materials but also protecting our Earth from serious environmental issues [1,41]. As a consequence, various approaches are being investigated worldwide to recycle valuable materials like cobalt from artificial ores besides natural ores [41]. Some of the most common approaches that are being used are discussed below.

16.3.1 Pyrometallurgical Approach

The pyrometallurgical process is a traditional approach for the recovery of valuable metals from both natural ores and artificial ores, which basically uses high-temperature oxidative/reductive conditions to impose physical-chemical transformation in the raw material, typically through smelting or other heat-based methods [40,42]. The pyrometallurgical process commonly involves calcining, roasting, smelting, and refining, and in these unit processes, the metals are separated according to their chemical and metallurgical characteristics. Nevertheless, the high energy requirement during metal recovery makes the process a bit expensive and usually produces polluting emissions, which causes the loss of metals from the scrap during combustion. However, nowadays, various strategies are made where greener reactants and renewable energy are effectively being used, and at the same time, efficient technologies are being introduced to mitigate the environmental impact of pyrometallurgical processes [43].

16.3.2 Hydrometallurgical Approach

The hydrometallurgical process is a type of extractive metallurgy for extracting valuable metals from both primary ores and artificial ores, which utilizes aqueous solutions during metal recovery [44,45]. Typically, the hydrometallurgical process involves three main steps: (1) metal dissolution (leaching), (2) purification of the solution, and (3) recovery of metal values. Among these, the most commonly used hydrometallurgical method is the leaching process, which is carried out in an aqueous solution containing a lixiviant. A lixiviant is a chemical compound, normally used in the hydrometallurgical process to selectively extract a particular metal from its ores, which acts like an interface between metal and solvent. The dissolution of metal values by the leaching process uses many chemicals, mostly acids, viz. mineral acids and organic acids and sometimes a mixture of acids. Some other lixiviants include cyanide, isocyanate, thiosulfate, thiourea, and iodide. The leaching process is followed by concentration and purification steps, which include precipitation, solvent extraction, and ion exchange. Finally, the purified metals from the aqueous solutions are recovered by either the precipitation method or the electrochemical method. In the current scenario, research focus is being given toward circular hydrometallurgy, i.e., the designing of energy-efficient processes that consume minimum chemical reagents and produce minimum waste materials for both primary and secondary resources [46].

16.3.3 Bio-Hydrometallurgical/Bioleaching Approach

Bio-hydrometallurgy, also known as bioleaching, is a process employed in the mining sector to recover valuable metals from low-quality minerals with the assistance of microorganisms [31,47]. Thus, it is also a type of extractive metallurgical process that couples metallurgical science with biological science and usually makes use of microbes and microbiological processes. The microbes can be bacteria (viz. Acidithiobacillus ferrooxidans, Leptospirillum ferrooxidans, Acidithiobacillus thiooxidans, and Sulfolobus sp.) or fungi (viz. Aspergillus niger and Penicillium) [48]. Moreover, each species of a microbe can have different selectivity and efficiency for a particular type of metal to be extracted [49]. Though the microbiological process is a very slow process, it is highly selective and is entirely based on the interaction of metals with microbes. The process works akin to natural biogeochemical cycles and is thus an environmentally benign process along with low investment and low energy requirement. Currently, the microbe-based leaching process gains much popularity due to the availability of a wide range of species of various microbes and the suitability of the process to recover valuable metals from low-quality minerals and industrial wastes with greater efficiency and selectivity [50].

16.4 IMPORTANCE OF BIOLEACHING IN THE PRESENT CONTEXT

Bioleaching is a process often hailed as an eco-friendly and sustainable technology that harnesses the metabolic abilities of microbes such as fungi, bacteria, and archaea to extract metals from low-quality minerals, metal concentrates, or electronic waste (e-waste) [51,52]. This approach starkly contrasts with traditional mining methods, which frequently result in substantial environmental harm, including habitat destruction, soil erosion, and water pollution. In bioleaching, microorganisms are employed to break down the mineral ores or waste materials, releasing the target metals into solution through various biochemical reactions [33,49]. These microorganisms possess enzymes that facilitate the dissolution of metal-containing compounds, thereby making the metals accessible for recovery. The process occurs under relatively mild conditions, typically at ambient temperatures and atmospheric pressure, reducing the need for energy-intensive procedures such as high-temperature smelting. One key advantage of bioleaching is its ability to extract metals from ores that would otherwise be economically unfeasible to mine using conventional methods due to their low metal content or complex mineralogy [48]. Additionally, bioleaching can be tailored to target specific metals, offering a selective approach to metal recovery [53–56]. Furthermore, bioleaching is often considered environmentally friendly as it minimizes the production of toxic waste and reduces the emission of greenhouse gases associated with conventional mining and metal extraction processes. By utilizing naturally occurring microorganisms and operating at moderate conditions, bioleaching presents a sustainable alternative that aligns with efforts to reduce the environmental footprint of industrial activities.

Though the bioleaching process is still slow, particularly when compared to the operations of pyrometallurgy and hydrometallurgy, a number of studies have been conducted on hybrid bioleaching in order to achieve better results from bioleaching experiments. By way of illustration, Sheel and Pant were able to recover 90% of the gold that was present in electronic waste by utilizing a mixture of ammonium thiosulfate and Lactobacillus acidophilus [57]. A unique biorecovery approach that was followed by electrochemical treatment was devised by Sinha et al. in order to recover copper [58]. They were successful in recovering 92.7% of the copper. By employing chemical-biological hybrid systems in the bioleaching process, Dolker et al. achieved a 25% boost in the amount of lithium leached and a 98% boost in the amount of cobalt biosorbed [59]. Electrodialytic remediation was coupled with microbial metabolism by Gomes et al. [60]. When both approaches were combined, a greater mobilization of metals occurred, resulting in higher concentrations of metal in both the electrode compartments, notably for copper and chromium. Each of these combinations displayed a significantly greater bioleaching potential compared to individual cultures.

During the fungal bioleaching process, the microbes generate a significant quantity of organic acids such as oxalic acid, citric acid, tartaric acid, and a variety of carboxylic acids. These acids play a crucial role in aiding the solubilization of metals from waste printed circuit boards (WPCBs) by regulating the medium acidity and their redox potential. Furthermore, mechanisms for bioleaching processes are well established in many literature studies [61]. Usually, this process occurs within a comparatively high pH range of 9.0–10.5 and the fungi show adaptability to high pulp densities of 10% (w/v) [62]. Several fungal and yeast species have been investigated for the purpose of bioleaching metals from solid industrial wastes [32]. Recent research indicates that the Penicillium chrysogenum strain KBS3 was able to obtain the highest possible solubilization of metals, viz. 55% Ni, 67% Cu, 69% Mg, 60% Co, and 65% Zn from mine tailings [23]. During the process of bioleaching metals from fly ash, an investigation into the role of Aspergillus niger was carried out [63]. It was possible to attain a recovery rate of 56.1% for copper, 15.7% for aluminum, 20.5% for lead, 49.5% for zinc, and 8.1% for tantalum.

16.5 BIOLEACHING PROS AND CONS

Bioleaching is a process of solubilizing precious metals from low-grade minerals and wastes utilizing microorganisms [26,28,30,49]. Furthermore, various modifications are being implemented to improve the solubilization and extraction of valuable metals from minerals and wastes. Despite its numerous advantages, this process is also plagued by several shortcomings.

16.5.1 Pros of Bioleaching

- Bioleaching is useful for both low- and high-grade ores as well as industrial/mining wastes, and thus, it is helpful in minimizing mine tailing sites and thus more effective than traditional mining.
- Bioleaching is useful to extract valuable minerals from low-grade ores, which are already processed by chemical metallurgical processes, and provides good recovery.
- Acid mine drainage poses a significant environmental threat, characterized by the creation and circulation of highly acidic water laden with heavy metals. This phenomenon arises primarily from the oxidation of metal sulfides. Sulfides produce sulfates, which subsequently get converted to sulfuric acid, resulting in the contamination of water resources downstream. Bioleaching is a process that can minimize sulfate toxins from mines without affecting the environment.
- The toxic sulfur dioxide (SO_2) is generated through the combustion of fossil fuels and the smelting of sulfur-containing mineral ores, posing a significant environmental challenge. In contrast, bioleaching is a natural biological process that does not emit any harmful sulfur dioxide. Furthermore, bioleaching proves to be more cost-effective than smelting processes, making it a more profitable option.
- Additionally, it is an environmentally friendly process with low operational costs and minimal energy requirements. Furthermore, it operates with minimal use of chemical reagents while achieving high efficiency even at low metal concentrations.
- Finally, the bioleaching process is not only confined to laboratory-scale research but also emerged as a viable approach for industrial-scale operation toward processing of all types of minerals.

16.5.2 Cons of Bioleaching

- Bioleaching is a very slow process compared to smelting and provides low pulp density, thereby limiting its commercial applications.

- Stringent reaction conditions need to be maintained for the survival of microorganisms along with a good metal recovery rate.
- The possibility of killing the microorganisms due to the heat generated during the process cannot be overlooked.
- There remains the possibility of the production of toxic chemicals as side products due to the microbial metabolic process.
- Unlike other processes, bioleaching cannot be quickly stopped once started.
- Proper disposal and/or storage of specific microbials after bioleaching still remain a major concern to stop biohazards to the environment.
- Furthermore, bioleaching exhibits slow process kinetics and reduced efficiency at higher pulp densities, making it less feasible in environments with high toxicity.

16.6 PROCESS OF COBALT DISSOLUTION THROUGH BIOLEACHING

Cobalt is one of the very important metals to be used in many industrial applications, like energy storage devices and electronic devices, which results in the continuous depletion of naturally occurring cobalt-bearing minerals due to their exhaustive use [1,36]. To compensate the demand, various industrial processes are being developed to recover cobalt not only from its ores but also from various secondary ores, like mine tailing, industrial wastes, electronic wastes, and low-grade ores. The best possible way is to recycle the process for the recovery of valuable materials, like cobalt [39,40,42]. Out of the many processes, bioprocess is a novel eco-friendly approach utilizing microorganisms to recover valuable metals of industrial importance that seems to be of great potential and contributes significantly toward the mining industry [22].

Basically, there are two approaches for cobalt recovery by bioleaching based on microbe-ore interaction: (1) Direct bioleaching, also known as contact leaching, involves the direct enzymatic action of microorganisms on minerals that readily undergo oxidation, thereby facilitating the separation of metals from the minerals, and (2) indirect bioleaching, also referred to as non-contact leaching, doesn't involve direct contact between microorganisms and minerals, but generates leaching agents such as acids or enzymes, which are then utilized for metal recovery [64–67]. The mechanism of contact leaching entails the acidophilic microbes to be adsorbed onto the sulfide ore surface, facilitated by their inherent electrostatic attraction or hydrophobicity, leading to the development of biofilms known as sessile cells. In aerobic conditions, these sessile cells of acidophilic microbes continuously oxidize sulfide ores via iron and sulfur oxidation processes, ultimately transferring the electrons produced by sulfide ores to O_2, as shown below:

$$MS + 2H^+ + 0.5O_2 \rightarrow M^{2+} + S^0 + H_2O$$

$$2H^+ + S^0 + 2O_2 \rightarrow H_2SO_4$$

The non-contact leaching mechanism does not entail direct attachment of acidophilic microbial cells to the sulfide ore surface. Instead, these cells, known as planktonic cells, remain suspended in a liquid phase. The planktonic cells of acidophilic microbes use the Fe^{2+} in the solution as their source of energy, which undergoes oxidation to form Fe^{3+}. Subsequently, the Fe^{3+} oxidizes and decomposes the sulfide ore, releasing Fe^{2+} and generating elemental sulfur. This process supplies energy to acidophilic microbes, allowing them to oxidize and decompose sulfide minerals continuously for metabolic growth. The typical reactions involved in the non-contact leaching mechanism are shown below:

$$2Fe^{2+} + 2H^{+} + 0.5O_2 \rightarrow 2Fe^{3+} + H_2O$$

$$MS + 2Fe^{3+} \rightarrow 2Fe^{2+} + M^{2+} + S^{0}$$

A third category known as cooperative leaching is sometimes envisaged where microbes attached to the surface containing metal values and free microbes cooperate, where the mechanisms of both contact leaching and non-contact leaching operate. A detailed schematic view of all the processes is shown in Figure 16.1.

In industrial settings, an indirect process is considered suitable for enhancing leaching efficiency. Generally, the benefit of employing an indirect bioleaching process lies in the autonomous production of lixiviant, which separates the bioprocess from the chemical process. This independence allows for the optimization of each process separately, thereby maximizing productivity [68]. This approach is used for minerals that do not have sufficient mineral components to sustain a viable microbial population. Moreover, it can handle superior concentrations of waste compared to the direct process, resulting in enhanced metal yields. In terms of cobalt bioleaching by bacteria and fungi, the indirect approach involves the generation of metabolic by-products; mostly, sulfuric acids are responsible for the dissolution of cobalt [22].

In a broad spectrum, the process of bioleaching for the recovery of metal values can be classified as acidolysis, redoxolysis, and complexolysis, as shown in Figure 16.2. This type of classification is based on the type of metabolic processes and their metabolic products as well as the types of carbon and energy sources used during the process [17,22,56]. The acidolysis process involves the production of bio-acids as metabolic products and the subsequent dissolution of metal values. The redoxolysis process involves the formation of extracellular polymeric substances (EPS) by microbes in contact with the metal surface through the formation of biofilm and the dissolution of metal values triggered as a result of electron transfer between the microbes and waste materials. The complexolysis process involves the formation of metabolic products, like bio-acids, and the subsequent dissolution of metal values by complexation/chelation reactions.

Typically, acidophilic bacteria (viz. sulfur-oxidizing and iron-oxidizing) are frequently employed in bioleaching reactions. Microorganisms belonging to these categories include Acidithiobacillus thiooxidans, Acidithiobacillus ferrooxidans, Sulfobacillus thermosulfidooxidans, and Leptospirillum ferrooxidans [49]. Mostly,

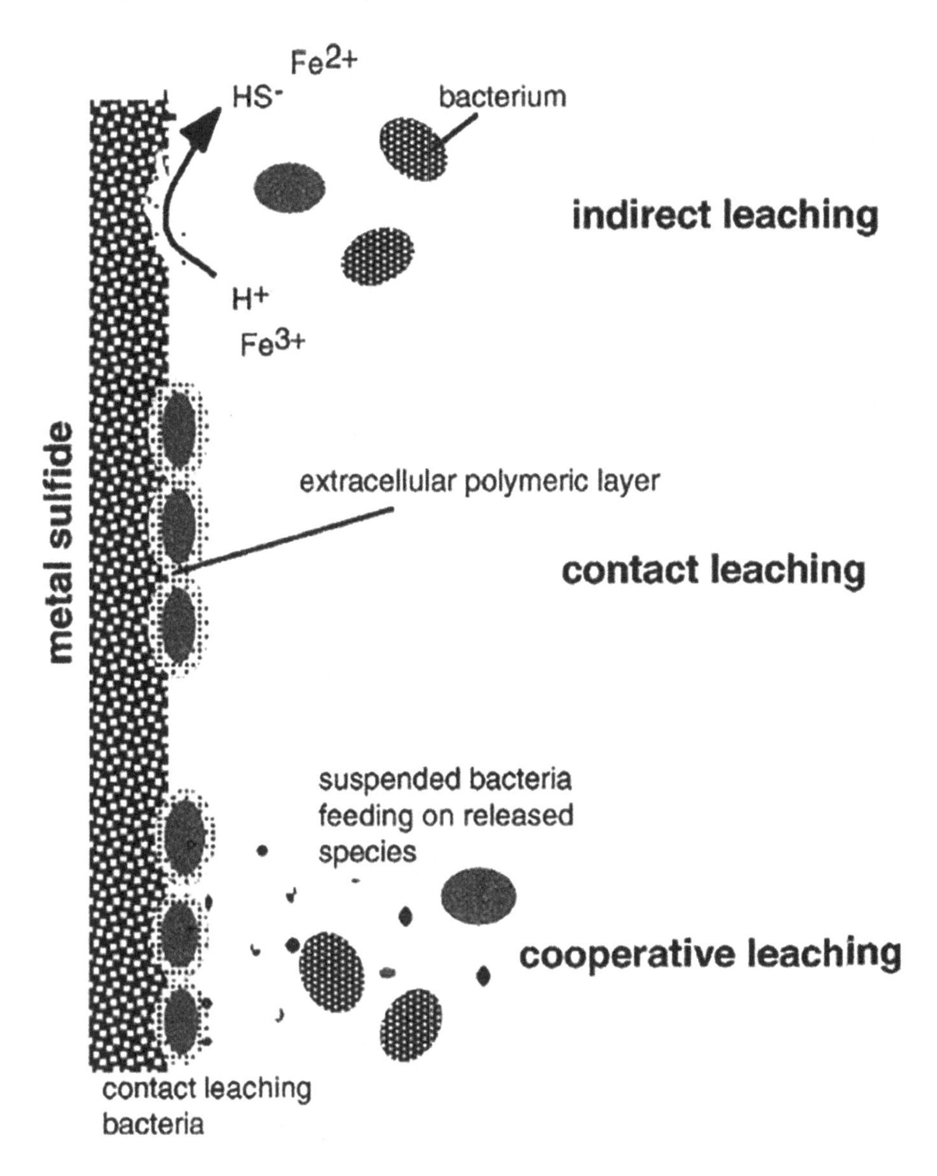

FIGURE 16.1 Schematic view of direct, indirect, and cooperative leaching for a sulfide mineral by bacteria [64].

the source of energy of these bacteria comes from the inorganic compounds (FeS_2, Fe^{2+}, and S^0) and facilitates the dissolution of metal through a series of biochemical reaction pathways, as shown below [22]:

$$2FeS_2 + 7O_2 + 2H_2O \rightarrow 2Fe^{2+} + 4SO_4^{2-} + 4H^+$$

$$4Fe^{2+} + 4H^+ + O_2 + \text{At. ferrooxidans} \rightarrow 4Fe^{3+} + 2H_2O$$

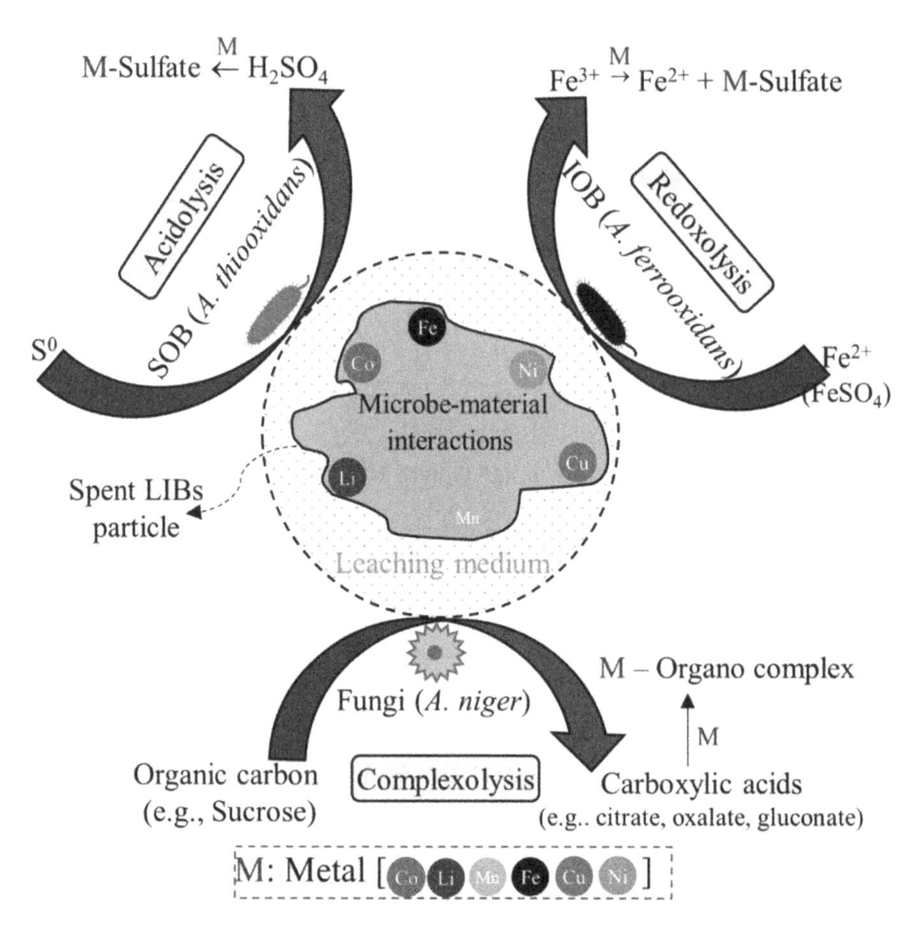

FIGURE 16.2 The mechanism of metal dissolution by microbes from spent lithium-ion battery [56].

$$2S^0 + 3O_2 + 2H_2O + \text{At. thiooxidans} \rightarrow 2H_2SO_4$$

$$Co_2O_3 + 6H^+ + 2e^- \rightarrow 2Co^{2+} + 3H_2O$$

$$2Fe^{2+} \rightarrow 2Fe^{3+} + 2e^-$$

In fact, the metabolites produced in the leaching medium by microbes during their growth, like bio-acids or biogenic acids, are responsible for the dissolution of metal values. Wu et al. proposed two basic mechanisms for the recovery of cobalt by bioleaching of the $LiCoO_2$ process, as shown in Figure 16.3 [33]. In the first mechanism, bio-oxidation of Fe^{2+} to Fe^{3+} on the bacterial cell leads to the release of electrons, which subsequently transfer through a series of biochemical reactions for the reduction of

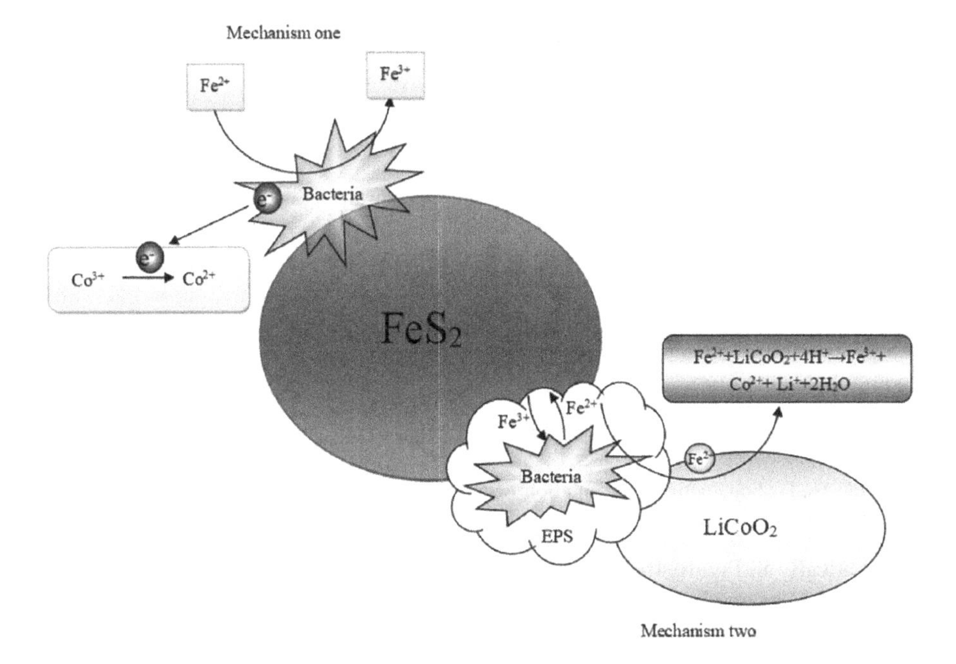

FIGURE 16.3 Proposed mechanism for the extraction of cobalt by bacteria from $LiCoO_2$ [33].

Co^{3+} to Co^{2+}. In the second mechanism, while bioleaching $LiCoO_2$, bacteria may secrete EPS. Compounds contained within these substances could aid in the extraction of metals from $LiCoO_2$. The majority of reported cobalt recovery through bioleaching processes typically involves either one or both of these mechanisms.

The fungi-assisted bioleaching of metals is due to the release of several organic acids, viz. formic, oxalic, acetic, citric, succinic, gluconic, lactic, and pyruvic acids [24,68]. Moreover, the relative acidity of an organic acid depends on its side-chain functionality and its electronic environment and is thus effective in the dissolution of cobalt. Furthermore, organic acids act like a chelating agent and make a complex with the Co^{2+} during the acidolysis reaction, thereby stabilizing them for further metal recovery [69]. Furthermore, metal values can sometimes get bio-accumulated within fungal biomass. In such cases, they are subsequently recovered as the functional moieties of the fungal cell wall are capable of binding metal ions to varying degrees. Thus, acidophilic bioleaching by bacteria proceeds through acidolysis and redoxolysis, whereas acidophilic bioleaching by fungi proceeds through acidolysis and complexolysis.

16.7 COBALT RECOVERY PROCESS THROUGH BIOLEACHING OF INDUSTRIAL WASTES

The leaching process to recover cobalt proceeds with the general step-by-step process except slight modifications based on the type of waste discards, as shown

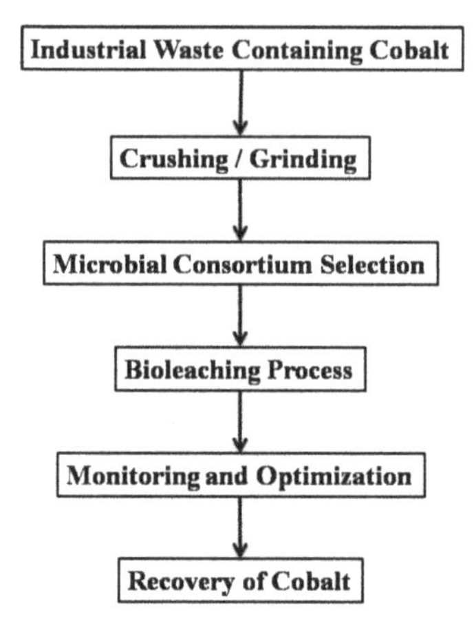

FIGURE 16.4 Schematic representation of the general process of bioleaching for cobalt recovery from industrial waste.

schematically in Figure 16.4 [66,67,70–72]. In general, industrial wastes containing cobalt may first undergo crushing or grinding to increase the surface area accessible to microorganisms and remove any contaminants or inhibitors that could interfere with the bioleaching process. A suitable microbial consortium or pure culture of microorganisms capable of leaching cobalt from the waste is selected. The waste materials are mixed with a nutrient medium and inoculated with the selected microbial consortium. The bioleaching process typically occurs under controlled conditions of temperature, pH, and oxygenation to optimize microbial activity. As the microorganisms grow and metabolize, they release organic acids or other compounds that dissolve cobalt from the waste matrix into the surrounding solution. The bioleaching process is monitored regularly to assess cobalt leaching efficiency, microbial growth, and the amount of cobalt in the leach solution. Adjustments to process parameters may be made to optimize cobalt recovery and minimize any potential environmental impacts. Once a sufficient concentration of cobalt has been leached into the solution, the cobalt can be extracted from the leachate using appropriate technology, such as solvent extraction, precipitation, and ion exchange. The remaining solid residue may undergo further treatment or be disposed of in an environmentally responsible manner. However, it's essential to consider factors such as microbial activity, process efficiency, and waste management to ensure the economic viability and environmental sustainability of bioleaching operations.

16.7.1 Cobalt Recovery from Low-Grade Ores and Mine Tailing

Tailings are basically mining wastes that remain after the extraction of valuable minerals from their ores, which still contain many elements of industrial importance. Mine tailings, some of the most extensive mining wastes globally, often cover extensive areas. They are produced during the processing of sulfide ores, and when they come into contact with water and oxygen, they often generate acid rock drainage (ARD), resulting in the pollution of ground and surface water [73,74]. Despite the detrimental environmental impacts of tailings, they also hold commercial potential because of the residual valuable metals they contain [22]. However, these mining wastes contain metal values in very low amounts and have a very complex composition; thus, alternative processes must be used to make the extraction process economically feasible [20,22,75]. While the utilization of tailings remains relatively underdeveloped, stakeholders in the mining industry recognize their potential as a future source of raw materials, particularly with the advent of new technologies. The ideal approach to managing tailings involves processing them to extract valuable metals and mitigate environmental impacts by converting reactive and hazardous minerals into stable forms [70].

Owing to the lack of independent cobalt deposits in nature, bioleaching is a promising method to convert complex mine tailings into secondary resources for cobalt. Bioleaching of low-grade ores to recover valuable metals has several advantages such as lower operational cost, low energy consumption, simplicity, and an environmentally benign process [20,73]. Processing of mine tailings not only recovers valuable minerals (although in lesser quantities) but also minimizes their negative environmental impacts, thereby maintaining resource conservation. The prime function of microbes in the bioleaching process is to produce some hydroxycarboxylic acids (viz. gluconic, citric, pyruvic, tartaric, and lactic) and related metabolites within the cultivation medium, which facilitates the dissolution of metal values [49]. Furthermore, the amount and purity of bio-acids produced depend on various factors, including microbial species, medium composition, pulp density, medium pH, temperature, aerobic environment, types of energy sources, and types of carbon sources [22].

Zhang et al. modified a mesophilic microbial consortium, predominantly composed of A. ferrooxidans and A. thiooxidans, achieving Co and Cu extractions of 91% and 57%, respectively, from bulk sulfidic mine tailings (containing 0.02% Co and 0.12% Cu) after 13 days in a stirred tank reactor [76]. Bioleaching experiments using a flotation concentrate from the tailings (with 0.06% Co and 0.57% Cu) resulted in 66% cobalt and 33% copper recovery. In another work, the same group reported the recovery of 74% Co and almost 100% Cu using a moderate thermophilic microbial consortium at 42°C in a 2-liter stirred tank reactor at 10% solid load [29]. Mäkinen et al. employed iron- and sulfur-oxidizing microorganisms to bioleach pyrite-rich tailings, aiming primarily to extract cobalt along with other valuable metals [77]. After microbial culture adaptation and batch bioleaching tests, a continuous-batch mode mini-pilot protocol was carried out using 10-L stirred tank reactors (maintained at 30°C with a solid content of 100 g/L). This approach resulted in significant leaching yields for the target metals (87% for Co, 100% for Zn, 67% for Ni, and 43% for Cu) in approximately 10 days of retention time. Sadeghieh et al. investigated the salinity impact on the bioleaching efficacy of a Cu-Ni-Co-bearing sulfidic tailing [78]. The results showed that the best extraction rates for the metals (87% Ni and 69% Co) were achieved under the following conditions: pH 1.8, 10% pulp density, Norris nutrient medium with 10 g/L of sodium chloride, and the use of moderately thermophilic microorganisms. Santos et al. conducted laboratory-scale tests on some Co-bearing minerals, viz. primary limonitic laterite ores and processing residues [79]. These materials underwent bioleaching in reducing conditions using a consortium of acidophilic bacteria (utilizing S as an electron donor) in stirred tank bioreactors maintained at pH 1.5 and a temperature of 35°C and successfully extracted 40%–50% of cobalt compared to some other metals within 30 days.

Thus, bioleaching provides a sustainable and economical approach for the extraction of cobalt from low-grade ores and mine tailings, contributing to the proficient utilization of resources and eco-friendly mining and processing of cobalt. However, like any mining operation, it's essential to implement proper management practices to mitigate environmental risks and ensure long-term sustainability.

16.7.2 Cobalt Recovery from Industrial Waste (Spent Catalyst and Spent Lithium-Ion Battery)

The refinery industries use a huge amount of metal catalysts, which contain a variety of metals, like Co, Fe, Ni, V, and Mo. After several cycles of use, the catalysts lose their efficiency and thus are discarded as waste. The metal values present in these wastes are very significant for the metal industries, and thus, several processes are being employed for their safe recovery, like bioleaching [39,47,80,81]. This eco-friendly approach offers a better prospect for the recovery of cobalt from spent catalysts. Gholami et al. employed A. ferrooxidans and A. thiooxidans to bioleach aluminum, cobalt, molybdenum, and nickel from hazardous spent catalysts [21]. The findings revealed that through a one-step bioleaching process with A. ferrooxidans, the maximum recovery attained was 63% Al, 96% Co, 84% Mo, and 99% Ni. However, utilizing A. thiooxidans yielded the highest removal efficiencies of 2.4%, 83%, 95%, and 16% for aluminum, cobalt, molybdenum, and nickel, respectively. Gholami et al. conducted bioleaching on a spent processing catalyst using A. ferrooxidans and A. thiooxidans to recover aluminum, cobalt, molybdenum, and nickel [82]. The findings indicated that the maximum recovery of 63% Al, 96% Co, 84% Mo, and 99% Ni was achieved in 30 days at pH 1.8–2.0 using A. ferrooxidans in the presence of ferrous sulfate. Conversely, the highest extractions of 2.4% Al, 83% Co, 95% Mo, and 16% Ni were achieved after 30 days at pH 3.9–4.4 using A. thiooxidans in the presence of sulfur.

The growth of the electronic industry directly impacts the battery industry, which generates a huge amount of spent lithium-ion batteries (LIBs), causing environmental pollution [83–85]. Cobalt is not only a major component in the cathode of LIBs but also a very minor component in the anodes of NiMH and NiCd batteries [86]. At the current time, $LiCoO_2$ is one of the most preferred cathode materials extensively used in portable electronic devices. Cobalt and lithium are the major constituents of spent LIBs besides others constituting ~30% Co and 10% Li [87]. Given the limited supply of cobalt and its growing demand, the mining industry faces significant pressure, as its supply cannot be sustained for long-term applications [86]. Therefore, to recover valuable metals such as cobalt, recycling processes are widely adopted, employing various approaches. Among these methods, bioleaching (bio-hydrometallurgy) has gained more focus in recent years because it employs specific microorganisms for the selective extraction of Co from spent LIBs [41,88,89]. Moreover, bioleaching proves to be economical, eco-friendly, and straightforward to operate and requires minimal energy [24,26,56,68].

In many studies, the extraction of Co from spent LIBs through bioleaching is conducted using acidophilic sulfur-oxidizing bacteria (SOB) and iron-oxidizing bacteria (IOB) [33,49,56,90]. Examples of SOB employed for this purpose include Acidithiobacillus thiooxidans, Acidithiobacillus caldus, Sulfobacillus thermosulfido-oxidans, and Alicyclobacillus sp., while examples of IOB include Acidithiobacillus ferrooxidans, Leptospirillum ferriphilum, Ferroplasma sp., and Sulfobacillus sp. Both SOB and IOB are a class of autotrophic microorganisms that get their energy by oxidizing inorganic compounds and are collectively known as chemolithoautotrophs [91]. The carbon source for these bacteria comes from the reductive fixation of atmospheric carbon dioxide. The energy source for SOB comes from inorganic compounds like elemental sulfur (S^0), whereas the energy source for IOB comes from ferrous ion (Fe^{2+}) [49,92]. For bacterial leaching, $FeSO_4$, Fe powder, and FeS_2 serve as the sources of iron and sulfur [89]. IOB oxidizes Fe^{2+} to Fe^{3+}, while SOB oxidizes S^0 to SO_4^{2-} [87]. Moreover, the advantage of using chemolithotrophic bacteria is their high level of resistance toward metal toxicity [30,91].

The working mechanism of heterotrophic microorganisms, like fungi, is a little different, the growth and metabolic activity of which depend on the carbon sources derived from organic carbon-based materials, like glucose and sucrose, which also act as energy sources [93]. Some of the fungal species used in the bioleaching process include Aspergillus tubingensis, Aspergillus niger, Penicillium chrysogenum, and Penicillium simplicissimum for the extraction of cobalt from secondary ores [56,94]. However, due to its ease of growth and high efficiency in metal dissolution, Aspergillus niger is mostly used in many bioleaching processes of LIBs [95]. It can better tolerate metal toxicity from spent LIBs, thereby demonstrating higher efficiency in metal bioleaching. Some of the advantages of fungal bioleaching over bacteria include (1) better tolerance to metal toxicity, (2) faster leaching rate, (3) capability to grow in both acid and alkaline media, and (4) having a shorter lag phase [24,68,96]. However, both bacterial-based bioleaching and fungal-based bioleaching result in higher lithium recovery compared to cobalt.

Recent research effort demonstrates that bioleaching of LIB using specific microbes results in efficient recovery of cobalt along with other metals. Alipanah et al. utilized Gluconobacter oxydans bacteria for the extraction of a biolixiviant from corn stover to bioleach the spent LIB cathode materials, where iron(II) served as a reducing agent to facilitate metal dissolution [97]. The life-cycle assessment demonstrated that bioleaching of spent LIBs could provide superior environmental sustainability when compared to alternative hydrometallurgical recovery methods like hydrochloric acid leaching. Specifically, bioleaching resulted in significantly lower CO_2 equivalent global warming potential per kilogram of recovered Co (16–19 kg CO_2 equivalent compared to 43–91 kg CO_2 equivalent). Figure 16.5 shows a schematic view of techno-economic analysis and life-cycle assessment of cobalt bioleaching from LIB. Noruzi et al. investigated the bioleaching of Co and Ni from spent LIB, utilizing A. ferrooxidans and A. thiooxidans through an innovative process catalyzed by silver ions [98]. The spent medium method achieved extraction rates of 45.2% Co and 71.5% Ni. Remarkably, a very high extraction yield of 99.95% for these metals was obtained in just 3 days using a two-step bioleaching approach. This approach involved gradually adding solid content and incorporating Ag^+ into the process. Zeng

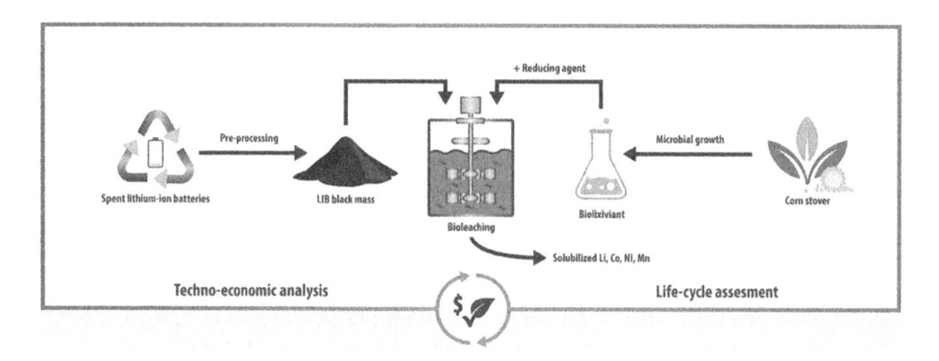

FIGURE 16.5 Schematic representation of techno-economic analysis and life-cycle assessment of cobalt bioleaching from spent LIB [97].

et al. also observed the impact of Ag^+ on the bioleaching process of Co extracted from spent LIB using A. ferrooxidans and observed the most significant effect at 0.02 g/L Ag^+ concentration, resulting in a remarkable cobalt leaching efficiency of 98.4% within a 7-day period, whereas only 43.1% Co was leached in the absence of Ag^+ [99]. The formation of a silver intermediate compound ($AgCoO_2$) on the particle surface, which readily dissolves, contributes to the enhancement of bioleaching efficiency. Once $AgCoO_2$ is oxidized by Fe^{3+}, the consumed Ag^+ is released for further catalytic action to recover Co^{2+} and Fe^{2+}. Further bacterial activity facilitates the oxidation of resulting Fe^{2+} back to Fe^{3+}. Zeng et al. further investigated Cu-catalyzed bioleaching as a method for recycling Co from spent LIB and demonstrated that 99.9% Co recovered in 6 days of bioleaching in the presence of 0.75g/L copper ions, whereas only 43.1% Co recovered in the absence of copper ions even after 10 days of reaction [100]. Hariyadi et al. isolated Aspergillus niger from candlenut and used it as a leaching agent to enhance detoxification and metal recovery from spent LIBs under different conditions and observed that the maximum recovery of both Co and Li in the leached liquor reached up to 72% [101]. Furthermore, bacterial strains derived from acid mine drainage also show promising efficiency toward extracting Co and Li in bioleaching processes. Putra et al. isolated the bacterium A. ferrooxidans from acid mine water to extract Co from LIBs, achieving a cobalt recovery rate of 74% over 14 days [102]. This was accomplished by adding 1 g of battery cathode per 100 mL under optimal conditions, which included 20% inoculum, pH 2–4, a temperature of 30°C, and the use of an aerator in the aeration system. Extracting cobalt through bioleaching at high pulp densities poses significant challenges, including microbial inhibition caused by elevated metal toxicity and substrate (iron) scarcity. Roy et al. explored the bioleaching of LIB at a high pulp density of 100 g/L employing the cost-efficient autotrophic bacterium A. ferrooxidans, achieving a recovery of 94% Co and 60% Li within 72 h [18,103]. Lobos et al. assessed the activity of A. niger, P. chrysogenum, and P. simplicissimum under varying concentrations of metal ions and discovered that pre-exposure to these metals allowed the fungi to build up tolerance to these metals [104]. Ghassa et al. explored the bioleaching of spent LIB under moderate thermophilic conditions of 45°C and observed the boost in Co recovery in the presence of $FeSO_4.7H_2O$, which served as a reducing agent. Remarkably, just 2 days of bioleaching in the presence of 24.25 g/L $FeSO_4.7H_2O$ resulted in the recovery of 99.9% Co, 99.7% Ni, and 84% Li [90].

Currently, electronic wastes are considered as secondary ores or artificial ores since they contain many precious metals (Au, Ag, and Pt), base metals (Al, Co, Cu, Ni, Zn, and Fe), rare earth metals (In, Nd, and Ta), and other metals (Be, Cd, Cr, Hg, Pb, Sb, Sn, and Ti) in larger quantities than those found in some natural ores [105]. Owing to their superb electrical conductivity and chemical resistance, these metals are used in various industries, such as electronics, automotive, jewelry, dentistry, and aviation. Thus, recycling of e-waste using the biological process for the recovery of valuable metals plays a significant role from both economic and environmental perspectives [71,72,105–107]. However, the main concern is the difficulties in treatment due to the heterogeneity of the materials present in the products, complex composition, and huge volumes.

16.8 PROSPECTS OF BIOLEACHING FOR THE COBALT INDUSTRY

Although the process of bioleaching is an effective way to get the metal values from low-grade ores, mine tailings, spent catalysts, spent LIBs, etc., there are many obstacles that remain, which hamper the efficiency and economy of the process. Thus, there is a need to improve the efficiency of the bioleaching process using modern sustainable technologies so that effective management of waste material can be made and at the same time recover valuable metals. This process will help in enhancing the leaching efficiency, minimizing process cost, and making use of resources effectively. Some of the measures that need particular attention are as follows:

1. Compared to traditional metal recovery, the use of biotechnology opens more options for the bioleaching process. The available genetic data can be used to synthesize specific microbes that are needed for industrial-scale recovery of cobalt from waste.
2. There is an utmost need to design genetically engineered microbes for selective and efficient leaching of cobalt from waste materials. Thus, future research efforts can enable the microbes to have genes with higher resistance to environmental factors to obtain a greater leaching rate of cobalt.
3. The slow kinetics can be improved by using a mixed system, where nano-catalyst can be employed along with the microbes in an optimum concentration range so that it will not hamper microbial growth.
4. The use of agricultural waste as a carbon and energy source for the microbial growth process may be explored for the development of eco-friendly technologies toward the recovery of valuable cobalt metal.
5. Since microbes are very sensitive to toxic level in their growth media, care should be taken to minimize the toxicity of waste material using some physico-chemical techniques before the actual bioleaching process.
6. The method of using the microbes for bioleaching may be revisited so that they can have a better tolerance level toward metal ion concentration, growth medium pH, temperature, and other environmental factors.

7. Innovative technologies can be implemented where other extraction methods can be coupled with the bioleaching process for the maximum recovery of cobalt from waste in a more economical way.
8. To further improve the metal recovery from waste materials, new strains have to be identified and isolated, implementing their effective use.
9. Last but not least, there is a need for commercial application of bioleaching at an industrial scale since the process is still on a laboratory scale.

16.9 CONCLUSION

In regard to metal recovery, both chemical and biological approaches have their own merits and demerits, but during their implementation, care should be taken on their efficiency and economic point of view. The need of the hour is to develop inexpensive extraction processes for the recovery of valuable metals, particularly from low-grade ores like waste materials. In this respect, the biological process emerges as one of the alternative processes for the recovery of cobalt without affecting much to the environment. In comparison with chemical leaching, the bioleaching process demonstrates higher cobalt removal efficiency from waste, albeit at a slower pace. However, bioleaching still requires testing of reactor configurations, development of upstream and downstream processes, scale-up considerations, assessment of operating expenses, capital expenditures, and product recovery methodologies. Moreover, developing nations should put more attention toward research and development for technologies based on biological processes, since the natural approaches to industrial processing are gaining popularity in many sectors. Overall, bioleaching offers a sustainable and cost-effective alternative for recovering cobalt from spent catalysts while reducing the environmental footprint associated with traditional extraction methods.

REFERENCES

1. A. Tisserant, S. Pauliuk, Matching global cobalt demand under different scenarios for co-production and mining attractiveness, *J. Econ. Struct.* 5 (2016) 4.
2. A. Julander, J. Kettelarij, C. Lidén, Cobalt. In: John, S., Johansen, J., Rustemeyer, T., Elsner, P., Maibach, H. (eds) *Kanerva's Occupational Dermatology*. Springer, Cham, (2020).
3. S. Luckeneder, S. Giljum, A. Schaffartzik, V. Maus, M. Tost, Surge in global metal mining threatens vulnerable ecosystems, *Glob. Environ. Chang.* 69 (2021) 102303.
4. T. Watari, K. Nansai, K. Nakajima, Major metals demand, supply, and environmental impacts to 2100: a critical review, *Resour. Conserv. Recycl.* 164 (2021) 105107.
5. R. Karn, N. Ojha, S. Abbas, S. Bhugra, A review on heavy metal contamination at mining sites and remedial techniques, *IOP Conf. Ser. Earth Environ. Sci.* 796 (2021) 012013.
6. I.V. Zabaikin, A.N. Lunkin, D.A. Lunkin, Disposal of mining waste and its impact on the environment, *Ecol. Eng. Environ. Technol.* 23 (2022) 7–14.
7. H.I. Abdel-Shafy, M.S.M. Mansour, Solid waste issue: sources, composition, disposal, recycling, and valorization, *Egypt. J. Pet.* 27 (2018) 1275–1290.
8. L.A. Guerrero, G. Maas, W. Hogland, Solid waste management challenges for cities in developing countries, *Waste Manag.* 33 (2013) 220–232.

9. S.D. Mancini, G.A. de Medeiros, M.X. Paes, B.O.S. de Oliveira, M.L.P. Antunes, R.G. de Souza, J.L. Ferraz, A.P. Bortoleto, J.A.P. de Oliveira, Circular economy and solid waste management: challenges and opportunities in Brazil, *Circ. Econ. Sustain.* 1 (2021) 261–282.
10. J. Cui, L. Zhang, Metallurgical recovery of metals from electronic waste: a review, *J. Hazard. Mater.* 158 (2008) 228–256.
11. R. Rautela, B.R. Yadav, S. Kumar, A review on technologies for recovery of metals from waste lithium-ion batteries, *J. Power Sources*. 580 (2023) 233428.
12. L. Brückner, J. Frank, T. Elwert, Industrial recycling of lithium-ion batteries – a critical review of metallurgical process routes, *Metals (Basel).* 10 (2020) 1107.
13. Z.J. Baum, R.E. Bird, X. Yu, J. Ma, Lithium-ion battery recycling – overview of techniques and trends, *ACS Energy Lett.* 7 (2022) 712–719.
14. O. Pourret, M.P. Faucon, Cobalt. In: White, W. (eds) *Encyclopedia of Geochemistry. Encyclopedia of Earth Sciences Series*. Springer, Cham (2017). https://doi.org/10.1007/978-3-319-39193-9_271-2.
15. W.M. Haynes, Abundance of elements in the earth's crust and in the sea references. In: *Handbook of Chemistry and Physics*, 97th ed., CRC Press, Boca Raton, FL, 1989: pp. 17–17.
16. Q. Dehaine, L.T. Tijsseling, H.J. Glass, T. Törmänen, A.R. Butcher, Geometallurgy of cobalt ores: a review, *Miner. Eng.* 160 (2021) 106656.
17. M. Sethurajan, S. Gaydardzhiev, Bioprocessing of spent lithium ion batteries for critical metals recovery – a review, *Resour. Conserv. Recycl.* 165 (2021) 105225.
18. J. Jegan Roy, M. Srinivasan, B. Cao, Bioleaching as an eco-friendly approach for metal recovery from spent NMC-based lithium-ion batteries at a high pulp density, *ACS Sustain. Chem. Eng.* 9 (2021) 3060–3069.
19. G.J. Olson, J.A. Brierley, C.L. Brierley, Bioleaching review part B: progress in bioleaching: applications of microbial processes by the minerals industries, *Appl. Microbiol. Biotechnol.* 63 (2003) 249–257.
20. S.K. Sarker, N. Haque, M. Bhuiyan, W. Bruckard, B.K. Pramanik, Recovery of strategically important critical minerals from mine tailings, *J. Environ. Chem. Eng.* 10 (2022) 107622.
21. R. Mafi Gholami, S.M. Borghei, S.M. Mousavi, Heavy metals recovery from spent catalyst using acidithiobacillus ferrooxidans and acidithiobacillus thiooxidans. In: ICCCE 2010-2010 International Conference on Chemistry and Chemical Engineering, Kyoto, Japan, 2010, pp. 331–335.
22. A.K. Saim, F.K. Darteh, A Comprehensive review on cobalt bioleaching from primary and tailings sources, *Miner. Process. Extr. Metall. Rev.* 344 (2023) 118511. DOI: 10.1080/08827508.2023.2181346.
23. S. Ilyas, R.A. Chi, J.C. Lee, Fungal bioleaching of metals from mine tailing, *Miner. Process. Extr. Metall. Rev.* 34 (2013) 185–194.
24. L. Dusengemungu, G. Kasali, C. Gwanama, B. Mubemba, Overview of fungal bioleaching of metals, *Environ. Adv.* 5 (2021) 100083.
25. F. Moosakazemi, S. Ghassa, M. Jafari, S.C. Chelgani, Bioleaching for recovery of metals from spent batteries – a review, *Miner. Process. Extr. Metall. Rev.* 4 (2022) 511.
26. I. Tezyapar Kara, K. Kremser, S.T. Wagland, F. Coulon, Bioleaching metal-bearing wastes and by-products for resource recovery: a review, *Environ. Chem. Lett.* 21 (2023) 3329.
27. H. Srichandan, R.K. Mohapatra, P.K. Parhi, S. Mishra, Bioleaching approach for extraction of metal values from secondary solid wastes: a critical review, *Hydrometallurgy*. 189 (2019) 105122.
28. G. Kour, R. Kothari, H.M. Singh, D. Pathania, S. Dhar, Microbial leaching for valuable metals harvesting: versatility for the bioeconomy, *Environ. Sustain.* 4 (2021) 215.

29. R. Zhang, A. Schippers, Stirred-tank bioleaching of copper and cobalt from mine tailings in Chile, *Miner. Eng.* 180 (2022) 107514.
30. K. Bosecker, Bioleaching: metal solubilization by microorganisms, *FEMS Microbiol. Rev.* 20 (1997) 591–604.
31. F. Anjum, M. Shahid, A. Akcil, Biohydrometallurgy techniques of low grade ores: a review on black shale, *Hydrometallurgy*. 117–118 (2012) 1–12.
32. D. Mishra, Y.H. Rhee, Microbial leaching of metals from solid industrial wastes, *J. Microbiol.* 52 (2014) 1–7.
33. W. Wu, X. Liu, X. Zhang, X. Li, Y. Qiu, M. Zhu, W. Tan, Mechanism underlying the bioleaching process of LiCoO2 by sulfur-oxidizing and iron-oxidizing bacteria, *J. Biosci. Bioeng.* 128 (2019) 344–354.
34. A. Pattanaik, L.B. Sukla, D. Pradhan, D.P. Krishna Samal, Microbial mechanism of metal sulfide dissolution, *Mater. Today Proc.* 30, 2020, 326–331.
35. M. Hapke, G. Hilt, Introduction to cobalt chemistry and catalysis, *Cobalt Catal. Org. Synth. Methods React.* 661, 2019: pp. 1–23.
36. Y. Yildiz, General aspects of the cobalt chemistry. In: Khan, M. (Ed.) *Cobalt.* IntechOpen, London (2017).
37. S.H. Farjana, N. Huda, M.A.P. Mahmud, Life cycle assessment of cobalt extraction process, *J. Sustain. Min.* 18 (2019) 150–161.
38. USGS, Cobalt Statistics and Information | U.S. Geological Survey (2023). https://www.usgs.gov/centers/national-minerals-information-center/cobalt-statistics-and-information.
39. D. Małolepsza, M. Rzelewska-Piekut, M. Emmons-Burzyńska, M. Regel-Rosocka, Waste-to-resources: leaching of cobalt from spent cobalt oxide catalyst, *Catalysts.* 13 (2023) 952.
40. S. Stopić, B. Friedrich, Recovery of cobalt from primary and secondary materials: an overiew, Vojnoteh. *GLAS.* 68 (2020) 321–337.
41. S. Krishnan, N.S. Zulkapli, H. Kamyab, S.M. Taib, M.F.B.M. Din, Z.A. Majid, S. Chaiprapat, I. Kenzo, Y. Ichikawa, M. Nasrullah, S. Chelliapan, N. Othman, Current technologies for recovery of metals from industrial wastes: an overview, *Environ. Technol. Innov.* 22 (2021) 101525.
42. M. Chandra, D. Yu, Q. Tian, X. Guo, Recovery of Cobalt from secondary resources: a comprehensive review, *Miner. Process. Extr. Metall. Rev.* 43 (2022) 679–700.
43. J.P. Harvey, W. Courchesne, M.D. Vo, K. Oishi, C. Robelin, U. Mahue, P. Leclerc, A. Al-Haiek, Greener reactants, renewable energies and environmental impact mitigation strategies in pyrometallurgical processes: a review, *MRS Energy Sustain.* 9 (2022) 212–247.
44. V. Gunarathne, A.U. Rajapaksha, M. Vithanage, D.S. Alessi, R. Selvasembian, M. Naushad, S. You, P. Oleszczuk, Y.S. Ok, Hydrometallurgical processes for heavy metals recovery from industrial sludges, *Crit. Rev. Environ. Sci. Technol.* 52 (2022) 1022–1062.
45. C.G. Anderson, H. Cui, Advances in mineral processing and hydrometallurgy, *Metals.* 11 (2021) 1393.
46. K. Binnemans, P.T. Jones, The twelve principles of circular hydrometallurgy, *J. Sustain. Metall.* 9 (2023) 1–25.
47. C. Erüst, A. Akcil, C.S. Gahan, A. Tuncuk, H. Deveci, Biohydrometallurgy of secondary metal resources: A potential alternative approach for metal recovery, *J. Chem. Technol. Biotechnol.* 88 (2013) 2115–2132.
48. M. Vera, A. Schippers, S. Hedrich, W. Sand, Progress in bioleaching: fundamentals and mechanisms of microbial metal sulfide oxidation – part A, *Appl. Microbiol. Biotechnol.* 106 (2022) 6933–6952.
49. S. Jones, J.M. Santini, Mechanisms of bioleaching: iron and sulfur oxidation by acidophilic microorganisms, *Essays Biochem.* 67 (2023) 685–699.

50. T. Naseri, V. Beiki, S.M. Mousavi, S. Farnaud, A comprehensive review of bioleaching optimization by statistical approaches: recycling mechanisms, factors affecting, challenges, and sustainability, *RSC Adv.* 13 (2023) 23570–23589.
51. W. Wu, X. Liu, X. Zhang, M. Zhu, W. Tan, Bioleaching of copper from waste printed circuit boards by bacteria-free cultural supernatant of iron–sulfur-oxidizing bacteria, *Bioresour. Bioprocess.* 5 (2018) 10.
52. D. Dutta, R. Rautela, L.K.S. Gujjala, D. Kundu, P. Sharma, M. Tembhare, S. Kumar, A review on recovery processes of metals from E-waste: a green perspective, *Sci. Total Environ.* 859 (2023) 160391.
53. S.K. Chaerun, R. Winarko, P.P. Butarbutar, Selective dissolution of magnesium from ferronickel slag by sulfur-oxidizing mixotrophic bacteria at room temperature, *J. Sustain. Metall.* 8 (2022) 1014–1025.
54. N.R. Nicomel, L. Otero-Gonzalez, A. Williamson, Y.S. Ok, P. Van Der Voort, T. Hennebel, G. Du Laing, Selective copper recovery from ammoniacal waste streams using a systematic biosorption process, *Chemosphere.* 286 (2022) 131935.
55. M.A. Askari Zamani, R. Vaghar, M. Oliazadeh, Selective copper dissolution during bioleaching of molybdenite concentrate, *Int. J. Miner. Process.* 81 (2006) 105–112.
56. B.K. Biswal, R. Balasubramanian, Recovery of valuable metals from spent lithium-ion batteries using microbial agents for bioleaching: a review, *Front. Microbiol.* 14 (2023) 1197081.
57. A. Sheel, D. Pant, Recovery of gold from electronic waste using chemical assisted microbial biosorption (hybrid) technique, *Bioresour. Technol.* 247 (2018) 1189–1192.
58. R. Sinha, G. Chauhan, A. Singh, A. Kumar, S. Acharya, A novel eco-friendly hybrid approach for recovery and reuse of copper from electronic waste, *J. Environ. Chem. Eng.* 6 (2018) 1053–1061.
59. T. Dolker, D. Pant, Chemical-biological hybrid systems for the metal recovery from waste lithium ion battery, *J. Environ. Manage.* 248 (2019) 109270.
60. H.I. Gomes, V. Funari, E. Dinelli, F. Soavi, Enhanced electrodialytic bioleaching of fly ashes of municipal solid waste incineration for metal recovery, *Electrochim. Acta.* 345 (2020) 136188.
61. A. Işıldar, J. van de Vossenberg, E.R. Rene, E.D. van Hullebusch, P.N.L. Lens, Biorecovery of metals from electronic waste. In: Rene, E., Sahinkaya, E., Lewis, A., Lens, P. (eds) *Sustainable Heavy Metal Remediation. Environmental Chemistry for a Sustainable World.* Springer, Cham (2017: pp. 241–278).
62. H. Brandl, R. Bosshard, M. Wegmann, Computer-munching microbes: metal leaching from electronic scrap by bacteria and fungi, *Process Metall.* 9 (1999) 569–576.
63. H.Y. Wu, Y.P. Ting, Metal extraction from municipal solid waste (MSW) incinerator fly ash - chemical leaching and fungal bioleaching, *Enzyme Microb. Technol.* 38 (2006) 839–847.
64. H. Tributsch, Direct versus indirect bioleaching, *Hydrometallurgy.* 59 (2001) 177–185.
65. D. Mishra, D.J. Kim, J.G. Ahn, Y.H. Rhee, Bioleaching: a microbial process of metal recovery; a review, *Met. Mater. Int.* 11 (2005) 249–256.
66. I. Asghari, S.M. Mousavi, F. Amiri, S. Tavassoli, Bioleaching of spent refinery catalysts: a review, *J. Ind. Eng. Chem.* 19 (2013) 1069–1081.
67. X. Zhang, H. Shi, N. Tan, M. Zhu, W. Tan, D. Daramola, T. Gu, Advances in bioleaching of waste lithium batteries under metal ion stress, *Bioresour. Bioprocess.* 10 (2023) 19.
68. V. Liapun, M. Motola, Current overview and future perspective in fungal biorecovery of metals from secondary sources, *J. Environ. Manage.* 332 (2023) 117345.
69. B.K. Biswal, U.U. Jadhav, M. Madhaiyan, L. Ji, E.H. Yang, B. Cao, Biological leaching and chemical precipitation methods for recovery of Co and Li from spent lithium-ion batteries, *ACS Sustain. Chem. Eng.* 6 (2018) 12343–12352.

70. X. Gao, L. Jiang, Y. Mao, B. Yao, P. Jiang, Progress, challenges, and perspectives of bioleaching for recovering heavy metals from mine tailings, *Adsorpt. Sci. Technol.* 2021 (2021) 9941979.
71. P.R. Yaashikaa, B. Priyanka, P. Senthil Kumar, S. Karishma, S. Jeevanantham, S. Indraganti, A review on recent advancements in recovery of valuable and toxic metals from e-waste using bioleaching approach, *Chemosphere.* 287 (2022) 132230.
72. A.I. Adetunji, P.J. Oberholster, M. Erasmus, Bioleaching of metals from e-waste using microorganisms: a review, *Minerals.* 13 (2023) 828.
73. M. Si, Y. Chen, C. Li, Y. Lin, J. Huang, F. Zhu, S. Tian, Q. Zhao, Recent advances and future prospects on the tailing covering technology for oxidation prevention of sulfide tailings, *Toxics.* 11 (2023) 11.
74. F. Krampah, G. Lartey-Young, P.O. Sanful, O. Dawohoso, A. Asare, Hydrochemistry of surface and groundwater in the vicinity of a mine waste rock dump: assessing impact of acid rock drainage (ARD), *J. Geosci. Environ. Prot.* 07 (2019) 52–67.
75. C. Falagán, B.M. Grail, D.B. Johnson, New approaches for extracting and recovering metals from mine tailings, *Miner. Eng.* 106 (2017) 71–78.
76. R. Zhang, S. Hedrich, F. Römer, D. Goldmann, A. Schippers, Bioleaching of cobalt from Cu/Co-rich sulfidic mine tailings from the polymetallic Rammelsberg mine, Germany, *Hydrometallurgy.* 197 (2020) 105443.
77. J. Mäkinen, M. Salo, M. Khoshkhoo, J.E. Sundkvist, P. Kinnunen, Bioleaching of cobalt from sulfide mining tailings; a mini-pilot study, *Hydrometallurgy.* 196 (2020) 105418.
78. S.M. Sadeghieh, A. Ahmadi, M.R. Hosseini, Effect of water salinity on the bioleaching of copper, nickel and cobalt from the sulphidic tailing of Golgohar Iron Mine, Iran, *Hydrometallurgy.* 198 (2020) 105503.
79. A.L. Santos, A. Dybowska, P.F. Schofield, R.J. Herrington, D.B. Johnson, Sulfur-enhanced reductive bioprocessing of cobalt-bearing materials for base metals recovery, *Hydrometallurgy.* 195 (2020) 105396.
80. Z. Wiecka, M. Rzelewska-Piekut, R. Cierpiszewski, K. Staszak, M. Regel-Rosocka, Hydrometallurgical recovery of cobalt(II) from spent industrial catalysts, *Catalysts.* 10 (2020) 61.
81. A. Akcil, F. Vegliò, F. Ferella, M.D. Okudan, A. Tuncuk, A review of metal recovery from spent petroleum catalysts and ash, *Waste Manag.* 45 (2015) 420.
82. R.M. Gholami, S.M. Borghei, S.M. Mousavi, Bacterial leaching of a spent Mo-Co-Ni refinery catalyst using Acidithiobacillus ferrooxidans and Acidithiobacillus thiooxidans, *Hydrometallurgy.* 106 (2011) 26–31.
83. W. Mrozik, M.A. Rajaeifar, O. Heidrich, P. Christensen, Environmental impacts, pollution sources and pathways of spent lithium-ion batteries, *Energy Environ. Sci.* 14 (2021) 6099–6121.
84. X. Wu, G. Ji, J. Wang, G. Zhou, Z. Liang, Toward sustainable all solid-state Li–metal batteries: perspectives on battery technology and recycling processes, *Adv. Mater.* 35 (2023) 2301540.
85. E. Asadi Dalini, G. Karimi, S. Zandevakili, M. Goodarzi, A review on environmental, economic and hydrometallurgical processes of recycling spent lithium-ion batteries, *Miner. Process. Extr. Metall. Rev.* 47 (2020) 1–22.
86. E.A. Olivetti, G. Ceder, G.G. Gaustad, X. Fu, Lithium-ion battery supply chain considerations: analysis of potential bottlenecks in critical metals, *Joule.* 1 (2017) 229–243.
87. A. Heydarian, S.M. Mousavi, F. Vakilchap, M. Baniasadi, Application of a mixed culture of adapted acidophilic bacteria in two-step bioleaching of spent lithium-ion laptop batteries, *J. Power Sources.* 378 (2018) 19–30.
88. A.B. Botelho Junior, S. Stopic, B. Friedrich, J.A.S. Tenório, D.C.R. Espinosa, Cobalt recovery from Li-ion battery recycling: a critical review, *Metals (Basel).* 11 (2021) 1999.

89. R. Golmohammadzadeh, F. Faraji, B. Jong, C. Pozo-Gonzalo, P.C. Banerjee, Current challenges and future opportunities toward recycling of spent lithium-ion batteries, *Renew. Sustain. Energy Rev.* 159 (2022) 112202.
90. S. Ghassa, A. Farzanegan, M. Gharabaghi, H. Abdollahi, Novel bioleaching of waste lithium ion batteries by mixed moderate thermophilic microorganisms, using iron scrap as energy source and reducing agent, *Hydrometallurgy.* 197 (2020) 105465.
91. H.L. Ehrlich, Inorganic energy sources for chemolithotrophic and mixotrophic bacteria, *Geomicrobiol. J.* 1 (1978) 65–83.
92. Y. Hong, M. Valix, Bioleaching of electronic waste using acidophilic sulfur oxidising bacteria, *J. Clean. Prod.* 65 (2014) 465–472.
93. T. Naseri, V. Beigi, A. Namdar, A. Keikavousi Behbahan, S.M. Mousavi, Biohydrometallurgical recycling approaches for returning valuable metals to the battery production cycle, *Nano Technol. Batter. Recycl. Remanuf. Reusing.* 2022 (2022) 217–246.
94. L. An (Ed.), *Recycling of Spent Lithium-Ion Batteries: Processing Methods and Environmental Impacts.* Springer, Cham (2019).
95. J.J. Roy, B. Cao, S. Madhavi, A review on the recycling of spent lithium-ion batteries (LIBs) by the bioleaching approach, *Chemosphere.* 282 (2021) 130944.
96. F. Dell'anno, E. Rastelli, E. Buschi, G. Barone, F. Beolchini, A. Dell'anno, Fungi can be more effective than bacteria for the bioremediation of marine sediments highly contaminated with heavy metals, *Microorganisms.* 10 (2022) 993.
97. M. Alipanah, D. Reed, V. Thompson, Y. Fujita, H. Jin, Sustainable bioleaching of lithium-ion batteries for critical materials recovery, *J. Clean. Prod.* 382 (2023) 135274.
98. F. Noruzi, N. Nasirpour, F. Vakilchap, S.M. Mousavi, Complete bioleaching of Co and Ni from spent batteries by a novel silver ion catalyzed process, *Appl. Microbiol. Biotechnol.* 106 (2022) 5301–5316.
99. G. Zeng, S. Luo, X. Deng, L. Li, C. Au, Influence of silver ions on bioleaching of cobalt from spent lithium batteries, *Miner. Eng.* 49 (2013) 40–44.
100. G. Zeng, X. Deng, S. Luo, X. Luo, J. Zou, A copper-catalyzed bioleaching process for enhancement of cobalt dissolution from spent lithium-ion batteries, *J. Hazard. Mater.* 199–200 (2012) 164–169.
101. A. Hariyadi, A.R. Masago, R. Febrianur, D. Rahmawati, Optimization fungal leaching of cobalt and lithium from spent Li-ion batteries using waste spices candlenut, *Key Eng. Mater.* 938 (2022) 177–182.
102. R.A. Putra, I. Al Fajri, A. Hariyadi, Metal bioleaching of used lithium-ion battery using acidophilic ferrooxidans isolated from acid mine drainage, *Key Eng. Mater.* 937 (2022) 193–200.
103. J.J. Roy, S. Madhavi, B. Cao, Metal extraction from spent lithium-ion batteries (LIBs) at high pulp density by environmentally friendly bioleaching process, *J. Clean. Prod.* 280 (2021) 124242.
104. A. Lobos, V.J. Harwood, K.M. Scott, J.A. Cunningham, Tolerance of three fungal species to lithium and cobalt: Implications for bioleaching of spent rechargeable Li-ion batteries, *J. Appl. Microbiol.* 131 (2021) 743–755.
105. S. Manikandan, D. Inbakandan, C. Valli Nachiyar, S. Karthick Raja Namasivayam, Towards sustainable metal recovery from e-waste: a mini review, *Sustain. Chem. Environ.* 2 (2023) 100001.
106. M. Baniasadi, F. Vakilchap, N. Bahaloo-Horeh, S.M. Mousavi, S. Farnaud, Advances in bioleaching as a sustainable method for metal recovery from e-waste: a review, *J. Ind. Eng. Chem.* 76 (2019) 75–90.
107. N. Nagarajan, P. Panchatcharam, A comprehensive review on sustainable metal recovery from e-waste based on physiochemical and biotechnological methods, *Eng. Sci.* 22 (2023) 844.

17 Bioleaching, an Alternative Tool for Platinum Group Metal Recovery from Industrial Waste

An Overview

Soumya Banerjee, Sankha Chakrabortty, Prasenjit Chakraborty, and Jayato Nayak

17.1 INTRODUCTION

The need for electronic equipment has increased dramatically alongside the increasing population and the rapid advancements in life. Massive amounts of electronic garbage have been generated at this concerning rate. Demand for e-waste has been on the rise in recent times, and e-waste generation and management have risen to the status of major environmental concerns. Electrical and electronic trash refers to any obsolete equipment that functions by electrical currents or electromagnetic fields. It is also known as "waste electrical and electronic equipment (WEEE)" or "e-waste." According to Hazra et al. [1], e-waste accounts for about 5% of the total solid trash produced each year, totalling around 40 million metric tonnes. The term "e-waste" refers to a wide variety of items, including but not limited to electronic appliances (both large and tiny), lighting, toys, electronic tools, medical equipment, monitoring systems, and automatic dispensers [1]. Likewise, it encompasses devices used for information and communication technology. Batteries, capacitors, cathode-ray tubes, and motherboards are all examples of electrical equipment that contains internal components. Illegal and careless disposal of e-waste into landfills, such as placing it in abandoned water sources or marshes without treating the soil for leachates, is a common problem. People are exposed to potentially harmful substances due to these dangerous and inappropriate techniques of disposal and recycling (Figure 17.1). Both official and illegal recycling processes expose people to e-waste, and people in the environment can come into contact with potentially harmful e-waste compounds. Developing nations such as Vietnam, China, Nigeria, India, and Thailand are more likely to practise informal recycling [2]. Disassembling old electronics in order to

DOI: 10.1201/9781003415541-17

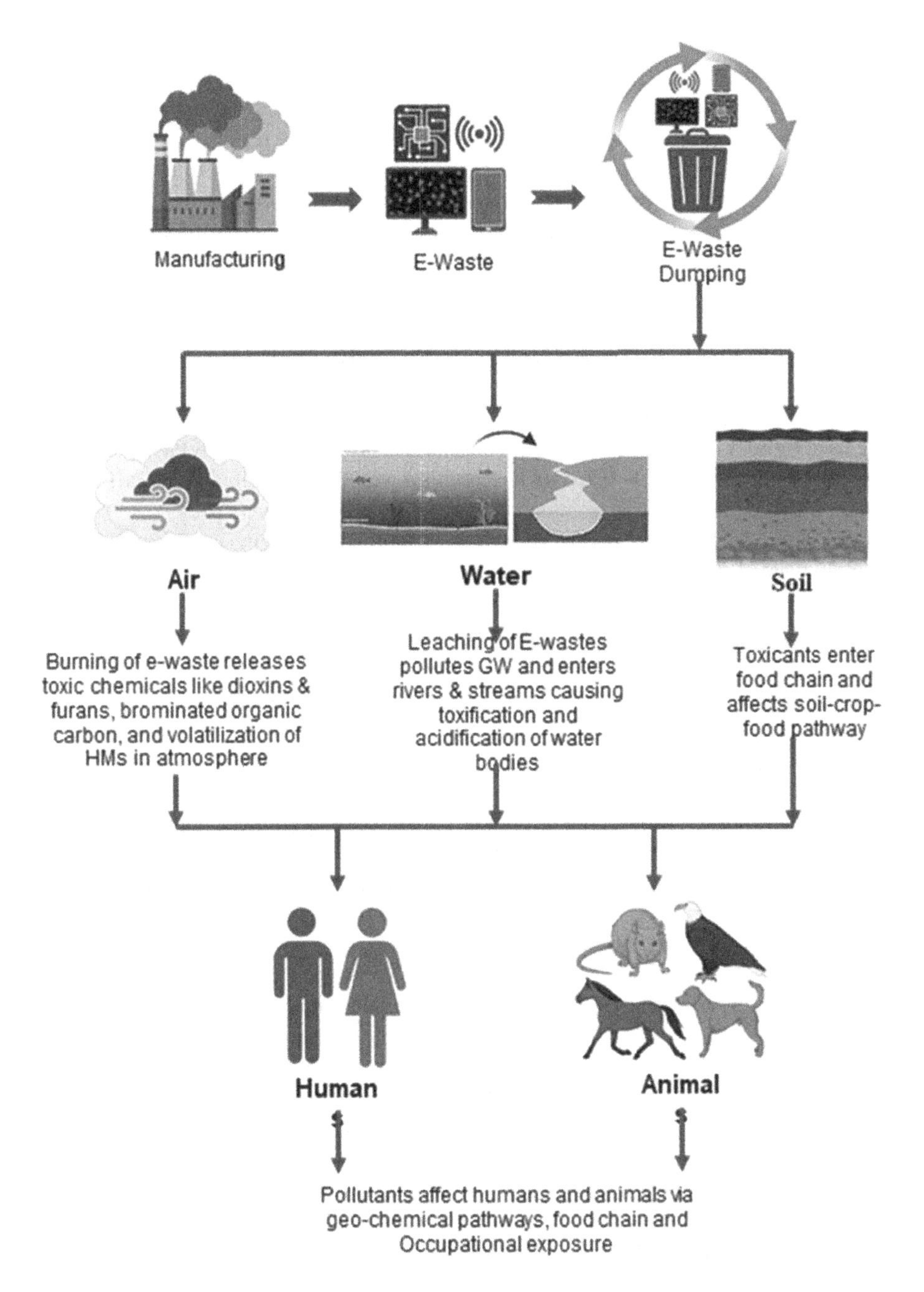

FIGURE 17.1 Diagram showing how e-waste affects the world. Adopted from Ref. [21].

salvage usable parts utilizing impromptu techniques is one example of an informal approach to recovering electrical trash.

Platinum group metals (PGMs), which include platinum, palladium, rhodium, iridium, osmium, and ruthenium, are critical components in various electronic devices, making them valuable materials found in electronic waste (e-waste). These metals are prized for their excellent catalytic properties, corrosion resistance, and electrical conductivity, making them integral to components such as hard drives, circuit boards, and catalytic converters in automobiles. E-waste represents a significant secondary source of PGMs due to their high demand and limited natural availability. Recycling e-waste not only recovers these precious metals, mitigating the need for environmentally disruptive mining, but also helps in managing hazardous waste and promoting sustainable material use. Efficient recovery and recycling of PGMs from e-waste can thus play a crucial role in supporting the circular economy and reducing the environmental footprint of electronic products.

The numerous industrial applications of PGMs stem from their many desirable properties, including a high melting point, excellent catalytic activity, corrosion resistance, and good selectivity. The majority of PGMs produced on a global scale are used by ACC [3–5]. The two most common PGMs are Pt and Pd. One of the most important components of ACC is Rh. Rhodium is a rare and valuable metal, but its low abundance in natural ores makes it expensive and restricts its use.

There has been a lot of recent interest in PGM recovery from secondary sources, namely wasted catalyst [4,6]. The percentage of PGMs in spent automotive catalyst (SAC) is relatively high when compared to natural ores. Industrial operations utilize pyrometallurgical and hydrometallurgical processes, which are among the PGM recovery strategies that result in high yield [7–10]. However, these procedures aren't without their share of drawbacks. As an example, pyrometallurgical processing produces slag, generates wastewater, and emits dangerous gaseous emissions during its high-temperature operation [11]. Hydrometallurgical procedures involve soaking the spent catalyst in an appropriate solvent or leaching agent to recover PGM. According to Cui and Zhang [11], La Brooy et al. [12], and Tesfaye et al. [13], these processes produce secondary (liquid) wastes that might be harmful to the environment. Additionally, they require particular equipment due to the strong corrosive acids they use. Recovering PGM from SAC requires new methods that are less harmful to the environment and more sustainable.

Many people believe that bioleaching is a safe way to recover metals without harming the environment. Biological oxidation and complexation mechanisms can mobilize metal ions from solid materials through bioleaching [14]. There seems to be a lack of research on the bioleaching of precious and PGMs, despite the considerable study of heavy metal bioleaching from secondary sources [14–17]. So yet, there is a lack of comprehensive data. Anecdotal evidence suggests that cyanide-forming bacteria may be useful in secondary source gold and PGM extraction [18–20]. Bacteria engage in biological cyanidation when they oxidatively decarboxylate glycine to form the secondary metabolite cyanide. In solutions, cyanide can be found as both the free cyanide anion (CN–) and the non-dissociated hydrocyanic acid (HCN).

In this chapter, a comprehensive review of various methods for recovering PGMs from electronic waste (e-waste) is presented, with a particular focus on bioleaching. Special emphasis is placed on the recovery of rhodium, one of the most valuable and challenging PGMs to extract due to its scarcity and chemical properties. This chapter begins by outlining the significance of PGMs in modern technology and their growing presence in e-waste, highlighting the need for efficient recovery methods to meet rising demand and reduce environmental impact. Conventional recovery processes, such as pyrometallurgy and hydrometallurgy, are explored in detail. Pyrometallurgy involves high-temperature treatments to extract metals, while hydrometallurgy utilizes aqueous chemistry for metal dissolution and recovery. Despite their effectiveness, these methods are energy-intensive and generate significant waste and greenhouse gas emissions. This chapter then transitions to an in-depth discussion on bioleaching, a burgeoning green technology that employs microorganisms to leach metals from e-waste. The bioleaching process is examined, focusing on its application to rhodium recovery. Various microorganisms, such as acidophilic bacteria and fungi, which can solubilize rhodium from e-waste, are discussed. Factors influencing bioleaching efficiency, including pH levels, temperature, and the presence of other metals, are analysed. Comparative studies between conventional methods and bioleaching are reviewed, highlighting the advantages of bioleaching, such as lower environmental impact, reduced energy consumption, and the potential for selective recovery of rhodium. This chapter concludes by identifying challenges in bioleaching, such as the need for optimizing microbial strains and process conditions, and proposes directions for future research to enhance the feasibility and efficiency of bioleaching for rhodium and other PGMs from e-waste. This review underscores the importance of advancing sustainable recovery technologies to promote the circular economy and mitigate the environmental burden of electronic waste.

17.2 CIRCULATION AND VALUE RECOVERY OF RESOURCES

For a sustainable future, a principle has to be fabricated to balance out waste and pollution with products and materials in use. This could develop a natural system for maintaining a circular economy. Thus, such an ecological-industrial based design was initially projected by industrial metabolism and industrial ecology. Unfortunately, such an idea on the recovery of best from waste was discarded.

It has been evident that before the circular economy was established other models existed and those were considered as the source of the circular economy. Industrial ecology and industrial metabolism during their inception were considered to be promising ones since they provided the idea of recirculating waste as intra- and inter-company secondary raw materials. Around the 1970s, this idea was first demonstrated by a Danish company called Kalundborg, which later conceived the concept of industrial symbiosis. This interdependence among participating organizations made a bilateral supply of resources and energy, commercial viability, and symbolic impression of "short mental distance" between authorities [22,23].

Classically, mining and metallurgical activities could be considered as a unidirectional process where resources are extracted, produced, used, and then dumped without consideration for reuse or recycling. As a result, more sustainable technologies are being used to convert these production processes into circulatory models that will primarily target the mitigation of impacts of any kind. Again, reinstating these secondary technologies for processing minerals needs reformation in the entire metallurgical system, which would establish a circular economy. Such an economy would then create a window that would decrease the dependency of primary resources by increasing the use of secondary resources, and this would target landfills to a greater extent. It has been reported that the global secondary reserve of aluminium was around 413 Mt among the United States, and China was found to contribute around 85 and 65 Mt, respectively [24].

Materials obtained as waste could be classified based on their source like textile and leather, metallurgy, petroleum, and mining and minerals, or they could be categorized based on their secondary usage like waste to resource or waste to energy. Various authors have pointed out that without proper terminology there could be complexity between the idea perceived in a laboratory and its use in society. Again, it has been observed that not all materials could be categorized in such a simplified way since a single material could be used for both energy and commodity production. Also, wastes coming out of a single source could contain various kinds of materials, which could vary in many ways like being toxic and non-toxic. Thus, a proper nomenclature has to be made that could be easily used to identify the type and nature of the waste.

Recovery of secondary material value could be categorized based on its economic importance as raw material via grouping them as basic or common, precious, and critical or rare. The primary materials are obtained from nature, whereas the secondary ones are obtained from recycling methods. This recycling technique involves physical, chemical, and biological methods for transformation. In one such study, it was reported that primary resources in comparison with secondary resources are more specific based on their previous anthropogenic use and lost value. As a result, products or materials with no potential value are considered as waste, which could be observed in the case of primary mining. Conversely, materials obtained from secondary mining are considered valuable, and secondary mining dedicates itself to using every possible form of waste which is dumped unscientifically in nature and causes pollution [25].

17.3 URBAN MINING

Over the past couple of decades, the concept of urban mining has been taking shape gradually, with a great emphasis on the circular economy. Although this could be considered as a recent concept, it has been a result of a constant demand for alternative resources that would create a lesser burden on the upstream extraction and production chain. In short, urban mining could be defined as the process that concentrates on the recovery of turning waste into value that could help anthropogenic activities as both material and energy.

One of the important aspects behind the conceptualization of urban mining could be attributed to the increase in waste which has been accumulating in natural openings, leading to catastrophic changes in unit ecosystems. Over the years, it has been observed that e-wastes are increasing, which is causing alarming concern among environmentalists and economists globally. Thus, urban mining could be considered as a way of managing e-wastes due to its diversity and high concentration of metal of around 60.2%. This form of non-conventional mining could be made possible with a particular type of protocol, which includes identifying potential urban mines, logistics, collection and classification of waste, sorting and disassembling, comminution and separation, and metal recovery.

Recovery of metal ions from secondary sources could be conducted via pyrometallurgy, hydrometallurgy, biohydrometallurgy, bioleaching, biosorption, and a combination of processes.

The term urban mining mostly concentrates on an idea that describes the recovery of value-added materials from anthropogenic wastes rather than urban space dumped with waste. One of the important aspects of urban mining is locating such mines similarly to that of primary mining, which are identified based on their deposits and exploitation potential. Similarly, in the case of urban mining wastes, maximum value-added potentials are located. Globally, various hotspots have been identified, which could be used for secondary mining. It has been observed that the amount of e-waste generated is directly linked to an updated lifestyle or to a population with an easier approach to electronic gadgets. One of the serious problems that occurs with urban mining is the unscientific sorting of wastes at the sources and then the availability of proper facilities for converting such waste to the best. Although countries like the United States, Norway, Australia, and China have been working hard on waste management, in India, the idea lacks both legislative and social aspects [26].

17.4 AN OVERVIEW OF PLATINUM-GRADE METALS

Platinum, palladium, rhodium, ruthenium, osmium, and platinum are the elements that make up the PGMs, which have long held a lot of respect in many fields. The significant PGM deposits found in countries like Russia, the United States, Canada, and South Africa from the 18th century to the 20th century solidified platinum's central position in the world's industrial landscape, as documented in historical accounts [27]. Over the past two centuries, platinum has shown to be remarkably versatile, finding employment as a catalyst in a wide variety of applications outside its traditional role in jewellery [28]. According to Figure 17.2 [27], there is a considerable demand for platinum, palladium, and rhodium in the industry market. There has been a meteoric rise in the usage of PGMs as catalysts for treating exhaust gases in the automotive sector since the 1970s [29].

The use of PGMs has skyrocketed in recent years, driven by the need to meet strict environmental regulations like the Euro standards [31]. Traditional PGM extraction from ores involves labour-intensive techniques, which highlights the need for more convenient and ecologically friendly methods to obtain catalytic systems. For this reason, a lot of people think that recycling catalytic material is the best option. This includes components like catalytic converters. The purpose of this paper is to offer a comprehensive review of the recent improvements in recycling catalytic converters for application in novel catalytic systems. Due to their exceptional chemical stability under different environmental conditions and harsh temperatures, PGMs are well-suited for industrial applications, especially as catalysts (Figure 17.3). These PGMs have the ability to speed up chemical reactions and progress certain chemical processes. Their prominence is well-documented in the chemical and chemical engineering communities [27].

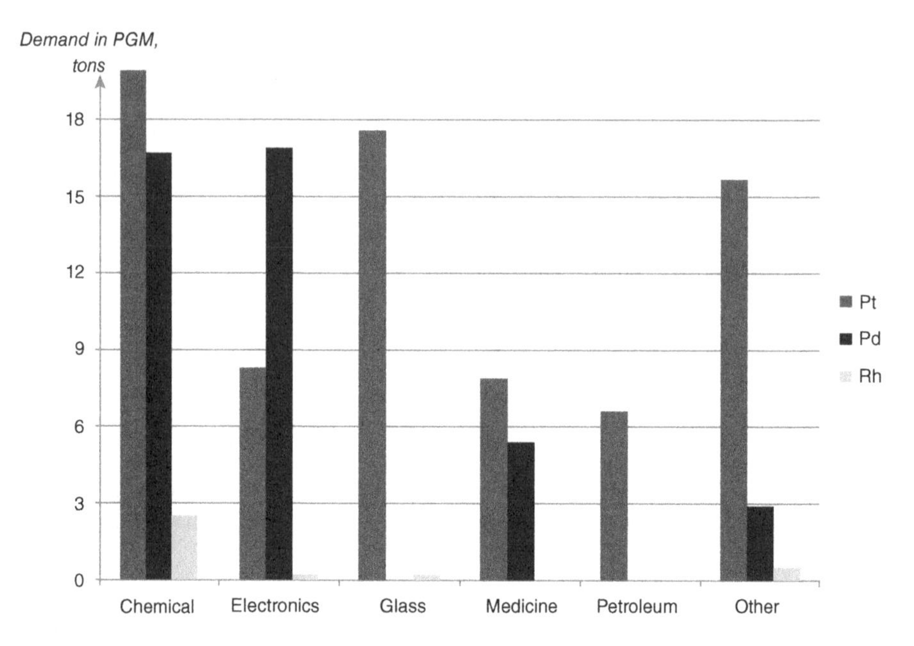

FIGURE 17.2 Industrial PGM demand in tonnes, 2023. Adopted from Ref. [30].

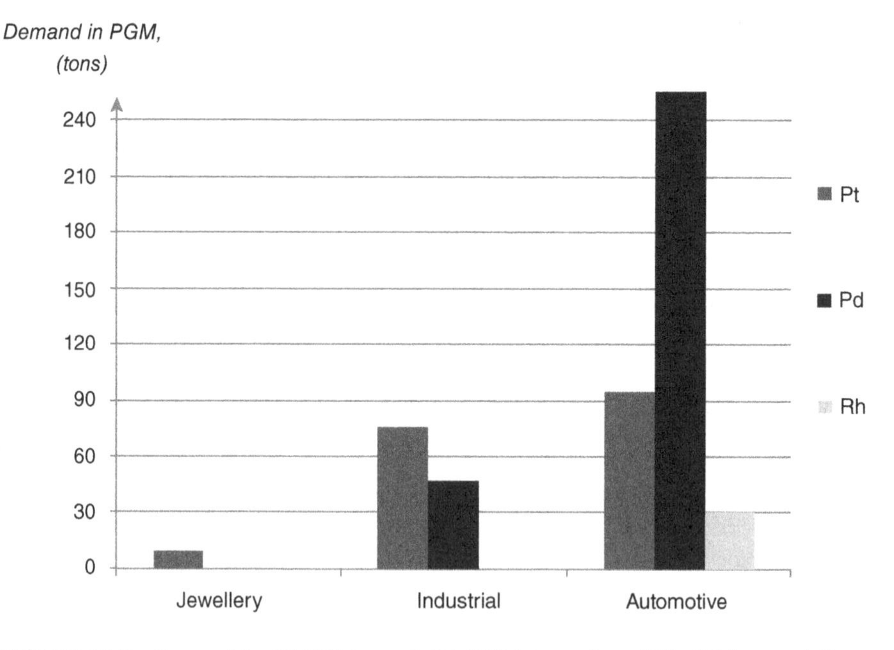

FIGURE 17.3 Demand for Pt (178 tonnes), Pd (303.5 tonnes), and Rh (29.5 tonnes) from different businesses. Adopted from Ref. [30].

17.5 AN OVERVIEW OF METAL RECOVERY FROM CONCENTRATES

The siderophilic state contains platinum metals in their pure form or alloy form, while the chalcophile state contains platinum metals in their ore form. Canada, Russia, South Africa, the United States, and Zimbabwe are home to some of the world's greatest ore deposits of PGM, which are located in igneous rock formations that are either ultramafic or mafic [32–34].

17.5.1 Recovery of PGMs from Sulphide Ores

Most PGMs come from sulphide ores; however, some PGMs come from metallic platinum placer deposits in Russia. However, the biggest resource of these ores in the world is located in South Africa [35]. South African ore typically contains 3–4 gms of platinum group members per metric tonne. Although recovering PGMs is a costly procedure, optimizing extraction efficiency to obtain the concentrate is of the utmost importance. One soluble PGM-containing component of the ore deposit is pentlandite [(Ni,Fe)9S8], whereas one unique mineral grain is braggite [(Pt,Pd)S]. Chalcocite, pyrrhotite, and pentlandite are common inclusions in PGM-containing minerals [36–38]. Because this process of isolating PGMs is so expensive, it is absolutely necessary to achieve maximum extraction efficacy into the concentrate. The multi-step technique for extracting PGMs entails the following:

1. Ore crushing and grinding
2. Froth flotation to obtain a platinum-rich sulphide concentrate
3. Melting and converting the concentrate to matte
4. Hydrometallurgical techniques (adsorption, chelation, ion exchange, leaching, precipitation, solvent extraction, etc.) to purify PGMs from the matte

17.5.2 Recovery of PGMs Using Gravity Separation

In contrast to the host rock and iron, the densities of minerals containing platinum are much higher. Example: Pt3Fe isoferroplatinum has a density of 18 g/cm^3, while PtFe tetraferroplatinum has a density of 16 g/cm^3. Graphite (Fe, Mg)Cr_2O_4, pyroxene (Mg, Fe)$_2Si_2O_6$, quartz SiO_2, chromite (Ni, Fe)$_9S_8$, and 5 g/cm^3 are the densities of the corresponding substances. Using this density contrast, a gravity concentrate with around 20% PGM is obtained. Additional flotation procedures are used on the leftover ore.

The extensive usage of PGMs in various industries has led to their presence in industrial sources being more concentrated than in natural ores [39,40]. The high monetary and ecological worth of these metals has piqued a lot of people's interest in finding sustainable ways to recycle waste products like catalytic converters [41–48].

17.5.3 Recovery of PGMs Using Hydrometallurgy Process

Because γ-Al_2O_3 exhibits amphoteric qualities, catalysts such as Pt-Re/γ-Al_2O_3, Pt-Sn/γAl_2O_3, and Pd/γ-Al_2O_3 can undergo dissolution treatment. After dissolving the finely ground catalyst in sulphuric acid, the residual solution contains a concentration of palladium (Pd) and platinum (Pt). Applying the cementation process with aluminium powder as a reducing agent helps recover part of the dissolved PGMs [49]. The research found that the catalytic converter was processed via autoclave leaching with sulphuric acid. One alternative is the Bayer process, which involves dissolving γ-Al_2O_3 in either NaOH or KOH [50]. This process does have some drawbacks, such as requiring complex equipment, high temperatures and pressures for γ-Al_2O_3 leaching, and significant energy usage.

Hydrochloric acid and oxidizing agents such as HNO_3, Cl_2, NaClO, $NaClO_3$, and H_2O_2 might facilitate the leaching of PGMs found in catalysts [51–53]. The chlorine that is produced has a strong oxidizing power and a lot of chloride ions, which are useful components of this leaching system. Hence, the PGMs can be found as chloro complexes, specifically $[PtCl6]^{2-}$, $[PdCl4]^{2-}$, and $[RhCl6]^{3}$.

17.5.4 A Small Amount of PGMs is Typically Produced by the Procedure

Hence, steps like pressure leaching, reduction, roasting, and fine grinding are necessary for pre-treatment. In order to remove organic molecules from the surface of the catalyst, numerous studies [54–57] have investigated different pre-treatment methods, such as oxidative roasting, reductive roasting, and pre-leaching. The research has shown that when Pt(IV) is mixed with a solution containing 30% HCl and different amounts of H_2O_2, the leaching efficiency of the metal shows an increasing trend [57]. The improvement in efficiency becomes clearly apparent as the roasting temperature of the catalyst is raised from 700°C to 900°C, with a constant concentration of H_2O_2. Furthermore, the study discovered that treating the catalyst greatly enhances the Pt(IV) leaching efficacy, specifically increasing the extraction rate from 84% to 100%. This is a notable improvement. Furthermore, leaching PGMs requires the use of acid or alkali solutions according to the specific circumstances. Some examples of such solutions are sodium chlorate, sulphuric acid, sodium hydroxide, and aqua regia [58–63]. Leaching is followed by metal concentration through precipitation or liquid extraction and, finally, purification.

The low extraction efficiency, particularly for rhodium, has stymied the progress made in hydrometallurgical procedures, which have received a lot of attention. Also, a lot of hazardous wastes are produced and the processing timeframes are long for these procedures. Furthermore, owing to their inherent rarity, PGMs present additional challenges when extracted from catalytic converters.

17.5.5 Recovery of PGMs Using Pyrometallurgical Process

One of the main ways to process PGMs from catalyst waste is the use of pyrometallurgical techniques, as shown in Figure 17.4 [39,64–67].

These procedures incorporate both chlorination and melting techniques for PGM recovery. In the smelting concentration step, the catalyst is melted alongside a flux, an auxiliary metal, and a reducing agent. The process culminates in the formation of an auxiliary metal-PGM alloy, which is further refined to remove platinum group

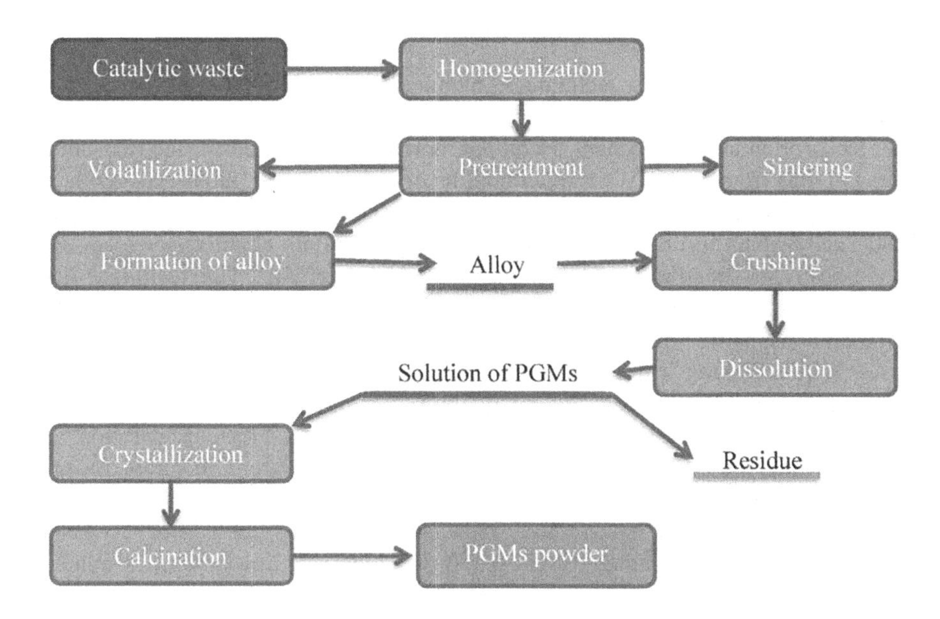

FIGURE 17.4 Process flow diagram for pyrometallurgy of PGM. Adopted from Ref. [30].

elements [68]. It is important to select an auxiliary metal with great care, taking into account its chemical compatibility, melting point, and mutual solubility with the PGMs. Iron, nickel, copper, and lead are common auxiliary metals that are famously good at this. Crushing, pelletizing, smelting, and separation are all parts of the pyrometallurgical process, which is the standard way for recovering PGMs from recycled catalysts [66,68].

Methods such as lead metal gathering were among the earliest [69]. Among its many advantages are its low melting point, ease of application, and straightforward purifying process. Regrettably, it has a low rhodium yield (around 70%–80%), and it releases lead oxide, a pollutant that can be harmful to humans and the environment.

The copper metal collection process involves heating the catalyst in an electric arc furnace with fluxes (SiO_2 and CaO), a collector ($CuCO_3$ or CuO), and a reducing agent. A mild reducing environment and relatively low temperatures are necessary for the successful collection of PGMs. The Serbian Institute of Mining and Metallurgy has utilized this technique to reduce PGMs [47]. It combines electrolytic refining with pyrometallurgical phases. Refining is the last phase after crushing, grinding, homogenizing, granulating, drying, melting, and electrolysis. Submerging the PGMs in a solution of pre-melted copper transports them from the dry pellets to the copper metal phase. An anode plate is made from electrolysed copper; the residue from this process comprises 20%–25% PGMs [47].

Smelting is a useful method for working with PGMs, but pyrometallurgy has limited practical applications. The major cause for this constraint is the high cost of equipment corrosion throughout the evaporation process. Chlorine and carbon monoxide are two more harmful gases that can damage employees and contribute to air pollution [64].

17.5.6 Recovery of PGMs Using Biometallurgy

Regarding metal recovery techniques, biometallurgy could be considered as a nascent method which is still in use at the laboratory scale and uses approaches like biosorption and bioleaching. Here, microorganisms are used for the extraction of metal ions from waste. Wastewater coming out of municipal, industrial, geothermal, and electrochemical sectors contains a considerable amount of such precious metal ions, which in the case of other technologies are not effective for recovery. Interestingly, biometallurgical techniques have been effective. This technique has been branded as a greener technology since it generates a lesser carbon footprint, minimal consumption of solvents and energy, and negligible emission of hazardous gases. While working with microorganisms for the extraction of PGMs, one of the major challenges that arises is the compatibility of PGMs with respect to the process parameters like pH, temperature, ionic strength, and metal ion concentration. At the same time, these PGMs are found as complex in wastewaters; therefore, proper pre-treatment of the wastewater is crucial. Bacterial species like *Pseudomonas fluorescens*, *Acidithiobacillus thiooxidans,* and *Chromobacterium violaceum* have been reported to have metal sequestering capability, which could even work in diluted metal solutions.

However, metal recovery has been found to be promising via biosorption. In this case, metal ions are removed by absorption precipitation, solvent extraction, and ionic exchange that supports the interaction of metal ions which charge microbial cell surfaces. Microbial species like *Desulfovibrio desulfuricans*, *D. vulgaris*, and *Chromobacterium violaceum* have been found to be effective for the biosorption of PGM ions from wastewater solutions.

The term "biometallurgy" refers to any biotechnological process in which microorganisms interact with metallic compounds or minerals that contain metals. In recent years, bioleaching, an essential component of biometallurgy, has been subjected to extensive research and application across the globe [70–72]. It is necessary for the recovery of metals from mining resources and waste materials. Here, microorganisms and the waste products of their metabolic processes are utilized as instruments for the purpose of separating metals from spent catalysts. This biologically mediated extraction technology really shines if it is paired with other leaching strategies because of the fact that it is both inexpensive and environmentally friendly [70].

This biological phenomenon depends completely on environmental conditions like pH, temperature, moisture, and O_2/CO_2 balance, making it a slower process. As a result, commercially different bioleaching approaches are being used based on the convenience of the process, which includes slope, heap, and in situ bioleaching. Under conventional condition, mineral ores are first crushed into finer particles. The ground particles are then dumped into a sloping-like condition forming a huge leaching mountain. The mountain is sprinkled with microbes containing water. The water will then be collected at the bottom where metal extraction will be executed. The technique of slope leaching has been found to be effective due to the reusability of the water for consecutive processes.

However, in the case of heap bioleaching, the metal is extracted in the same way as in slope bioleaching, but the only difference resides in the position of the ore. Here, the ore is placed in a heap and the metal along with the water oozes down followed by the extraction of metal from the leached solution. Lastly, in the case of in situ bioleaching, the microbial solution is piped directly down to the ore in the mining area. After the metals are leached, they will be piped out from the mining area. Thus, all these methods have been found to be effective in the removal of metals from ores without causing any temporary or permanent damage to their surroundings like in the case of traditional mining methods.

The interaction of microbes with metal ions during bioleaching could be broadly classified into two classes of mechanisms, viz. direct and indirect modes of leaching.

Direct leaching is conducted for metals which could be oxidized easily. In this type of interaction, the microorganisms stay in direct contact with metal ores where they separate the metal catalytically via oxidation. Microorganisms are sulphur-reducing bacteria that use this technique where they attach themselves to the ore by forming a biofilm. During the formation of biofilm, the bacterial cells attach to the ores via membrane interaction, leading to enzymatic actions and resulting in the separation of metal ions. This biofilm is formed in a very short amount of time during which they produce an extracellular polysaccharide layer, which acts as the layer between the cell and the surrounding. This layer provides a protective shield by regulating metabolic activities, maintaining pH balance, and taking part in oxidation processes.

The area of bioleaching encompasses both one-step and two-step direct methods. The simultaneous fermentation of microbes and the leaching of metals occur in the one-step process [73]. Although both steps of the two-step direct method take place in the same container, the order of culture and leaching in the former is different. However, in the case of the indirect technique, it is imperative that the leaching process transpire in a medium that has been fermented to eliminate microorganisms [74].

Researchers in a number of investigations [75–78] ground electronic waste and automobile catalytic converters into incredibly small particles. To remove the metals from these materials, the next step was to apply powerful mineral acids. The produced leachates were not entirely suitable for use by living things due to their low pH and high concentration of metals. This needs some discussion. Because of this, a crucial step that needed to be done was to dilute the solution or change the pH [77]. To avoid direct contact with these poisonous leachates, a second option was to premetallize the biomass or utilize the off-gases produced by bacterial fermentation. These methods, when combined with an autocatalytic reduction process, allowed for the recovery of valuable metals, including gold, copper, palladium, and platinum. The most remarkable result of this study is the capacity to selectively and sequentially collect PGMs, while also recovering almost all of the metal species. When compared to traditional chemical recovery methods, the biometallurgy process clearly excels in differentiating between PGMs. In the paper [79], the authors conducted an experimental investigation to recover platinum using the bioleaching technique. This study explored different methods of bioleaching. Direct bioleaching, which may be either a one-step process or a two-step process, and indirect bioleaching, which could involve wasted medium bioleaching with or without pH control, were among these approaches. Pt recovery was enhanced by acid-base management, according to their study's results. A 37% recovery rate of platinum was achieved through the

TABLE 17.1
A List of Microorganisms Used to Extract Metals from Their Sources

Metals	Sources	Organisms Used for Bioleaching
Lithium (Li)	Petalite	*Aspergillus niger* and *Penicillium purpurogenum*
Platinum (Pt)	Hexachloroplatinate	*Pseudomonas* sp., *Shewanella algae*, and *Desulfovibrio desulfuricans*
Iron (Fe^{3+})	Pyrite	*Chlorella vulgaris* and *Microcystis* sp.
Nickel (Ni)	Iron ore limonite	*Spirulina* spp. and *Aulosira fertilissima*
Zinc (Zn)	Zinc carbonate	*Aspergillus niger* and *Pseudomonas* sp.
Uranium (U)	Uranium nitrate	*Paenibacillus* sp. and *Pseudomonas putida*
Cobalt (Co)	Cobaltite	*Chlamydomonas reinhardtii* and *Spirogyra hyalina*
Copper (Cu^{2+})	Chalcocite, chalcopyrite, and covellite	*Chlorella* spp., *Phormidium* spp., and *Microcystis* spp.
Silver (Ag)	Silver nitrate	*Escherichia coli*, *Klebsiella pneumoniae*, and *Lactobacillus fermentum*
Gold (Au)	Chloroauric acid	*Bacillus megaterium*, *Bacillus subtilis*, and *Escherichia coli*

process of identifying the optimal operating parameters. The necessary parameters for the bioleaching process were a pH of 0.5, a pulp content of 1% by weight, and a temperature of 70°C.

When compared to other, more traditional recovery methods, the amount of PGMs recovered during bioleaching is usually fairly low. A new option that is good for the environment and energy efficient is bioleaching, and its advantages should be highlighted [78].

In Table 17.1, a list of microorganisms used in extracting different metals from their sources has been given.

17.6 CONCLUSIONS

In conclusion, the exponential rise in electronic equipment usage, driven by population growth and technological advancements, has led to a significant increase in electronic waste (e-waste). E-waste, which includes a wide range of obsolete electronic devices and components, poses a substantial environmental challenge due to improper disposal and recycling practices, especially in developing countries. The presence of valuable materials, such as PGMs, in e-waste underscores the importance of efficient recovery and recycling processes. Traditional recovery methods like pyrometallurgy and hydrometallurgy, while effective, have notable environmental drawbacks, including the generation of hazardous waste and emissions. In contrast, emerging techniques such as bioleaching offer a promising alternative by using microorganisms to recover metals with lower environmental impact. This review highlights the critical need for sustainable e-waste management strategies that not only recover valuable resources but also minimize ecological harm, thereby supporting a circular economy. Moving forward, further research and development in environmentally friendly recovery technologies, particularly bioleaching, are essential to address the growing e-waste problem and harness its potential as a valuable secondary resource.

REFERENCES

1. A. Hazra, S. Das, A. Ganguly, P. Das, P. Chatterjee, N. Murmu, P. Banerjee, *Plasma Arc Technology: A Potential Solution Toward Waste to Energy Conversion, and of GHGs Mitigation, Waste Valorisation and Recycling.* Springer, Chem, pp. 203–217 (2019).
2. K. Grant, F.C. Goldizen, P.D. Sly, M.N. Brune, M. Neira, M. van den Berg, R. E Norman, Health consequences of exposure to e-waste: a systematic review, *Lancet Glob. Health* 1 (6) (2013) 350–361.
3. S. Karim, Y.-P. Ting, Ultrasound-assisted nitric acid pretreatment for enhanced biorecovery of platinum group metals from spent automotive catalyst, *J. Clean. Prod.* 255 (2020) 120199.
4. S. Karim, Y.-P. Ting, Recycling pathways for platinum group metals from spent automotive catalyst: a review on conventional approaches and bio-processes, *Res. Conserv. Recycl.* 170 (2021) 105588.
5. D. Shin, J. Park, J. Jeong, B.-S. Kim, A biological cyanide production and accumulation system and the recovery of platinum-group metals from spent automotive catalysts by biogenic cyanide, *Hydrometallurgy* 158 (2015) 10–18.
6. A.N. Nikoloski, K.-L. Ang, D. Li, Recovery of platinum, palladium and rhodium from acidic chloride leach solution using ion exchange resins, *Hydrometallurgy* 152 (2015) 20–32.
7. R.J. Dawson, G.H. Kelsall, Recovery of platinum from secondary materials: electrochemical reactor for platinum deposition from aqueous iodide solutions, *J. Appl. Electrochem.* 46 (12) (2016) 1221–1236.
8. C.-H. Kim, S.I. Woo, S.H. Jeon, Recovery of platinum-group metals from recycled automotive catalytic converters by carbochlorination, *Ind. Eng. Chem. Res.* 39 (5) (2000) 1185–1192.
9. Y. Li, N. Kawashima, J. Li, A.P. Chandra, A.R. Gerson, A review of the structure, and fundamental mechanisms and kinetics of the leaching of chalcopyrite, *Adv. Coll. Intf. Sci.* 197–198 (2013) 1–32.
10. S. Wang, A. Chen, Z. Zhang, J. Peng, Leaching of palladium and rhodium from spent automobile catalysts by microwave roasting, *Environ. Prog. Sustain. Eng.* 33 (3) (2014) 913–917.
11. J. Cui, L. Zhang, Metallurgical recovery of metals from electronic waste: a review, *J. Hazard. Mater.* 158 (2) (2008) 228–256.
12. S.R. La Brooy, H.G. Linge, G.S. Walker, Review of gold extraction from ores, *Miner. Eng.* 7 (10) (1994) 1213–1241.
13. F. Tesfaye, D. Lindberg, J. Hamuyuni, Valuable metals and energy recovery from electronic waste streams. In: Zhang, L., Drelich, J.W., Neelameggham, N.R., Guillen, D.P., Haque, N., Zhu, J., Sun, Z., Wang, T., Howarter, J.A., Tesfaye, F., Ikhmayies, S., Olivetti, E., Kennedy, M.W. (Eds.), *Energy Technology 2017: Carbon Dioxide Management and Other Technologies.* Springer International, Cham, pp. 103–116 (2017).
14. S. Ilyas, J.-C. Lee, Biometallurgical recovery of metals from waste electrical and electronic equipment: a review, *Chem. Bio. Eng. Revs.* 1 (4) (2014) 148–169.
15. S. Vyas, Y.-P. Ting, Microbial leaching of heavy metals using Escherichia coli and evaluation of bioleaching mechanism, *Bioresour. Technol. Rep.* 9 (2020) 100368.
16. S. Vyas, S. Das, Y.-P. Ting, Predictive modeling and response analysis of spent catalyst bioleaching using artificial neural network, *Bioresour. Technol. Rep.* 9 (2020) 100389.

17. Z. Yu, H. Han, P. Feng, S. Zhao, T. Zhou, A. Kakade, S. Kulshrestha, S. Majeed, X. Li, Recent advances in the recovery of metals from waste through biological processes, *Bioresour. Technol.* 297 (2020) 122416.
18. G. Natarajan, Y.P. Ting, Gold biorecovery from e-waste: an improved strategy through spent medium leaching with pH modification, *Chemosphere* 136 (2015) 232–238.
19. H. Brandl, S. Lehmann, M.A. Faramarzi, D. Martinelli, Biomobilization of silver, gold, and platinum from solid waste materials by HCN-forming microorganisms, *Hydrometallurgy* 94 (1) (2008) 14–17.
20. D. Shin, J. Jeong, S. Lee, B.D. Pandey, J.-C. Lee, Evaluation of bioleaching factors on gold recovery from ore by cyanide-producing bacteria, *Miner. Eng.* 48 (2013) 20–24.
21. M. Jaiswal, S. Srivastava, A review on sustainable approach of bioleaching of precious metals from electronic wastes, *J. Hazard. Mater. Adv.* 14 (2024) 100435.
22. R. A. Frosch, N. Gallopoulos, Strategies for manufacturing, *Sci. Am.* 261 (3) (1989) 144–152.
23. J. R. Ehrenfeld, Industrial ecology: paradigm shift or illusion? *J. Indust. Ecol.* 8 (1–2) (2004) 1–6.
24. K. N. Maung, T. Yoshida, G. Liu, C. M. Lwin, D. B. Muller, S. Hashimoto, Assessment of secondary aluminum reserves of nations, *Resour. Conserv. Recycl.* 126 (2017) 34–41, https://doi.org/10.1016/j.resconrec.2017.06.016.
25. M. A. Reuter, R. B. Ashford, Material criticality for the montan sector: evaluating risk, opportunity, and stewardship, *J. Sustain. Metall.* 1 (1) (2012) 5–18.
26. S. K. Ghosh (Ed.). *Urban Mining and Sustainable Waste Management.* Springer, Singapore, 2020. https://doi.org/10.1007/978-981-15-0532-4.
27. A.E. Hughes, N. Haque, S.A. Northey, S. Giddey, Platinum group metals: a review of resources, production and usage with a focus on catalysts, *Resources* 10 (2021) 1–40, https://doi.org/10.3390/resources10090093.
28. F. Gervilla, J. García-Guinea, L.F. Capit´ an-Vallvey, Platina in the 18th century: mineralogy of the crude concentrate used in the first modern attempts at refining platinum, *Mineral. Mag.* 84 (2020) 289–299, https://doi.org/10.1180/mgm.2020.3.
29. R. F. Schulte, Minerals Yearbook, PLATINUM-GROUP METALS, U.S. Department of the Interior U.S. Geological Survey, 2019, 1–13, https://pubs.usgs.gov/myb/vol1/2019/myb1-2019-platinum-group.pdf.
30. I. Chidunchi, M. Kulikov, R. Safarov, E. Kopishev, Extraction of platinum group metals from catalytic converters, *Heliyon* 2024 (2024) e25283, https://doi.org/10.1016/j.heliyon.2024.e25283.
31. A. Gritsenko, V. Shepelev, G. Salimonenko, Y. Cherkassov, P. Buyvol, Environmental control and test dynamic control of the engine output parameters, *FME Trans.* 48 (2020) 889–898, https://doi.org/10.5937/fme2004889G.
32. G.M. Mudd, S.M. Jowitt, T.T. Werner, Global platinum group element resources, reserves and mining – a critical assessment, *Sci. Total Environ.* 614–625 (2018) 622–623, https://doi.org/10.1016/j.scitotenv.2017.11.350.
33. C. Hagelüken, Markets for the catalyst metals platinum, palladium and rhodium, *Metall* 60 (2006) 31–42.
34. A. Vymazalov´, F. Zaccarini, G. Garuti, F. Laufek, D. Mauro, C.J. Stanley, C. Biagioni, Bowlesite, PtSnS, a new platinum group mineral (PGM) from the merensky reef of the bushveld complex, South Africa, *Mineral. Mag.* 84 (2020) 468–476, https://doi.org/10.1180/mgm.2020.32.
35. L.A. Cramer, The extractive metallurgy of South Africa's platinum ores, *JOM* 53 (2001) 14–18, https://doi.org/10.1007/s11837-001-0048-1.
36. C.F. Vermaak, L.P. Hendriks, A review of the mineralogy of the Merensky Reef, with specific reference to new data on the precious metal mineralogy, *Econ. Geol.* 71 (1976) 1244–1269, https://doi.org/10.2113/gsecongeo.71.7.1244.

37. C.J. Penberthy, E.J. Oosthuyzen, R.K.W. Merkle, The recovery of platinum-group elements from the UG-2 chromitite, Bushveld complex - a mineralogical perspective, *Mineral. Petrol.* 68 (2000) 213–222, https://doi.org/10.1007/s007100050010.
38. R. Lukpanov, D. Dyusembinov, Z. Shakhmov, D. Tsygulov, Y. Aibuldinov, N.I. Vatin, Impregnating compound for cement-concrete road pavement, *Crystals* 12 (2022) 161, https://doi.org/10.3390/cryst12020161.
39. M.K. Jha, J.C. Lee, M.S. Kim, J. Jeong, B.S. Kim, V. Kumar, Hydrometallurgical recovery/recycling of platinum by the leaching of spent catalysts: a review, *Hydrometallurgy* 133 (2013) 23–32, https://doi.org/10.1016/j.hydromet.2012.11.012.
40. K. Othmer, *Kirk-Othmer Encyclopedia of Chemical Technology*, 5th ed., 2006. Wiley, New York.
41. T.N. Angelidis, E. Skouraki, Preliminary studies of platinum dissolution from a spent industrial catalyst, *Appl. Catal. Gen.* 142 (1996) 387–395, https://doi. org/10.1016/0926-860X(96)00088-9.
42. C. Nowottny, W. Halwachs, K. Schügerl, Recovery of platinum, palladium and rhodium from industrial process leaching solutions by reactive extraction, *Sep. Purif. Technol.* 12 (1997) 135–144, https://doi.org/10.1016/S1383-5866(97)00041-5.
43. M. Baghalha, H. Khosravian Gh, H.R. Mortaheb, Kinetics of platinum extraction from spent reforming catalysts in aqua-regia solutions, *Hydrometallurgy* 95 (2009) 247–253, https://doi.org/10.1016/j.hydromet.2008.06.003.
44. Y.-K. Taninouchi, T.H. Okabe, Recovery of platinum group metals from spent catalysts using iron chloride vapor treatment, *Metall. Mater. Trans. B Process Metall. Mater. Process. Sci.* 49 (2018) 1781–1793, https://doi.org/10.1007/s11663-018-1269-9.
45. Y.-K. Taninouchi, T. Watanabe, T.H. Okabe, Magnetic concentration of platinum group metals from catalyst scraps using iron deposition pretreatment, *Metall. Mater. Trans. B Process Metall. Mater. Process. Sci.* 48 (2017) 2027–2036, https://doi.org/10.1007/s11663-017-0999-4.
46. S. Suzuki, M. Ogino, T. Matsumoto, Recovery of platinum group metals at Nippon PGM Co., ltd, *J. MMIJ.* 123 (2007) 734–736, https://doi.org/10.2473/ournalofmmij.123.734.
47. C. Liu, S. Sun, X. Zhu, G. Tu, Metals smelting-collection method for recycling of platinum group metals from waste catalysts: a mini review, *Waste Manag. Res.* 39 (2021) 43–52, https://doi.org/10.1177/0734242X20969795.
48. N. Omirzak, R.S. Yerkasov, Quantum chemical study of structural properties for copper halide complexes with protonated acetamides, *Bull. L.N. Gumilyov Eurasian Natl. Univ. Chem. Geogr. Ecol. Ser.* 138 (2022) 7–17, https://doi.org/10.32523/2616-6771-2022-138-1-7-17.
49. M.S. Kim, E.Y. Kim, J. Jeong, J.C. Lee, W. Kim, Recovery of platinum and palladium from the spent petroleum catalysts by substrate dissolution in sulfuric acid, *Mater. Trans.* 51 (2010) 1927–1933, https://doi.org/10.2320/matertrans.M2010218.
50. R.K. Mishra, PGM recoveries by atmospheric and autoclave leaching of alumina bead catalyst, *Precious Met.* 1987 (1987) 177–195.
51. R.S. Marinho, C.N. da Silva, J.C. Afonso, J.W.S.D. da Cunha, 2011. Recovery of platinum, tin and indium from spent catalysts in chloride medium using strong basic anion exchange resins, *J. Hazard. Mater.* 192, 1155–1160, https://doi.org/10.1016/j.jhazmat.2011.06.021.
52. P.P. Sun, M.S. Lee, Separation of Pt from hydrochloric acid leaching solution of spent catalysts by solvent extraction and ion exchange, *Hydrometallurgy* 110 (2011) 91–98, https://doi.org/10.1016/j.hydromet.2011.09.002.
53. A.A. De S´ a Pinheiro, T.S. De Lima, P.C. Campos, J.C. Afonso, Recovery of platinum from spent catalysts in a fluoride-containing medium, *Hydrometallurgy* 74 (2004) 77–84, https://doi.org/10.1016/j.hydromet.2004.01.001.

54. D. Jimenez De Aberasturi, R. Pinedo, I. Ruiz De Larramendi, J.I. Ruiz De Larramendi, T. Rojo, Recovery by hydrometallurgical extraction of the platinum-group metals from car catalytic converters, *Miner. Eng.* 24 (2011) 505–513, https://doi.org/10.1016/j.mineng.2010.12.009
55. A. Fornalczyk, M. Saternus, Removal of platinum group metals from the used auto catalytic converter, *Metalurgija* 48 (2009) 133–136.
56. I.E. Suleimenov, O. Guven, G.A. Mun, C. Uzun, O.A. Gabrielyan, S.B. Kabdushev, L. Agibaeva, A. Nurtazin, Hysteresis effects during the phase transition in solutions of temperature sensitive polymers, *Eurasian Chem. J.* 19 (2017) 41, https://doi.org/10.18321/ectj501.
57. P.P. Sun, M.S. Lee, Recovery of platinum from chloride leaching solution of spent catalysts by solvent extraction, *Mater. Trans.* 54 (2013) 74–80, https://doi.org/10.2320/matertrans.M2012320.
58. J.Y. Lee, B. Raju, B.N. Kumar, J.R. Kumar, H.K. Park, B.R. Reddy, Solvent extraction separation and recovery of palladium and platinum from chloride leach liquors of spent automobile catalyst, *Sep. Purif. Technol.* 73 (2010) 213–218, https://doi.org/10.1016/j.seppur.2010.04.003.
59. A.S. Kirichenko, A.N. Seregin, A.I. Volkov, Developing a technology for recycling automotive exhaust-gas catalysts, *Metallurgist* 58 (2014) 250–255, https://doi.org/10.1007/s11015-014-9897-z.
60. R.S. Marinho, J.C. Afonso, J.W.S.D. da Cunha, Recovery of platinum from spent catalysts by liquid-liquid extraction in chloride medium, *J. Hazard. Mater.* 179 (2010) 488–494, https://doi.org/10.1016/j.jhazmat.2010.03.029.
61. Z. Wiecka, M. Rzelewska-Piekut, M. Regel-Rosocka, Recovery of platinum group metals from spent automotive converters by leaching with organic and inorganic acids and extraction with quaternary phosphonium salts, *Sep. Purif. Technol.* 280 (2022), 119933, https://doi.org/10.1016/j.seppur.2021.119933.
62. A.P. Paiva, F.V. Piedras, P.G. Rodrigues, C.A. Nogueira, Hydrometallurgical recovery of platinum-group metals from spent auto-catalysts – focus on leaching and solvent extraction, *Sep. Purif. Technol.* 286 (2022), 120474, https://doi.org/10.1016/j.seppur.2022.120474.
63. A. M´endez, C.A. Nogueira, A.P. Paiva, Recovery of platinum from a spent automotive catalyst through chloride leaching and solvent extraction, *Recycling* 6 (2021), 27, https://doi.org/10.3390/recycling6020027.
64. H. Dong, J. Zhao, J. Chen, Y. Wu, B. Li, Recovery of platinum group metals from spent catalysts: a review, *Int. J. Miner. Process.* 145 (2015) 108–113, https://doi.org/10.1016/j.minpro.2015.06.009.
65. Z. Peng, Z. Li, X. Lin, H. Tang, L. Ye, Y. Ma, M. Rao, Y. Zhang, G. Li, T. Jiang, Pyrometallurgical recovery of platinum group metals from spent catalysts, *JOM* 69 (2017) 1553–1562, https://doi.org/10.1007/s11837-017-2450-3.
66. A. Fornalczyk, M. Saternus, Vapour treatment method against other pyro- and hydrometallurgical processes applied to recover platinum from used auto catalytic converters, *Acta Metall. Sin. (English Lett.)* 26 (2013) 247–256, https://doi.org/10.1007/s40195-012-0125-1.
67. J. Willner, A. Fornalczyk, J. Cebulski, K. Janiszewski, Preliminary studies on simultaneous recovery of precious metals from different waste materials by pyrometallurgical method, *Arch. Metall. Mater.* 59 (2014) 801–804, https://doi.org/10.2478/amm-2014-0136.
68. G. Liu, A. Tokumaru, S. Owada, Concentration of PGMs from automobile catalyst by combining surface grinding and quenching, *Eur. Metall. Conf. EMC* 2013 (2013) 235–254.

69. M. Benson, C.R. Bennett, J.E. Harry, M.K. Patel, M. Cross, The recovery mechanism of platinum group metals from catalytic converters in spent automotive exhaust systems, *Resour. Conserv. Recycl.* 31 (2000) 1–7, https://doi.org/10.1016/S0921-3449(00)00062-8.
70. T. Hennebel, N. Boon, S. Maes, M. Lenz, Biotechnologies for critical raw material recovery from primary and secondary sources: R& D priorities and future perspectives, *N. Biotech.* 32 (2015) 121–127, https://doi.org/10.1016/j.nbt.2013.08.004.
71. D.B. Johnson, Biomining-biotechnologies for extracting and recovering metals from ores and waste materials, *Curr. Opin. Biotechnol.* 30 (2014) 24–31, https://doi.org/10.1016/j.copbio.2014.04.008.
72. D. Gauthier, L.S. Sabjerg, K.M. Jensen, A.T. Lindhardt, M. Bunge, K. Finster, R.L. Meyer, T. Skrydstrup, Environmentally benign recovery and reactivation of palladium from industrial waste by using gram-negative bacteria, *ChemSusChem* 3 (2010) 1036–1039, https://doi.org/10.1002/cssc.201000091.
73. W.-Q. Zhuang, J.P. Fitts, C.M. Ajo-Franklin, S. Maes, L. Alvarez-Cohen, T. Hennebel, Recovery of critical metals using biometallurgy, *Curr. Opin. Biotechnol.* 33 (2015) 327–335, https://doi.org/10.1016/j.copbio.2015.03.019.
74. I. Asghari, S.M. Mousavi, F. Amiri, S. Tavassoli, Bioleaching of spent refinery catalysts: a review, *J. Ind. Eng. Chem.* 19 (2013) 1069–1081, https://doi.org/10.1016/j.jiec.2012.12.005.
75. D. Santhiya, Y.-P. Ting, Use of adapted Aspergillus Niger in the bioleaching of spent refinery processing catalyst, *J. Biotechnol.* 121 (2006) 62–74, https://doi.org/10.1016/j.jbiotec.2005.07.002.
76. G. Colica, S. Caparrotta, R. De Philippis, Selective biosorption and recovery of Ruthenium from industrial effluents with Rhodopseudomonas palustris strains, *Appl. Microbiol. Biotechnol.* 95 (2012) 381–387, https://doi.org/10.1007/s00253-012-4053-9.
77. L.E. Macaskie, N.J. Creamer, A.M.M. Essa, N.L. Brown, A new approach for the recovery of precious metals from solution and from leachates derived from electronic scrap, *Biotechnol. Bioeng.* 96 (2007) 631–639, https://doi.org/10.1002/bit.21108.
78. S.K. Padamata, A.S. Yasinskiy, P.V. Polyakov, E.A. Pavlov, D.Y. Varyukhin, Recovery of noble metals from spent catalysts: a review, *Metall. Mater. Trans. B Process Metall. Mater. Process. Sci.* 51 (2020) 2413–2435, https://doi.org/10.1007/s11663-020-01913-w.
79. Y. Wang, P. Ziemkiewicz, A. Noble, A hybrid experimental and theoretical approach to optimize recovery of rare earth elements from acid mine drainage precipitates by oxalic acid precipitation, *Minerals* 12 (2022), 236, https://doi.org/10.3390/min12020236.

18 Bioleaching of Laterites for the Extraction of Valuable Metal Ions

A Microbial Approach for Sustainable Recovery

Tulsi Kumar, Jyoti Gulia, Yashika Rani, Amit Lath, Preeti Kumari, Nater Pal Singh, and Anita Rani Santal

18.1 INTRODUCTION

18.1.1 DEFINITION OF LATERITES

Laterites are iron-rich, weathered soil deposits found abundantly in tropical and subtropical regions worldwide. These residual soils are enriched with various valuable metal ions, including nickel, cobalt, copper, and manganese, owing to the intensive leaching of other soluble elements during the laterization process (Bustillo Revuelta, 2018). The extraction of these metals from lateritic ores has garnered significant interest due to their widespread applications in diverse industries, such as steel production, battery manufacturing, and catalysis (Abdollahi et al., 2024).

Conventional methods for metal recovery from laterites involve energy-intensive pyrometallurgical processes or environmentally detrimental hydrometallurgical techniques that rely on harsh chemicals and generate substantial amounts of toxic waste (Nkuna et al., 2022). In recent years, bioleaching has emerged as a promising alternative for the sustainable extraction of valuable metals from lateritic ores. This process harnesses the metabolic activities of naturally occurring microorganisms to solubilize and recover metals from low-grade ores or mineral wastes (Singh, 2023).

Lateritic bioleaching is a complicated interaction between several microbial populations and the mineral matrix. The chemolithotrophs (*Acidithiobacillus* sp., *Leptospirillum* sp., and *Sulfobacillus* sp.) are the most common microbes responsible for metal solubilization in lateritic ores (Rathna & Nakkeeran, 2020).

DOI: 10.1201/9781003415541-18

Acidithiobacillus ferrooxidans and *Acidithiobacillus thiooxidans* are two of the most widely studied and utilized bacteria in bioleaching operations. These acidophilic microorganisms derive energy by oxidizing ferrous iron (Fe^{2+}) and reduced sulfur compounds, respectively, leading to the production of sulfuric acid and Fe^{3+} ions (Sukla et al., 2013). The generated acid facilitates the dissolution of valuable metals from the lateritic matrix, making them available for subsequent recovery (Abdollahi et al., 2024). In addition to these primary bioleaching agents, other microorganisms play crucial roles in the bioleaching process. Heterotrophic bacteria, such as *Acidiphilium* sp. and *Acidocella* sp., can contribute to the solubilization of metals through the production of organic acids and chelating agents (Kanekar & Kanekar, 2022). The Archaea, particularly members of the order Thermoplasmatales, have also been identified in bioleaching environments and may contribute to metal dissolution through their metabolic activities (Vera et al., 2022). A comprehensive dataset on the organic acid production capabilities of four distinct microbial isolates has been shown in Table 18.1. These isolates are well-known for their extraordinary capacity to synthesize a wide range of organic acids, which has important commercial and biotechnological uses. *Aspergillus niger*, a filamentous fungus, stands out as a prolific producer of both citric acid and gluconic acid, with remarkable concentrations of 9,960 and 14,400 mg/L, respectively. *Pseudomonas putida*, a versatile bacterial species, exhibits a diverse organic acid production profile, synthesizing citric acid (1,340 mg/L), gluconic acid (12,940 mg/L), and oxalic acid (1,050 mg/L). *Pseudomonas koreensis*, another bacterial isolate, demonstrates a remarkable ability to produce gluconic acid at a high concentration of 10,800 mg/L. Additionally, it is capable of synthesizing oxalic acid, at a lower concentration of 70 mg/L while *Penicillium bilaji*, a fungal isolate, synthesizes citric acid (6,230 mg/L), gluconic acid (4,480 mg/L), and oxalic acid (400 mg/L). This broad spectrum of organic acid production could make *Penicillium bilaji* an attractive candidate for acidification in the process of bioleaching (Saleh et al., 2019).

TABLE 18.1
Concentration and Type of the Acids Formed by Bacterial and Fungal Isolates (Saleh et al., 2019)

Sr.No.	Isolates	Acid Types	Acid Concentration (mg/L)
1	*Aspergillus niger*	Citric acid	9,960
		Gluconic acid	14,400
2	*Pseudomonas putida*	Citric acid	1,340
		Gluconic acid	12,940
		Oxalic acid	1,050
3	*Pseudomonas koreensis*	Gluconic acid	10,800
		Oxalic acid	70
4	*Penicillium bilaji*	Citric acid	6,230
		Gluconic acid	4,480
		Oxalic acid	400

18.1.2 Importance of Laterites in the Mining Industry

Laterites are an important source of various valuable metal ions, such as nickel, cobalt, chromium, manganese, and aluminum-bearing ores like bauxite, which are essential for a wide range of industrial applications, including the production of stainless steel, batteries, and catalysts (Sukla et al., 2021). Nickel is crucial for stainless steel, batteries, and industrial applications. Cobalt is essential for lithium-ion batteries. Aluminum is a primary metal for various industries, including construction and transportation. As the demand for these metals continues to increase, the mining industry has turned its attention to the exploitation of lateritic ores as a sustainable alternative to traditional metal sources (Akhmetov et al., 2023).

18.1.3 Traditional Methods of Metal Extraction from Laterites

Traditionally, the extraction of valuable metal ions from lateritic ores has been accomplished through conventional pyrometallurgical and hydrometallurgical techniques, such as roasting, smelting, and acid leaching (Pandey et al., 2023; Whitworth et al., 2022). These methods, however, have several drawbacks, including high energy consumption, the generation of hazardous waste, and the release of harmful emissions, which have led to increasing environmental concerns (Pandey et al., 2023).

18.1.4 Environmental Concerns with Conventional Extraction

The traditional methods of metal extraction from lateritic ores have been criticized for their significant environmental impact (Stanković et al., 2020). Deforestation due to land clearing disrupts habitats, leading to biodiversity loss (Caetano et al., 2024). Chemicals used in the extraction process can contaminate soil and water, threatening ecosystems and human health. Emissions from smelting contribute to air pollution and acid rain. Additionally, the energy-intensive nature of these methods results in high carbon emissions, exacerbating climate change as a result, there is a growing need for more sustainable and environmentally friendly approaches to the extraction of valuable metals from lateritic ores (Agboola et al., 2020).

18.2 LATERITE FORMATION AND COMPOSITION

18.2.1 Geological Formation of Laterites

The formation of laterites is a complex and extensive process that occurs primarily in tropical and subtropical regions, where the environmental conditions are characterized by high temperatures, high rainfall, and well-drained landscapes. The geological formation of laterites is the result of the intense weathering and leaching of underlying parent rocks, which are typically igneous or metamorphic in nature (Tardy, 1997). The process begins with the weathering of the parent rock, which is driven by the combined effects of physical, chemical, and biological factors. Physical weathering, such as freeze-thaw cycles, mechanical abrasion, and the expansion of mineral grains, helps to break down the parent rock into smaller fragments (Thorne et al., 2012). Chemical weathering, on the other hand, involves the dissolution and transformation of the mineral components of the parent rock through the action of water, carbon dioxide, and various organic and inorganic acids (Birkeland, 1984; Sarkar et al., 2020). The leaching of silica and other mobile elements, such as calcium, magnesium, and sodium, is a crucial step in the formation of laterites. As the weathering process progresses, these elements are preferentially removed from the parent rock, leaving behind the less mobile elements, such as iron and aluminum (Kruger, 2019). This selective leaching leads to the relative enrichment of iron and aluminum oxides, which are the primary constituents of lateritic soils and rocks (Borra et al., 2016). The formation of specific laterite types, such as bauxitic laterites, ferruginous laterites, and nickeliferous laterites, is influenced by the composition of the parent rock, the local climate, and the duration and intensity of the weathering process (Freyssinet et al., 2005). For example, the formation of bauxitic laterites is favored by the weathering of aluminosilicate-rich rocks, such as granites and gneisses, while the formation of nickeliferous laterites is associated with the weathering of ultramafic rocks, such as serpentinized peridotites (Slukin et al., 2019). The final stage of laterite formation involves the accumulation and redistribution of the remaining iron and aluminum oxides, often in the form of goethite, hematite, and gibbsite, which can form distinct horizons or layers within the lateritic profile (Ghosh & Guchhait, 2020; Gilkes, 1992). These processes can lead to the development of a complex and heterogeneous lateritic landscape, with varying degrees of weathering and mineral composition. The geological formation of laterites is a dynamic and ongoing process, and the characteristics of these deposits can be influenced by a wide range of factors, including the regional climate, the nature of the parent rock, the duration of the weathering process, and the local topography and hydrology (Daramola et al., 2024; Zhao et al., 2024).

18.2.2 Types of Laterites

Laterites can be classified into several different types, depending on their composition and the degree of weathering. The main types of laterites include bauxitic laterites, ferruginous laterites, and nickeliferous laterites, each of which is characterized by its specific mineral assemblage and the predominant metal ions present (Butt & Cluzel, 2013; Freyssinet et al., 2005).

18.2.2.1 Bauxitic Laterites

Bauxitic laterites are characterized by their high content of aluminum oxides, primarily in the form of the mineral gibbsite ($Al(OH)_3$). These laterites are typically formed through the intense weathering of aluminosilicate-rich parent rocks, such as granites, gneisses, and bauxites (Stoops & Marcelino, 2018). The leaching of silica and other elements from the parent rock leads to the relative enrichment of aluminum oxides, resulting in the formation of these aluminum-rich lateritic deposits.

18.2.2.2 Ferruginous Laterites

Ferruginous laterites are dominated by iron oxides, with the primary mineral phases being goethite (α-FeOOH) and hematite (Fe_2O_3). These laterites are often associated with the weathering of iron-rich parent rocks, such as basalts, andesites, and ferruginous schists (Levett et al., 2020). The selective leaching of other elements, such as silica and aluminum, leads to the concentration of iron oxides within the lateritic profile (Stoops & Marcelino, 2018).

18.2.2.3 Nickeliferous Laterites

Nickeliferous laterites are characterized by their elevated concentrations of nickel, often accompanied by cobalt and chromium (Herrington et al., 2016). These laterites are typically formed through the weathering of ultramafic rocks, such as serpentinized peridotites and dunites, which are naturally enriched in nickel-bearing minerals (Dalvi et al., 2004). The weathering and leaching processes concentrate the nickel, cobalt, and chromium within the lateritic profile, resulting in the formation of these economically important deposits (Marsh & Anderson, 2011). Table 18.2 depicts the percentage of nickel present in various laterite ores that are mined from different parts of the world. These are mainly used in the formation of batteries.

It is important to note that the boundaries between these different laterite types are not always clear-cut, and transitional or mixed-type laterites can also occur. The specific mineral assemblage and the relative proportions of the various metal oxides within a lateritic deposit are influenced by a range of factors, including the composition of the parent rock, the intensity and duration of the weathering process, and the local environmental conditions (Gleeson et al., 2003). The classification of laterites into these distinct types is essential for understanding their geochemical characteristics, their potential for the extraction of valuable metal ions, and the appropriate processing and utilization strategies to be employed (Suparta & Pujiono, 2023).

TABLE 18.2
Types of Ores Use in Laterite Batteries (Mudd, 2010; Naseri et al., 2022)

Sr.No.	Types of Ores	Ni%
1	Caldag lateritic nickel ore	1.22
2	Laterite ore Tubay region, Mindanao, Philippines	1.28
3	Lateritic nickel ore from mid-Anatolia region	1.37
4	Nickel silicate ore, Rudinici	1.29
5	Limonitic, Indonesia	1.22
6	Chromite overburden	0.87
7	Saprolite ore	2.70
8	Sulawesi saprolitic ore	1.76
9	Halmahera saprolitic ore	1.28
10	Low-grade hematitic laterite ore	0.73
11	Garnieritic type	4.23
12	Lizardite	4.33
13	South African nickel laterite	0.02

18.2.3 Composition of Lateritic Ores

Lateritic ores are complex and heterogeneous materials, comprising a wide range of mineral phases and chemical constituents. The exact composition of a lateritic ore can vary significantly depending on the nature of the parent rock, the intensity and duration of the weathering processes, and the local environmental conditions (Duzgoren-Aydin et al., 2002; Sergeev, 2023).

The primary mineral phases found in lateritic ores are typically iron and aluminum oxides, with the specific mineral assemblage being closely related to the type of laterite deposit. Table 18.3 provides us a data on various mineral groups present in laterite ores and their chemical formulas. In bauxitic laterites, the dominant mineral phase is usually gibbsite Al $(OH)_3$, while in ferruginous laterites, the iron oxides goethite (α-FeOOH) and hematite (Fe_2O_3) are the primary constituents. Nickeliferous laterites, on the other hand, often contain a combination of iron oxides, nickel-bearing silicates, and minor amounts of cobalt and chromium-bearing minerals (Diop et al., 2024; Hakim et al., 2023; Herrington et al., 2016). In addition to the principal iron

TABLE 18.3
Minerals Groups Present in Laterite Ores and Their Chemical Formula (Watling et al., 2011; Wells & Chia, 2011)

Mineral Group	Minerals	Chemical Formulas
Kaolin	Kaolinite, halloysite	$Al_2Si_2O_5(OH)_4$
Opal	Opal	$SiO_2 \cdot nH_2O$
Smectite	Montmorillonite,	$(Na,Ca)_{0.33}(Al,Mg)_2(Si_4O_{10})(OH)_2 \cdot nH_2O$
	Nontronite	$Na_{0.3}Fe_2Si_3AlO_{10}(OH)_2.4H_2O$
Mg-clays	Saponite, palygorskite	$Mg_6(Si_8O_{20})(OH)_4 \cdot 4H_2O$
Chlorite	Fe/Mg/intermediate chlorite types	$(Mg,Fe)_3(Si,Al)_4O_{10}(OH)_2 \cdot (Mg,Fe)_3(OH)_6$
	Chlorite	$[(Fe,Mg)5Al](AlSi_3)O_{10}OH)_8$
Chromite	Chromite	$Fe^{2+}Cr_2O_4$
Talc	Talc	$Mg_3Si_4O_{10}(OH)_2$
Serpentine	Lizardite	$(Mg,Fe)_3Si_2O_5(OH)_4$
Primary silicates	Hornblende, riebeckite	$(Ca,Na,K)_2$-$3(Mg,Fe,Al)_5(Si,Al)_8O_{22}(OH)_2$
Carbonates	Dolomite	$CaMg(CO_3)_2$,
Mn oxides	Magnesite	$MgCO_3$
Fe oxides	Hematite and maghemite	Fe_2O_3
	Goethite	FeO(OH)
Quartz	Quartz	SiO_2
Albite	Albite	$NaAlSi_3O_8$
Magnesiohornblende	Magnesiohornblende	$Ca_2[Mg_4(Al,Fe^{3+})]Si_7AlO_{22}(OH)_2$
Accessory phases	Strontianite	$SrCO_3$
	Artinite	$Mg_2(CO_3)(OH)_2 \cdot 3H_2O$
	Jarosite	$KFe_3(SO_4)_2(OH)_6$
	Pyroaurite	$Mg_6Fe_2^{3+}(CO_3)(OH)_{16} \cdot 4H_2O$

and aluminum oxides, lateritic ores may also contain variable amounts of other mineral constituents, such as silica (SiO_2), kaolinite $Al_2Si_2O_5(OH)_4$, quartz (SiO_2), and residual primary minerals from the parent rock. The presence and proportions of these secondary mineral phases can significantly influence the physical and chemical properties of the lateritic ore, as well as its behavior during extraction and processing (Freyssinet et al., 2005; Meshram et al., 2019). The chemical composition of lateritic ores is typically dominated by the major elements of iron, aluminum, silicon, and magnesium, with varying concentrations of valuable metal ions such as nickel, cobalt, chromium, and manganese (Mudd, 2010). The relative abundance of these metal ions is a key factor in determining the economic potential of a lateritic deposit and the most appropriate extraction and processing strategies (Pintowantoro & Abdul, 2019).

18.3 MICROBIAL MECHANISMS IN BIOLEACHING

18.3.1 Role of Microorganisms in Metal Extraction

Bioleaching, or the technique of recovering precious metal ions from lateritic ores with microorganisms, is a well-established and rapidly expanding technology in the field of sustainable metal recovery. This microbial technique for metal extraction is based on the ability of certain microbes to solubilize and mobilize metal-bearing materials (Abdollahi et al., 2024). The key players in bioleaching are a diverse group of microorganisms, primarily acidophilic bacteria and archaea, which thrive in the low-pH environments typically associated with lateritic ores (Rathna & Nakkeeran, 2020). These microorganisms possess unique metabolic capabilities that allow them to catalyze the dissolution of metal-bearing minerals through various mechanisms (Sukla et al., 2021). One of the primary mechanisms employed by bioleaching microorganisms is the production of organic acids, such as sulfuric acid and oxalic acid, which can effectively dissolve the mineral matrix and release the desired metal ions (Nasab et al., 2020). For example, the acidophilic bacterium *Acidithiobacillus ferrooxidans* is known to generate sulfuric acid as a byproduct of its metabolism, which can then be used to solubilize iron and other metal-bearing minerals (Giese, 2021). In addition to the production of organic acids, bioleaching microorganisms can also facilitate metal extraction through the secretion of chelating agents, such as siderophores, which can form stable complexes with metal ions and enhance their solubility (Sarkodie et al., 2022). These chelating agents can selectively target specific metal ions, allowing for more selective and efficient metal recovery (Naseri et al., 2022). Furthermore, certain microorganisms can directly oxidize or reduce metal-bearing minerals, using them as electron donors or acceptors in their metabolic processes (Rawlings, 2002). This direct interaction with the mineral matrix can lead to the mobilization of the desired metal ions, contributing to the overall bioleaching efficiency (Dulay et al., 2020). The success of bioleaching processes is heavily dependent on the specific environmental conditions that support the growth and metabolic activity of the leaching microorganisms. Factors such as pH, temperature, oxygen availability, nutrient supply, and the presence of inhibitory substances can all influence the efficiency and selectivity of the bioleaching process (Nkuna et al., 2022). By leveraging the unique metabolic capabilities of these specialized microorganisms, bioleaching offers a promising and environmentally friendly approach to the extraction of valuable metal ions from lateritic ores, paving the way for more sustainable mining and metal recovery practices (Olson et al., 2003).

18.3.2 Factors Influencing Microbial Activity

The success and efficiency of bioleaching processes for the extraction of valuable metal ions from lateritic ores are heavily dependent on the environmental conditions that support the growth and metabolic activity of the microorganisms involved. A thorough understanding of the key factors influencing microbial activity is crucial for the optimization and effective implementation of bioleaching technologies (Behera & Mulaba-Bafubiandi, 2015).

18.3.2.1 pH

The pH of the bioleaching is a critical factor, as the majority of the microorganisms employed in these processes are acidophilic, thriving in low-pH conditions. The optimal pH range for most bioleaching microorganisms, such as *Acidithiobacillus ferrooxidans* and *Leptospirillum ferrooxidans*, is typically between 2.0 and 3.0. At higher pH levels, the solubility of metal ions decreases, and the activity of the leaching microorganisms is inhibited (Potysz et al., 2018).

18.3.2.2 Temperature

Temperature also plays a significant role in the growth and metabolic activity of bioleaching microorganisms. Most acidophilic bacteria and archaea involved in bioleaching processes are mesophilic, with optimal growth temperatures ranging from 30°C to 40°C. However, some thermophilic microorganisms, such as *Sulfobacillus* sp. and *Acidianus* sp., can thrive at higher temperatures, making them suitable for bioleaching operations in regions with elevated ambient temperatures (Potysz et al., 2018; Whitworth et al., 2022).

18.3.2.3 Oxygen Availability

Aeration and the availability of dissolved oxygen are crucial for the growth and metabolic activity of aerobic bioleaching microorganisms. These microorganisms require oxygen as an electron acceptor in their respiratory processes, and the lack of sufficient oxygen can limit their ability to solubilize and extract metal ions from the lateritic ore (Rathna & Nakkeeran, 2020; Slukin et al., 2019).

18.3.2.4 Nutrient Supply

The availability of essential nutrients, such as nitrogen, phosphorus, and trace elements, can also influence the performance of bioleaching microorganisms. Nutrient-deficient conditions can lead to slower growth rates and reduced metabolic activity, thereby limiting the efficiency of the bioleaching process (Vera et al., 2022).

18.3.2.5 Inhibitory Substances

Certain ions, compounds, or elements present in the lateritic ore or the surrounding environment can act as inhibitors, interfering with the growth and metabolic activity of the bioleaching microorganisms. For example, the presence of heavy metals, such as copper or mercury, or the accumulation of toxic metabolic byproducts can negatively impact the performance of the leaching microorganisms (Giese et al., 2019).

18.4 BIOLEACHING TECHNIQUES

The bioleaching of lateritic ores for the extraction of valuable metal ions can be accomplished through a variety of techniques, each with its advantages, limitations, and specific operational requirements. These bioleaching approaches have been developed and refined over the years to optimize the efficiency, selectivity, and environmental sustainability of the metal extraction process (Figure 18.1).

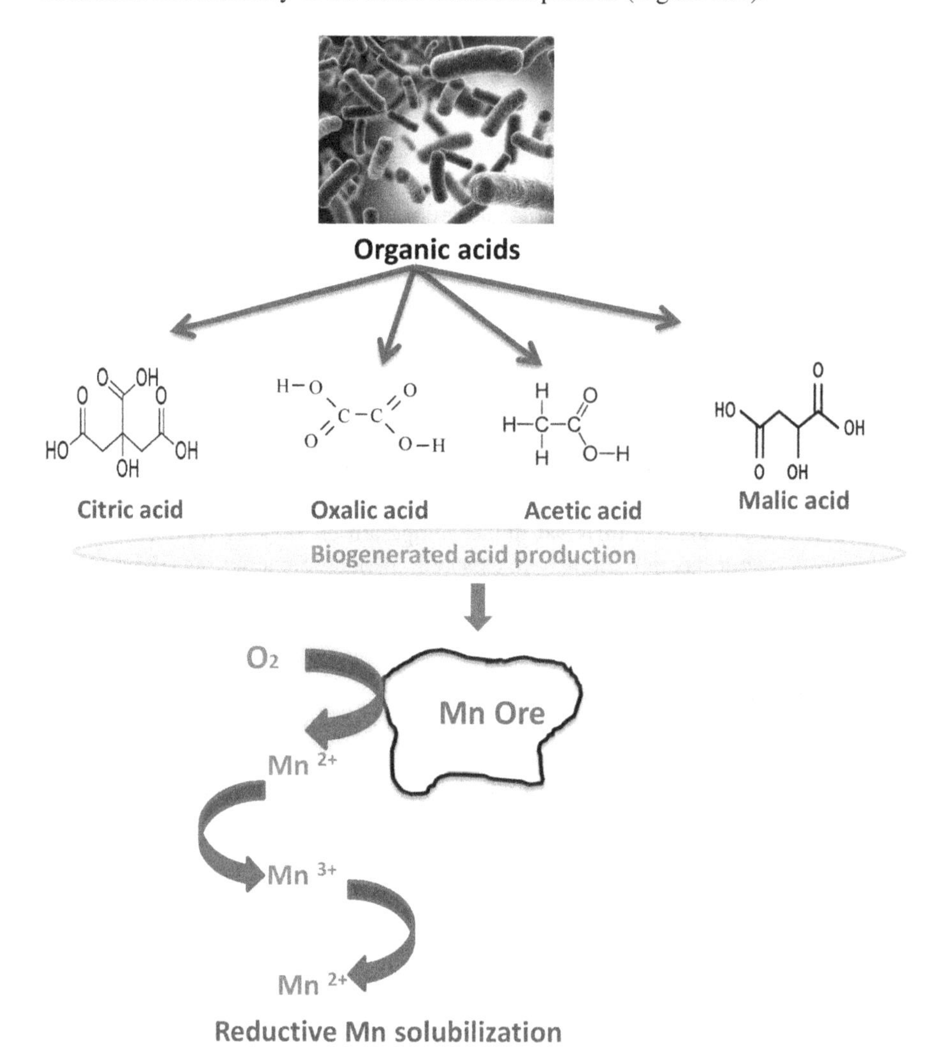

FIGURE 18.1 Systematic representation of bioleaching.

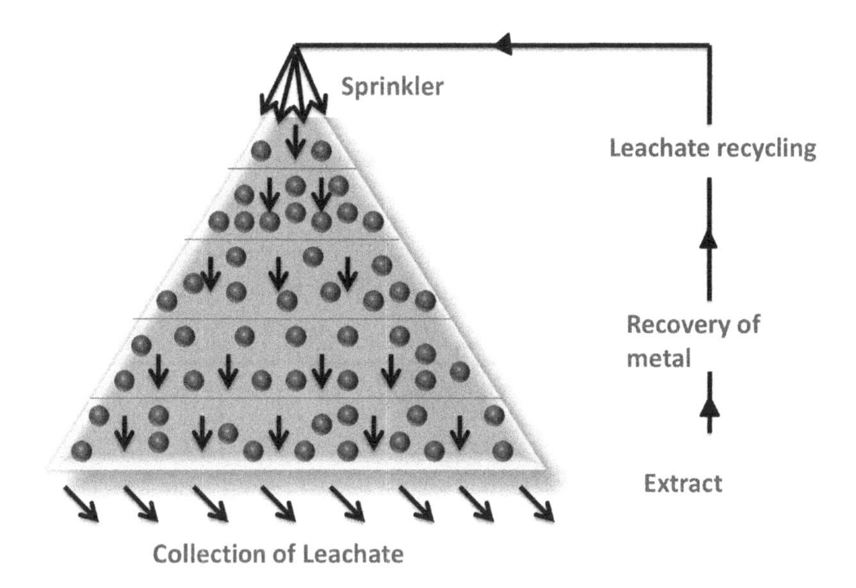

FIGURE 18.2 Heap bioleaching.

18.4.1 Heap Bioleaching

Heap bioleaching is a widely adopted technique for the extraction of metals from lateritic ores. In this method, the crushed and agglomerated lateritic ore is stacked in large, irrigated heaps, which are then inoculated with the desired leaching microorganisms (Thenepalli et al., 2019). The lixiviant solution, containing the necessary nutrients and microbial inoculum, is continuously circulated through the heap, facilitating the dissolution and mobilization of the target metal ions. The advantages of heap bioleaching include relatively low capital and operating costs, the ability to process large volumes of ore, and the potential for in situ metal recovery (Figure 18.2) (Watling et al., 2011).

18.4.2 Dump Bioleaching

It includes the heaped uncrushed waste rock. It includes 0.1%–0.5% of copper, which is too low and cannot be recovered by traditional means. It includes a larger amount of dump having a greater length in comparison to width this stature is known as finger dump having $800 \times 35 \times 200$ m measurements. Dump bioleaching involves acid solution permeation via mineralized particles over mass. Here, local microorganisms undergo multiplication in the ideal conditions. Ultimately, the natural solubilization of metals and gathering of profluent happens at a base that has metal richness. For the proper draining, the action of microbes is done at different temperature conditions (Figure 18.3) (Kumar & Yaashikaa, 2020).

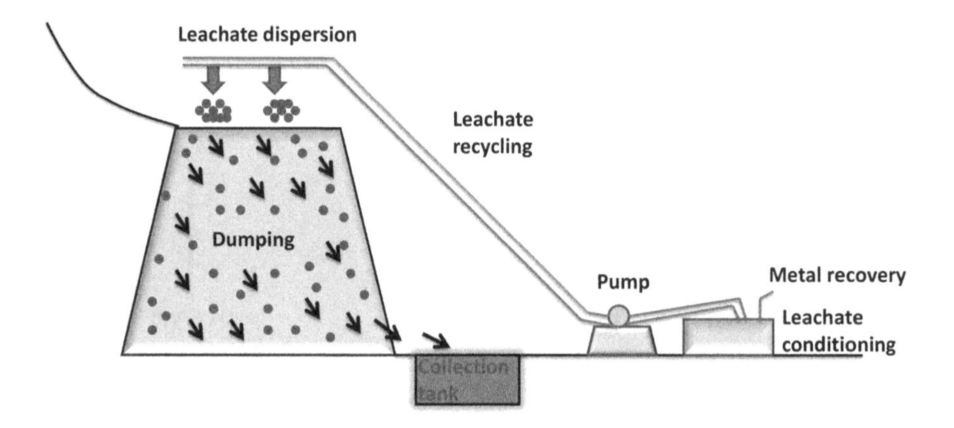

FIGURE 18.3 Dump bioleaching.

18.4.3 Tank Bioleaching

Tank bioleaching utilizes agitated, aerated, and pH-controlled bioreactors to carry out the leaching process. This approach allows for better control over the environmental conditions, such as pH, temperature, and oxygen availability, which can lead to improved metal recovery rates and more efficient metal extraction. Tank bioleaching is particularly suitable for the processing of finely ground lateritic ores, as the intimate contact between the ore, the lixiviant, and the microorganisms can enhance the leaching efficiency (Mishra et al., 2023; Zhang, 2022).

18.4.3.1 In Situ Bioleaching

In situ bioleaching involves the injection of a lixiviant solution containing the necessary microorganisms and nutrients directly into the lateritic ore deposit, without physically excavating the ore (Earley III, 2020). This technique allows for the leaching of the ore in its natural geological formation, minimizing the environmental impact and the need for extensive ore handling and processing. However, the success of in situ bioleaching is heavily dependent on the hydrogeological characteristics of the deposit and the ability to effectively distribute the lixiviant solution throughout the ore body (Figure 18.4) (Olson et al., 2003).

18.4.3.2 Stirred Tank Bioleaching

Stirred tank bioleaching utilizes agitated reactors to facilitate the interaction between the lateritic ore, the leaching solution, and the microorganisms (Rasoulnia et al., 2021). This approach can enhance the mass transfer of nutrients and dissolved metal ions, leading to improved metal extraction efficiency. The controlled environment of the stirred tank reactor also allows for the optimization of parameters such as pH, temperature, and aeration, contributing to the overall performance of the bioleaching process (Srichandan et al., 2020). Each of these bioleaching techniques has its own

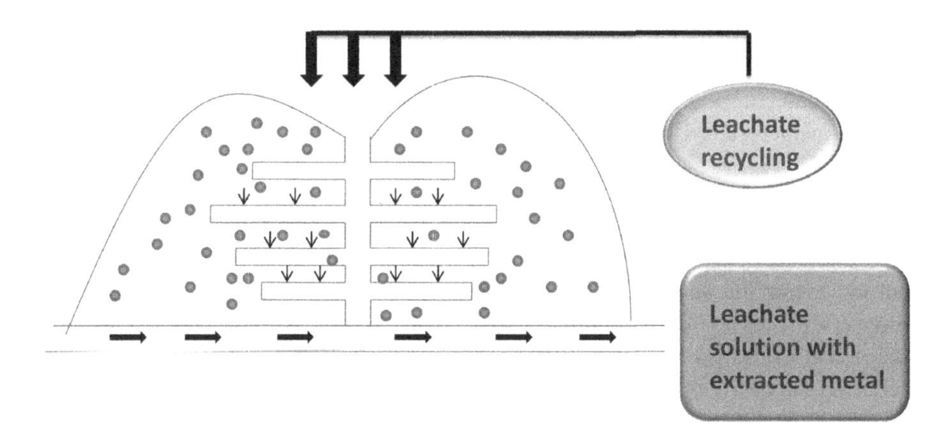

FIGURE 18.4 In situ bioleaching.

set of advantages and limitations, and the selection of the most appropriate approach will depend on factors such as the characteristics of the lateritic ore, the target metal ions, the available infrastructure, and the specific environmental and economic constraints of the operation. Ongoing research and technological advancements continue to improve the efficiency, selectivity, and sustainability of bioleaching processes for the extraction of valuable metal ions from lateritic ores (Olson et al., 2003).

18.5 ADVANCEMENTS IN BIOLEACHING TECHNOLOGY

Significant progress has been made in the development of more efficient and innovative bioleaching technologies for the extraction of valuable metal ions from lateritic ores. These advancements have the potential to revolutionize the way metal extraction is carried out, making the process more environmentally friendly, economically viable, and sustainable in the long term.

18.5.1 Genetic Engineering and Enzyme Technology

The application of genetic engineering and enzyme technology has shown promising results in enhancing bioleaching efficiency and selectivity. Researchers have successfully engineered microorganisms with improved metal-solubilizing capabilities, increased tolerance to environmental stressors, and the ability to selectively target specific metal ions. Additionally, the use of engineered enzymes can contribute to the more efficient dissolution and mobilization of metal-bearing minerals (Gao et al., 2021; Sukla et al., 2021).

18.5.2 Biofilm-Based Bioleaching

The utilization of microbial biofilms has demonstrated improved metal recovery rates in bioleaching processes (Li et al., 2022). Biofilms are complex communities of microorganisms that can form stable attachments to mineral surfaces, enhancing their ability to solubilize and extract the desired metal ions. The properties of biofilms, such as increased resistance to environmental stressors and improved mass transfer, can contribute to the overall efficiency and stability of the bioleaching operation (Kachieng'a & Unuofin, 2021).

18.5.3 Integrated Bioleaching and Downstream Processing

The integration of bioleaching with other hydrometallurgical techniques, such as solvent extraction and ion exchange, can lead to more efficient and holistic metal recovery processes. By combining bioleaching with downstream processing steps, the overall metal extraction and purification efficiency can be improved, while also reducing the environmental impact and waste generation associated with the process.

18.5.4 Bioleaching of Refractory Ores

Advancements in bioleaching technology have enabled the processing of more complex and refractory lateritic ores, which were previously considered unsuitable for conventional extraction methods. The ability to effectively leach and recover metals from these challenging ores expands the potential sources of valuable metal ions and contributes to the diversification of the mining industry's feedstock (Olson et al., 2003).

18.5.5 Optimization of Operational Parameters

Ongoing research on the optimization of bioleaching parameters, such as pH, temperature, nutrient supply, and microbial consortia composition, has led to significant improvements in metal extraction efficiency and process sustainability. The careful optimization of these operational parameters can enhance the selectivity, recovery rates, and environmental performance of the bioleaching process, making it a more attractive option for metal extraction. These advancements in bioleaching technology, combined with the inherent environmental benefits of microbial-based metal extraction, have the potential to transform the mining industry's approach to the recovery of valuable metal ions from lateritic ores. By leveraging these innovative techniques, the industry can strive toward more sustainable and eco-friendly metal production, contributing to the development of a circular economy and a more resource-efficient future (Stanković et al., 2024).

18.6 ECONOMIC AND ENVIRONMENTAL CONSIDERATIONS

The successful implementation of bioleaching techniques for the extraction of valuable metal ions from lateritic ores requires careful consideration of both economic and environmental factors. These two aspects are closely intertwined and play a crucial role in the long-term sustainability and viability of the bioleaching process. From an economic perspective, the feasibility of bioleaching processes depends on a range of factors, including the capital and operating costs, the metal recovery rates, the market prices of the target metal ions, and the potential for the recovery and reuse of the leaching solutions. The capital costs associated with the implementation of bioleaching technology can be influenced by factors such as the scale of the operation, the complexity of the ore processing, and the integration with downstream processing steps (Abdollahi et al., 2024). Operational costs, on the other hand, are typically driven by the energy requirements, the cost of nutrients and other reagents, and the management of the resulting waste streams. The metal recovery rates achieved through bioleaching processes directly impact the economic feasibility of the operation. Maximizing the extraction efficiency of the target metal ions, while minimizing the losses, is crucial for enhancing the profitability of the bioleaching project. Additionally, the potential for the recovery and reuse of the leaching solutions can contribute to the overall economic viability of the process. By implementing closed-loop or recycling systems, the consumption of reagents and the generation of waste can be reduced, leading to cost savings and improved resource efficiency (Roberto et al., 2022; Shah et al., 2020). The environmental considerations are equally important when evaluating the viability of bioleaching technologies. Compared to traditional pyrometallurgical and hydrometallurgical extraction methods, bioleaching processes typically generate less hazardous waste, consume less energy, and have a lower carbon footprint, making them a more sustainable option for the mining industry (Shah et al., 2020). The reduced environmental impact of bioleaching can be attributed to factors such as lower energy requirements, the potential for the treatment and reuse of effluent streams, and the minimization of harmful emissions and waste generation. By adopting bioleaching techniques, mining operations can contribute to the reduction of their environmental footprint and the preservation of local ecosystems. Furthermore, the potential for the recovery and utilization of residual mineral resources, such as iron and aluminum oxides, can enhance the overall sustainability of the bioleaching process. By maximizing the valorization of the lateritic ore, the environmental and economic benefits of the metal extraction operation can be further amplified (Nkuna et al., 2022).

18.7 CONCLUSIONS

The extraction of valuable metal ions from lateritic ores through conventional pyrometallurgical and hydrometallurgical techniques has been associated with significant environmental concerns, including high energy consumption, the generation of hazardous waste, and the release of harmful emissions. In response to these challenges, the utilization of microbial bioleaching has emerged as a promising alternative approach that can offer a more sustainable and environmentally friendly solution. Bioleaching processes leverage the metabolic capabilities of various microorganisms, such as acidophilic bacteria and archaea, to solubilize the metal-bearing minerals and facilitate the extraction of the desired metal ions. The success of these processes is heavily influenced by the environmental conditions that support the growth and activity of the leaching microorganisms, as well as the specific characteristics of the lateritic ore being processed. Significant advancements have been made in bioleaching technology, including using genetically engineered microorganisms, developing biofilm-based systems, and integrating bioleaching with downstream processing techniques. These innovations have contributed to improved metal recovery rates, increased process efficiency, and enhanced environmental sustainability. As the demand for valuable metal ions continues to grow, the implementation of bioleaching as a sustainable and environmentally responsible approach to laterite processing can play a crucial role in meeting the needs of various industries while minimizing the environmental impact of mining operations. By carefully balancing the economic and environmental considerations, bioleaching can emerge as a viable and promising solution for the extraction of valuable metal ions from lateritic ores, contributing to the development of a more sustainable and resource-efficient future.

REFERENCES

Abdollahi, H., Hosseini Nasab, M., & Yadollahi, A. (2024). Bioleaching of lateritic nickel ores. In *Biotechnological Innovations in the Mineral-Metal Industry* (pp. 41–66). Cham: Springer International Publishing.

Agboola, O., Babatunde, D. E., Fayomi, O. S. I., Sadiku, E. R., Popoola, P., Moropeng, L., & Mamudu, O. A. (2020). A review on the impact of mining operation: Monitoring, assessment and management. *Results in Engineering*, *8*, 100181.

Akhmetov, N., Manakhov, A., & Al-Qasim, A. S. (2023). Li-ion battery cathode recycling: An emerging response to growing metal demand and accumulating battery waste. *Electronics*, *12*(5), 1152.

Behera, S. K., & Mulaba-Bafubiandi, A. F. (2015). Advances in microbial leaching processes for nickel extraction from lateritic minerals – A review. *Korean Journal of Chemical Engineering*, *32*, 1447–1454.

Birkeland, P. W. (1984). *Soils and Geomorphology* (pp. 372). Cambridge: Cambridge University Press.

Borra, C. R., Blanpain, B., Pontikes, Y., Binnemans, K., & Van Gerven, T. (2016). Recovery of rare earths and other valuable metals from bauxite residue (red mud): A review. *Journal of Sustainable Metallurgy*, *2*, 365–386.

Bustillo Revuelta, M., & Bustillo Revuelta, M. (2018). Mineral deposits: Types and geology. In *Mineral Resources: From Exploration to Sustainability Assessment* (pp. 49–119). Cham: Springer International Publishing.

Butt, C. R., & Cluzel, D. (2013). Nickel laterite ore deposits: Weathered serpentinites. *Elements*, *9*(2), 123–128.

Caetano, G. C., Ostroski, I. C., & de Barros, M. A. S. D. (2024). Lateritic nickel and cobalt recovery routes: Strategic technologies. *Mineral Processing and Extractive Metallurgy Review*, *2024*, 1–15.

Dalvi, A. D., Bacon, W. G., & Osborne, R. C. (2004). The past and the future of nickel laterites. In *PDAC 2004 International Convention, Trade Show & Investors Exchange, North Carolina*, USA (pp. 1–27).

Daramola, S. O., Hingston, E. D. C., & Demlie, M. (2024). A review of lateritic soils and their use as landfill liners. *Environmental Earth Sciences*, *83*(3), 118.

Diop, T., Gaudin, A., & Lebeau, T. (2024). Sorption of Arsenic (V) by Senegalese Laterites for the Treatment of Arsenic-Contaminated Mining Waters. Available at SSRN 4747560. http://dx.doi.org/10.2139/ssrn.4747560

Dulay, H., Tabares, M., Kashefi, K., & Reguera, G. (2020). Cobalt resistance via detoxification and mineralization in the iron-reducing bacterium *Geobacter sulfurreducens*. *Frontiers in Microbiology*, *11*, 600463.

Duzgoren-Aydin, N. S., Aydin, A., & Malpas, J. (2002). Re-assessment of chemical weathering indices: Case study on pyroclastic rocks of Hong Kong. *Engineering Geology*, *63*(1–2), 99–119.

Earley III, D. (2020). *In Situ Recovery & Remediation of Metals*. Englewood, CO: Society for Mining, Metallurgy & Exploration.

Freyssinet, P., Butt, C. R. M., Morris, R. C., & Piantone, P. (2005). Ore-forming processes related to lateritic weathering. In *One Hundredth Anniversary Volume*. McLean, VA: Society of Economic Geologists.

Gao, X., Jiang, L., Mao, Y., Yao, B., & Jiang, P. (2021). Progress, challenges, and perspectives of bioleaching for recovering heavy metals from mine tailings. *Adsorption Science & Technology*, 2021, 1–13.

Ghosh, S., & Guchhait, S. K. (2020). *Laterites of the Bengal Basin: Characterization, Geochronology and Evolution (p. Basel)*. Switzerland: Springer.

Giese, E. C. (2021). Influence of organic acids on pentlandite bioleaching by *Acidithiobacillus ferrooxidans* LR. *3 Biotech*, *11*(4), 165.

Giese, E. C., Carpen, H. L., Bertolino, L. C., & Schneider, C. L. (2019). Characterization and bioleaching of nickel laterite ore using Bacillus subtilis strain. *Biotechnology Progress*, *35*(6), e2860.

Gilkes, R. J. (1992). Introduction to the petrology of soils and chemical weathering. *Soil Science*, *154*(2), 169.

Gleeson, S. A., Butt, C. R. M., & Elias, M. (2003). Nickel laterites: A review. *SEG Newsletter*, *54*, 1–18.

Hakim, A. Y. A., Sunjaya, D., Hede, A. N. H., Indriati, T., & Hidayat, T. (2023). Critical raw materials associated with the lateritic bauxite and red mud in West Kalimantan, Indonesia. *Geochemistry: Exploration, Environment, Analysis*, *23*(3), 022–064.

Herrington, R., Mondillo, N., Boni, M., Thorne, R., & Tavlan, M. (2016). Bauxite and nickel-cobalt lateritic deposits of the Tethyan belt. In *Economic Geology Special Publication* (Vol. 19). McLean, VA: Society of Economic Geologists.

Kachieng'a, L. O., & Unuofin, J. O. (2021). The potentials of biofilm reactor as recourse for the recuperation of rare earth metals/elements from wastewater: A review. *Environmental Science and Pollution Research*, *28*(33), 44755–44767.

Kanekar, P. P., & Kanekar, S. P. (2022). Acidophilic microorganisms. In *Diversity and Biotechnology of Extremophilic Microorganisms from India* (pp. 155–185). Singapore: Springer Nature Singapore.

Kruger, J. (2019). The behaviour of chromium and its isotopes in nickel laterites. (Doctoral dissertation, University of Southampton).

Kumar, P. S., & Yaashikaa, P. R. (2020). Recent trends and challenges in bioleaching technologies. *Biovalorisation of Wastes to Renewable Chemicals and Biofuels*, 373–388.

Levett, A., Vasconcelos, P. M., Gagen, E. J., Rintoul, L., Spier, C., Guagliardo, P., & Southam, G. (2020). Microbial weathering signatures in lateritic ferruginous duricrusts. *Earth and Planetary Science Letters*, *538*, 116209.

Li, Z., Wang, X., Wang, J., Yuan, X., Jiang, X., Wang, Y., & Wang, F. (2022). Bacterial biofilms as platforms engineered for diverse applications. *Biotechnology Advances*, *57*, 107932.

Marsh, E. E., Anderson, E. D., & Gray, F. (2011). *Ni-Co Laterites: A Deposit Model*. Denver, CO: US Department of the Interior, US Geological Survey.

Meshram, R., Bhondwe, A., Jawadand, S., Raut, T., Dandekar, S., Joshi, V., & Randive, K. (2019). Formation of lateritic kaolin deposit over hoskote-Kolar granodiorite: A case study from Nandigudi and Bavanhalli localities, Kolar District, Karnataka. *Journal of Applied Geochemistry*, *21*(2), 201–211.

Mishra, S., Panda, S., Akcil, A., & Dembele, S. (2023). Biotechnological avenues in mineral processing: Fundamentals, applications and advances in bioleaching and bio-beneficiation. *Mineral Processing and Extractive Metallurgy Review*, *44*(1), 22–51.

Mudd, G. M. (2010). Global trends and environmental issues in nickel mining: Sulfides versus laterites. *Ore Geology Reviews*, *38*(1–2), 9–26.

Nasab, M. H., Noaparast, M., Abdollahi, H., & Amoozegar, M. A. (2020). Indirect bioleaching of Co and Ni from iron rich laterite ore, using metabolic carboxylic acids generated by *P. putida*, *P. koreensis*, *P. bilaji* and *A. niger*. *Hydrometallurgy*, 193, 105309.

Naseri, T., Pourhossein, F., Mousavi, S. M., Kaksonen, A. H., & Kuchta, K. (2022). Manganese bioleaching: An emerging approach for manganese recovery from spent batteries. *Reviews in Environmental Science and Bio/Technology*, *21*(2), 447–468.

Nkuna, R., Ijoma, G. N., Matambo, T. S., & Chimwani, N. (2022). Accessing metals from low-grade ores and the environmental impact considerations: A review of the perspectives of conventional versus bioleaching strategies. *Minerals*, *12*(5), 506.

Olson, G. J., Brierley, J. A., & Brierley, C. L. (2003). Bioleaching review part B: Progress in bioleaching: applications of microbial processes by the minerals industries. *Applied Microbiology and Biotechnology*, *63*(3), 249–257.

Pandey, N., Tripathy, S. K., Patra, S. K., & Jha, G. (2023). Recent progress in hydrometallurgical processing of nickel lateritic ore. *Transactions of the Indian Institute of Metals*, *76*(1), 11–30.

Pintowantoro, S., & Abdul, F. (2019). Selective reduction of laterite nickel ore. *Materials Transactions*, *60*(11), 2245–2254.

Potysz, A., van Hullebusch, E. D., & Kierczak, J. (2018). Perspectives regarding the use of metallurgical slags as secondary metal resources – A review of bioleaching approaches. *Journal of Environmental Management*, *219*, 138–152.

Rasoulnia, P., Barthen, R., & Lakaniemi, A. M. (2021). A critical review of bioleaching of rare earth elements: The mechanisms and effect of process parameters. *Critical Reviews in Environmental Science and Technology*, *51*(4), 378–427.

Rathna, R., & Nakkeeran, E. (2020). Biological treatment for the recovery of minerals from low-grade ores. In *Current Developments in Biotechnology and Bioengineering* (pp. 437–458). Amsterdam, The Netherlands: Elsevier.

Rawlings, D. E. (2002). Heavy metal mining using microbes. *Annual Reviews in Microbiology*, *56*(1), 65–91.

Roberto, F. F., & Schippers, A. (2022). Progress in bioleaching: Part B, applications of microbial processes by the minerals industries. *Applied Microbiology and Biotechnology*, *106*(18), 5913–5928.

Saleh, D. K., Abdollahi, H., Noaparast, M., Nosratabad, A. F., & Tuovinen, O. H. (2019). Dissolution of Al from metakaolin with carboxylic acids produced by *Aspergillus niger, Penicillium bilaji, Pseudomonas putida,* and *Pseudomonas koreensis*. *Hydrometallurgy*, *186*, 235–243.

Sarkar, T., Mishra, M., & Chatterjee, S. (2020). On detailed field-based observations of laterite and laterization: A study in the Paschim Medinipur lateritic upland of India. *Journal of Sedimentary Environments*, *5*(2), 219–245.

Sarkodie, E. K., Jiang, L., Li, K., Yang, J., Guo, Z., Shi, J., & Liu, X. (2022). A review on the bioleaching of toxic metal (loid) s from contaminated soil: Insight into the mechanism of action and the role of influencing factors. *Frontiers in Microbiology*, *13*, 1049277.

Sergeev, N. (2023). Quantifying weathering intensity using chemical proxies: A weathering index AFB. *Australian Journal of Earth Sciences*, *70*(2), 260–284.

Shah, S. S., Palmieri, M. C., Sponchiado, S. R. P., & Bevilaqua, D. (2020). Environmentally sustainable and cost-effective bioleaching of aluminum from low-grade bauxite ore using marine-derived *Aspergillus niger*. *Hydrometallurgy*, *195*, 105368.

Singh, H. R. (2023). Bioremediation of heavy metals-a microbial approach. In *Seed Certification, Organic Farming and Horticultural Practices* (pp.83). Maharashtra, India: Bhumi publishing.

Slukin, A. D., Boeva, N. M., Zhegallo, E. A., & Bortnikov, N. S. (2019). Biogenic dissolution of quartz during formation of laterite bauxites (according to the results of electron microscopic study). *Doklady Earth Sciences, 486*, 541–544.

Srichandan, H., Mohapatra, R. K., Singh, P. K., Mishra, S., Parhi, P. K., & Naik, K. (2020). Column bioleaching applications, process development, mechanism, parametric effect and modelling: A review. *Journal of Industrial and Engineering Chemistry*, *90*, 1–16.

Stanković, S., Goldmann, S., Kraemer, D., Ufer, K., & Schippers, A. (2024). Bioleaching of a lateritic ore (Piauí, Brazil) in percolators. *Hydrometallurgy*, *224*, 106262.

Stanković, S., Stopić, S., Sokić, M., Marković, B., & Friedrich, B. (2020). Review of the past, present, and future of the hydrometallurgical production of nickel and cobalt from lateritic ores. *Metallurgical & Materials Engineering*, *26*(2), 199–208.

Stoops, G., & Marcelino, V. (2018). Lateritic and bauxitic materials. In *Interpretation of micromorphological features of soils and regoliths* (pp. 691–720). Amsterdam, The Netherlands: Elsevier.

Sukla, L. B., Behera, S. K., & Pradhan, N. (2013). Microbial recovery of nickel from lateritic (oxidic) nickel ore: A review. *Geomicrobiology and Biogeochemistry*, *39*, 137–151.

Sukla, L. B., Pattanaik, A., Samal, D. K., & Pradhan, D. (2021). Microbial leaching for recovery of nickel and cobalt from lateritic ore: A review. In *Ni-Co 2021: The 5th International Symposium on Nickel and Cobalt* (pp. 207–217). Cham: Springer International Publishing.

Suparta, W., & Pujiono, J. (2023). Characteristics of nickel laterite in Langgikima, Southeast Sulawesi: Potential of mineral resources and their significance in mining extraction. *AIP Conference Proceedings*, 2983(1), 030009.

Tardy, Y. (1997). *Petrology of Laterites and Tropical Soils*. Rotterdam: AA Balkema.

Thenepalli, T., Chilakala, R., Habte, L., Tuan, L. Q., & Kim, C. S. (2019). A brief note on the heap leaching technologies for the recovery of valuable metals. *Sustainability*, *11*(12), 3347.

Thorne, R. L., Roberts, S., & Herrington, R. (2012). Climate change and the formation of nickel laterite deposits. *Geology*, *40*(4), 331–334.

Vera, M., Schippers, A., Hedrich, S., & Sand, W. (2022). Progress in bioleaching: Fundamentals and mechanisms of microbial metal sulfide oxidation – part A. *Applied Microbiology and Biotechnology*, *106*(21), 6933–6952.

Watling, H. R., Elliot, A. D., Fletcher, H. M., Robinson, D. J., & Sully, D. M. (2011). Ore mineralogy of nickel laterites: Controls on processing characteristics under simulated heap-leach conditions. *Australian Journal of Earth Sciences*, *58*(7), 725–744.

Wells, M. A., & Chia, J. (2011). Quantification of Ni laterite mineralogy and composition: A new approach. *Australian Journal of Earth Sciences*, *58*(7), 711–724.

Whitworth, A. J., Vaughan, J., Southam, G., van der Ent, A., Nkrumah, P. N., Ma, X., & Parbhakar-Fox, A. (2022). Review on metal extraction technologies suitable for critical metal recovery from mining and processing wastes. *Minerals Engineering*, *182*, 107537.

Zhang, M. (2022). Physicochemical and biological processes in iron ore bioprocessing. In *Iron Ores Bioprocessing* (pp. 89–110). Cham: Springer International Publishing.

Zhao, L., Niu, S., Zhou, S., Li, L., Huang, F., Wang, Y., Niu, X., Chen, T., Mo, L., & Zhang, M. (2024). New insight into genesis of the maojun laterite Fe–Mn deposit in the Lanshan area, Hunan Province, South China: Evidence from detailed mineralogical and geochemical studies. *Ore Geology Reviews*, *165*, 105900.

Index

Printed in the United States
by Baker & Taylor Publisher Services